Handbook of Genetic Diagnostic Technologies in Reproductive Medicine

Improving Patient Success Rates and Infant Health

Second Edition

Edited by

Carlos Simón MD

Professor of Obstetrics and Gynecology at the University of Valencia
Senior Lecturer, Department of Obstetrics and Gynecology
Beth Israel Deaconess Medical Center, Harvard University Medical School
Boston, Massachusetts, USA
Head of the Scientific Advisory Board of Igenomix
Valencia, Spain

Carmen Rubio PhD

Research Senior Director for Embryo Chromosomal Studies, Igenomix
Scientist at Igenomix Foundation, INCLIVA, University of Valencia
Valencia, Spain

CRC Press
Taylor & Francis Group
Boca Raton London New York

CRC Press is an imprint of the
Taylor & Francis Group, an **informa** business

First edition published 2022
by CRC Press
6000 Broken Sound Parkway NW, Suite 300, Boca Raton, FL 33487-2742

and by CRC Press
2 Park Square, Milton Park, Abingdon, Oxon, OX14 4RN

ISBN: 978-0-367-45718-1 (hbk)
ISBN: 978-1-032-24490-7 (pbk)
ISBN: 978-1-003-02494-1 (ebk)

DOI: 10.1201/9781003024941

Typeset in Times
by Deanta Global Publishing Services, Chennai, India

Contents

Contributors vii
Preface xiii

1 **The Brave New World of Genomics** 1
Sandra García Herrero, Blanca Simon Frances, Cristian Perez-Garcia, and Javier Garcia-Planells

2 **Genetics at the Cell Level: The Human Cell Atlas** 11
Valentina Lorenzi and Roser Vento-Tormo

3 **The Genetic Risk of a Couple Aiming to Conceive** 24
Joe Leigh Simpson, Svetlana Rechitsky, and Anver Kuliev

4 **Carrier Screening for Single-Gene Disorders** 36
Julio Martin, Arantxa Hervas, Ana Bover, Laura Santa, and Ana Cervero

5 **Meiotic Abnormalities in Infertile Males** 42
Mireia Solé, Francesca Vidal, Joan Blanco, and Zaida Sarrate

6 **Chromosomal Analysis of Sperm** 51
Lorena Rodrigo Vivó, Tantra Martínez Benito, and Azarina Ferro Barbero

7 **Epigenetics in Sperm, Epigenetic Diagnostics, and Transgenerational Inheritance** 61
Jennifer L. M. Thorson, Millissia Ben Maamar, and Michael K. Skinner

8 **Aneuploidy in Human Oocytes and Preimplantation Embryos** 72
Eva R. Hoffmann

9 **Epigenetics from Oocytes to Embryos** 92
Dagnė Daškevičiūtė, Marta Sanchez-Delgado, and David Monk

10 **Embryo Kinetics and Aneuploidy** 103
Fernando Meseguer, Noelia Ramírez, and Marcos Meseguer

11 **Preimplantation Genetic Testing of Aneuploidies (PGT-A)** 117
Daniela N. Bakalova, Darren K. Griffin, Maria E. Póo, and Alan R. Thornhill

12 **Preimplantation Genetic Testing for Structural Rearrangements** 136
Inmaculada Campos-Galindo and Vanessa Peinado

13 **Mosaicism in Preimplantation Embryos** 151
Maurizio Poli and Antonio Capalbo

14 **Embryo Cell-Free DNA in the Culture Medium and Its Potential for Non-Invasive Aneuploidy Testing** 160
Carmen Rubio, Luis Navarro-Sánchez, and Carmen M. García-Pascual

15 **Mitochondria and Embryo Viability** 173
Irene Corachan Garcia, Laura Iñiguez Quiles, and Antonio Diez-Juan

16 **Preimplantation Genetic Diagnosis for Single Gene Disorders** 185
Ana Cervero, Jose Antonio Martínez-Conejero, Lucía Sanz-Salvador, Claudia Gil-Sanchís, Maribel Sánchez-Piris, and Laura Iñiguez Quiles

17 **Preimplantation Genetic Testing for Polygenic Disorders** 195
Nathan R. Treff, Diego Marin, and Laurent C. A. M. Tellier

18 **Should Genome Editing Replace Embryo Selection Following PGT?** 202
Nada Kubikova and Dagan Wells

19 **Molecular Diagnosis of Endometrial Receptivity** 213
Maria Ruiz-Alonso, Diana Valbuena, and Carlos Simón

20 **The Vaginal Microbiome** 225
Shahriar Mowla, Phillip R. Bennett, and David A. MacIntyre

21 **The Uterine Microbiota** 235
Jonah Bardos, Carlos Simón, and Inmaculada Moreno

22 **Endometritis: New Times, New Opportunities** 247
Ettore Cicinelli, Rossana Cicinelli, Carla Mariaflavia Santarsiero, and Amerigo Vitagliano

23 **Decidualization Resistance: A New Condition Identified in Severe Preeclampsia** 256
Irene Muñoz-Blat, Nerea Castillo-Marco, Teresa Cordero, Carlos Simón, and Tamara Garrido-Gómez

24 **Advanced Cell Therapy for Asherman's Syndrome** 267
Jordi Ventura and Xavier Santamaria

25 **Differential Genetic Diagnosis between Leiomyoma and Leiomyosarcoma** 276
Alba Machado-Lopez and Aymara Mas

26 **Non-Invasive Prenatal Testing (NIPT)** 284
Nuria Balaguer, Emilia Mateu-Brull, and Miguel Milán

27 **Advances in Non-Invasive Diagnosis of Single-Gene Disorders and Fetal Exome Sequencing** 301
Liesbeth Vossaert, Roni Zemet, and Ignatia B. Van den Veyver

28 Chromosome Abnormalities in Human Pregnancy Loss: A Review of Cytogenetic and Molecular Analyses **315**
Kathy Hardy and Terry Hassold

29 Products of Conception: Current Methodologies and Clinical Applications **330**
Nasser Al-Asmar and Marcia Riboldi

30 Mother and Embryo Cross Communication during Conception **345**
Anna Idelevich, Andrea Peralta, and Felipe Vilella

31 Perinatal and Pediatric Outcome of Pregnancies Following PGT-M/SR/A **361**
Malou Heijligers and Christine de Die-Smulders

32 Genetic Counseling in Assisted Reproductive Technology **368**
Alyssa Snider and Juliana Cuzzi

Index 387

Contributors

Nasser Al-Asmar
Igenomix
Valencia, Spain

Daniela N. Bakalova
School of Biosciences
University of Kent
Canterbury, UK
and
Igenomix UK Limited
Guildford, UK

Nuria Balaguer
Igenomix
Valencia, Spain

Jonah Bardos
Department of Obstetrics and Gynecology and Department of Human Genetics
Miller School of Medicine
University of Miami
Miami, FL

Millissia Ben Maamar
Center for Reproductive Biology
School of Biological Sciences
Washington State University
Pullman, WA

Phillip R. Bennett
Institute of Reproductive and Developmental Biology
Department of Metabolism, Digestion and Reproduction
and
March of Dimes Prematurity Research Centre
Imperial College London
London, UK

Joan Blanco
Genetics of Male Fertility Group
Cellular Biology
Universitat Autònoma de Barcelona
Bellaterra, Spain

Ana Bover
Carrier Screening Research Group
Igenomix
Valencia, Spain

Inmaculada Campos-Galindo
Igenomix
Valencia, Spain

Antonio Capalbo
Igenomix
Marostica, Italy
and
Valencia, Spain

Nerea Castillo-Marco
INCLIVA Instituto de Investigación Sanitaria
Igenomix Foundation
Valencia, Spain

Ana Cervero
Igenomix
Valencia, Spain

Ettore Cicinelli
Department of Biomedical and Human Oncological Science (DIMO)
and 2nd Unit of Obstetrics and Gynecology
University of Bari
Bari, Italy

Rossana Cicinelli
Department of Biomedical and Human Oncological Science (DIMO)
and
2nd Unit of Obstetrics and Gynecology
University of Bari
Bari, Italy

Irene Corachan Garcia
Hospital Clínico Universitario
Instituto de Investigación INCLIVA
and
School of Medicine
University of Valencia
Valencia, Spain

Teresa Cordero
INCLIVA Instituto de Investigación Sanitaria
Igenomix Foundation
Valencia, Spain

Juliana Cuzzi
Igenomix USA & Canada
Miami, FL

Dagnė Daškevičiūtė
Biomedical Research Centre
University of East Anglia
Norwich Research Park
Norwich, UK

Christine de Die-Smulders
Department of Clinical Genetics
Maastricht University
Maastricht, the Netherlands

Antonio Diez-Juan
Igenomix
Valencia, Spain

Azarina Ferro Barbero
PGD Molecular Cytogenetics
Igenomix
Valencia, Spain

Sandra García Herrero
Genomic Precision Diagnostic Unit
Igenomix
Valencia, Spain

Carmen M. García-Pascual
R&D Department
Igenomix
Valencia, Spain
and
INCLIVA Instituto de Investigación Sanitaria
Igenomix Foundation
Valencia, Spain

Javier Garcia-Planells
Genomic Precision Diagnostic Unit
Igenomix
Valencia, Spain

Tamara Garrido-Gómez
INCLIVA Instituto de Investigación Sanitaria
Igenomix Foundation
Valencia, Spain

Claudia Gil-Sanchís
Igenomix
Valencia, Spain

Darren K. Griffin
School of Biosciences
University of Kent
Canterbury, UK

Kathy Hardy
CytoLabs
Bentley, Australia

Terry Hassold
School of Molecular Biosciences
Washington State University
Pullman, WA

Malou Heijligers
Department of Clinical Genetics
Maastricht University
Maastricht, the Netherlands

Arantxa Hervas
CGT
Igenomix
Valencia, Spain

Eva R. Hoffmann
DNRF Center for Chromosome Stability
Department of Cellular and Molecular Medicine
University of Copenhagen
Copenhagen, Denmark

Anna Idelevich
INCLIVA Instituto de Investigación Sanitaria
Igenomix Foundation
Valencia, Spain

Laura Iñiguez Quiles
Igenomix
Valencia, Spain

Nada Kubikova
Nuffield Department of Women's and Reproductive Health
University of Oxford
John Radcliffe Hospital
and
Jesus College
University of Oxford
Oxford, UK

Anver Kuliev
Department of Human and Medical Genetics
Herbert Wertheim College of Medicine
Florida International University
Miami, FL
and
Reproductive Genetic Innovation (RGI)
Northbrook, IL

Valentina Lorenzi
Wellcome Sanger Institute
Hinxton, UK

Alba Machado-Lopez
INCLIVA Instituto de Investigación Sanitaria
Igenomix Foundation
Valencia, Spain

David A. MacIntyre
Institute of Reproductive and Developmental Biology
Department of Metabolism, Digestion and Reproduction
and
March of Dimes Prematurity Research Centre
Imperial College London
London, UK

Diego Marin
Genomic Prediction Inc.
North Brunswick, NJ

Julio Martin
Clinical Applied Development
Single-Gene Disorders Unit
Igenomix
Valencia, Spain

Tantra Martínez Benito
PGD Molecular Cytogenetics
Igenomix
Valencia, Spain

Jose Antonio Martínez-Conejero
Igenomix
Valencia, Spain

Aymara Mas
INCLIVA Instituto de Investigación Sanitaria
Igenomix Foundation
Valencia, Spain

Emilia Mateu-Brull
Igenomix
Valencia, Spain

Fernando Meseguer
Research Laboratory
IVIRMA Global
Valencia, Spain

Marcos Meseguer
IVF Unit
IVIRMA Global
Valencia, Spain

Miguel Milán
Igenomix
Valencia, Spain

David Monk
Biomedical Research Centre
University of East Anglia
Norwich Research Park
Norwich, UK

Inmaculada Moreno
INCLIVA Instituto de Investigación Sanitaria
Igenomix Foundation
Valencia, Spain

Shahriar Mowla
Institute of Reproductive and Developmental Biology
Department of Metabolism, Digestion and Reproduction
Imperial College London
London, UK

Irene Muñoz-Blat
INCLIVA Instituto de Investigación Sanitaria
Igenomix Foundation
Valencia, Spain

Luis Navarro-Sánchez
R&D Department
Igenomix
Valencia, Spain

Vanessa Peinado
Igenomix
Valencia, Spain

Andrea Peralta
INCLIVA Instituto de Investigación Sanitaria
Igenomix Foundation
Valencia, Spain

Cristian Perez-Garcia
Genomic Precision Diagnostic Unit
Igenomix
Valencia, Spain

Maurizio Poli
Igenomix
Marostica, Italy
and
REPROOMICS
Amsterdam, the Netherlands

Maria E. Póo
Igenomix Mexico
Ciudad de México, Mexico

Noelia Ramírez
Research Laboratory
IVIRMA Global
Valencia, Spain

Svetlana Rechitsky
Department of Human and Medical Genetics
Herbert Wertheim College of Medicine
Florida International University
Miami, FL
and
Reproductive Genetic Innovation (RGI)
Northbrook, IL

Marcia Riboldi
Igenomix Brazil and Argentina
São Paulo, Brazil

Lorena Rodrigo Vivó
PGD Molecular Cytogenetics
Igenomix
Valencia, Spain

Maria Ruiz-Alonso
ERA
Igenomix
Valencia, Spain

Marta Sanchez-Delgado
ZMBH-Zentrum für Molekulare Biologie
Universität Heidelberg
Heidelberg, Germany

Maribel Sánchez-Piris
Igenomix
Valencia, Spain

Laura Santa
CGT
Igenomix
Valencia, Spain

Xavier Santamaria
Asherman Syndrome
Igenomix
Valencia, Spain
and
Ob/Gyn Department
Vall Hebron Institut (VHIR)
Barcelona, Spain

Carla Mariaflavia Santarsiero
Department of Biomedical and Human Oncological Science (DIMO)
and
2nd Unit of Obstetrics and Gynecology
University of Bari
Bari, Italy

Lucía Sanz-Salvador
Igenomix
Valencia, Spain

Zaida Sarrate
Genetics of Male Fertility Group
Cellular Biology
Universitat Autònoma de Barcelona
Bellaterra, Spain

Blanca Simon Frances
Genomic Precision Diagnostic Unit
Igenomix
Valencia, Spain
and
MedStar Washington Hospital Center
Washington, DC

Joe Leigh Simpson
Department of Human and Medical Genetics and Department of Obstetrics and Gynecology
Herbert Wertheim College of Medicine
Florida International University
Miami, FL
and
Reproductive Genetic Innovation (RGI)
Northbrook, IL

Michael K. Skinner
Center for Reproductive Biology
School of Biological Sciences
Washington State University
Pullman, WA

Alyssa Snider
Clinical Genetic Services
Igenomix USA and Canada
Torrance, CA

Mireia Solé
Genetics of Male Fertility Group
Cellular Biology
Universitat Autònoma de Barcelona
Bellaterra, Spain

Laurent C. A. M. Tellier
Genomic Prediction Inc.
North Brunswick
NJ, USA

Alan R Thornhill
School of Biosciences
University of Kent
Canterbury, UK
and
Igenomix UK Limited
Guildford, UK

Jennifer L. M. Thorson
Center for Reproductive Biology
School of Biological Sciences
Washington State University
Pullman, WA

Nathan R. Treff
Genomic Prediction Inc.
North Brunswick, NJ

Diana Valbuena
Igenomix
Valencia, Spain

Ignatia B. Van den Veyver
Departments of Obstetrics and Gynecology and Molecular & Human Genetics
Baylor College of Medicine
Texas Children's Hospital Pavilion for Women
Duncan Neurological Research Institute at Texas Children's Hospital
Houston, TX

Roser Vento-Tormo
Wellcome Sanger Institute
Hinxton, UK

Jordi Ventura
University of Valencia
and
Igenomix
Valencia, Spain

Francesca Vidal
Genetics of Male Fertility Group
Cellular Biology
Universitat Autònoma de Barcelona
Bellaterra, Spain

Felipe Vilella
INCLIVA Instituto de Investigación Sanitaria
Igenomix Foundation
Valencia, Spain

Amerigo Vitagliano
Department of Biomedical and Human Oncological Science (DIMO)
and
2nd Unit of Obstetrics and Gynecology
University of Bari
Bari, Italy
and
Department of Women and Children's Health
University of Padua
Padua, Italy

Liesbeth Vossaert
Department of Molecular & Human Genetics
Baylor College of Medicine
Houston, TX

Dagan Wells
Nuffield Department of Women's and Reproductive Health
University of Oxford
John Radcliffe Hospital
and
Juno Genetics
Oxford, UK

Roni Zemet
Department of Molecular & Human Genetics
Baylor College of Medicine
Houston, TX

Preface

Infertility affects 15% of couples at reproductive age worldwide seeking to become parents. However, having a baby is just the first challenge to overcome; the next and most important is to obtain a healthy newborn free of preventable genetic conditions. Humans have a reproductive efficiency of about 30% in terms of live birth, and this statistic does not improve much after IVF. In addition, genetic disorders affect 1% of live births and are responsible for 20% of pediatric hospitalizations and 20% of infant mortality.

In this handbook, we cover the genetic reproductive journey for patients and doctors in the genomic era, from preconception to childhood. This journey capitalizes on next-generation sequencing (NGS) as a precision genomic diagnostic tool to understand the mechanisms underlying genetic conditions and to prevent disease or physiological impairments. We are indebted to our 91 colleagues who, using their unique expertise and experience, have crafted 32 chapters to take you on this genetic reproductive journey.

Genetic technologies are applied at different stages of the reproductive lifecycle: from preconception carrier screening to couples aiming to conceive, the genetics of the sperm and the embryo, to prenatal and postnatal testing; also, from invasive to non-invasive strategies, such as analysis of fetal cell-free DNA in plasma or embryo cell-free DNA in spent blastocyst media; nor to forget the maternal endometrium to identify the personalized window of endometrial receptivity and the existence of a healthy, dysbiotic, or pathological microbiome that might impair embryo implantation and pregnancy; and the study of cross-communication between the endometrium and the embryo, as well as the early detection of pregnancy complications such as preeclampsia. In this new and ever-burgeoning world of genetics and IVF, genetic counseling and perinatal and pediatric follow-up are crucial supports for positive reproductive outcomes.

With this text, we wish to bring awareness of the importance of genetics in our field and foster a comprehensive knowledge about what genetic tests are available, when they can be prescribed, and how results should be interpreted for clinical management to help advance the goal of achieving healthy newborns.

Carlos Simón
Carmen Rubio

The Brave New World of Genomics

1

Sandra García Herrero, Blanca Simon Frances, Cristian Perez-Garcia, and Javier Garcia-Planells

INTRODUCTION

Today, many genomic laboratories can obtain the complete sequence of a human genome in a few days at a low cost. To reach this point, an exciting revolution has had to happen. Something very similar to a scientific and technological Big Bang.

In April 2003, just a couple of decades ago, the first draft sequence of the human genome was published. To achieve this milestone, an intense international collaboration of almost 3,000 researchers worldwide was necessary, which spanned about 15 years and required an investment of close to 3 billion dollars. This initiative culminated in 2000 with the sequencing of the first human genome over 15 months at a cost of 300 million US dollars.

Only 18 years later, genome sequencing is available to almost anyone who wants it. We even have initiatives that provide us with our genome for free, speaking in economic terms (https://www.personalgenomes.org/), although with high intangible, barely noticeable costs, in exchange for our privacy and access to our personal, highly sensitive, and valuable heritage, genomic data. Heritage is unique not only to us as individuals but also to our relatives as it is the genetic currency that is passed on from generation to generation. This data exchange entails important ethical and regulatory dilemmas that will be featured in another story and that Misha Angrist describes perfectly in his book *Here Is a Human Being: At the Dawn of Personal Genomics* [1].

Scientists are generally not prone to effusiveness, but we can affirm that genomics is one of the areas in science where reality exceeds our dreams and expectations. This impressive genomic revolution in which we are involved allows for the continuous characterization and discovery of genes and molecular causes of many genetic diseases in ever shorter times, increasing the diagnostic yield of many diseases, facilitating the differential diagnosis of complex and genetically heterogeneous diseases to ultimately develop increasingly powerful and sensitive diagnostic methods.

One of the main levers of change has been the development of the technologies for genome sequencing, also known as next-generation sequencing (NGS), with different waves of technology improvements. This technology provides millions of sequences read in only one assay, allowing us to obtain the complete sequence of an individual's genome in a few days and at an astonishing cost of under 1,000 US dollars. The progressive application of NGS technologies in research has boosted our clinical knowledge, increasing the discovery rate of genes associated with Mendelian disorders [2], allowing the establishment

DOI: 10.1201/9781003024941-1

of robust associations between genes and genetic markers in complex diseases to obtain polygenic risk scores, and learning about the intricate molecular basis underlying genetic diseases, as well as the etiology of the transformation of normal cells into malignant cancer cells. All these technological and scientific advances are paving the way towards a precise and personalized medicine.

Whole-exome sequencing (WES) is being used in the clinical setting to obtain a personalized diagnosis of more than 30% of individuals with intellectual disabilities, developmental delay, or congenital abnormalities [3] and allowing for the diagnosis of genetically heterogeneous diseases or those with a complex differential diagnosis that previously represented an unsolvable, diagnostic maze. Currently, NGS technologies provide: i) complete genetic analysis of a newborn admitted to the ICU in a few days to provide the most accurate treatment; ii) analysis of fetal DNA in a few hours from a maternal blood sample without the need for unnecessarily increasing the risk of miscarriage through the use of invasive tests; iii) identification of the genetic profile of a tumor in order to accurately tailor the treatment, providing a precise and personalized treatment for the patient and the disease; iv) identification of mutations and categorization of tumor profiles by liquid biopsy, useful for follow-up and monitoring of the evolution of the disease in a previously treated patient; v) recognition of variants accountable for increased predisposition to common diseases; vi) prediction of individual patient response to certain drugs, avoiding their adverse effects; and vii) the anticipation and prevention of clinical scenarios, one of the great challenges in medicine in developed societies.

Given the impact of the genetic background in the development of rare, common, chronic, and multifactorial diseases, genomic medicine must have the maximum precision to provide the greatest diagnostic yield to direct the treatment. For that, we highlight the importance of incorporating genetic medicine into clinical decisions and treatment algorithms. On the way to personalized medicine, genomic precision medicine can be defined as the centerpiece of the evolution of genomic knowledge. Its main aim is to gather all the information from translational genomic research and direct it towards patient risk stratification of common genetic and biological characteristics, ultimately leading to an accurate diagnosis and, as a byproduct, improving clinical management with personalized treatments and the establishment of accurate prognoses.

This first draft sequence of the human genome provides us with an instructions manual of life, written in a very extensive, complex format with more than 3300 million characters [4]. A reference map on which to base new projects aimed at identifying the genes associated with biological processes to better understand the genetic basis of human health and of the pathology of human diseases. The challenges faced by scientists, researchers, and clinicians are diverse for many reasons, one of them being the ability to read the genomic information necessary to extract conclusions, based on clinical and scientific evidence, in a timely and cost-effective manner. Another challenge is the ability to establish a link between known biological processes and genetic conditions to make a clinical decision that will guide treatment and prognosis of a wide range of diseases with different pathophysiologies on physiologically diverse individuals. As we will see throughout this work, some of these challenges are being addressed in a satisfactory way, whereas most of them are yet to be solved, given that they require important technological, scientific, clinical, and ethical advances.

Although it is an indisputable fact that the emergence of molecular genetic techniques such as NGS has allowed us to standardize pathways to achieve an accurate diagnosis, it has also translated into an number of different genetic tests, where the high level of uncertainty is a challenge to handle and decipher.

WES (whole-exome sequencing) and WGS (whole-genome sequencing) are the most recent technologies based on NGS developed for genomic diagnostic purposes. Undoubtedly, genomic tools, especially WES and WGS, give us a global view of a disease, increasing the knowledge of the underlying mechanisms encoded by the genetic condition. At the same time, such tools also increase the probabilities of finding variants associated with other health conditions, which may or may not be medically actionable, and, being unexpected, are so-called "secondary" or "incidental" findings. The way of reporting and managing these secondary findings, as well as the way of communicating and giving proper clinical recommendations to patients, are some of the main genomic diagnostic challenges. Genomic, high-resolution, high-throughput technologies have opened up new diagnostic pathways, introducing a personalized

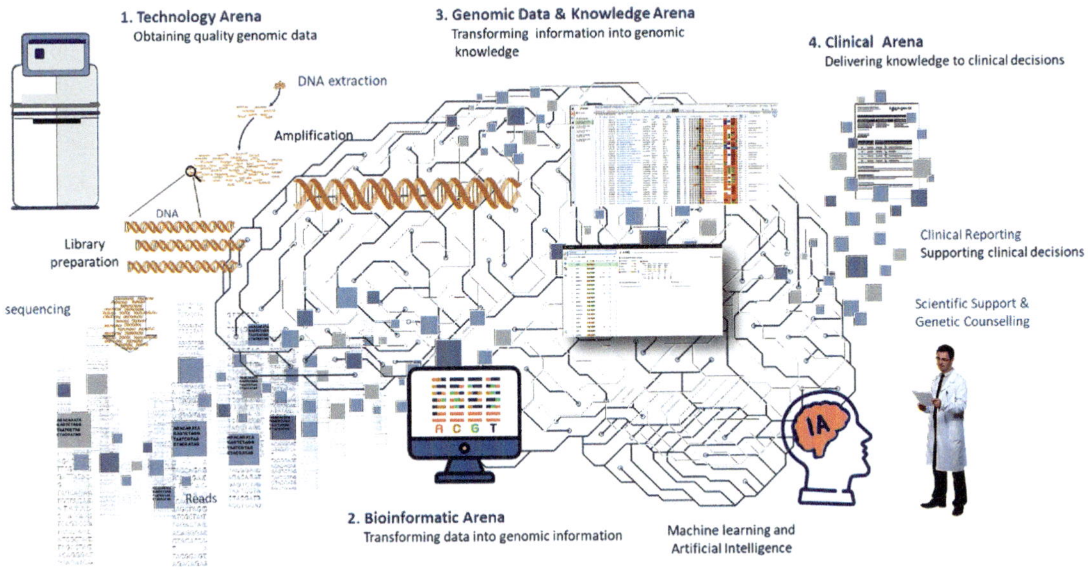

FIGURE 1.1 Integrating technology, bioinformatics, data and knowledge management, and clinical management into the genomic diagnostic process.

clinical management but one paired with new challenges as to the use, management, and interpretation of generated databases, needing great bioinformatic support to conduct a final genomic diagnosis.

The diagnostic process comprises a set of steps, rules, and algorithms that aim to identify a candidate variant underlying a genetic condition. For this purpose, a huge number of databases, computational (*in silico*) predictive programs, and scientific literature must be handled by trained diagnostic personnel in charge, making the final interpretation extremely context dependent (Figure 1.1).

In order to safely progress on this journey through the brave new world of genomics, it is imperative to overcome some challenges, for which it is necessary to be comprehensively trained and to become an expert in the field. It is crucial that the clinical data obtained through this diagnostic process is supervised and approved by a clinician-scientist or physician specialized in genetic diseases to ultimately provide an ethically and clinically appropriate recommendation within the context of the disease and the treatment resources available for that particular disease.

TECHNOLOGY ARENA

The impressive and continual development of genomic technologies in recent years has been leading this genomic revolution. Since 2005, the year in which Roche began to commercialize the first NGS sequencer, an important race started to reach the goal of clinical diagnosis. Two years later, a new technology, Illumina, and thereafter other competitors such as SOLID and Ion Torrent, currently marketed by Thermo Fisher, joined the bandwagon [5]. Third-generation sequencing technologies are based on single-molecule sequencing, a new approach that avoids the bias of PCR amplification and produces longer sequences. Although a clear advantage is associated with this new technology, the caveat is the loss of sensitivity, a key aspect in diagnosis [6–8]. The recently reported and promising technology, fourth-generation sequencing, centered around the *in situ* reading of nucleic acid sequences within tissues and cells, exploits previous NGS chemistries [9]. This technology pledges to become an essential tool for the implementation of personalized medicine [10, 11].

This genomic technology race promises an exciting and challenging future, becoming a reality for research. However, further developments to improve sensitivity, specificity, and clinical validations still need to be established before it becomes a reality in the clinical setting.

Currently, Illumina sequencers are leading this race, given the quality of its sequencing technology, its versatility, and affordable price. Thermo Fisher sequencers follow closely thanks to their integrative solutions and clinical applications, and finally Roche sequencers, which unfortunately could not keep up with their competitors despite starting with the advantage of longer reads.

This amazing competition has allowed a greater accessibility of these diagnostic technologies, expanding the number and quality of their applications. This has led to new approaches and innovative strategies, increasing the accuracy, and reducing the time and costs involved.

Undoubtedly, in the coming years, new players will join this challenging race. Promising new sequencing systems have been developed but have not been able to compete on price or sensitivity, including new fourth-generation sequencing systems. In any case, only those sequencing systems that can provide competitive advantages, such as price or simplicity, without compromising on the quality provided by current sequencing systems, will be able to enter this competitive race of genomics diagnostics.

It is important to highlight that being able to buy the latest generation sequencer does not guarantee better or more accurate diagnosis. Before the acquisition of NGS machines, many issues must be dealt with, and skills must be acquired. The clinical applications of these technologies require extensive technical and molecular biology knowledge to be able to generate genomic data. But we still have a long way to go until this raw data can be converted into clinically useful information.

The days when decisions were based on technical aspects, such as pyrosequencing, sequencing by synthesis or by ligation, hybridization capture or amplicon sequencing gene panels, mean or minimum coverage, vertical (depth) or horizontal coverage, etc., are well over. NGS platforms and technologies are constantly changing and must be adapted and validated before being used in clinical diagnosis. For any NGS platform or enrichment method of regions of interest (ROI), we need to establish the performance and diagnostic yield of our approach in our clinical setting. Currently, any NGS clinical approach does not guarantee a 100% accurate analysis of ROI. This may be due to technical reasons or even intrinsic features of the DNA structure itself. This raises the importance of calculating the analytical sensitivity and specificity and defining the limits of detection, according to the clinical goals of the test [12].

Thanks to this outstanding technological development, we have learned how to obtain the whole genomic sequence of a human being. Currently, we can generate the instructions of life based on the genetic makeup at extraordinarily low costs and short times with great precision and analytical sensitivity.

But are we able to understand these instructions? Can we make clinical decisions based on these data? Can we accurately manage all the information of an individual in an adequate time? How can we manage this uncertainty and at the same time trust in the final diagnosis?

Important questions that raise new challenges must be solved with the aid of different professional profiles and computing technologies, introducing us to the exciting world of Big Data.

BIOINFORMATIC ARENA

First of all, it is important to remember that the human genome consists of 3.3 billion nucleotides packed into each cell. This overwhelming amount of information is impossible to read or understand without the help of the latest developments in computational technologies [13]. Additionally, the complexity level increases given that our body is made up of 30 billion cells grouped into tissues, in which, genes are differentially expressed based on their individual functions [14], while many cells also acquire a differential number of mutations in different genes throughout their life [15]. Each of these analyses adds multiple layers of complexity to the pursuit of biological knowledge, raising the dimensionality of the problem. Variables accumulate, driving the analysis to what tends to be called the curse of dimensionality, meaning that, as complexity increases, so does the sparsity of the data, so that reaching statistical conclusions becomes essentially unachievable.

Therefore, the use of powerful computational tools is essential to be able to process these massive amounts of data at the germinal/hereditary level, but, remarkably, also at the somatic level.

Due to the general idea of bioinformatic tools being seen as black boxes for clinicians focused on the diagnostic divisions, during the past few years there has been an enormous effort to share, not only new software and methods to the ecosystem, but also technical knowledge through courses and tutorials, with the final objective of teaching the basics and the advanced statistical and computational techniques to the non-bioinformatician staff. Just as equipment is connected in a factory, concatenating bioinformatic tools gives birth to a bioinformatic pipeline, and, thanks to the open-source way of thinking of the community, everyone can reach that knowledge and skills at no cost.

In this regard, bioinformatic pipelines must be an essential part of the diagnostic process as they will have a great impact on disease management and patient care.

The main challenges to face at this stage are the following:

- The availability of numerous bioinformatic algorithms, with their strengths and weaknesses, require considerable expertise to decide which is the most suitable combination of them.
- Bioinformatic pipelines should be customized based on laboratory needs, the sequencing platform, and the expertise of the bioinformatician teams.
- A validation process is necessary with every change that may be introduced in the pipelines, as the absence of these processes may have negative consequences for patient care.
- Variant annotation and classification, according to clinical evidence and state of the art, require a periodic check and update of each of the databases.
- Updates at the level of the reference genome being used on the process must be maintained, attaining higher quality in the process
- Incorporation of automatic classification algorithms, based on machine learning and artificial intelligence, should be deeply validated, as trained models work only with previously seen scenarios and might not assess novel data correctly.

If we continue the diagnostic process from where we left off in the previous section, sequencing technologies provide massive amounts of data that require multiple computationally intensive steps for an appropriate analysis. It goes without saying that the computational analysis of the data is a key point in the diagnostic process. A poor computational processing of the data can ruin the technically best sequencing process and provide a misdiagnosis or inaccurate result.

As today, the routine bioinformatic pipeline used on whole-exome sequencing and custom panel analysis must contain the following steps:

- Base calling and demultiplexing, steps followed to assign a quality score to each of the bases in the sequence, and generation of raw FASTQ files.
- Sequence alignment with the human reference genome, providing the genomic context to each sequence with a mapping quality score, followed by duplicate read marking or removal. To reduce its size, these sequence alignment data are stored in a Binary Alignment Map (BAM) file.
- Variant calling to identify variants or differences between the sample and the reference genome sequence with the greatest accuracy. We can identify different types of variants by means of specific algorithmic procedures: single nucleotide variants (SNV) or small insertion and deletions (indels). These data are stored in a variant call format (VCF) file.
- Variant annotation, understood as querying different databases to provide context about chromosomal and gene location, potential amino acid variant, allele frequency, etc. Using this information, we are able to classify and prioritize the variants depending on the clinical significance, level of evidence, pathogenicity, etc.

- Complex events calling, where the presence of a different type of event, such as structural variations (SVs) or copy number variations (CNVs) is evaluated. Although a standardized way exists for storing these data in the VCF format and naming the events, it is unfortunately not always followed.
- Metric calculation and revision should be followed as a quality standard on each of the samples. Data as read duplication, on-target reads, mean coverage, etc. provides valuable information with which to detect issues regarding the state of the sample or to the lab process.

All these processes are subject to constant optimization and automation improvements to obtain the lowest analytical and computational times without a quality loss of analysis. The overall objective is to provide the maximum level of information without overwhelming the clinician with the massive numbers of available databases found nowadays, favoring the most informative ones that help to identify and classify all the variants of clinical interest with the greatest accuracy.

This is the reason why bioinformatic analysis is of great importance at the diagnostic level, paying special attention to documenting each process, its limitations, and the level of database updating, as well as analytical and clinical validation [16].

At the end of the bioinformatic analysis, we must procure a list of the variants identified in each individual and classify them based on relevant features, such as the quality of sequence (excluding sequencing artifacts), biological effect, pathogenic prediction, and potential relationship with the disease or clinical characteristic that we intend to diagnose. International guidelines provide guidance for the interpretation of genetic variants according to their potential pathogenicity and relationship with the disease [17], harmonizing the classification criteria for all laboratories performing genetic diagnosis, and becoming the basis for the quality standards for laboratory accreditation and genomic analysis software platforms.

Nowadays, the amount of genetic data and information on variants provided by the different sequencing technologies is very high, especially when whole-genome sequencing is performed, but the quality and level of clinical and scientific evidence of those variants is still limited, making the clinical interpretation process difficult. For example, if we re-analyze the whole exome data of an undiagnosed individual 12 months after the first analysis, the overall diagnosis rate increases from 30 to 41% [18], thanks to the advances in research and knowledge over the 12-month period as well as the improvement in classification processes.

This gap between genomic sequencing capabilities and associated clinical knowledge reveals a large number of variants of uncertain significance (VUSs), variants whose clinical significance is not still known due to the lack of scientific evidence. The accurate classification of these VUSs, which account for about 40% of all variants, is one of the main challenges in the area of genomic diagnostics.

GENOMIC DATA AND KNOWLEDGE ARENA

If we assume that we start from high-quality technical processes and bioinformatic algorithms, after having carried out an intensive analytical and clinical validation of them, the main arena is currently focused on the management of biological, clinical, and scientific data, their level of update, and the ease of reaching practical conclusions that provide a clear answer to the clinical indication from our genomic test.

As with everything in science, it is crucial to stay updated on the latest evidence as new genes are discovered every day and so variants may be reclassified, and expanded phenotypic information becomes available. The ascending spiral of knowledge and the unstoppable discovery of disease-causing genes and variants make it difficult for geneticists and physicians to master the updated information.

Enormous efforts are being made by many laboratories and scientific consortiums around the world to generate genetic data, annotate and classify variants, and translate those data into clinical actionable findings. International data sharing is an efficient solution to increasing genetic medical knowledge and increasing the level of evidence for the largest number of genetic variants, facilitating the process of interpretation and genetic diagnosis, but not enough.

Despite all international collaborative efforts, genomic and biological data are too large, too complex, and are stored in different databases that must be mined using different approaches. In addition, clinical data are recorded without following standardized phenotypic formats and clinical definitions that ensure the contextual meaning of the patient record.

For this reason, genomics has become a data-driven science, which requires complex, computational models and the application of machine-learning techniques and artificial intelligence to handle such quantities of data from a variety of sources while maintaining a high level of precision and ever-shorter analysis times.

The application of machine-learning algorithms and artificial intelligence (AI) in genomics helps to refine the interpretation of genomic data according to the clinical context by the development of computer systems that replicate the complex diagnostic processes carried out by genomic experts.

However, the quality and performance of these processes strongly depend on the accuracy of the data and its classification throughout the multiple sources of available information. For this, it is necessary to clean the data by removing errors, irregularities, and missing information coming from heterogeneous data. This process of data cleaning can consume about 80% of the time spent on data analysis.

As new technological innovations keep appearing in the data science area and its applications to a practical routine seem to come closer, there are tiny, take-home messages that each of us should know by now, as data surrounds us in almost every aspect of our daily actions. Briefly, the fact that highlights why structuring and contrasting the data is important is the time spent by a data scientist in data cleaning, taking up to 80%, leaving only 20% for analysis. Specifying the variables and their values before storing them is one of the many solutions to this problem, as is a structured organization of the data.

Scientists are heavily dependent on the quality and quantity of the data, and role models will only be built after all the previous recommendations have been taken into account, otherwise the valuable time and money will be wasted on the creation of a model that, although providing a result, may not be close to the real value, and we will fail in the use of these powerful techniques.

Understanding and applying the knowledge in the data science field requires a multidisciplinary team, including experts not only in the areas of mathematics and data engineering, but also in probably the most complex and specialized field, the one used to create models. The lack of specialized professionals in this field together with a deficient communication between the different disciplines creates an enticing environment in the language of data in which getting involved will eventually save money and pave a solid ground to a successful project. AI should be understood as a mathematical tool to discover advanced patterns that are not seen by our regular eyesight, even being hard or impossible to extract from the model generated by the computer. The field is divided into two different kinds of models: supervised and unsupervised models. The supervised model works by training a model with labeled data and allowing it to generalize the problem to new data, just as a variant can be classified as pathogenic or benign. The unsupervised model, on the other hand, is a tool to associate or aggregate data without previous knowledge, just like clustering cell data by its tissue of origin without actually having to label it from the beginning and leaving the interpretation to the scientist.

CLINICAL ARENA

Global traditional clinical genetics has changed dramatically in only a few decades from traditional non-molecular technologies, based around visual microscope-based techniques (i.e., fluorescence *in situ* hybridization (FISH) or G-banding karyotype), to the latest molecular high-throughput techniques as in next-generation sequencing (NGS).

Up to the beginning of molecular genetics, traditional clinical genetics was focused on identifying monogenic disorders, with the use of artisanal technologies tailored for a specific gene or mutation and previously oriented clinically with a detailed evaluation of the patient's medical and family histories.

The "explosion" of next-generation sequencing technologies and the ability to obtain large amounts of genomic information through a single assay has made it easier for us to use a large number of clinical applications that a few years ago were unapproachable. From the diagnostic point of view, there has been an important paradigm shift when approaching the analysis of many diseases, which can be caused by different genes or where differential diagnosis is very complex to establish, or for the resolution of complex phenotypes or non-specific conditions through the simultaneous analysis of a number of genes, or even the whole exome, encompassing the complete set of genes of an individual. Currently, the use of gene panels is the most cost-effective strategy, compared to the classical approach of defining a diagnostic algorithm and sequentially analyzing the different genes involved.

In addition, genomic technologies facilitate the development of new applications and developments in the field of screening of people who do not feel ill or before any symptoms become noticeable. Some of the most widely used NGS-based screening tests currently are focused on carrier testing to establish at a preconception level the risk for a couple of transmitting a genetic disease to their offspring. Another example is the genetic newborn screening tests to promptly identify those neonates who may develop a genetic disease for which there is an effective treatment or clinical management, as a complementary test to the biochemical screening of the newborn.

In general, screening tests are characterized by being a proactive offer of services to a generally healthy population, where early clinical intervention would benefit them in preventing the development of a disease, facilitating its clinical management, the access to treatment or the application of preventive measures. This definition of a screening test is very important as the temptation to provide patients with as much information as possible is present, thanks to the high level of data production from genomic technologies. But sometimes less is more, and an excess of non-actionable information will generate more uncertainty and patient anxiety in an effort to provide a primary prevention strategy. There is a large number of variants of uncertain significance identified (40%) and a small number of genes associated with Mendelian phenotypes (24%: 4500 genes out of a total of 19,000 human genes [19]) which accounts for this large body of information. A screening test, given its preventive and preclinical nature, the cost of uncertainty may be harmful to the patient due to the lack of evidence supporting their gene–disease association, and actionability in terms of risk prevention, reproductive approaches, or clinical management in the case of newborn screening.

In the case of diagnostic tests, the uncertainty related to the interpretation of variants and genes without sufficient evidence tends to have a lower impact since we know that the patient is affected by a clinical condition. Although the genetic test does not provide a precise genetic diagnosis of the disease or condition, the clinical diagnosis and evolution of the patient will not vary in the face of an uncertain result of the genetic test, whereas the test allows us to rule out the main known genetic causes of the disease.

In essence, the ultimate goal of genomic diagnosis, as with many other diagnostic tests, is to be able to provide a precise answer to a clinical question posed with respect to our patient, with the maximum level of detection for those variants that can define the phenotype and with the least degree of uncertainty. These aspects, in terms of quality, can be translated by clinical utility, sensitivity, and specificity.

The end point of the genomic diagnostic process is the elaboration of a report collecting all the clinically useful information obtained from the data analyzed and interpreted according to the clinical indications. The diagnostic process implies a clear description of all the stages that we have discussed: the sensitivity, specificity, and limits of detection of the technologies used, the algorithms and updating levels of the databases used, and the levels of evidence that support the clinical interpretation of the results.

Reporting is probably the most important part of this complex process because it involves a written proof of the quality of the whole procedure, the means of describing and communicating complex information, and a commitment to the patient that the entire process has been accomplished according to the highest-quality standards. It is convenient to describe the main elements that a report must contain and, therefore, which we must include when we prepare a report or demand a report when we are the recipients of it.

According to Song et al. [20], a report must be:

1. *Accurate*:
   WGS test results will have profound clinical and personal implications for patients and family members.
2. *Unambiguous*:
   The results of these types of studies are usually of lifelong relevance and are a reliable resource for the patient, physician, and other family members. It is common for this information to be used in the future by other health care professionals or laboratories. For this reason, any ambiguity or inaccuracy could have serious repercussions at a patient level as well as for the rest of the family. This applies both for positive and negative results.
3. *Clear*:
   Results must be presented in a clear fashion, preferably in a table format using standardized nomenclature, for it to be understood and interpreted by different health care professionals and laboratories. A clear report of the results is crucial in ensuring clinical decision making.
4. *Evidence-based interpretation*:
   The report should list supporting evidence and criteria applied to reach the conclusion and cite the source of information.
   The results, or data, must be interpreted using scientific and clinical-based evidence by properly trained personnel. This interpretation may vary with time, given the incorporation of new evidence or discoveries, which is why the interpretation of all conclusions must be referenced accordingly, with up-to-date evidence coinciding with the date of the report. The report should list supporting evidence and criteria applied to reach the final conclusion.
5. *Oriented towards the clinical question*:
   We should not forget that the ultimate goal of these tests is to answer a clinical question, which should be reflected in the section "Clinical Indication."
   Occasionally, given the vast amount of information obtained with these technologies, we focus more on reporting and sometimes forget the real reason of this test, the clinical enigma that we want to answer as well as the patient's concerns. The patient will have signed an informed consent form, regarding the type of information we must provide, and we have to keep in mind that we don't have the patient's consent to release additional unnecessary information regarding other aspects, and we must protect the principle of autonomy of the patient not knowing certain information.
6. *Transparent*:
   This aspect is of great importance given the number of technologies and the variety of approaches that we have at our disposition, each one with its advantages and limitations, as well as the pivotal evolution of each of them. The type of technology together with a bioinformatic pipeline, databases at different stages of update and when that analysis has been carried out plays a crucial role in the transparency of a result. Any change in any of these variables could yield a totally different result, potentially leading to an incorrect diagnosis, and thus changing clinical management.
   For those reasons, the methodology used, the databases that are consulted, and the level of update achieved should all be reflected in the report, as well as any in-house validation of the procedures, with its corresponding sensitivity, specificity, and detection limits.

It is clear that the genomic revolution is changing the paradigms of a growing number of medical specialties. At the pediatric and neonatal level, the application of next-generation sequencing analysis is providing us with diagnostic yields greater than 40% in many complex and heterogeneous diseases that were previously difficult to diagnose. This precise diagnosis allows early access to new and sophisticated treatments, as well as improved clinical care of an increasing number of patients. In the field of prenatal diagnosis, we can increase the diagnostic capabilities, in terms of both sensitivity and time to response. This is especially true for those cases where ultrasound findings have been detected and for which it is difficult to establish a differential clinical diagnosis. But also, at the level of aneuploidy screening and other common genetic diseases, these techniques allow us to analyze the fetus through maternal blood without the risks associated with invasive tests. In oncology and pathological anatomy, we are able to establish the profile of a tumor in order to identify genetic markers that narrow down the diagnosis, its origin, to guide therapy, to establish prognosis, to predict adverse responses or toxicities, and include the patient in relevant clinical trials, increasing not only their survival, but also their quality of life. Genomics also has a profound impact in the prevention of more common diseases, such as cardiovascular disease, including sudden death. Disciplines such as pharmacogenomics lead to an increase in scientific knowledge and clinical evidence which is necessary to pave the way towards personalized medicine.

There is also an important area of personal genomics, in which the use of genetic markers related to phenotypic characteristics, abilities, or even attitudes, will generate a lot of concern and interest. This is a field in which ethical and regulatory aspects will be decisive in clearly defining where the clinical benefits end and the individual´s interests begin.

A promising future will see the incorporation of even more powerful technologies, such as whole-genome sequencing, that increase the amount of data and complexity by several orders of magnitude and will provide us with a series of advantages when it comes to detecting new types of mutations. These will in turn require corresponding technical and analytical validation processes and the development of powerful tools for analysis, data updating, and management of the increasing body of information.

Undoubtedly, a complex process of constant learning, in which our knowledge and experience, as well as our awareness of our capacities and limitations, will allow us to slowly implement the process, with a higher level of complexity, and raise it to the higher level, the Brave New World of Genomics.

REFERENCES

1. Angrist M. *Here Is a Human Being: At the Dawn of Personal Genomics*. Harper Collins, 2010.
2. Bamshad MJ, et al. *Am J Hum Genet*. 2019;105(3):448–455.
3. Manickam K, et al. *Genet Med*. 2021;23(11):2029–2037.
4. Collins FS. *The Language of Life: DNA and the Revolution in Personalized Medicine*. Harper Perennial. 2011.
5. Liu L, et al. *J Biomed Biotechnol*. 2012;2012:251364.
6. Eid J, et al. *Science*. 2009;323(5910):133–138.
7. Branton D, et al. *Nat. Biotechnol*. 2008;26(10):1146–1153.
8. Timp W, et al. *IEEE Trans Nanotechnol*. 2010;9(3):281–294.
9. Mignardi, M., et al. *Genome Med*. 2014;6:31.
10. Ke R, et al. *Nat Methods*. 2013;10:857–860.
11. Lee JH, et al. *Science*. 2014;343:1360–1363.
12. Matthijs G, et al. *Eur J Hum Genet*. 2016;24(1):2–5. Erratum in: *Eur J Hum Genet*. 2016;24(10):1515.
13. Krassowski M, et al. *Front Genet*. 2020;11:610798.
14. Costa-Silva J, et al. *PLoS ONE*. 2017;12(12):e0190152.
15. García-Nieto PE, et al. *Genome Biol*. 2019;20:298.
16. Roy S, et al. *J Mol Diagn*. 2018;20(1):4–27.
17. Richards, S., et al. *Genet Med*. 2015;17:405–423.
18. Ewans LJ, et al. *Genet Med*. 2018;20(12):1564–1574.
19. Online Mendelian Inheritance in Man, OMIM®. McKusick-Nathans Institute of Genetic Medicine, Johns Hopkins University (Baltimore, MD), {10 September 2021}.
20. Song C, et al. *Curr Protoc Hum Genet*. 2018;6:e61.

Genetics at the Cell Level

The Human Cell Atlas

2

Valentina Lorenzi and Roser Vento-Tormo

UNCOVERING CELLULAR HETEROGENEITY IN THE HUMAN BODY

The Human Cell Atlas (HCA) is an international consortium established at the end of 2016 with the mission of mapping and characterizing all cells in the human body in terms of their distinctive patterns of gene expression, physiological states, and location (Rozenblatt-Rosen *et al.*, 2017); (Regev *et al.*, 2017) (www.humancellatlas.org). It is an open and collaborative initiative, bringing together experts across multiple disciplines, and is meant to progress in phases. Recently, the first maps focused on specific organs and tissues (Ordovas-Montanes *et al.*, 2018; Vento-Tormo *et al.*, 2018; Popescu *et al.*, 2019; Ramachandran *et al.*, 2019; Smillie *et al.*, 2019; Stewart *et al.*, 2019; Vieira Braga *et al.*, 2019) have laid the foundations for further work aimed at completing the atlas to include at least ten billion cells that fully represent the world's diversity.

The desire to comprehensively characterize and classify cells into distinct types is not new. Research has long been focused on cataloging cells based on their shape, location, biological function, and molecular components with increasing levels of detail. Only recently, however, have advances in single-cell genomic technologies made it possible to undertake the high-resolution, unbiased, and systematic characterization of cells in the human body which is at the heart of the HCA (Svensson, Vento-Tormo, and Teichmann, 2018). Such revolutionary technologies allow us to profile the genome and the genome products—including chromatin architecture, RNA transcripts, and proteins from single cells (Lander, 1996).

Genomic profiling technologies have been used for many years to describe ensembles of cells in a tissue, called bulk tissue samples, but it was not until the beginning of the last decade that they could be employed in the characterization of individual cells. This represents a dramatic step forward in the study of cellular heterogeneity and has led to the discovery of new cell types, thus resonating with the HCA's main goal of creating a comprehensive, integrated catalog of all the cell types present in the human body.

DOI: 10.1201/9781003024941-2

From Bulk to Single Cells

No two cells in a body are exactly the same. Although the set of genetic instructions is shared across all the cells in an individual, the form and function of each cell, even within the same cell type, are unique. Such uniqueness is accomplished by finely regulating gene expression, meaning the quantity of RNAs and proteins that are produced from each gene in a given cell.

Traditionally, gene expression has been studied by taking thousands of cells in a tissue and treating them as a single, homogeneous entity. This approach measures the *aggregated* expression level for each gene across the population of input cells (Wang, Gerstein, and Snyder, 2009). Although very useful for comparing expression signatures of a tissue across different conditions, such as health and disease, bulk studies inevitably miss out on what makes each cell unique (Raj and van Oudenaarden, 2008).

The last decade has witnessed the rise of new, single-cell sequencing technologies that allow researchers to capture the cell diversity within a tissue, a feature which is lost in bulk populations (Kulkarni *et al.*, 2019). These technologies can measure the *distribution* of expression levels for each gene in each cell across a population, thus providing answers to new biological questions in which cell-specific changes in gene expression play a major role. The ability to discern individual cells within a tissue can, for instance, lead to the identification of new cell types and improve our understanding of organ development and homeostasis (Figure 2.1).

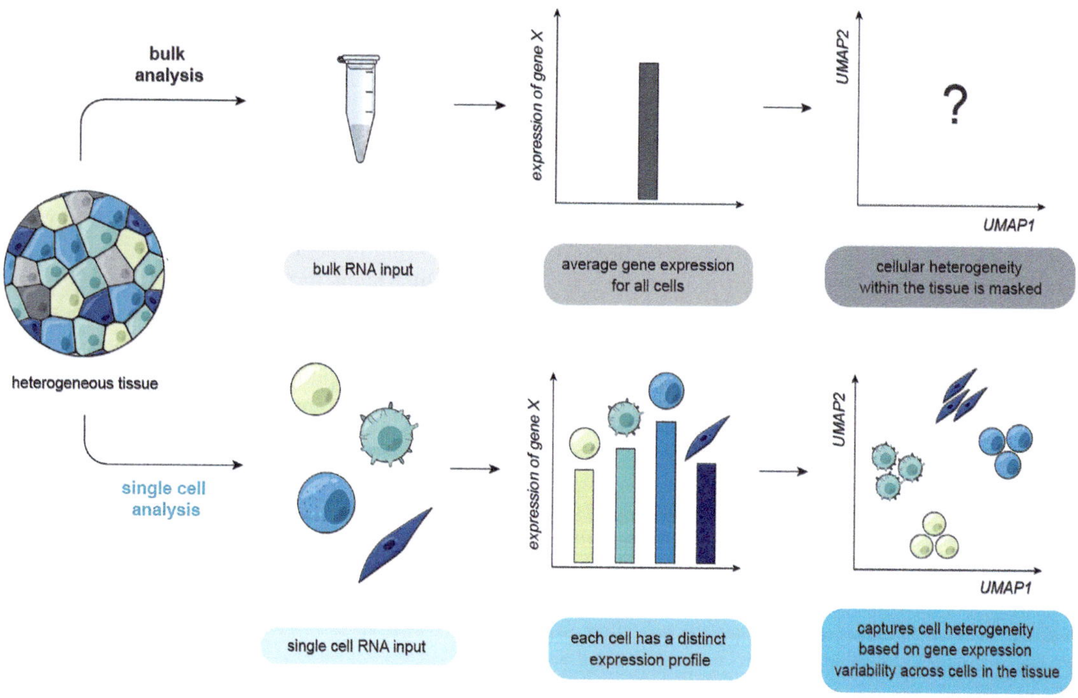

FIGURE 2.1 Schematic representation of the major differences between bulk and single-cell transcriptomic analysis. Bulk transcriptomic analysis measures the average expression for each gene across all cells in a tissue, making it unfit for unraveling cellular heterogeneity. In contrast, single-cell transcriptomic analysis outputs the gene expression profile specific to each cell, thus capturing cellular heterogeneity within a tissue.

Finding New Cell Types

Our knowledge about the types of cells present in the human body is still limited and discovering new, often rare, cell types poses a significant challenge. Indeed, the rarity of these cell types makes them hard to identify *via* bulk transcriptomic technologies, whereas targeted, single-cell molecular techniques that interrogate gene expression of a few selected cells are also ill suited because these new cell types are, by definition, unknown. This is where single-cell sequencing technologies come into play, offering an unbiased and novel understanding of the cellular composition of tissues.

Notably, in 2018, transcriptomic profiling of human bronchial epithelial cells with single-cell sequencing technologies led to the identification of a rare lung cell type that plays a major role in cystic fibrosis (Plasschaert *et al.*, 2018). Named the pulmonary ionocyte, this cell type expresses high levels of the cystic fibrosis transmembrane conductance regulator (CFTR) gene, which codes for a cell membrane protein that allows the outward flow of chloride ions and which is mutated in patients with cystic fibrosis, causing pathogenic accumulation of mucus in the airways. Currently available drugs for treating cystic fibrosis are able to target specific mutated versions of the CFTR protein, and knowing which cells are contributing to the production of the protein could be conducive to more precise delivery of the treatment.

Several projects within the HCA have led to the discovery of new cell subtypes that play an important role during human development (Vento-Tormo *et al.*, 2018; Popescu *et al.*, 2019; Park *et al.*, 2020). A recent study profiling the uterine-placental interface at single-cell resolution identified three distinct subsets of decidual natural killer cells (dNK) and decidual stromal cells with potential roles in early pregnancy in humans (Vento-Tormo *et al.*, 2018). The novel dNK subsets found in the tissue are thought to be key to maintaining the immunomodulatory environment during early pregnancy, helping in the establishment of placentation through invasion of the placental trophoblast. This study has advanced our understanding of the uterine environment during the early stages of pregnancy and will have a direct impact on pregnancy-related pathologies, such as preeclampsia or fetal growth restriction (Figure 2.2).

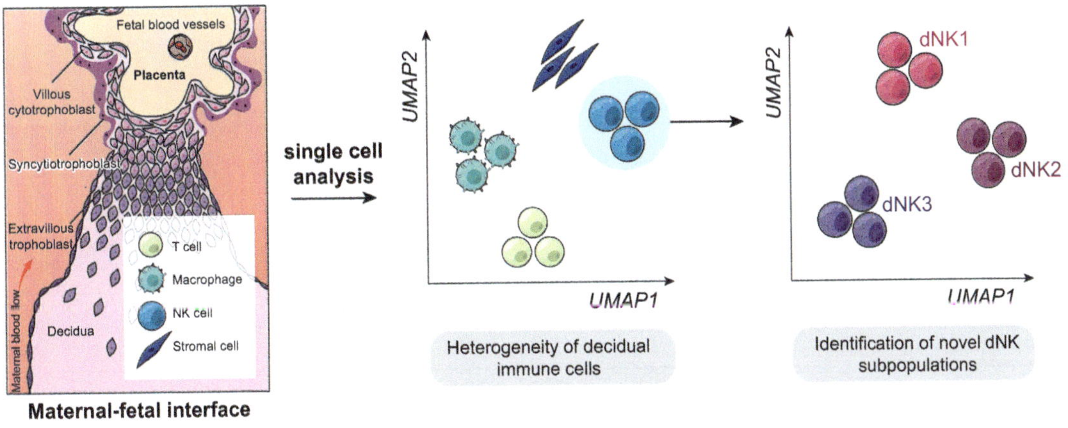

FIGURE 2.2 Single-cell profiling of the maternal–fetal interface during early pregnancy in humans has led to the discovery of three novel subsets of decidual natural killer cells (dNKs). Each subpopulation of dNK was found to have a specific function in the establishment of placentation.

Creation of a Comprehensive Body of Knowledge

Each single-cell study carried out within the framework of the HCA initiative provides invaluable information on how cells are organized into functional organs. However, in order to fully exploit the newly acquired knowledge for further basic research as well as for medical and translational purposes, it is necessary to integrate it into a single, coherent data resource. Moreover, having an atlas means that meta-analyses can be performed across cells, tissues, and organs, thus gaining a comprehensive understanding that goes well beyond what can be gathered from the individual studies.

In more practical terms, the integrated maps of the human body in the HCA allow researchers and practitioners around the world to query organs, tissues, and cells to obtain information about their molecular and organizational features. The analogy used in the White Paper (Regen *et al.*, 2018) to describe the purpose of the HCA is that of Google maps: the atlas allows its users to navigate the human body at various levels of resolution to identify patterns and interactions among its fundamental elements, zooming in and out, depending on the research goals (Figure 2.3).

Building an integrative resource comes with its own challenges. New computational tools are being developed to integrate datasets from distinct batches, including technologies and protocols for processing or sequencing platforms (Lopez *et al.*, 2018; Korsunsky *et al.*, 2019; Polański *et al.*, 2019). In addition, data have to be released on an open-access basis to ensure its accessibility. The Data Coordination Platform (DCP) seeks to coordinate data accessibility through its deposition in a cloud-based platform and integrate it into a user-friendly portal.

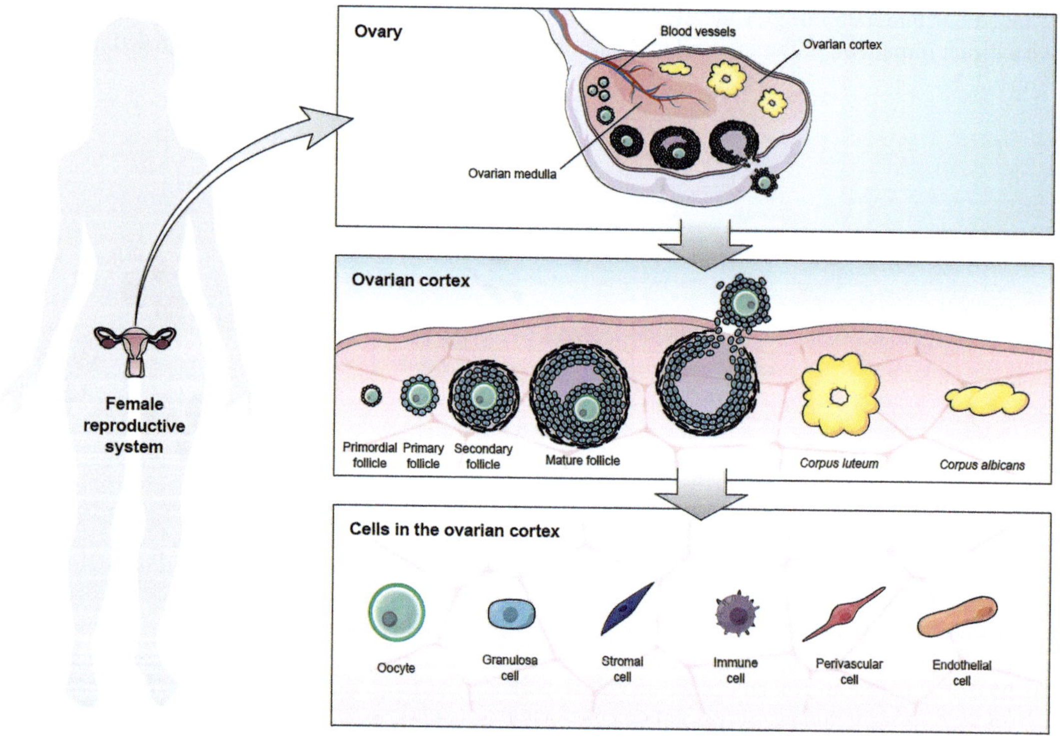

FIGURE 2.3 Example of an integrated map of the human ovary. Such maps can be queried by researchers and practitioners at the organ, tissue, and cell levels to obtain information about the molecular and organizational features of the structure of interest.

SINGLE-CELL RNA SEQUENCING TECHNOLOGIES

Single-cell sequencing is a relatively recent technological development which was named Method of the Year in 2013 by the journal *Nature Methods* (Nawy, 2014). The earliest single-cell sequencing method applicable on a large scale was transcriptome analysis *via* RNA-Seq (single-cell RNA sequencing, scRNA-seq) (Tang *et al.*, 2009). scRNA-seq marked an enormous progression in the definition of cell identities because it offered a comprehensive and unbiased characterization of the cell at the molecular level (Trapnell, 2015).

Currently available scRNA-seq protocols are based on technologies that vary greatly in cost and sensitivity. Individual strengths and weaknesses of scRNA-seq protocols need to be assessed in light of the biological questions of interest (Papalexi and Satija, 2018). Regardless of the underlying technology, they all involve a workflow which can be broken down into three main steps: physical separation of the cells; reverse transcription and PCR amplification of the polyadenylated RNAs; and sequencing of the PCR products. Depending on their approach to cell isolation and transcript amplification, the technologies can be broadly divided into plate-based, droplet-based, and combinatorial indexing protocols. Moreover, scRNA-seq data can now be complemented with spatial data to study the exact location of cells within a tissue.

Plate-Based Protocols

The first protocols for the unbiased quantification of the transcriptome of single cells were developed in a single tube containing lysis buffer (Tang *et al.*, 2009). Shortly after, multiplexing and robotics enabled the processing of hundreds of cells, which meant that cells could be processed in 96-well or 384-well plates (Baugh *et al.*, 2001; Islam *et al.*, 2011; Ramsköld *et al.*, 2012; Jaitin *et al.*, 2014; Zeisel *et al.*, 2015). Whereas the first protocols isolated cells *via* micro-pipetting and placed them into a plate containing lysis buffer, it is now common to isolate them by fluorescent activated cell sorting (FACS). This method allows the user to isolate cells by protein surface markers selected on the basis of prior information about the cell type. Following lysis of the cells, conversion of RNA into cDNA and subsequent cDNA amplification are performed separately on each cell and the PCR products are sequenced *via* next-generation sequencing (NGS) (Figure 2.4, top panel).

One example of plate-based technology that is broadly used is Smart-Seq2 (Picelli *et al.*, 2013, 2014). The advantage of this method is that it enables the quantification of full-length transcriptomes. This information has proven useful for the reconstruction of highly variable T- and B-cell receptors required to control B and T cell clonal expansion in response to antigens (Stubbington *et al.*, 2016; Lindeman *et al.*, 2018). Smart-seq2 has also been used to reconstruct haplotypes of KIR (killer-cell immunoglobulin-like receptor), which are key receptors involved in the activation and inhibitory states of NK cells (Vento-Tormo *et al.*, 2018). Smart-Seq3 is a newly defined protocol that offers even greater sensitivity than its predecessors, enabling isoform identification (Hagemann-Jensen *et al.*, 2020).

Droplet-Based Protocols

Instead of using wells to carry out the reverse transcription and PCR reactions, droplet-based methods, such as Drop-seq (Macosko *et al.*, 2015) and InDrop (Klein *et al.*, 2015), employ microdroplets which encapsulate a single cell and a gel bead covered with barcodes. After reverse transcription and amplification, the mRNAs inside each droplet are pooled together, significantly increasing the multiplexing of the methodology, and sequenced in parallel *via* NGS. This technology has been recently adapted and commercialized by 10x Genomics, increasing the accessibility of the method across the scientific community.

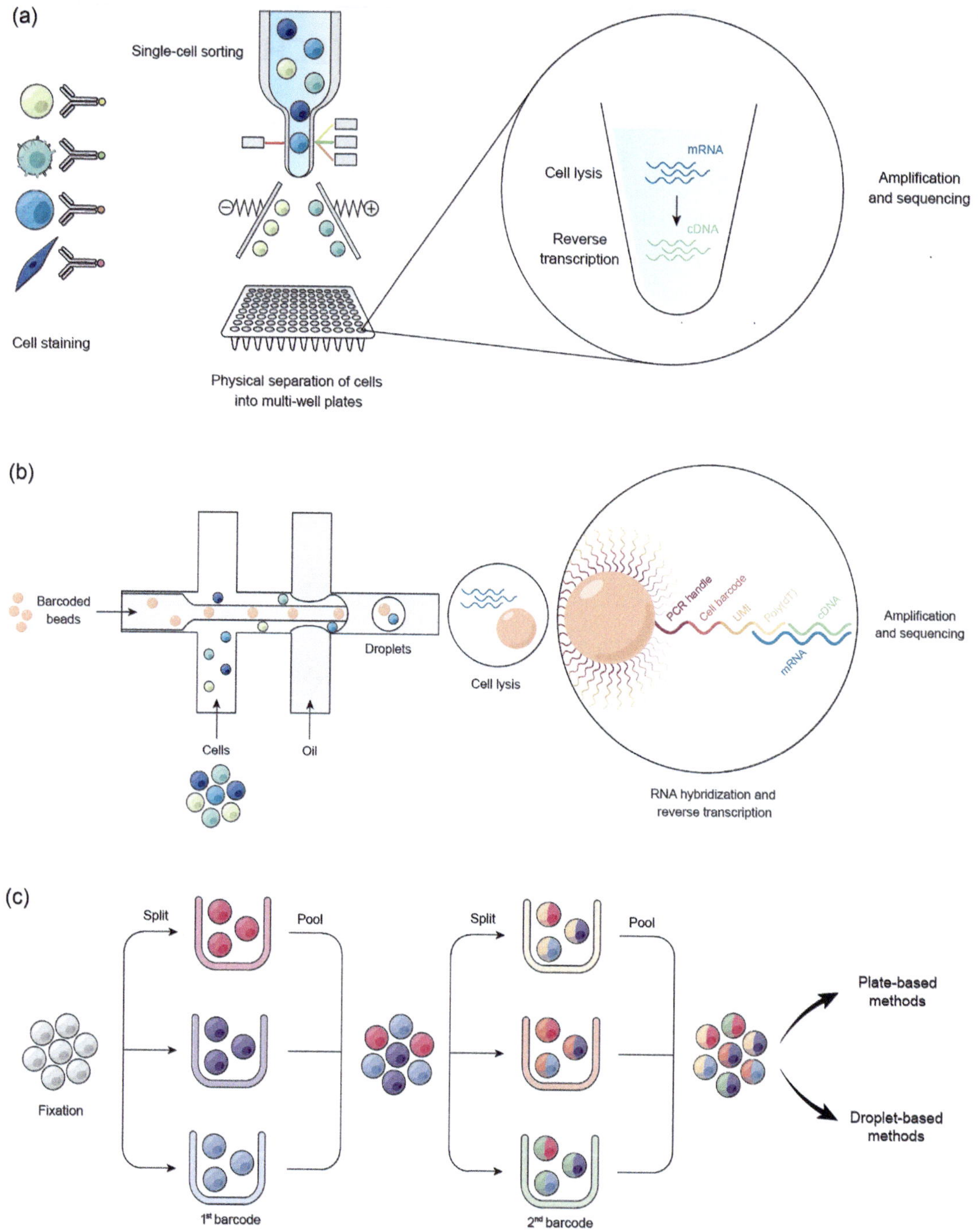

FIGURE 2.4 (a) In plate-based scRNA-seq protocols, cells are physically isolated into a multi-well plate using a cell-sorting technique. Cell lysis, reverse transcription, and amplification are carried out separately within each well. (b) Droplet-based scRNA-seq protocols employ a microfluidics device that enables the formation of microdroplets encapsulating both a single cell and a gel bead covered with barcodes. After reverse transcription and amplification, the mRNAs inside each droplet are pooled together for parallel sequencing. (c) Schematic representation of the principle behind combinatorial indexing protocols. Fixed cells are randomly split into a multi-well plate, where each well contains a unique barcode. After being labeled with the first barcode, the cells are pooled, shuffled, and split again randomly into the same set of wells. This so-called *split-pool* cycle is iterated until the number of combinations of possible barcodes is much higher than the number of cells being profiled.

In these protocols, individual barcoded gel beads and cells are flowed into a microfluidics device (Guo *et al.*, 2012) where a mixture of aqueous and oil flows allows for their compartmentalization into a single droplet. Each barcode is composed of three segments: a PCR handle to initiate the PCR reaction; a cell barcode that is unique to each bead; and a unique molecular identifier (UMI) which is different for each cellular mRNA molecule. Cell barcodes and UMIs are generated by randomly assembling nucleotides. UMIs provide a means to remove PCR artifacts during the step of RNA amplification. Once the beads and cells are trapped in the droplets, a lysis buffer breaks the cell membrane and releases the mRNAs, which come into contact with the barcodes and ligate to them. This results in every mRNA molecule present inside the droplet being labeled with the same cell barcode and a different UMI. It is thus possible to subsequently employ massive parallel sequencing without losing the information about the cell of origin of each mRNA (Figure 2.4, central panel).

The main benefit of droplet-based protocols lies in the large number of cells that can be profiled in parallel in a single experiment, the reduction in reagent volumes, and multiplexing, which directly translates into a much lower cost per cell. On the other hand, the sensitivity is often lower than with plate-based methods and only one end of the transcript is sequenced. Recent studies have shown that the definition of cell identity is highly dependent on the number of cells sequenced (Svensson, da Veiga Beltrame, and Pachter, 2019). Therefore, droplet-based methods are the strategies-of-choice for studying heterogeneous cell populations and discovering rare cell types.

Plate-Based Combinatorial Indexing Protocols

Building upon the concept of cellular barcoding, plate-based combinatorial indexing scRNA-seq protocols, such as single-cell combinatorial indexing RNA sequencing (sci-RNA-seq) (Cao *et al.*, 2017) and split-pool ligation-based transcriptome sequencing (SPLiT-seq) (Rosenberg *et al.*, 2018) allow the profiling of thousands of cells without having to physically isolate each cell. In these methods, the cells are fixed, and their mRNA is manipulated *in situ* by sequential addition of random barcodes in a combinatorial fashion. These strategies overcome the cell isolation inefficiency associated with the droplet-based methods and therefore reduce the pool of barcodes required (Figure 2.4, bottom panel).

In SPLiT-seq, formaldehyde-fixed cells are randomly split into a 96-well plate with unique barcodes for each well. All cells are labeled with the first barcode, and reverse transcription is carried out: at this stage; the chance of two cells having the same barcode is 1/96. The cells are then pooled, shuffled, and split again randomly into the same set of wells. After the second barcode is added, the probability that two cells have the same barcode is 1/9216 (96*96). This procedure can be iterated until the number of combinations of possible barcodes is much higher than the number of cells being profiled, and this is what overcomes the need for physically isolating each cell. The cells are eventually lysed, and their labeled cDNAs are PCR amplified and sequenced.

Droplet-Based Combinatorial Indexing Protocols

The inefficiency of droplet-based methods limits the number of cell barcodes that can be designed, and, consequently, limits their multiplexing capacity. On the other hand, the potential for massive-scale profiling of droplet-based methods still surpasses that of plate-based methods. Newly developed, droplet-based, combinatorial indexing protocols, such as single-cell combinatorial fluidic indexing (scifi-RNA-seq) (Datlinger *et al.*, 2019) and droplet single-cell assay for transposase-accessible chromatin using sequencing (dscATAC-seq) (Lareau *et al.*, 2019), apply the idea behind plate-based combinatorial indexing protocols to overcome the limitation of the cell isolation inefficiency of droplet-based methods and are thus able to fully exploit their potential for massive-scale profiling.

In scifi-RNA-seq, permeabilized cells are first pre-indexed by reverse transcription in microwell plates, similarly to plate-based combinatorial indexing protocols. Then, the pre-indexed cells are pooled, randomly mixed, and encapsulated with a microfluidic droplet generator. The encapsulation is carried out such that most droplets are filled with potentially more than one cell, and, inside the droplets, the transcripts are labeled with a second barcode. Because the cells are randomly mixed between the first (plate well) and the second (droplet) round of barcoding, the combination of the two barcodes enables the unique identification of individual cells. Moreover, by allowing each droplet to encapsulate multiple cells, scifi-RNA-seq solves the cell isolation inefficiency inherent to droplet-based scRNA-seq.

The combination of both methods results in an exponential increase in the number of cells that can be sequenced in each reaction and a significant reduction in reagent usage and hence costs. However, the higher number of cells also increases sequencing demands which have not experienced the exponential reduction in costs observed for the reagents. Therefore, it is important to select the method of choice based on the user's budget and the scientific question to address.

Integration of Sequencing and Spatial Data

Because single-cell sequencing methods require cells to be dissociated, information about the spatial location of each cell within the tissue of origin is inevitably lost. Knowledge of how the cells are spatially organized in the tissue, however, is crucial to fully understanding cellular identity. Indeed, both the "absolute" location of a cell within the tissue and its relative position compared to other cells are often linked to the cell's function. Combining such information with single-cell gene expression data, measured by RNA-seq, also allows the inference of complex interactions, such as cell-cell communication. Currently, scRNA-seq data can be integrated with either image-based or sequencing-based spatial data.

A commonly used image-based technique is fluorescence *in situ* hybridization (FISH). Originally designed for detecting chromosomal abnormalities in diseases, it was employed to visualize specific chromosomal locations by using fluorescent oligonucleotide probes complementary to the region of interest (Gall and Pardue, 1969). The application of the original molecular principle to image mRNA in an intact tissue is known as single-molecule FISH (smFISH) and allows for the visualization of individual cells (including their subcellular structure) along with their spatial coordinates within the tissue (Figure 2.5, top panel). Spatial technologies based on smFISH vary greatly in scale, ranging from tens of imaged mRNAs/cell, such as in RNAscope (Wang *et al.*, 2012) and SABER-FISH (Kishi *et al.*, 2019), to hundreds or thousands of imaged mRNAs/cell *via* the use of imageable barcodes, such as in MERFISH (Xia *et al.*, 2019) and seqFISH (Qian *et al.*, 2020). By integrating scRNA-seq and smFISH data, it is possible to transfer cell-type annotations derived from gene expression measurements of the entire transcriptome to spatially resolved cells. Once cell types are mapped onto the tissue, the architecture of the tissue can then be reconstructed in terms of both its overall cellular composition and the spatial relationships among cell types.

More recently, sequencing-based methods, such as the spatial transcriptomic technology (Visium) commercialized by 10x Genomics, allow the whole transcriptome to be mapped directly onto the tissue (Ståhl *et al.*, 2016). In Visium, the tissue slide is first placed onto a chip used to measure gene expression, which is made up of capture areas containing thousands of barcoded spots. Each spot consists of millions of oligonucleotides with unique spatial barcodes. After permeabilization of the tissue, mRNAs can bind to the barcoded oligonucleotides on the underlying spot by diffusion. Reverse transcription is carried out and cDNAs are pooled and sequenced *via* NGS as in massive parallel scRNA-seq protocols (Figure 2.5, bottom panel). Although greatly promising, the Visium technology does not reach single-cell resolution when measuring transcriptome-wide gene expression because the size of each spot (~50 μ) is several times the average size of a cell (approximately 10–50 cells/spot). To overcome this limitation, matched scRNA-seq from the same sample, can be used to deconvolve the Visium data from spatially resolved spots to spatially resolved cells (Kleshchevnikov *et al.*, 2022; Andersson *et al.*, 2019, 2020). More recent, sequencing-based spatial technologies include Slide-seq (Rodriques *et al.*, 2019), which employs barcoded beads spatially indexed by SOLiD sequencing, and high-definition spatial transcriptomics (Vickovic *et al.*, 2019), which uses a high-density bead array and decodes each bead's location *via* rounds of hybridization with complementary, labeled, decoder oligonucleotides.

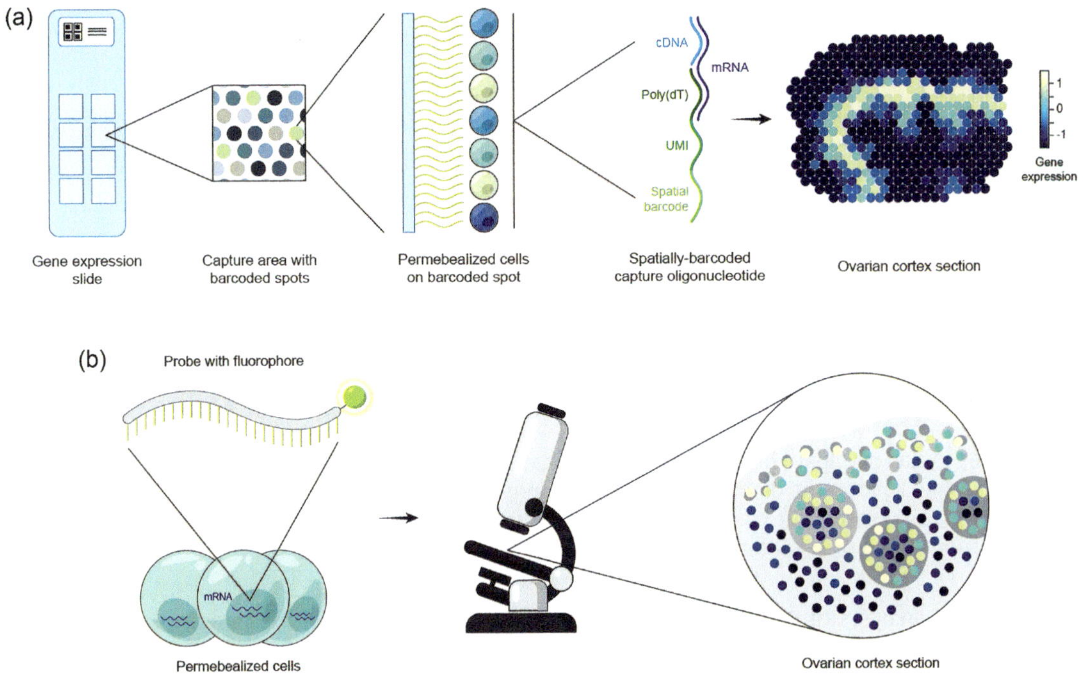

FIGURE 2.5 (a) Schematic representation of the main steps in the workflow for Visium, as an example of sequencing-based spatial technologies. The tissue is first loaded onto the gene expression slide and then permeabilized so that all cellular mRNAs can bind to the barcoded oligonucleotides on the slide. (b) Schematic representation of the main steps in the workflow for imaging-based spatial technologies. Fluorescent oligonucleotides are incubated with the permeabilized tissue, allowing for their hybridization with cellular mRNAs of interest and subsequent imaging under the microscope.

TOWARD SINGLE-CELL REPRODUCTIVE MEDICINE

The potential applications of the HCA to both basic and translational research are limitless. By taking our understanding of human biology to a higher level of resolution, an atlas can help uncover the cellular basis of disease as well as lead to the development of improved *in vitro* models, revolutionizing drug testing. The aforementioned applications are particularly relevant in reproductive medicine, where the physiology of the tissues and many widespread diseases are still poorly understood. In addition, there is a need for better *in vitro* models to accelerate the process of drug development and testing (Figure 2.6).

Single-Cell Atlases of Healthy Tissue

Before we can study the cellular dysregulation brought about by disease, it is imperative to have a comprehensive reference map of the molecular state of the cells in the healthy human tissue. The human reproductive system is responsible for producing gametes and hormones and for accommodating and nurturing the fetus, all of which are highly regulated functions that require timely activation of distinct cellular phenotypes. Single-cell transcriptomics has already helped shed light on the diversity of cells present in the human reproductive system, including primary and secondary reproductive organs across various stages of development, and have hinted at the cellular mechanisms potentially underlying many pathological conditions.

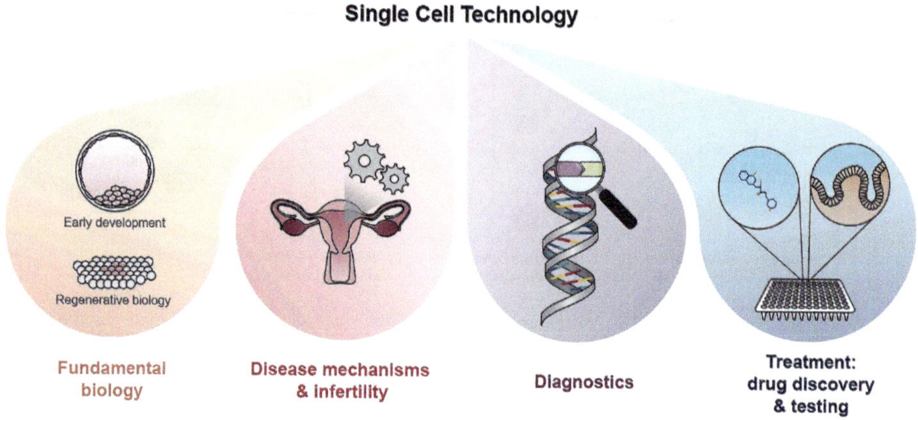

FIGURE 2.6 Areas of impact of single-cell technologies in reproductive medicine. Single-cell technologies have already shed light on the biological mechanisms underlying the functioning of the healthy reproductive system as well as several disease conditions. Moreover, they are instrumental in the development of new tools for testing and treating such conditions.

Single-cell analyses of the neonatal, adolescent, and adult male primary reproductive organs elucidated the transcriptomic and developmental stages of human spermatogenesis and investigated the cellular crosstalk between germline and somatic cells present in the testes (Guo *et al.*, 2018; Sohni *et al.*, 2019; Shami *et al.*, 2020; Xia *et al.*, 2020). Similarly, scRNA-seq profiling of the adult ovaries provided a map of the molecular signatures of the cell types in the inner and outer ovarian cortices at the different stages of follicular development (Fan *et al.*, 2019; Wagner *et al.*, 2020). Altogether, these datasets constitute an invaluable resource with which to dissect the potential mechanisms of male and female infertility as well as other pathologies associated with the reproductive system, such as cancer, and to develop new treatments and assisted reproductive technologies.

Other studies include atlasing of secondary female reproductive organs, such as the endometrium (Suryawanshi *et al.*, 2018; Vento-Tormo *et al.*, 2018; Wang *et al.*, 2018, 2020; Lucas *et al.*, 2020) and the fallopian tube (Dinh *et al.*, 2021; Hu *et al.*, 2020). Single-cell transcriptomic analysis of endometrial biopsies across the menstrual cycle revealed a high degree of heterogeneity in the cellular composition of the tissue, with characteristic signatures for each cell type and phase of endometrial transformation (Suryawanshi *et al.*, 2018; Vento-Tormo *et al.*, 2018; Wang *et al.*, 2018, 2020; Lucas *et al.*, 2020). The fallopian tube also undergoes structural changes in response to the menstrual cycle and is thought to harbor the cell-of-origin for many high-grade serous ovarian cancers (HGSOCs). Droplet-based scRNA-seq of fallopian tubes from healthy individuals revealed the transcriptional programs underlying different epithelial cell populations (Dinh *et al.*, 2020; Hu *et al.*, 2020). Furthermore, computational deconvolution of HGSOCs based on the transcriptional signatures of the epithelial populations present in the healthy tissue revealed that early secretory epithelial cells from the fallopian tubes are likely to be the precursor state for many HGSOCs (Dinh *et al.*, 2020; Hu *et al.*, 2020).

Single-Cell Atlases of Disease

Disease involves the disruption of normal cellular functions and interactions. Initial efforts toward building reference single-cell maps of non-physiological conditions in the field of reproductive medicine, such as ovarian cancer, have already provided invaluable insights. They have helped uncover heterogeneity within these conditions and provided mechanistic insights into their development.

Among the plethora of malignancies affecting the ovary, HGSOC predominates in the clinical setting and is infamous for its high fatality rate and poor prognosis (Lisio *et al.*, 2019). At the time of diagnosis, one-third of patients with HGSOC present ascites, which act as a reservoir of cell types that provide a tumor-promoting microenvironment for cancer cells (Ahmed and Stenvers, 2013). A recent analysis of ascites samples from HGSOC patients, using droplet-based and plate-based scRNA-seq, resolved the expression profiles of cancer, immune, and stromal cells. It uncovered inter-patient and intra-patient heterogeneity within cancer-associated fibroblasts (CAFs) and macrophages found in ascites and suggested a role for JAK-STAT signaling in both malignant cells and CAFs. It also helped redefine the immunoreactive and mesenchymal subtypes of HGSOCs described by The Cancer Genome Atlas (TCGA), finding that they are derived from macrophages and CAFs, respectively, rather than from malignant cells (Izar *et al.*, 2020).

Improving *In Vitro* Models

Mapping the development of malignant as well as non-malignant cells is important to understanding the origin and progression of conditions such as HGSOCs *in vivo* lineage tracing experiments have provided invaluable insights into how cells differentiate in mice although translating these findings to humans is often challenging. A solution to this challenge is offered by *in vitro* models, which allow researchers to study and manipulate developmental processes. Then, scRNA-seq can be used to obtain an unbiased and comprehensive read-out of the changes occurring at the transcriptome level as cells develop or are perturbed in the dish.

Prompted by the *in vivo* findings about the potential role of JAK-STAT signaling in the ascites of HGSOC patients, Izar *et al.* (2020) used primary HGSOC cell lines and patient ascites-derived xenograft models to test the effects of JAK-STAT signaling inhibition. They performed a drug screen with compounds targeting different nodes of the JAK-STAT signaling pathway and identified one compound, JSI-124, as having potent anti-tumor activity. Taken together, the results from the *in vivo* and *in vitro* analyses revealed that inhibition of the JAK-STAT signaling pathway may be a therapeutic option for HGSOC patients (Izar *et al.*, 2020).

Other noteworthy studies employed scRNA-seq to disentangle the heterogeneity of endometrial organoids (Fitzgerald *et al.*, 2019) and cultures of primary decidualizing endometrial stromal cells (Lucas *et al.*, 2020). Future quantitative comparisons of *in vitro* models and their corresponding *in vivo* single-cell references will prove useful in improving the efficiency and accuracy of current *in vitro* models (Boretto *et al.*, 2017; Turco *et al.*, 2017). In addition, such comparisons will help us develop more accurate *in vitro* models that mirror the fundamental developmental events that give rise to the formation and differentiation of reproductive organs.

CONCLUSION AND FUTURE PERSPECTIVES

While scRNA-seq remains the most widely used single-cell sequencing approach, it is now also possible to achieve single-cell resolution when measuring chromatin accessibility, DNA methylation, cell surface proteins, histone modifications, and chromosomal information (Stuart and Satija, 2019). Moreover, as described in this chapter, image-based and sequencing-based spatial technologies can currently be used to locate cells anatomically within a tissue. Integrating the aforementioned data types or simultaneously measuring multiple modalities is invaluable when defining cell identities. Unsurprisingly, single-cell multimodal omics (i.e., a combination of distinct single-cell genomic methods) was named Method of the Year in 2019 ("Method of the Year 2019: Single-cell multimodal omics," Teichmann, 2020).

Creating single-cell references of the human reproductive system that encompass information from multiple layers of regulation of the genome will prove crucial in further dissecting the cellular mechanisms underlying both physiological and pathological conditions. Moreover, such multi-omic atlases will serve as a framework for the development of better *in vitro* models. Altogether, single-cell sequencing technologies are taking us towards a comprehensive understanding of the complexity of our body, by—to put it in the words of the HCA—"mapping the human body one cell at a time."

ACKNOWLEDGMENTS

We thank Antonio Garcia, scientific illustrator from Bio-Graphics, for help with the illustrations and Charlotte Wright, Luz Garcia-Alonso, and Regina Hoo for insightful comments. R.V-T is supported by Wellcome Sanger core funding (grant number WT206194), and V.L is supported by the European Union's Horizon 2020 Research and Innovation Programme under grant agreement No. 874741.

AUTHORS' CONTRIBUTIONS

V.L. and R.V.-T .wrote and edited the manuscript.

BIBLIOGRAPHY

Ahmed, N., *et al.* (2013) *Frontiers in Oncology*, 3, p. 256.
Andersson, A., *et al.* (2019) *bioRxiv*. Available at: https://www.biorxiv.org/content/10.1101/2019.12.13.874495v1.full
Andersson, A. *et al.* (2020) *Communications Biology*, 3(1), p. 565.
Baugh, L. R., *et al.* (2001) *Nucleic Acids Research*, 29(5), p. E29.
Boretto, M., *et al.* (2017) *Development*, 144(10), pp. 1775–1786.
Cao, J., *et al.* (2017) *Science*, 357(6352), pp. 661–667.
Datlinger, P., *et al.* (2019) *bioRxiv*. Available at: https://www.biorxiv.org/content/10.1101/2019.12.17.879304v1.abstract.
Dinh, H. Q., *et al.* (2021) *Cell Reports*, 35(2), p. 108978.
Fan, X., *et al.* (2019) *Nature Communications*, 10(1), p. 3164.
Fitzgerald, H. C., *et al.* (2019) *Proceedings of the National Academy of Sciences of the United States of America*, 116(46), pp. 23132–23142.
Gall, J. G., et al. (1969) *Proceedings of the National Academy of Sciences of the United States of America*, 63(2), pp. 378–383.
Guo, J., *et al.* (2018) *Cell Research*, 28(12), pp. 1141–1157.
Guo, M. T., *et al.* (2012) *Lab on a Chip*, p. 2146.
Hagemann-Jensen, M., *et al.* (2020) *Nature Biotechnology*.
Hu, Z., *et al.* (2020) *Cancer Cell*, 37(2), pp. 226–242.e7.
Islam, S., *et al.* (2011) *Genome Research*, 21(7), pp. 1160–1167.
Izar, B., *et al.* (2020) *Nature Medicine*, 26, pp. 1271–1279.
Jaitin, D. A., *et al.* (2014) *Science*, 343(6172), pp. 776–779.
Kishi, J. Y., *et al.* (2019) *Nature Methods*, 16(6), pp. 533–544.
Klein, A. M., *et al.* (2015) *Cell*, 161(5), pp. 1187–1201.
Kleshchevnikov, V., *et al.* (2022) *Nature Biotechnology*.
Korsunsky, I., *et al.* (2019) *Nature Methods*, 16(12), pp. 1289–1296.
Kulkarni, A., *et al.* (2019) *Current Opinion in Biotechnology*, pp. 129–136.
Lander, E. S. (1996) *Science*, 274(5287), pp. 536–539.

Lareau, C. A., *et al.* (2019) *Nature Biotechnology*, pp. 916–924.
Lindeman, I., *et al.* (2018) *Nature Methods*, 15(8), pp. 563–565.
Lisio, M.-A., *et al.* (2019) *International Journal of Molecular Sciences*, 20(4).
Lopez, R., *et al.* (2018) *Nature Methods*, 15(12), pp. 1053–1058.
Lucas, E. S., *et al.* (2020) *Communications Biology*, 3(1), p. 37.
Macosko, E. Z., *et al.* (2015) *Cell*, 161(5), pp. 1202–1214.
Nawy, T. (2014) *Nature Methods*, 11(1), p. 18.
Ordovas-Montanes, J., *et al.* (2018) *Nature*, 560(7720), pp. 649–654.
Papalexi, E., et al. (2018) *Nature Reviews. Immunology*, 18(1), pp. 35–45.
Park, J.-E., *et al.* (2020) *Science*, 367(6480).
Picelli, S., *et al.* (2013) *Nature Methods*, 10(11), pp. 1096–1098.
Picelli, S., *et al.* (2014) *Nature Protocols*, 9(1), pp. 171–181.
Plasschaert, L. W., *et al.* (2018) *Nature*, 560(7718), pp. 377–381.
Polański, K., *et al.* (2020) *Bioinformatics*, 36(3), pp. 964–965.
Popescu, D.-M., *et al.* (2019) *Nature*, 574(7778), pp. 365–371.
Qian, X., *et al.* (2020) *Nature Methods*, 17(1), pp. 101–106.
Raj, A., et al. (2008) *Cell*, pp. 216–226.
Ramachandran, P., *et al.* (2019) *Nature*, 575(7783), pp. 512–518.
Ramsköld, D., *et al.* (2012) *Nature Biotechnology*, 30(8), pp. 777–782.
Regev, A., *et al.* (2017) *eLife*, 6.
Regev, A., *et al.* (2018) *ArXiv*. Available at https://arxiv.org/abs/1810.05192.
Rodriques, S. G., *et al.* (2019) *Science*, 363(6434), pp. 1463–1467.
Rosenberg, A. B., *et al.* (2018), *Science*, 360(6385), pp. 176–182.
Rozenblatt-Rosen, O., *et al.* (2017) *Nature News*, 550(7677), p. 451.
Shami, A. N., *et al.* (2020) *Developmental Cell*, 54(4), pp. 529–548.e12.
Smillie, C. S., *et al.* (2019) *Cell*, 178(3), pp. 714–730.e22.
Sohni, A., *et al.* (2019) *Cell Reports*, 26(6), pp. 1501–1517.e4.
Ståhl, P. L., *et al.* (2016) *Science*, 353(6294), pp. 78–82.
Stewart, B. J., *et al.* (2019) *Science*, 365(6460), pp. 1461–1466.
Stuart, T., et al. (2019) *Nature Reviews Genetics*, pp. 257–272.
Stubbington, M. J. T., *et al.* (2016) *Nature Methods*, 13(4), pp. 329–332.
Suryawanshi, H., *et al.* (2018) *Science Advances*, 4(10), p. eaau4788.
Svensson, V., *et al.* (2020) *Database: the journal of biological databases and curation.*
Svensson, V., *et al.* (2018) *Nature Protocols*, 13(4), pp. 599–604.
Tang, F. *et al.* (2009) *Nature Methods*, 6(5), pp. 377–382.
Teichmann, S. (2020) *Nature Methods*, 17(1), p. 1.
Trapnell, C. (2015) *Genome Research*, 25(10), pp. 1491–1498.
Turco, M. Y., *et al.* (2017) *Nature Cell Biology*, 19(5), pp. 568–577.
Vento-Tormo, R., *et al.* (2018) *Nature*, 563(7731), pp. 347–353.
Vickovic, S., *et al.* (2019) *Nature Methods*, 16(10), pp. 987–990.
Vieira Braga, F. A., *et al.* (2019) *Nature Medicine*, 25(7), pp. 1153–1163.
Wagner, M., *et al.* (2020) *Nature Communications*, 11(1), p. 1147.
Wang, F., *et al.* (2012), *The Journal of Molecular Diagnostics*, 14(1), pp. 22–29.
Wang, W., *et al.* (2018) *Fertility and Sterility*, p. e2.
Wang, W., *et al.* (2020) *Nature Medicine*, 26(10), pp. 1644–1653.
Wang, Z., *et al.* (2009) *Nature Reviews. Genetics*, 10(1), pp. 57–63.
Xia, B., *et al.* (2020), *Cell*, 180(2), pp. 248–262.e21.
Xia, C., *et al.* (2019) *Proceedings of the National Academy of Sciences of the United States of America*, 116(39), pp. 19490–19499.
Zeisel, A., *et al.* (2015) *Science*, 347(6226), pp. 1138–1142.

The Genetic Risk of a Couple Aiming to Conceive

3

Joe Leigh Simpson, Svetlana Rechitsky, and Anver Kuliev

INTRODUCTION

For over 50 years, ethnicity-based screening has identified asymptomatic individuals who are heterozygous for selected disorders. Tay-Sachs disease and hemoglobinopathies were the initial targets of screening, meeting the requisite requirement for a protein-based gene product in which it was possible to distinguish between heterozygous and homozygous asymptomatic individuals. In recent years, it has become clear that the clinical usefulness of ethnicity-based screening was less than predicted, in part because not all individuals of a given ethnic group are aware of their own ancestry. Pan-ethnic screening of carriers, based on DNA sequencing, is now pursued, increasing the number of at-risk couples identified. The genetic basis for at least 5,000 rare Mendelian traits has been determined, so the potential benefit is enormous (1). Moreover, at least 100,000 robust associations have been identified between genomic regions and common genetic diseases (1). Although not yet practicable, aspirational goals include DNA sequencing of every embryo or every fetus in every pregnancy. Chromosomal abnormalities associated with advanced maternal age can be detected non-invasively during pregnancy. Advanced paternal age is not associated with chromosomal abnormalities but is correlated with increased frequency of *de novo* mutations causing single-gene disorders.

Rationale for Determining Increased Risk

Couples wanting to conceive naturally wish to learn whether their children will be normal and healthy. This applies to those able to conceive naturally as well as to those being treated for infertility. Concern is genuine because 2–3% of liveborn infants have a major congenital anomaly or a disorder of genetic etiology. Half of these are evident at birth, as a result of a chromosome abnormality (0.6%), a single-gene disorder (1%), or a trait showing complex (polygenic) inheritance (1%).

DOI: 10.1201/9781003024941-3

Ideally, pre-conception identification of risk could lead directly to the primary prevention of occurrence of affected offspring. An excellent example is maternal administration of a vitamin (folic acid) pharmaceutical agent (2). Maternal administration of folic acid reduces the incidence of spina bifida by 75%, but few other examples have had such an impact. A major impediment to this approach is that, in the United States, half of all pregnancies are unplanned. Even when a pregnancy is planned, aspiring parents rarely query in advance with regard to applicable, preventative measures. Before implantation, preimplantation genetic testing (PGT) can select a euploid (PGT-A) embryo or an embryo lacking a mutant monogenic genotype (PGT-M).

Once a clinical pregnancy is established, several approaches can be used to maximize the normal outcome of a liveborn pregnancy. The specific approach depends on the etiology of the disorder of concern and the stage of gestation. In the first trimester, chorionic villus sampling provides villi to assess. In the second trimester, amniocentesis provides amniotic fluid cells. DNA obtained allows sequencing of targeted regions or even the entire exome (whole-exome sequencing, WES) or genome (whole-genome sequencing, WGS). "Non-invasive" prenatal testing (NIPT) utilizes cell-free DNA in the maternal plasma to identify chromosomal abnormalities and selected monogenic disorders.

Family History

Family history helps define risks for occurrence of chromosomal abnormalities, monogenic disorders, and polygenic disorders in a given family. This information in turn will direct the health provider to identify specific genetic tests that assess concerns over risks of the condition in question. A genetic history involves inquiring into the health status of first-degree relatives (siblings, parents, or offspring), second-degree relatives (nephews, nieces, aunts, uncles, grandparents, or grandchildren), and third-degree relatives (maternal or paternal first cousins). Similarly, genetic history is warranted for family members of sperm or oocyte donors. A thorough history will frequently raise potential concerns, but relatively few concerns actually warrant formal clinical genetic referral for diagnostic validation and counseling. Most findings are sufficiently straightforward that a well-informed reproductive medicine specialist can give advice concerning specific risk. If the risk is sufficiently low, no invasive procedures or other testing may be necessary. Only rarely will a birth defect in a third-degree relative warrant concern. A cardiac defect of polygenic etiology in a second- or third-degree relative confers little increased risk over that in the general population. However, a second-degree relative (e.g., grandparent) having certain autosomal dominant disorders may warrant physical examination of his or her grandchildren, who may display physical features suggesting the diagnosis of a transmissible mutation. An example is multiple *café-au-lait* spots indicating neurofibromatosis, especially if located in the axillary region.

Family history might reveal recurrent miscarriages, stillbirths, or anomalous liveborn infants that require counseling and laboratory testing. A woman having recurrent miscarriages is at increased risk of structural chromosomal rearrangements, especially if she is < 25 years old and has a sibling who also had miscarriages (3). Management recommendations need to be provided, and specific diagnostic approaches outlined.

Parental ages should always be recorded. Advanced *maternal* age confers increased risk of aneuploid offspring (4). Increased *paternal* age confers little increased risk for aneuploid offspring, although sperm with cytological abnormalities (e.g., breakage) may be associated with adverse pregnancy outcomes that result from defective DNA repair. The risk of a *de novo* single- gene mutation increases, as noted, with advanced paternal age. There is, however, no specific maternal or paternal age threshold beyond which a sperm or oocyte donor should categorically be excluded. Each reproductive medicine unit should develop and adhere to its own consistent policy.

Ethnicity should always be recorded. It has already been noted that some autosomal recessive disorders are sufficiently common that universal carrier screening is recommended in asymptomatic individuals. Discussed later in this chapter, and in Chapter 4, pan-ethnic screening now allows hundreds of single-gene disorders to be screened to detect carriers to identify at-risk pregnancies.

An individual presenting for infertility treatment is usually assumed to have no heritable condition, but this must be confirmed. An ostensibly healthy but infertile adult may actually have features indicating mild expression of a single-gene disorder. Irrespective of mild severity (expressivity), these individuals may have a mutant gene that is potentially transmissible to offspring. A chromosomal disorder may be recognized for the first time at infertility consultation, for example 47, XXY Klinefelter syndrome having azoospermia. A male with congenital bilateral absence of vas deferens (CBAVD) may have a pathogenic variant of the cystic fibrosis gene. He may have one ΔF508 allele and one abnormal 5T allele. If the clinically unaffected partner is heterozygous for ΔF508, the risk of homozygous ΔF508 cystic fibrosis in the offspring is one in four.

Morphological or developmental abnormalities may result from teratogenic factors (prescription drugs, chemotherapeutic agents, radiation) interfering with development of an embryo/fetus that otherwise would have been normal. Although the subject of considerable media attention and parental concern, there are actually few proven teratogens. Standard obstetrics and pediatric texts list teratogenic drugs, pathogenic viruses, and environmental toxicants, the latter of which are often poorly defined. The most commonly encountered detrimental environmental toxicants are mercury, heavy metals, phthalates, and pesticides. Environmental exposure levels of these compounds usually do not rise to the level of clinical concern, but this might not be so if exposure were under occupational circumstances. Alteration in job assignments during pregnancy may be necessary.

Primary and Secondary Prevention of Chromosomal Abnormalities and Birth Defects

Primary prevention of aneuploid offspring is not generally possible, although pre-conceptional analysis of the first polar body (PBI) technically provides this option. At this stage, fertilization has not occurred and metaphase II oocytes can be recovered. Liveborn outcomes can be predicted by analysis of sequential first and second polar body (PB2), which is expelled before fusion of maternal and paternal pronuclei (syngamy) to generate the zygote. Pre-conception testing to take into account the uncommon (up to 5%) paternal aneuploidy is not possible. Another form of primary prevention is PGT-A (5, 6). Embryo selection excludes affected embryos; thus, implantation of affected embryos can be avoided without clinical termination of pregnancy.

Secondary prevention is standard obstetrical practice. Interrogation of an established pregnancy traditionally follows an invasive prenatal diagnostic procedure (amniocentesis or chorionic villus sampling, CVS). Chromosomal abnormalities exist in 1 in 200 liveborn infants. Attracting the greatest clinical attention are the three autosomal trisomies compatible with livebirth (7–9). All are associated with advanced maternal age (10–12), and usually arise during maternal meiosis I. Liveborn risk for trisomy 21 at maternal age 35 years is 0.30% (3/1000). The prevalence at mid-trimester (assessed by amniocentesis) is 0.33%, whereas prevalence in the first trimester (assessed by CVS) is 0.39% (11). Higher rates earlier in gestation reflect ongoing natural selection against aneuploid embryos and fetuses (12).

Gestation-specific prevalences apply to total chromosomal abnormalities as well as for specific trisomies. For trisomies other than trisomy 21, decreases in prevalence proceed more rapidly as gestation advances (11). Chromosomal abnormalities are responsible for 50–60% of first-trimester miscarriages, only 5–10% of stillborn infants, and only 0.6% of liveborn infants.

Recurrent Aneuploidy

After one aneuploid offspring (liveborn or miscarried), the risk of having another aneuploid offspring increases. The likelihood of liveborn trisomy recurring is approximately 1% (12), for which reason prenatal genetic diagnosis is routinely offered. That aneuploidy can be recurrent is shown by tracking chromosomal outcome in sequential miscarriages. In successive pregnancies, chromosomal status is not random.

Rather, miscarriages in a single individual are more likely to be either recurrent aneuploidy or recurrent euploidy (13–15). That is, if the chromosomal complement of the first miscarriage is aneuploid, the likelihood is increased that the complement of the second miscarriage will also be aneuploid. Recurrent aneuploidy may involve not only autosomal trisomy, but other numerical chromosomal abnormalities as well. The significance of this observation is that the initial aneuploid event may have resulted in a lethal chromosome (e.g., trisomy 20), but a subsequent miscarriage may involve a chromosome that is compatible with livebirth (e.g., trisomy 21). Thus, an invasive prenatal genetic diagnosis or PGT-A should be offered after any aneuploid miscarriage.

Plausible biological explanations exist for recurrent aneuploidy. Meiosis and mitosis are under genetic control. Mutations may result from faulty DNA repair, failed chromosomal synapsis, or perturbed homologous recombination. Chromosomal disjunction can result in embryos having chromosomal abnormalities that manifest as either trisomic miscarriages or trisomic liveborns. Chromosomal status of miscarriages can be determined through so-called non-invasive 24-chromosome cell-free DNA analysis recovered from maternal plasma. This methodology is used routinely to assess liveborn trisomies 13, 18, and 21 in ongoing pregnancies, but can be applied to all chromosome in all pregnancies, including those having resulted in miscarriages.

Chromosomal Translocation

Structural chromosomal rearrangements may be ascertained in the context of karyotypes undertaken following invasive prenatal diagnostic procedures or analysis of products of conception (miscarriage). In a translocation, regions of different chromosomes have been interchanged. No maternal age effect exists. There may or may not be associated phenotypic abnormalities. A pregnancy resulting in a child with a *de novo* structural chromosome alteration (i.e., where neither parent has a rearrangement) has a 1% empirical risk of another affected offspring in any subsequent pregnancy. This is best explained by germline mosaicism, the abnormal index case presumably arising in a single parental gamete. Balanced translocations are most often ascertained clinically in a couple having had either recurrent miscarriages or a child with birth defects. Likelihood of recurrence is much higher than the *de novo* 1%.

Reciprocal translocations involve interchange of chromosomal material between two or more chromosomes. Breaks occur at various locations in the long or short arms; the phenotype depends on the specific chromosomal segments duplicated or deficient. If no phenotypic consequences are evident, the translocation is said to be balanced. Robertsonian translocations involve centric fusion between two acrocentric (13, 14, 15, 21, 22). Genomes of acrocentric chromosomes are predominantly concentrated on the long arms, with the very small, short arm coding only for ribosomal RNA. In Robertsonian translocations, two acrocentric chromosomes have thus effectively fused. In a balanced Robertsonian translocation having a normal phenotype, there are 45 chromosomes rather than the usual 46.

Parental karyotypes or array CGH are routinely recommended in couples experiencing recurrent miscarriages. If detected, likelihood of a successful pregnancy in a given ART cycle is decreased considerably because chromosomal segregation results in a disproportionate number of unbalanced embryos that cannot result in a full-term pregnancy. Table 3.1 shows one cohort of embryos in a couple in which one partner had a translocation that necessitated embryo selection by PGT-SR.

Frequency and management of translocation carriers

In 2% of couples experiencing repeated miscarriages, one partner shows a balanced translocation (7, 16). Stephenson and Sierra (16) studied 1893 couples and found 51 with a structural rearrangement. Among the 51 were 28 with a reciprocal translocation, 12 with a Robertsonian translocation, seven with an inversion, and four with other rearrangements (e.g., excessive centromeric heterochromatin). Frequency of a balanced translocation is higher in females than males, and higher still if there exists a family history of a stillborn or abnormal liveborn. Translocation heterozygosity is not correlated with advanced maternal age

TABLE 3.1 Preimplantation Genetic Testing for Aneuploidy (PGT-A) and for Chromosomal Structural Rearrangements (PGT-SR) in Cohorts of Embryos from a Couple Having a Reciprocal Translocation Involving Chromosomes 4 and 17 (46, XY, del(4;17)(q35;q23))

CYCLE	*EMBRYO*	*RESULTS*	*INTERPRETATION*	*EMBRYO TRANSFER*
1	4	Complex chromosomal abnormalities	Unbalanced / Abnormal	No
2	1	46, XY, del(17)(q23.1; q25.3)	Unbalanced / Abnormal	No
	2	46, XY / NGS(1-22)x2,(XY)x1	Normal / Balanced Male	Yes
	4	45, XY, −21	Aneuploid	No
	5	Complex chromosomal abnormalities	Unbalanced / Abnormal	No
	10	45, XX, −14	Aneuploid	No

or with consecutive miscarriages. The likelihood of detecting a balanced translocation does not substantively differ after one, two, or three miscarriages.

Likelihood of abnormal liveborns resulting from a balanced translocation

As a result of normal segregation, offspring may be unbalanced or balanced if one parent has a balanced translocation. Management of couples with a balanced translocation varies, but all begin with detection. Parental karyotypes were once universally recommended after two or three miscarriages, consecutive or not. Alternatively, RCOG (Royal College of Obstetricians and Gynaecologists) (17), ESHRE (European Society of Human Reproduction and Embryology) (18) and Anton (19) reason that a) the likelihood of there being a balanced translocation is low (2%), and b) the likelihood of abnormal liveborns deriving from such a translocation is low as for an inversion (19). Thus, testing for parental karyotypes may not be cost effective. Franssen et al. (3) provide a useful table estimating the likelihood of detecting a balanced translocation in a couple experiencing recurrent miscarriages, taking into account maternal age, number of previous miscarriages, and presence or absence of a sib who had a miscarriage. The likelihood of a balanced translocation being present is highest in a couple of maternal age under 25 years old, who has experienced multiple losses and also has a sib having experienced miscarriages.

Theoretical versus empirical risks

Theoretical risks for a parent with a balanced reciprocal translocation of having unbalanced translocation offspring can be calculated, but more useful are empirical data. Theoretical risks are greater than empirical risks (20), and less clinically applicable. However, empirical data are not always available for a specific reciprocal translocation. Even if the same two chromosomes are involved, even if also the same arms (short arm or long arm), the precise chromosomal breakpoints very likely will differ. Thus, only generalizations are possible, derived from pooled data compiled from different translocations.

Empirical data exist more often for balanced Robertsonian translocations. For a female carrying a t(14q;21q) translocation, the empirical risk of having a genetically unbalanced translocation offspring is approximately 10%. Risk is less than 2% for offspring of a male carrying such a translocation. Robertsonian translocations involving chromosomes other than chromosome 21 show lower empirical liveborn risks. In t(13q;14q), the liveborn risk for trisomy 13 is 1% or less. The lower risk compared to t(14q;21q) presumably reflects the greater lethality of trisomies or monosomies involving chromosomes 13, 14, 15, or 22 than those involving chromosome 21.

A variable affecting empirical counseling in a case with a parental balanced translocation is mode of identification. Risk of an unbalanced liveborn offspring is 3% if the parental balanced translocation was identified through recurrent miscarriage, but 20% when ascertained through an abnormal liveborn.

Cumulative likelihood of livebirths in translocation carriers

Distinct from concern over unbalanced segregants in a given pregnancy is the likelihood of achieving *any* liveborn offspring in a couple having a translocation ascertained because of multiple miscarriages. The cumulative likelihood of successful pregnancy (65–70%) is not substantively different from the livebirth rate in the general population (8, 16). However, time to achieve pregnancy is much longer (5–6 years) than in couples not having a translocation (21). Increased time to achieve pregnancy may be relatively less concerning to younger couples, but the mean delay of 5–6 years to achieve a liveborn conception can be a major impediment in older couples. The increased duration required reflects the high proportion of unbalanced embryos in a given cycle cohort. In a cohort tested by PGT-SR (structural) and PGT-A, it is not uncommon for all or nearly all embryos to be unbalanced, as illustrated in Table 3.1. PGT-SR is thus recommended in order to achieve a viable pregnancy with a euploid embryo.

Chromosomal Inversions

Chromosomal inversions differ mechanistically from translocations but exert similar roles clinically in terms of recurrent miscarriages or abnormal liveborns. Individuals having an inversion are phenotypically normal, but the order of their genes on a specific chromosome is reversed. This arises when a single chromosome undergoes two breaks. In pericentric inversions, one break occurs on the short arm and the other on the long arm. In paracentric inversions, both breaks occur on the same arm (i.e., short or long). In either inversion, DNA repair occurs, but in doing so the delineated chromosomal segment becomes inverted through 180°; thus, the sequence of genes on the chromosome involved is reversed (22). To illustrate, the original gene sequence of ABCDEFGH may become HGFEDCBA. Stephenson and Sierra (16) detected seven inversions among 1893 couples suffering recurrent miscarriages (0.37%), whereas Goddijn (8) detected 9 inversions among 1324 couples (0.68%). Inversions of sufficient length would be expected to be detected by karyotype, based on altered sequences of bands. Importantly, microarray-based comparative genomic hybridization (CGH) or next-generation sequencing (NGS) platforms, based solely on quantity of DNA, cannot identify a balanced inversion because DNA content alone remains unchanged. Analysis based on single nucleotide polymorphisms (karyomapping) can be informative.

The mechanism resulting in adverse clinical outcomes in inversion heterozygotes originates during cell division. In mitotic or meiotic cell division, crossing-over must occur to assure orderly chromosomal disjunction. If a sequence is inverted, this process must still occur (22). When it does, some chromosomal regions are duplicated whereas others are deficient. Based on pooled data involving different chromosomes, females having a pericentric inversion have a 7% risk of abnormal liveborns, whereas males have a 5% risk (9). Analogous to translocations, pericentric inversions ascertained through a phenotypically normal proband are more likely to result in live infants who are phenotypically normal than if pericentric inversions were ascertained through a phenotypically abnormal proband.

In paracentric inversions, recombinants would be expected to show large duplicated or deficient regions. There should theoretically be low likelihoods of unbalanced liveborns because most recombinants would be predicted to be lethal. The pooled risk of unbalanced viable offspring is reported to be 4% (19).

Risk of Single-Gene (Monogenic) Disorders

The protein-coding regions are defined as exons, and code for 21,000 genes in the human genome. This constitutes 1.5% of the genome. At least 5,000 Mendelian disorders are exonic in origin (1) The other 98.5% of the genome contains non-coding DNA, believed to be responsible for gene regulation. Non-coding regions (introns) may separate coding genes or be interspersed among exons of a clinically significant gene. At a single genetic locus, many different nuclear mutations may arise. In compound heterozygosity, two different alleles may show different base mutations, each deleterious. Causative mutant alleles influencing a single phenotype may exist in two different genetic loci (mixed heterozygosity). It follows that both genes are necessary for a normal phenotype.

Newborn screening

Couples at risk of monogenic disorders were once recognized only after birth of an affected child or after such an event in other family members. In the 1950s, identification of affected children began in certain regions through newborn screening programs. At present, such legislation is mandated in every US state for 35 or more inherited conditions (23). At least 5,000 asymptomatic US newborns are identified annually through newborn screening programs, resulting in 1 in 300 newborns being treated with specialized diets. The five most common conditions are 1) hearing loss, a highly heterogeneous condition, 2) congenital hypothyroidism, likewise heterogeneous, 3) cystic fibrosis, only one gene but molecularly heterogeneous, 4) sickle-cell anemia, one gene in African-Americans and, usually due to a single pathogenic mutation in codon 6 of the β-globin gene, and 5) medium-chain acyl-CoA dehydrogenase (MCAD) deficiency, heterogeneous as well. The latter three disorders are monogenic metabolic disorders, as are most of the 35 disorders being screened for. The notable exceptions to this are the spectrum of cardiac anomalies screened for by pulse oximetry (24).

Inborn errors of metabolism detected by newborn screening are based on detecting recognized abnormalities in protein-based gene products or secondary metabolites. An example is PKU (phenylketonuria), based on phenylpyruvate-detecting tandem mass spectrometry. Although many DNA mutations are responsible for most disorders subjected to newborn screening, DNA-based tests paradoxically are in part disadvantageous, compared with gene-product (protein) tests. The former rarely reaches 100% detection of the many causative DNA changes. By contrast, protein-based assessments (e.g., of Tay-Sachs disease) have only a single reliable summative quantitative end-point (hexosaminidase A activity for Tay-Sachs disease) which can be tested for directly or indirectly. DNA methodology is used to complement some protein-based diagnostic algorithms.

Ethnicity-Based Carrier Screening

It is obviously preferable for couples at risk of having offspring with a single-gene disorder to know their risk before conceiving. Causative mutations are known for most autosomal dominant, and X-linked recessive conditions. Identification of at-risk status before conception is based on parental DNA sequencing and is possible for at least 1,000 well-defined autosomal recessive disorders. Prenatal genetic diagnosis during pregnancy could be applied. Targeted or whole-genome sequencing (WGS) or whole-exome sequencing (WES) can be performed.

For over 50 years, identifying heterozygosity for a mutant autosomal recessive disorder in candidate asymptomatic individuals has been a goal. If both partners are heterozygous for a mutation for the same disorder, the risk is 25% for any given offspring being homozygous and, hence, affected. Programs to identify heterozygotes in carrier-screening programs achieve ≥ 95% detection rates. A negative screening test result greatly reduces the likelihood that an individual is a carrier.

Ideally, both partners will be screened concurrently, but carrier screening could also begin with testing the partner at greater risk (e.g., on the basis of family history or ethnicity). If one partner is heterozygous for a mutation in the key gene, the other partner should then be screened. If both partners are carriers, the couple should be informed of their options: invasive prenatal diagnosis (CVS or amniocentesis), PGT-M, donor gametes (eggs or sperm), or adoption. All these options exclude liveborn offspring with the at-risk disorder.

If a couple is shown to be at risk for an autosomal recessive disorder, other relatives may be at risk as well. Family members, especially sibs or offspring of prior partnerships, would benefit from being informed of the availability of carrier screening. The strategy is now testing because a family member is already known to be at potential risk. Importantly, permission from individuals known to have the mutant allele must be obtained before their genetic results are disclosed to others. Communication is best coming from the affected couple, who can explain the issue and rationale for them undergoing tests.

Initially, only protein-based (gene product) methodologies were applicable for genetic screening because DNA sequencing had not been developed. Prior to the availability of DNA strategies, the American College of Obstetricians and Gynecologists (ACOG) had recommended that only a limited number of disorders would be suitable for ethnicity-based carrier screening (25). The American College of Medical Genetics and Genomics (ACMG) recommended a similar, but not identical, set of disorders for screening.

The first carrier-screening program was for Tay-Sachs disease in individuals of Ashkenazi Jewish ancestry. Begun in the 1970s, heterozygote detection took advantage of the observation that the activity of hexosaminidase A, encoded by the causative gene, was quantitatively different among individuals who were homozygous normal (100% activity), heterozygous mutant (50%), or homozygous mutant (3% or less). Differentiation took advantage of the existence of a related but different hexosaminidase gene (encoding hexosaminidase B). Carrier detection was based on the relative ratios of the protein gene products hexosaminidase A and B.

The other carrier-screening program offered initially involved hemoglobinopathies (25). Sickle-cell disease is prevalent among individuals of African or African-American origin, 1 in 12 being heterozygotes for sickle-cell anemia (i.e., having the phenotype known as sickle-cell trait). Most affected individuals of African or African-American descent are the result of a missense mutation on codon 6 of the β-globin gene of hemoglobin, a single pathogenic variant resulting in a substitution of valine to glutamic acid in β-globin. A minority of cases from the Middle East have other hemoglobinopathy mutations. Irrespective of ethnic origin, heterozygosity can be identified through hemoglobin electrophoresis, assessing the motility of the protein gene product. Hemoglobin electrophoresis is also utilized for screening for carriers of β-thalassemias, hemoglobinopathies which are more frequent among Greeks, Italians (particularly Sicilians), Turks, Arabs, southern Iranians, and Asian Indians. Carrier detection is based on the presence of anemia which is not due to iron deficiency. The phenotype is further characterized by a mean corpuscular volume of less than 80 fL (microcytosis) and normal iron saturation levels. Hemoglobin electrophoresis shows decreased hemoglobin β but increased hemoglobin F as a characteristic for detecting β-thalassemia (26). DNA-based methods are now standard.

DNA-Based Expanded Pan-Ethnic Carrier Screening

Universal pan-ethnic carrier screening is now possible because screening can be DNA-based, and advances in molecular technologies have made DNA sequencing practicable. Both partners need not be of the same ethnicity or at similar risk (27). Hundreds of disorders can be screened concomitantly. ACOG, the National Society of Genetic Counselors (NSGC), ACMG, and the Perinatal Quality Foundation have all endorsed the concept of pan-ethnic screening (28, 29). ACOG explicitly states that "ethnic-specific, pan-ethnic, and expanded carrier screening are acceptable strategies for pregnancy and prenatal carrier screening" (28). These recommendations replace the ACOG Committee Opinions offered during the era of protein-based ethnicity carrier screening (26).

DNA-based screening actually began in the 1990s as the Human Genome Initiative was under development, albeit on a more limited basis than at present. Screening for carriers of Tay-Sachs disease was initially based solely on protein gene product, hexosaminidase A. However, only a few base mutations are responsible for most cases of Tay-Sachs in the Ashkenazim. Screening these mutations resulted in high heterozygote carrier frequencies. Other autosomal recessive disorders in the Ashkenazim could similarly be screened for. Carrier screening was thus recommended by ACOG for Canavan disease and for familial dysautonomia (25), with "consideration" given for other disorders common in the Ashkenazim: mucolipidosis IV; Niemann-Pick disease type A; Fanconi anemia types A, B, C; Bloom syndrome; Gaucher disease (GBA); Joubert syndrome; familial hyperinsulinemia (ABCC8); maple syrup urine integrated

(BCKDHA, BCKDHB, DBT) disease; and Usher syndrome. In addition to Canavan disease and familial dysautonomia, ACMG recommended screening for Niemann-Pick (type A) and Bloom syndrome. DNA panels for these disorders detect 95–99% of heterozygotes. The same approach was later applied in pan-ethnic couples carrying other mutant genes.

The stage was thus set for DNA-based screening to detect couples at risk of a plethora of genetic disorders. Cystic fibrosis was the first such condition to be screened solely on a panel of causative mutations. Cystic fibrosis varies in prevalence among ethnic groups: 1 in 25 Caucasians are heterozygotes and 1 in 24 Ashkenazi Jews. In African-Americans, only 1 in 65 are heterozygous, in Hispanic-Americans 1 in 46, and in Asian-Americans 1 in 94 (30). AMCG and ACOG jointly developed a pan-ethnic DNA panel consisting of the 23 most common mutations in CF carriers (30). Importantly, the panel did not attempt to encompass every CF mutation, which runs into hundreds.

Screening based on the 23-mutation panel was initially "offered" only to Caucasians and Ashkenazi Jewish populations. In these two ethnic groups, heterozygote detection rates based on the initial 23 mutation panel were 88% and 94%, respectively. In other ethnic groups, detection rates were lower, for which reason screening was only "made available." Recommendations were later updated by ACOG, ACMG, and NSGC, and now cystic fibrosis carrier screening is offered to all ethnicities (31–33).

Prior to the current era of expanded pan-ethnic screening, two other disorders where DNA screening was universally considered for screening by ACMG or ACOG were identified: spinal muscular atrophy (SMA) and fragile X syndrome (fragile X mental retardation [FMR1]). Heterozygote frequencies for autosomal recessive spinal muscular atrophy range from 1 in 40 to 1 in 60 in most populations but are lower in Hispanics (34). The causative gene is Survival Motor Neuron 1 (SMN1). ACMG has long recommended population screening (34), but ACOG recommends screening only for couples with a positive family history (28, 29). Screening for SMN1 carriers is complex because, on each copy of chromosome 5, the SMN1 gene is present in duplicate as two individual and functional copies of the SMN1 gene, totaling four functional copies per individual. Confusion may arise if DNA analysis reports a single individual having two copies of the SMN1. This could be explained either by each homologous chromosome having one copy of the mutant allele (*trans*) or by a single chromosome having two (both) copies of the mutant allele (*cis*) but with zero copies on the homolog. In the latter situation, the chromosome with no copies could result in an affected embryo if the other partner likewise contributes a chromosome lacking a wild-type copy of the SMN1 gene.

The other condition consistently considered is fragile X syndrome (FMR1), the most common inherited form of intellectual disability in males (1 in 3600). The molecular basis of fragile X involves expansion of the trinucleotide CGG repeat. In unaffected individuals, the mean number of CGG repeats is 32–33. If greater than 200 CGG repeats are present, mental retardation occurs in almost all males, and in many females. CGG expansion does not increase in male meiosis, but it does in female meiosis. In a female having fewer than 200 CGG repeats, the number may expand during meiosis and result in her male offspring inheriting an allele with more than 200 CGG repeats. ACOG, ACMG, and NSGC all recommend screening/testing for women having a family history of fragile X, but not screening in the general population. In the United States, the fragile X carrier frequency for women with no known risk factors is 1 in 257 (35). Population carrier screening for fragile X is recommended in Israel. Of reproductive relevance, women with 55–199 CTT repeats (premutation), but who do not show mental retardation, are at increased risk (15%) of premature ovarian failure.

Genetic Counseling in Pan-Ethnic Carrier Screening

Pan-ethnic expanded carrier screening (ECS) now encompasses so many genetic disorders that it is no longer practical to provide information on every disorder to be screened in every couple. Fortunately, vendors provide useful brochures and online information to accompany screening and explain test results. A practical approach is to defer detailed disease-specific counseling until a couple is shown to be at risk for a given disorder.

Chapter 4 provides details on platforms and molecular methodology available. Some 400 different genes are plausible candidates (36) and screening is offered by one or more platforms (vendors). Costs of DNA sequencing are no longer the main impediment in deciding the number of disorders to be screened for, nor even the number of causative alleles to be interrogated. Testing copy number variants (CNV) increases the carrier detection rates, especially in genes traditionally difficult to sequence, such as FMR1, SMN1, 21-hydroxylane deficiency congenital adrenal hyperplasia, and α-thalassemia. Heterozygote detection rates can approximate to almost 100% for cystic fibrosis and Duchenne/Becker muscular dystrophy (36, 37).

No specific number of disorders is recommended for a pan-ethnic panel. More disorders may not necessarily be better because as the number of disorders increases so does complexity. For example, in one study, the first 18 disorders screened accounted for 84% of the informative mutations, whereas the next 73 disorders added only an additional 11% (38). Moreover, increasing accumulation of detected variants of unknown significance (VUS) must be taken into account. Nevertheless, the value of expanded carrier screening is unquestioned. Hogan et al. (37) used a panel that detected 1 in 22 couples to be at risk of a detectable disorder; 1 in 300 fetuses would have been affected.

Expanded Carrier Screening and *In Vitro* Fertilization for Adult-Onset Disorders

The original goal of carrier screening was identifying pregnancies at risk of disorders of childhood onset. This remains the general expectation. However, expanded carrier screening to identify couples at risk of offspring with a homozygous childhood-onset disorder may unwittingly identify a heterozygous adult who is at risk for a different, adult-onset disorder. An example arises in screening for autosomal recessive Fanconi anemia (FAN) or autosomal recessive ataxia telangiectasia (ATM). Both of these childhood-onset conditions are caused by defects in DNA repair; only homozygous fetuses are at risk of these childhood disorders. However, a heterozygous female detected in the screening process is at increased risk of breast cancer. In a separate scenario, women under treatment for breast cancer may be tested by gene panels to identify the presence of cancer-susceptible genes. In 10–15% of breast cancer cases, the etiology is a heritable, cancer-susceptibility gene.

The NIH National Cancer Institute lists 31 familial cancer susceptibility syndromes (39), most being adult-onset and autosomal dominant. Of relevance to the practice of reproductive medicine and *in vitro* fertilization, interest is often kindled when a family member (e.g., a mother) proves to have cancer, and has undergone testing that identifies a causative and heritable mutation. Although the affected parent is usually beyond reproductive age, her offspring often are not and may wish to conceive. If the mutation is autosomal dominant (e.g., BRCA1), the daughter of the index case has a 50% likelihood of having inherited the mutation. Predictably, she may wish to avoid transmitting the familial mutant allele to her yet-to-be conceived offspring (grandchildren of the index case). PGT-M is the preferred prenatal option, chosen by half or more of at-risk couples. As for any single-gene disorder, chorionic villus sampling or amniocentesis can detect affected fetuses; however, management might involve clinical pregnancy termination. PGT-M avoids this by pre-conception selection of an unaffected embryo that can be transferred without fear of later needing termination of a clinical pregnancy. For this reason, the number of PGT-M cycles for heritable cancers has increased two-fold in recent years.

In families seeking PGT-M for adult-onset disorders, molecular heterogeneity often exists and must be taken into account. This contrasts with many childhood-onset conditions, especially disorders of high prevalence in certain ethnic groups (e.g., Tay Sachs disease). A limited number of mutations (pathogenic variants) may account for the great majority of carriers. In sickle-cell anemia in African-Americans, almost all cases involve a codon 6 of the β-globin gene encoding part of hemoglobin A. The situation is usually different in adult-onset cancers, which are typically inherited in autosomal dominant fashion. At Reproductive Genetic Innovation, we have performed PGT-M for 151 families with BRCA1 pathogenic variants and for 119 families with BRCA2 pathogenic variants. In the case of BRCA1, 71 of the 151 (47.0%) at-risk families showed a single deletion, 187delAG. In 14 other families (9.3%), the pathogenic variant was 5385insC. Six other variants were found in more than one family, but 44 other variants were found in only a single family. The spectrum was similar in families at risk of BRCA2. The most common BRCA2 pathogenic variant was 6174delT, present in 51 of 119 (42.9%) at-risk families. One other pathogenic BRCA2 variant was found in two different families but 46 others were detected in only a single family.

The same clinical diagnostic consequences arise in other adult-onset autosomal dominant disorders. Most commonly sought for PGT-M are the heritable cardiac conditions Long Q-T syndrome (LGT1, LGT2, LGT8), hypertrophic cardiomyopathy (CMH1, CMH4, CMH8), and dilated cardiomyopathy (Type 1A, 1DD, 1E,1G). In these autosomal dominant disorders, identification typically follows a family member being identified in the context of athletic exertion. Relatives of reproductive age often prefer prenatal genetic diagnosis through PGT-M.

Genetic screening for common complex conditions is increasingly being explored. Disorders of interest include not only structural birth defects involving a single organ system (e.g., cleft palate, hip dysplasia, cardiac anomaly). Recurrence risks in first-degree relatives are 2–5%. In certain adult-onset disorders (e.g., hypertension, diabetes mellitus), there is a 10–15% recurrence risk. These conditions are considered to be the result of the cumulative effect of many adverse genes. The odds ratio for risk of an individual anomaly has been based on presence or absence of at-risk alleles.

PGT-M is justified by a 10–15% increased risk for offspring at risk of long-standing life-threatening disorders. A pitfall in predicting outcome for an embryo at risk of a common complex condition is that the number of causative alleles remains largely unknown. One illustrative study involves Hirschsprung disease, a disorder in which counseling has traditionally been provided solely on the basis of empirical risks stratified by sex and by length of the involved colon (long segment or short segment). In 190 index cases and 47 first-degree affected relatives, a large number of at-risk alleles or CNVs was identified by Tiligman et al. (40). Prediction in a pregnancy at risk was generated by batching the most common causative variants to generate odds ratios that could be communicated to families. In these common complex disorders, stratifying embryos dichotomously into affected *versus* unaffected phenotype seems arguable; however, selecting the embryo least likely to be affected seems reasonable.

REFERENCES

1. Collins FS, *et al. N Engl J Med.* 2021;384:1–4.
2. Viswanathan M, *et al. JAMA.* 2017;317:190–203.
3. Franssen MT, *et al. BMJ.* 2005;331:137–41.
4. Morris JK, *et al. Prenat Diagn.* 2003;23:252–8.
5. Zegers-Hochschild F, *et al. Hum Reprod.* 2017;32:1786–801. *Feertil Steril.* 2017;108:393–406, (Simultaneous Publicatiosa).
6. Kuliev AM, *et al. Practical Preimplantation Genetic Diagnosis.* 3rd ed. London, Heidelberg, New York: Springer-Verlag; 2020.
7. Simpson JL, *et al. Fertil Seril.* 1981;36:584–90.
8. Goddijn M, *et al. Hum Reprod.* 2004;19:1013–7.
9. Pettenati MJ, *et al. Am J Med Genet.* 1995;55:171–87.
10. American College of Obstetricians and Gynecologists. *Obstet Gynecol.* 2016;127:e123–37.
11. Benn PA. In *Genetic Disorders and the Fetus.* 6th ed. (Milunsky A, Milunsky JM, eds.). Oxford, Chichester, Hoboken: Wiley-Blackwell; 2010;194–312.
12. Snijders RJ, Nicolaides KH. *Ultrasound Markers for Fetal Chromosomal Defects.* NewYork: Parthenon; 1996.
13. Warburton D, *et al. Am J Hum Genet.* 1987;41:465–83.
14. Warburton D, *et al. Am J Hum Genet.* 2004;75:376–85.

15. Simpson JL, *et al.* In *Obstetrics Normal and Problem Pregnancies* (Gabbe SG, Niebyl JR, Simpson JL, et al., eds.). 7th ed. 2017;578–94.
16. Stephenson MD. *Hum Reprod.* 2006;21:1076–82.
17. Royal College of Obstetricians and Gynaecologists. *The Management of Early Pregnancy Loss. Guidleine No. 25.* London: RCOG; 2011.
18. European Society of Human Reproduction and Embryology. *Recurrent Pregnancy Loss.* 2017. https://www.eshre.eu/Guidelines-and-Legal/Guidelines/Recurrent-pregnancy-loss.aspx.
19. Anton E, *et al. Fertil Steril.* 2006;85:661–6.
20. Daniel A, *et al. Am J Med Genet.* 1989;33:14–53.
21. Sugiura-Ogasawara, *et al. Fertil Steril.* 2004;81:367–73.
22. Gardner RJ, *et al. Chromosome Abnormalities and Genetic Counselling.* New York: Oxford; 2012.
23. American College of Obstetricians and Gynecologists. *Obstet Gynecol.* 2019;133:e357–61.
24. Grosse SD, *et al. MMWR Morn Mortal Wkly Rep* 2017;66:888–90.
25. Driscoll DA, *et al.* In *Obstetrics. Normal and Problem Preganacies.* 7th ed. (Gabbe SG, Niebyl JR, Simpson JL, et al., eds.). Philadelphia: Elsevier; 2017;193–211.
26. American College of Obstetricians and Gynecologists. *Obstet Gynecol.* 2007;109:229–37.
27. Edwards JG, *et al. Obstet Gynecol.* 2015;125:653–62.
28. American College of Obstetricians and Gynecologists. *Obstet Gynecol.* 2017;129:e35–e40.
29. American College of Obstetricians and Gynecologists. *Obstet Gynecol.* 2017;129:e41–e55.
30. Grody WW, *et al. Genet Med.* 2001;3:149–154.
31. Watson MS, *et al. Genet Med.* 2004;6:387–91.
32. American College of Obstetricians and Gynecologists. *Obstet Gynecol.* 2011;117:1028–31.
33. Langfelder-Schwind E, *et al. J Genet Couns.* 2014;23:5–15.
34. Prior TW, Professional Practice and Guidelines Committee. *Genet Med.* 2008;10:840–2.
35. Cronister A, *et al. Obstet Gynecol.* 2008;111:596–601.
36. Haque IS, *et al. JAMA*, 2016;316:734–42.
37. Hogan GJ, *et al. Clin Chem.* 2018;64:7.
38. Lazarin GA, *et al. Genet Med.* 2013;15:178–86.
39. NIH National Cancer Institute. *Cancer Genetics Overview (PDQ®) – Health Professional Version.* https://www.cancer.gov/about-cancer/causes-prevention/genetics/overview-pdq (Accessed 6/7/19).
40. Tiligman JM, *et al. N Engl J Med.* 2019;380:1421–32.

Carrier Screening for Single-Gene Disorders

4

Julio Martin, Arantxa Hervas, Ana Bover, Laura Santa, and Ana Cervero

INTRODUCTION

The role of genetic variation in human diseases is well known; some DNA changes or genotypes across a person's genome explain single-gene diseases as well as contributing to developing complex traits. The application of DNA sequencing and genetic studies to the practice of medicine has ramped up progress in variant detection and interpretation; nevertheless, significant work is still needed to prevent the birth of affected children and to improve the lives of affected people. The use of genetic testing, such as expanded carrier screening, has helped to reduce the frequency and impact of significant Mendelian diseases. These screening tests allow the identification of individuals and couples at risk of conceiving children who will be affected by diseases traceable to single-gene mutations.

CARRIER SCREENING IS AN IMPORTANT COMPONENT OF PRECONCEPTION AND PRENATAL CARE

Single-gene disorder screening was first proposed for condition-directed carrier testing for phenylketonuria in the 1960s, followed by sickle-cell anemia and Tay-Sachs diseases, both of which targeted high-risk ethnic populations.

Following the practice of screening for other conditions, like cystic fibrosis and the thalassemias, the prevalence of these conditions decreased significantly [1, 2]. Moreover, the continued increase in our genetic knowledge has allowed the decoding of close to 6,000 disorders with suspected Mendelian inheritance, approximately 2,000 of which being recessive diseases (https://www.omim.org/statistics/entry; last accessed in September 2021), opening up the possibility of a large pre-conception testing approach.

Mendelian disease prevention by genetic screening is clinically pertinent. Together, these conditions account for around 20% of infant mortalities and 10% of pediatric hospitalizations [3, 4]. With the advent of high-throughput next-generation sequencing (NGS) and bioinformatic advances, comprehensive pre-conception screening becomes feasible and practicable, allowing simultaneous, efficient, and affordable testing for a large number of genetic conditions that a family history would never detect. Using NGS technologies,

DOI: 10.1201/9781003024941-4

each researcher, geneticist, or clinician can determine the appropriate approach to sequencing DNA (or RNA) to analyze sequence variation [5]. In this chapter, we will discuss the use of NGS for comprehensive DNA sequencing of genes causing Mendelian disorders as applied to test individuals and used to characterize their carrier burden and the risk for their descendants, as the best approach to ensuring a healthy baby.

WHAT WE KNOW TODAY

Carrier screening (CS) denotes genetic testing performed on an otherwise healthy, asymptomatic individual to determine whether that person has a mutation within a given gene associated with a recessive disorder; in addition, females are also tested for X-linked genes. It can be performed to test for a single condition, a limited number of specific conditions, or hundreds of disorders. Testing for many conditions simultaneously is known as expanded carrier screening (ECS). Remarkably, ECS will identify most individuals and couples who are at risk of having affected children. Nevertheless, it must be noted that any individual who tests negative for a specific mutant allele still has a residual risk (RR) of being a true carrier. This RR will vary depending on the disease prevalence, detection rate, and ethnicity, among other variables.

The goal of CS is to provide individuals with meaningful information with which to consider their range of reproductive options, based on their personal values [6]. Historically, pre-conception CS has been recommended only for a handful of mutations targeting specific populations known to have high frequencies of these disorders. The diseases most frequently screened for were cystic fibrosis, hemoglobinopathies, conditions associated with Ashkenazi ethnicity, spinal muscular atrophy, and, in women, fragile X syndrome [7]. The list of recommended disorders to be included in CS tests is taken from the guidelines of the professional societies whose criteria are based on condition severity, race or ethnicity, prevalence, carrier frequency, detection rates, and RR. However, there is now a tendency toward the development of pan-ethnic genetic tests in consonance with current multiracial societies. Indeed, some authors believe that, with the tools and techniques available today, genetic analyses should test as many variants as possible [8]. Traditional and ethnicity-based tests were thought to have a higher mutation detection rate and be more cost effective, but they have proved unsuitable for patients of mixed or unknown ethnic background [8, 9].

The advantage of NGS-based genetic analysis is the possibility of designing a comprehensive assay to test all patients regardless of their clinical history and ethnicity [10]. Concerning inheritance, most disorders included on current expanded panels are autosomal recessive; some may be X-linked, tested in female samples mainly [11]. In addition, the promotion of pan-ethnic carrier testing has wide use in IVF programs, using gamete donors [11, 12].

WHAT GENETIC CONDITIONS TO SCREEN FOR?

Many studies have focused on describing and/or determining the criteria for selection of genetic conditions. Traditional methods have focused generally on conditions that significantly affect life expectancy or life quality, including cognitive or physical disabilities or a requirement for lifelong medical therapies, and those that have fetal, neonatal, or early childhood onset and a well-defined, typically severe, phenotype [11]. Therefore, ECS genetic content mostly follows clinical criteria and recommendations of professional societies for disease/gene inclusion, i.e., clinical utility, disease prevalence, disease severity, test accuracy and cost, and reporting variants with a highly penetrant phenotype. Indeed, there is good consensus regarding frequent and severe conditions, including those specific for certain ethnicities. However, there is a great variability for a significant part of the gene content in the different gene panels. For example, the inclusion criteria for multigene panels may consider conditions caused by the alteration of several genes and it is common to see the inclusion of some rare (low-prevalence) genes implicated in the pathogenesis of a relative frequent condition.

Similarly, digenic effect is another scenario to be considered so that certain, low-prevalence genes may also be included to fully test for a condition showing this type of inheritance. Also, conditions associated with an adult-onset phenotype may be excluded, and deep consideration should be given as to whether gene/variants with high allele frequencies and moderate phenotypes (for example, HFE-associated hemochromatosis, Serpinal/alpha-1 antitrypsin, and others) should be widely tested for. In part, this explains why some expanded panels also include rare genes for conditions with great variation in their clinical presentation. In addition, this may impact the calculation of the assay metrics, including limited knowledge about the carrier frequency and the detection rate. Therefore, calculation of RR after a negative screening result is limited for some tested genes. Interestingly, a recent study [13] presented a method to calculate these and other assay metrics. The authors suggested an approach that laboratories conducting CS should evaluate, i.e., to report the results and the methods used, thus facilitating objective comparisons between laboratories. There is also great consensus that CS tests should report only pathogenic and likely pathogenic variants, thus leaving out reports of variants of uncertain (or unknown) significance (VUS).

In summary, it is important that test providers follow guidelines to ensure that actionable information is offered to individuals and expecting families [14].

Early studies, using NGS-based approaches to screen for more than 400 severe autosomal and X-linked recessive childhood disease genes, mostly included severe diseases but also analyzed some genes associated with low-incidence and variable-severity conditions [15]. Later, other authors made calculations by modeling populations to demonstrate decreases in frequency of genetic diseases and health cost-reduction by using ECS [16]. A significant decrease of 61% in the incidence of affected children has been reported when results for NGS-based testing were compared with couples without testing. Although more long-term implications are still pending, CS NGS-based tests help to prevent genetic conditions not only in high-risk populations [17] but also in the general population.

METHODS AND VARIANT INTERPRETATION

Array hybridization was initially explored for expanded carrier detection [18]. It is a cost-efficient test, but it is also a fixed-content method with significant analytical limitations, such as limited lists of mutations per gene. On the contrary, variant interpretation is straightforward because only pathogenic variants were detected on the array. Due to its analytical design, the clinical utility of the array-based test is limited because a significant portion of carriers will not be detected.

Alternatively, targeted capture and NGS have shown efficacy and scalability for resequencing human exomes and genomes [15, 19, 20], with excellent analytical accuracy in terms of both sensitivity and specificity, including operational feasibility for CS [9, 12, 15, 21, 22].

Regarding variant interpretation, most providers today use their own criteria to determine the clinical impact of findings from sequencing, adapted from or following agreed standards and guidelines for variant categorization [23]. One possible scenario is to first choose pathogenic or likely pathogenic variants annotated in databases (i.e., ClinVar [24] and/or HGMD [25]). A curated list of these already known pathogenic variants should allow for automation of variant classification of a significant number of findings. These mutations typically correspond to variants reported in patients with solid medical evidence. A second step or filter could be evaluating allele frequency, implemented to classify detected variants as common or rare. Typically, variants with an allele frequency > 1% in dbSNP (www.ncbi.nlm.nih.gov/SNP) in the 1000 Genomes project (www.1000genomes.org) or in an in-house database are defined as common variants and are usually categorized as likely polymorphisms. Exceptions are made for well-described annotated pathogenic variants with allele frequency > 1%. Variants with a frequency < 1% are considered to be rare variants. Further filtering steps take into consideration the type of mutation and its functional impact, zygosity, disease prevalence, detection in patients vs. controls, etc. Regarding mutation types, rare missense single-nucleotide variants (SNVs) and in-frame small insertion or deletions (indels) coding sequences with an allele frequency lower than the estimated prevalence of the corresponding disease with no homozygous status ever detected in controls, but not reported in patients or reported but without clear evidence of being a disease-causing change, could or are normally classified as variants of unknown significance (VUS). Finally, rare variants—typically below 1%—with severe functional impact (frameshift deletions, nonsense SNVs, and splice site variants) and with allele frequency below the corresponding disease prevalence with homozygous status ever detected in controls are classified as likely pathogenic. The previous examples for variant interpretation must be taken with caution since they were enumerated here or represent general examples. Each analysis of variant interpretation will require more in-depth understanding before a final classification can be awarded.

CLINICAL RESULTS

Several preclinical and clinical validation reports using NGS-based carrier screening have been published [9, 12, 21, 22]. All studies are rather similar in terms of preclinical validation: a set of DNA samples previously characterized by a different method, usually Sanger sequencing, were re-analyzed. For example, in one of our earliest studies [12], the selected DNA samples were positive for mutations affecting 27 genes of interest. Overall, the analytical sensitivity found was > 99%, with an estimated clinical sensitivity of 98%.

Regarding clinical results, a different carrier burden may be found due to various considerations, such as gene/mutation content, method, decision tree to classify sequence variation, etc. Continuing with our initial study [12] on individuals undergoing carrier testing as pre-conception screening in fertility clinics, a total of 2,570 tests were performed, on both patients and gamete donors. In total, 1,796 unique pathogenic or likely pathogenic variants were detected, and 13,785 VUS variants were defined. Of the 2,570 patients investigated, 2,161 (84%) were positive for at least one pathogenic variant. The average carrier burden of recessive or X-linked conditions was 2.3 mutations per sample. In addition, for couples using their own gametes, 7 out of 138 turned out to be carriers of a pathogenic or likely pathogenic variant in the same gene. This accounts for 5% of the tested couples. In these cases, pre-implantation genetic testing (PGT-M) was recommended during post-test genetic counseling. Moreover, in 6 out of 287 female patients, positive results for X-linked disorders were obtained, which accounted for approximately 2% of the total cohort. PGT-M was recommended as well. Gamete donors who tested microbiologically negative were subjected simultaneously to karyotype analysis and fragile X (females only) investigation; individuals with abnormal karyotypes were not tested further and were banned as donors. An additional 18 female donors were excluded from the program because they carried a pathogenic or likely pathogenic variant in an X chromosome gene, including fragile X, representing 1.94% of the total tests requested. They received information on the adverse finding, including genetic counseling, and were discouraged from entering the donor program. The remaining donors were included in a blind match, informatically controlled database. By request, the match system always displayed a set of donors genetically compatible with the patient requesting gamete donation. The blinded system allowed us to include VUS variants as criteria for assigning a matched donor. The system discouraged the assignment of donors having VUS variants in the same gene where the patient was carrying a known pathogenic variant, or *vice versa*.

LIMITATIONS

Patients with genetic conditions constitute an important portion of the world population with special health care needs [4]. Together with advances in genetic knowledge and technologies, we can easily assume that by providing information about the carrier status we are helping individuals, families, and society in general. However, evidence addressing reproductive outcomes after ECS is still needed to demonstrate the clinical utility of all these progresses when offered to the general population. Moreover, we must foster worldwide epidemiological studies of inherited disorders, particularly relevant for developing countries with limited resources since most epidemiological research has been conducted with individuals from Europe or North America. Findings from those studies will allow the definition of genes of interest, associated carrier rates, and other metrics relevant for the implementation of CS programs. Another area of improvement is related to residual risk calculations [13]; accurate development of detection rate and residual risk is critical to providing adequate clinical utility and support to couples making reproductive choices. Finally, in recent years, we have seen great progress in new developments of bioinformatic tools related to search, aggregation, analysis, etc., including the sharing of knowledge related to variant interpretation; these tools have impacted positively on our understanding of human DNA variation [26]. Nevertheless, clinical interpretation for a significant part of genetic variation is still lacking, especially associated with clinically variable phenotypes or conditions showing incomplete penetrance. This presents a limitation or challenge to the geneticist and clinician as to how to counsel and guide the medical management of a patient/couple carrying an inconclusive test result. Together with technical differences in capturing and/or sequencing efficiency, variant interpretation across different studies is a limitation in providing definitive estimations of the carrier burden in a significant number of human genes.

CONCLUSIONS

Carrier screening is an important component of pre-conception care. It aims to identify couples at risk of passing on genetic conditions to their offspring. Today, high-throughput genotyping and sequencing approaches allow for efficient screening of many diseases simultaneously, namely ECS. The comprehensive

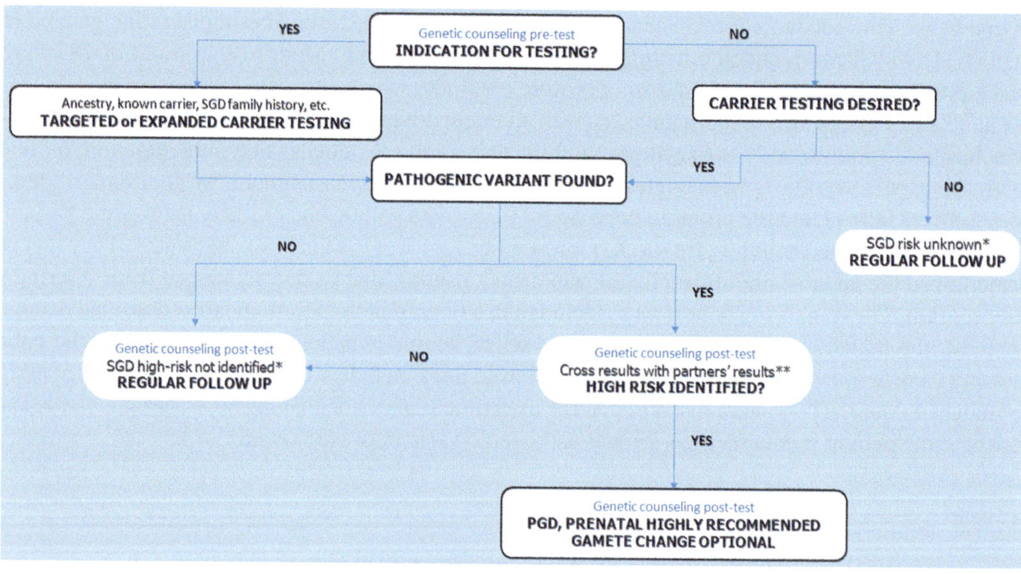

FIGURE 4.1 Algorithm for carrier screening. *Notes*: *Family wishes are assumed during all processes of this algorithm, which refers to the risk to future offspring regarding AR and XL conditions. **Testing partner is assumed. *Abbreviations*: AR, autosomal recessive condition; SGD, single-gene disorders; XL, X-linked condition.

or expanded carrier genetic test described here is built upon the many advantages offered by next-generation DNA sequencing platforms.

Appropriate and accurate pre- and post-test genetic counseling is of utmost importance. Individuals and couples must understand the purpose of the genetic test, the disorders analyzed and their severity, and the fact that, even for a negative test result, a residual risk remains. In addition, they must be advised about disorders that arise due to a *de novo* mutation, mutations not included in the test, or an untested rare disorder.

In the case of gamete donors, additional input may be needed. Gamete donors are typically young individuals attending the reproductive clinic for the purpose of donating, and they may not expect to be tested for genetic disorders. Even after pre-test counseling and giving consent for testing, the possibility of being a carrier for a certain mutation may not be on potential donors' minds at all. Therefore, in the event of a positive result, the genetic counselor may have to deal with the potential donor's shock at being excluded from the donation program. In such instances, post-test genetic counseling must emphasize that, generally, there is no clinical risk for the tested individual but at the same time must indicate the clinical relevance of the information for future family planning (Figure 4.1).

REFERENCES

1. Castellani C, et al. *JAMA*. 2009;302:2573–2579.
2. Bozkurt G. *Hemoglobin*. 2007;31:257–264.
3. Costa T, et al. *Am J Med Genet*. 1985;21:243–255.
4. Kumar P, et al. *Mayo Clin Proc*. 2001;76:777–783.
5. Martin J, et al. *Fertil Steril*. 2013;99(4):1054–1061.
6. ACMG Committee on Genetics. *Obstet Gynecol*. 2017;129:e35–e40.
7. Beaudet AL. *Genome Med*. 2015;7:79.
8. Perreault-Micale C, et al. *Mol Genet Genom Med*. 2015;3(4):363–373.
9. Hallam S, et al. *J Mol Diagn*. 2014;16:180–189.
10. Lazarin GA, et al. *Genet Med*. 2013;15(3):178–186.
11. Edwards JG, et al. *Obstet Gynecol*. 2015;125(3):653–662.
12. Martin J, et al. *Fertil Steril*. 2015;104(5):1286–1293.
13. Leung ML, et al. *J Mol Diag*. 2021;23(1):91–102.
14. Grody W, et al. *Genet Med*. 2013;15:482–483.
15. Bell CJ, et al. *Sci Transl Med*. 2011;3(65):65ra4.
16. Azimi M, et al. *Mol Genet Genom Med*. 2016;4(3):292–302.
17. Kaback MM. *Eur J Pediatr* 2000;159:S192–S195.
18. Srinivasan BS, et al. *Reprod Biomed Online*. 2010;21:537–551.
19. Ng SB, et al. *Nat Genet*. 2010;42:30–35.
20. Roach JC, et al. *Science*. 2010;328(5978):636–639.
21. Umbarger MA, et al. *Genet Med*. 2014;16(2):132–140.
22. Abulí A, et al. *Hum Mutat*. 2016;37(6):516–523.
23. Richards S, et al. *Genet Med*. 2015;17:405–423.
24. Landrum MJ, et al. *Nucleic Acids Res*. 2016;44(D1):D862–D868.
25. Stenson PD, et al. *Hum Genet*. 2017;136:665–677.
26. Kopanos C, et al. *Bioinformatics*. 2019;35(11):1978–1980.

Meiotic Abnormalities in Infertile Males

5

Mireia Solé, Francesca Vidal,
Joan Blanco, and Zaida Sarrate

OVERVIEW OF MEIOSIS

Meiosis is a key process in spermatogenesis, with the final goal of the production of gametes with haploid genetic content. A significant number of exhaustive revisions of this process have been published in the specialized literature [1–16]. In meiosis, DNA replication is followed by two consecutive cell divisions that reduce the number of chromosomes by half. The first meiotic division is reductional and involves the segregation of homologous chromosomes to opposite poles, while in the second meiotic division sister chromatid segregation occurs (Figure 5.1). Prophase of the first meiotic division includes highly complex events and consists of different substages (leptotene, zygotene, pachytene, diplotene, and diakinesis). During prophase I, homologous chromosomes pair; synaptonemal complexes are assembled and disassembled; and reciprocal exchanges between homologous chromatids take place, a phenomenon called meiotic recombination. In male meiosis, specifically at the zygotene and pachytene stages, chromosomes X and Y form a particular structure called a sex vesicle. At the end of prophase I, chromosome bivalents show chiasmata, which should be interpreted as the visible manifestation of recombination. Next, at metaphase I, bivalents reach maximum condensation, chiasmata are still visible, and bivalents appear arranged at the equatorial plate to segregate the homologous chromosomes to the opposite poles during anaphase I. As a result of the first meiotic division, two haploid cells are produced (each chromosome with two sister chromatids). During the second meiotic division, which starts without previous DNA replication, the pericentromeric cohesion between sister chromatids is lost, leading to the formation of haploid cells in where each chromosome shows a single chromatid.

The meiotic process is highly regulated, involving different cell control mechanisms (checkpoints) designed to detect anomalies in chromosome recombination and segregation. The recombination checkpoint prevents cells from overcoming this stage if chromosome pairing anomalies are present [2, 17], and spermatogenic arrest at this stage has been associated with defects in recombination and/or in homologous chromosome synapsis [8, 18]. The spindle assembly checkpoint regulates the transition from metaphase to anaphase and acts to retain cells at metaphase I or II until all bivalents or chromosomes are properly orientated to the spindle [19–21], which is a prerequisite for correct segregation.

DOI: 10.1201/9781003024941-5

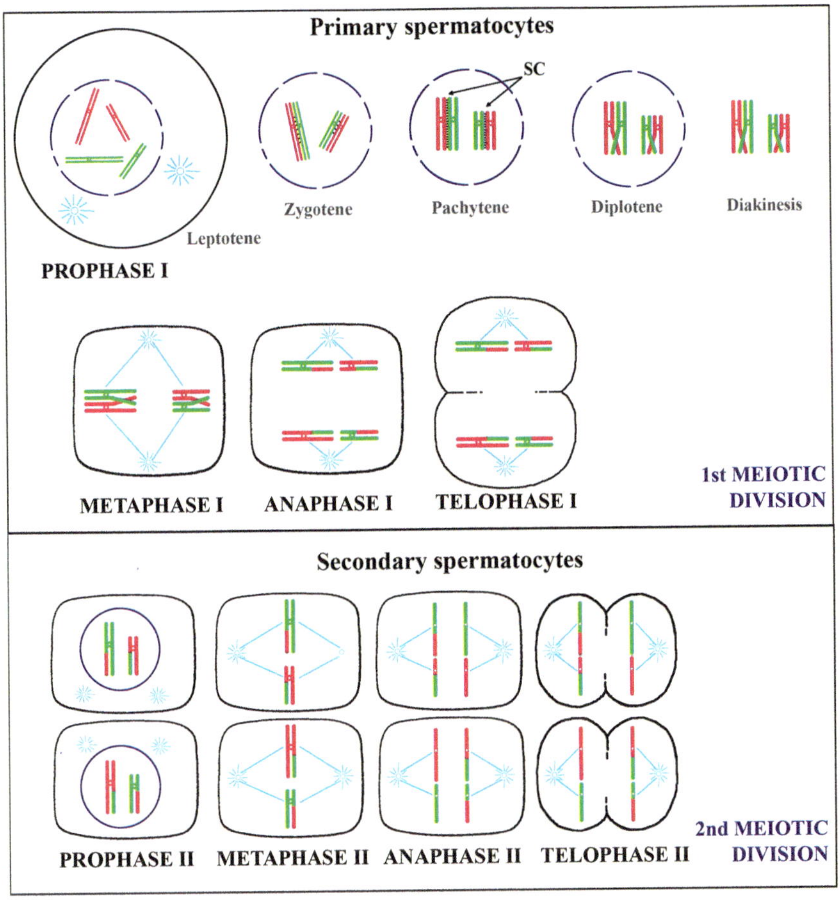

FIGURE 5.1 Schematic overview of the meiotic process (SC, synaptonemal complex).

METHODS OF STUDY

Meiotic cytogenetic studies address the detection of abnormalities that exclusively affect the germline. Thus, studies are performed in testicular tissue cells. Biopsies are usually obtained under local anesthesia and are disaggregated using the appropriate protocol according to the type of subsequent analysis performed. Since the 1980s, several cytogenetic techniques have been implemented. Essentially, these techniques address the study of synaptonemal complexes and the analysis of the meiotic chromosomes [22].

Immunostaining of synaptonemal complex elements and MLH1 recombination foci (Figure 5.2) has been used to evaluate synapsis and chromosome recombination at the pachytene stage [23]. Moreover, individual identification of each synaptonemal complex using centromere-specific multicolor fluorescence *in situ* hybridization (cenM-FISH) or subtelomere labeling [24–26] has contributed to a better understanding of the synaptic process in normal and abnormal scenarios. However, these studies have limited application in clinical contexts and will not be discussed in this chapter.

In contrast, studies of meiotic chromosomes using uniform staining have been implemented in some laboratories. Briefly, this protocol consists of mechanical disaggregation of the testicular tissue in a hypotonic solution (0.075 M KCl), followed by cell fixation using methanol:acetic acid (3:1). The cell suspension obtained is dropped onto slides, and samples are stained with Leishman stain (20%) for

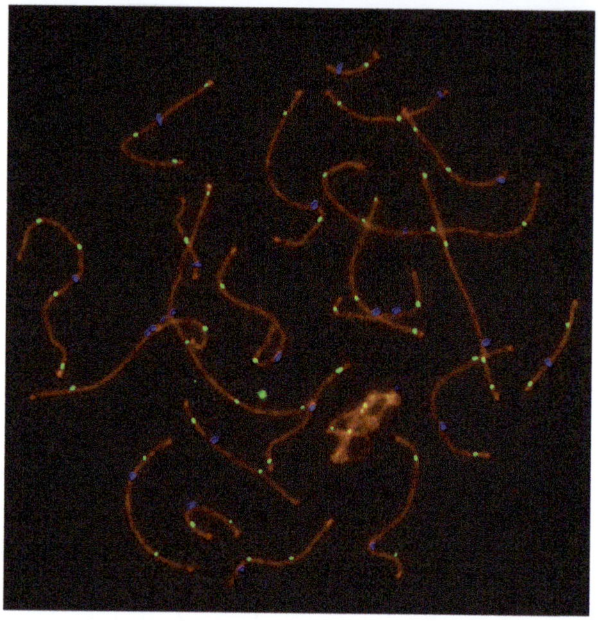

FIGURE 5.2 Human pachytene spermatocyte. Synaptonemal complexes are immunolabelled in red (SYCP3), centromeres in blue (CREST), and MLH1 recombination foci in green. (Image kindly provided by V. Peinado.)

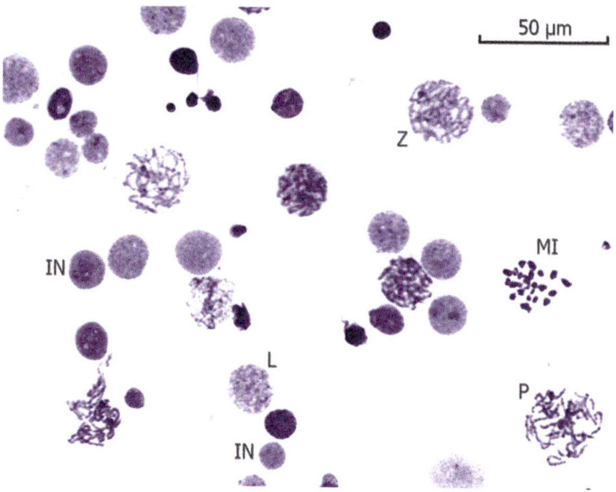

FIGURE 5.3 Leishman stained spermatogenic cells obtained from human testicular tissue. IN: Interphase nucleus; L: Leptotene; Z: Zygotene; P: Pachytene; MI: Metaphase I.

evaluation with an optical microscope (Figure 5.3). This methodology allows the analysis of chromosome features at prophase I, metaphase I, and metaphase II, identifying meiotic abnormalities in these stages (Figure 5.4).

The implementation of multiplex fluorescence *in situ* hybridization protocols (M-FISH) has allowed the unequivocal identification of chromosomes at metaphase I and metaphase II [27] (Figure 5.5). This methodology, limited to the research field due to the disproportionate relationship between cost and information obtained, has provided new data on the characterization of chromosome meiotic abnormalities described using uniform staining [28, 29].

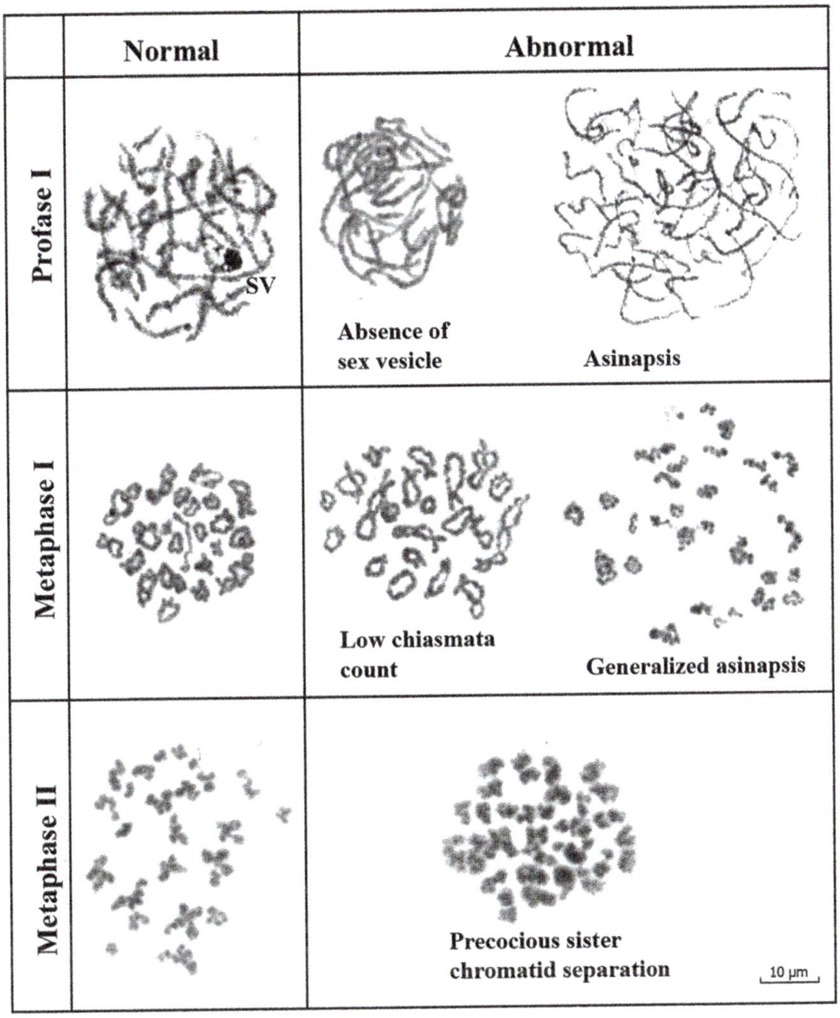

FIGURE 5.4 Leishman stained meiotic figures from human testicular biopsies. SV: sex vesicle.

CLASSIFICATION OF MEIOTIC ANOMALIES

Meiotic anomalies are usually classified into two categories: asynapsis and desynapsis. Asynapsis is the abnormal pairing of chromosomes from early prophase I with no formation of the sex vesicle, anomalies in synaptonemal complexes, and large reductions in the number of chiasmata in metaphase I [30, 31]. Desynapsis is the apparently normal pairing of chromosomes up to the pachytene stage, displaying sex vesicle formation, but with visible anomalies in some synaptonemal complexes and low chiasmata count in metaphase I [31, 32]. Focusing on bivalents in metaphase I, abnormalities may affect one, several, or most of them. Furthermore, these anomalies may involve all cells or coexist with normal cell lines [22].

In a clinical scenario, the analysis of uniform staining preparations enables us to reach a diagnosis of "normal meiosis" or "altered meiosis." In normal meiosis, all spermatogenic stages are present in normal proportions, showing normal cell features. In abnormal meiosis, a proportion of the spermatogenic cells is abnormal (meiotic arrest), usually due to an accumulation of prophase I stages or metaphase I cells.

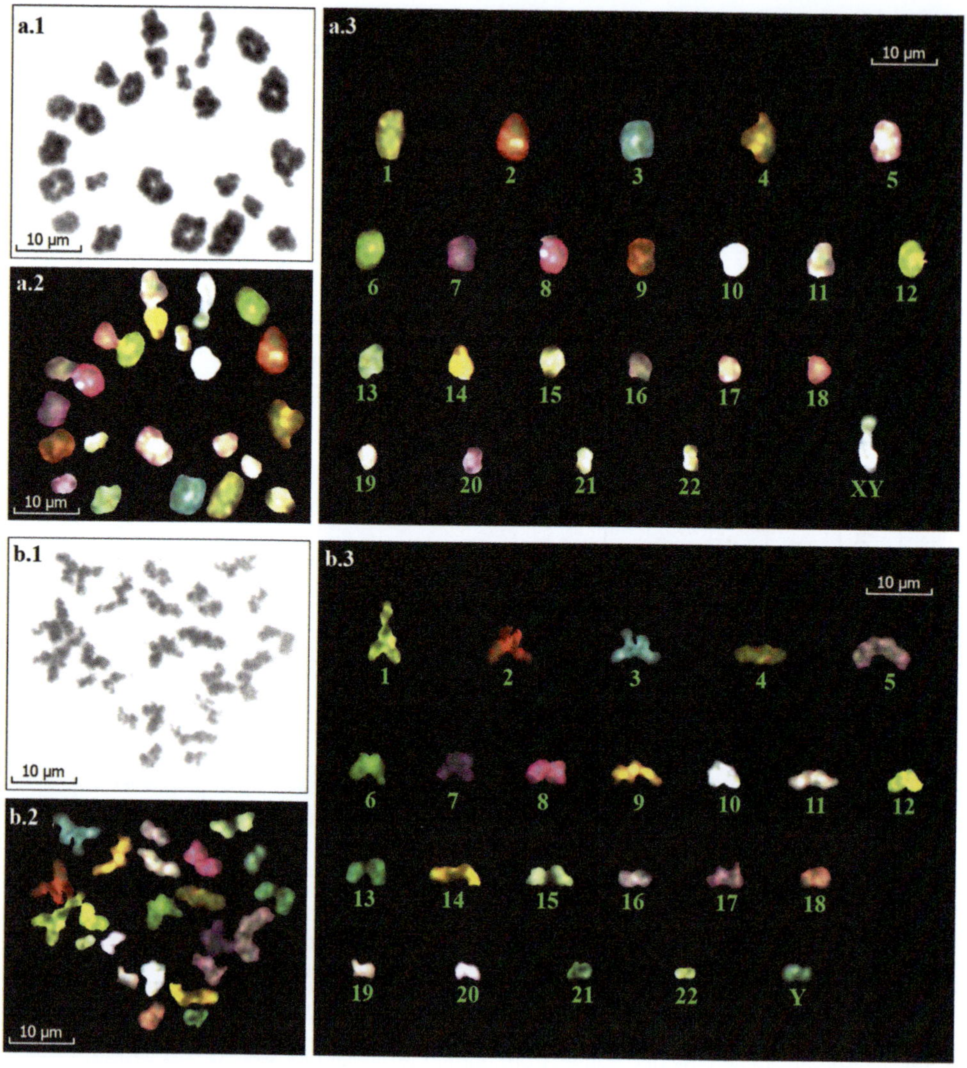

FIGURE 5.5 Chromosome identification at metaphase I and metaphase II using sequential Leishman stain and M-FISH procedures. (a) Metaphase I. (b) Metaphase II. a.1, b.1: Leishman staining. a.2, b.2: M-FISH images. a.3, b.3: M-FISH karyotypes.

Abnormal meiosis also arises if anomalies such as low chiasmata count or presence of univalents are observed.

Low Chiasmata Count

The reduction in the number of chiasmata per bivalent can influence their orientation in the metaphase plate and therefore affect the segregation of homologous chromosomes during anaphase I [1]. This phenomenon, combined with failures in the control mechanisms, could increase the proportion of chromosomal abnormalities in spermatozoa [33].

The relationship between male infertility and low chiasmata count has been described by different authors [22, 34–40], affecting the reproductive fitness of these individuals [41–44]. The use of M-FISH techniques has established that bivalents formed by medium and large chromosomes are the most susceptible to this phenomenon [28]. Despite this, since these chromosomes have a basal number of chiasmata higher than two [35], the reduction rarely leads to the presence of univalents. Accordingly, sperm FISH studies do not show preferred increased values of chromosomal abnormalities in medium and large chromosomes [40]. Therefore, the presence of at least one chiasma seems to guarantee the correct segregation of chromosomes during meiosis in most cases.

Presence of Univalents

The lack of chiasmata between homologous chromosomes compromises their orientation at metaphase I, resulting in segregation errors and leading to chromosomally abnormal gametes [45].

Several studies have reported the presence of univalents in infertile individuals [22, 35, 40]. This abnormality, which results from an extreme reduction in the number of chiasmata, is the most frequently observed anomaly in spermatocytes and usually involves sex chromosomes and small autosomes [28]. It is expected that defects in the processes of pairing, synapsis, and/or recombination of homologous chromosomes during prophase I could lead to the production of achiasmatic chromosomes [5, 7]. Since small autosomes (F and G group) and sex chromosomes often show a single chiasma [35], it is not surprising that these chromosomes were the ones observed more frequently as univalents [38]. This meiotic behavior is also consistent with sperm FISH studies that described the highest aneuploidy rates in sex chromosomes and G group chromosomes (especially chromosome 21) [40].

Meiotic Arrest

Meiotic arrest, characterized by the presence of unexpected proportions of some spermatogenic stages, has been described by different authors in infertile patients [22, 46]. The activation of checkpoints at pachytene and metaphase I/anaphase I would arrest and remove cells with meiotic abnormalities [47]. Depending on the severity of the alterations and the effectiveness of control mechanisms, a total or partial arrest of spermatogenesis could occur, resulting in azoospermia or more or less severe oligozoospermia. It is expected that a low number of spermatozoa is related to an increased rate of chromosome abnormalities, probably due to the inefficiencies in control mechanisms.

RELATIONSHIP BETWEEN MEIOTIC ANOMALIES AND MALE INFERTILITY

Published data show that meiotic abnormalities often occur in infertile males [22]. Studies of diakinesis/metaphase I in a series of 1,100 patients with different seminal alterations (from azoospermia to normozoospermia) and normal karyotype described synaptic abnormalities in 6–8% of males [48]. The incidence of meiotic anomalies was higher (17.5%) in a study performed on 103 infertile males with severe oligoasthenozoospermia [49]. Another study also described synaptic anomalies in 27% of the 60 normozoospermic males with a long history of sterility or with previous IVF failures evaluated [22]. Moreover, a study of 31 infertile males that incorporated M-FISH to characterize meiotic abnormalities suggested that these figures could reach an incidence of 48.4% and are especially associated with oligozoospermia [28].

According to the pieces of evidence provided in the preceding paragraphs, the best candidates for a meiotic study would be infertile males with a normal karyotype and unexplained infertility. Among them, males with normozoospermia and long-term sterility or with IVF failures (embryonic factor, no fertilization, repeated IVF failures) or infertile males with severe oligozoospermia [48].

MEIOTIC STUDIES IN THE CLINICAL DIAGNOSIS OF MALE INFERTILITY

Limitations of the Technique

Although the technique is cheap and fast, the analysis of meiotic images is tricky and requires experienced personnel. Moreover, sample features often result in some limitations due to the small amount of material obtained, the few cells under division, or, in the case of partial arrest during prophase I, the small numbers of metaphase I and II spermatocytes available for analysis. Besides, uniform chromosome staining procedures, together with the characteristic appearance of metaphase I and metaphase II chromosomes, make it difficult to identify the chromosomes affected by a given anomaly. Another limitation is that, occasionally, the characteristics of the preparations obtained are not compatible with cytogenetic analysis.

Finally, it is important to note that, to ensure accurate analysis results, it is strongly recommended to establish internal laboratory limits of normality for the different meiotic alterations.

Interpretation of the Results

Meiotic studies allow for determining whether the meiotic phase of spermatogenesis progresses correctly, but anomalies observed in this process do not always correlate with an increased rate of abnormalities in sperm. In this sense, when meiotic and sperm FISH studies were compared, a clear reduction in abnormal cells across was observed in 74% of patients [40], probably due to the activation of meiotic checkpoints that would selectively eliminate cells showing synaptic anomalies [50, 51]. Even so, an altered meiosis diagnosis would have to be interpreted as evidence of abnormalities in the pairing, recombination, and/ or segregation of meiotic chromosomes, indicating that spermatogenesis is compromised.

Taking all of this into consideration, individuals diagnosed as having altered meiosis should be advised about their reproductive risk, and it is highly recommended they have a sperm FISH study performed (see Chapter 6, this volume) to establish the outcome of chromosome abnormality at the end of the spermatogenic process.

NEW INSIGHTS INTO THE STUDY OF MALE INFERTILITY

The use of the three-dimensional fluorescent *in situ* hybridization (3D-FISH) [52] and the *in situ* chromosome conformation capture sequencing (Hi-C) techniques (which consist of determine and quantify interactions between loci that are physically close) [53] have demonstrated the functional relevance of genomic architecture. The non-random chromosomal positioning within the nucleus allows a spatiotemporal control of gene transcription and regulation, cell development, and division-fundamental biological processes for cell functionality. In this context, there is a growing idea that chromosomal territoriality could become a potential useful diagnostic tool for different diseases [54] and even infertility [55, 56].

An increasing number of articles define alterations in the chromosome positioning of male germ cells and spermatozoa from individuals with parameters related to infertility [57]. For instance, it has been observed that there is an altered distribution of specific chromosomes in sperm nuclei from infertile patients with: compromised semen parameters [58], increased aneuploidy frequencies [59], increased aneuploidy frequencies and oligoozospermia [60], immotile sperm populations [61], and reciprocal translocation carriers [62, 63]. Moreover, an altered distribution of bivalents at metaphase I is observed in infertile patients carriers of a Robertsonian translocation [64, 65] and in infertile patients with the presence of sex-chromosome univalents [66]. Thus, this evidence seems to relate a correct positioning of chromosomes with a correct development of the spermatogenesis process. Not only that but, it has been suggested that a proper chromosomal architecture in sperm nucleus is critical for early embryo development [67–71]. After fertilization, the nuclear membrane decomposes and sperm chromatin initiates decondensation by interacting with the maternal cytoplasmic environment and advancing in chromatin remodeling [72]. Therefore, the longitudinal positioning of chromosomes in the sperm nucleus could determine the pace of chromatin domains that are decondensed and remodeled [67–71]. Likewise, the relative position of chromosomes would facilitate or prevent specific interchromosomal contacts [73]. Consequently, the positioning of chromosomes in the sperm nucleus would become an additional epigenetic control, it would influence the gene expression pattern in the early stages of the embryo [67–71], and it may play a part in the *de novo* formation of anomalies observed in ICSI (intracytoplasmic sperm injection) babies [74].

Overall, in the face of the close relationship between chromosomal positioning and male infertility, an abnormal genome organization in male germ cells and sperm could be a detectable parameter that would help to explain infertility. However, the molecular mechanisms that regulate the spatial distribution of the genome and the involvement of this organization for spermatogenesis development are still unknown. Nevertheless, many studies are being carried out to understand the rules governing its operation. Therefore, the concept of 3D genome architecture as a diagnostic tool is just an emerging idea and there is a long way to go before its application.

ACKNOWLEDGMENTS

This study was supported by Project GJ515013 (Universitat Autònoma de Barcelona, Spain) and PI21/00564 (Instituto de Salud Carlos III, Ministerio de Ciencia e Innovación, Gobierno de España).

REFERENCES

1. Petronczki M, et al. *Cell.* 2003;112:423–40.
2. Burgoyne PS, et al. *Nat Rev Genet.* 2009;10:207–16.
3. Ding DQ, et al. *FEBS J.* 2009;277:565–70.
4. Longhese MP, et al. *DNA Repair.* 2009;8:1127–38.
5. Handel MA, et al. *Nat Rev Genet.* 2010;11:124–36.
6. Maiato H, et al. *Cell Mol Life Sci.* 2010; 67:2251–69.
7. Székvölgyi L, et al. *FEBS J.* 2010;277:571–89.
8. Kurahashi H, et al. *Front Genet.* 2012;3:112.
9. Watanabe Y. *Nat Rev Mol Cell Biol.* 2013;13:370–382.
10. Baudat F, et al. *Nat Rev Genet.* 2013;14:794–806.
11. Miller MP, et al. *Curr Opin Cell Biol.* 2013;25:687–96.
12. Hunter N. *Cold Spring Harb Perspect Biol.* 2015;7(12):1–35.
13. Manhart CM, et al. *DNA Repair.* 2016;38:84–93.
14. Cahoon CK, et al. *Nat Struct Mol Biol.* 2016;23:369–77.
15. Zickler D, et al. *Semin Cell Dev Biol.* 2016;54:135–48.
16. Hillers KJ, et al. *Meiosis. WormBook.* 2017;4:1–43.
17. Subramanian VV, et al. *Cold Spring Harb Perspect Biol.* 2014;6:a016675.
18. Li XC, et al. *Heredity.* 2009;102:24–30.
19. Vogt E, et al. *Mutat Res.* 2008;651:14–29.

20. Kops GJPL, et al. *Curr Biol.* 2020;18;30(10):589–602.
21. Gorbsky GJ. *FEBS J.* 2015;282:2471–87.
22. Egozcue J, et al. *Cytogenet Genome Res.* 2005;111:337–42.
23. Barlow AL, et al. *Chromosome Res.* 1996;4:562–73.
24. Oliver-Bonet M, et al. *Eur J Hum Genet.* 2003;11:879–83.
25. Codina-Pascual M, et al. *Cytogenet Genome Res.* 2004;107:18–21.
26. Sun F, et al. *Am J Hum Genet.* 2004;74:521–31.
27. Sarrate Z, et al. *Fertil Steril.* 2004;82:712–17.
28. Sarrate Z, et al. *Asian J Androl.* 2014;16:838–44.
29. Sarrate Z, et al. *Asian J Androl.* 2017;19:1–2.
30. Hultén M, et al. *Hereditas.* 1970;65:285–90.
31. Egozcue J. In *Diagnóstico de la Esterilidad Masculina* (Marina, S, ed.). Barcelona: ASESA; 1983;167–177.
32. Pearson PL, et al. *Cytogenetics.* 1970;9:460–67.
33. Ioannou D, et al. *Adv Exp Med Biol.* 2015;868:1–21.
34. McDermott A. *Ann Hum Genet.* 1973;37:13–20.
35. Skakkebaek NE, et al. *J Reprod Fertil.* 1973;35:23–36.
36. Chandley AC, et al. *Ann Hum Genet.* 1976;40:165–76.
37. Lamont MA, Faed MJ, and Baxby K. *J Clin Pathol.* 1981;34:145–50.
38. Laurie DA, et al. *Ann Hum Genet.* 1985;49:189–201.
39. Uroz L, et al. *Fertil Steril.* 2011;95:141–6.
40. Sarrate Z, et al. *Fertil Steril.* 2010;93:1892–902.
41. Rubio C, et al. *Hum Reprod.* 2001;16:2084–92.
42. Carrell DT, et al. *Arch Androl.* 2003;49:49–55.
43. Vialard F, et al. *J Assist Reprod Genet.* 2008;25:539–45.
44. Caseiro AL, et al. *Reprod Biomed Online.* 2015;31:481–5.
45. Hassold T, et al. *Nat Rev Gene.* 2001;2:280–91.
46. Martin RH. *Reprod Toxicol.* 2006; 22:142–7.
47. Hunt PA, et al. *Science.* 2002;296:2181–83.
48. Egozcue J, et al. *Hum Genet.* 1983;65:185–8.
49. Vendrell JM, et al. *Hum Reprod.* 1999;14:375–8.
50. Blanco J, et al. *Fertil Steril.* 2003;79:913–8.
51. Buwe A, et al. *Cytogenet Genome Res.* 2005;111:213–28.
52. Cremer M, et al. In *The Nucleus* (R Hancock ed.) Totowa, NJ: Humana Press; 2008;205–39.
53. Lieberman-Aiden E, et al. *Science.* 2009;326:289–93.
54. Meaburn KJ. *Front Genet.* 2016;7:134.
55. Mudrack O, et al. In *Male Infertility: Diagnosis and Treatment* (SC Oehninger and TF Kruger, eds.). London: Informa; 2007;73–83.
56. Ioannou D, et al. *Cytogen Genome Res.* 2011;133:269–79.
57. Sarrate Z, et al. *J Assist Reprod Genet.* 2018;35:1929–38.
58. Finch KA, et al. *Hum Reprod.* 2008;23:1263–70.
59. Olszewska M, et al.. *Chromosome Res.* 2008;16(6):875–90.
60. Ioannou D, et al. *Cytogenet Genome Res.* 2011;133:269–79.
61. Alladin N, et al. *Syst Biol Reprod Med.* 2013;59:146–52.
62. Wiland E, et al. *Chromosom Res.* 2008;16:291–305.
63. Mebrek, ML, et al. *J Assist Reprod Genet.* 2020;37:803–9.
64. Solé M, et al. *J Assist Reprod Genet.* 2017;34:131–8.
65. Wiland, E., et al. *Sci Rep.* 2019;9.
66. Mayans C, et al. *Asian J Andrology.* 2018;20(6):626–8.
67. Greaves IK, et al. *Chromosom Res.* 2003;11:503–12.
68. Foster HA, et al. *J Cell Sci.* 2005;118:1811–20.
69. Mudrak O, et al. In *Male Infertility: Diagnosis and Treatment* (Oehninger S and Kruger TF eds.). 2007;73–85.
70. Zalensky A, et al. *Biochem Soc Trans.* 2007;35:609–11.
71. Mudrak OS, et al. In *Human Interphase Chromosomes* (Yurov Y, Vorsanova S, Iourov I eds.). New York: Springer; 2013.
72. Wright SJ. *Curr Top Dev Biol.* 1999;46:133–78.
73. Cavalli G. *Curr Opin Genet Dev.* 2007;17: 443–50.
74. Van Steirteghem A, et al. *Hum Reprod Update.* 2002;8:111–6.

Chromosomal Analysis of Sperm

6

Lorena Rodrigo Vivó, Tantra Martínez Benito, and Azarina Ferro Barbero

INTRODUCTION

Infertility affects 10–15% of reproductive-age couples. In 1992, the World Health Organization determined that male factor infertility accounts for 10–30% of cases (1). Most of the factors associated with male infertility have a genetic component, e.g., chromosomal abnormalities and genetic defects, that affect the reproductive system or spermatogenesis. Errors in synapsis, recombination, and DNA repair during male meiosis can generate abnormal segregation of homologous chromosomes at meiosis I or sister chromatids at meiosis II, leading to spermatozoa with numerical chromosome abnormalities, e.g., aneuploidies or diploidies.

Checkpoint mechanisms regulating the different stages of male meiosis can eliminate defective germ cells to ensure spermatozoa have a normal chromosome content. Thus, spermatogenesis can be partially arrested in any of the maturation stages, resulting in *oligozoospermia* (reduced sperm number), or can be completely arrested, resulting in *azoospermia* (absence of sperm). However, if these control mechanisms are deficient, abnormal cell lineages can escape the checkpoints and give rise to chromosomally abnormal spermatozoa.

In couples with severe male factor infertility, intracytoplasmic sperm injection (ICSI) increases the chance of pregnancy. However, and importantly, prenatal testing from ICSI pregnancies has indicated an increased incidence of *de novo* sex chromosome aneuploidies and structural rearrangements (2, 3). Most chromosomal aberrations in these pregnancies are of paternal origin (4, 5) and are attributed to the sperm quality of the infertile males (6, 7). Between 2 and 26% of infertile men with a normal karyotype exhibit cytogenetic anomalies confined to the germ cell line (8, 9), which makes sperm chromosome studies particularly relevant.

In 1970, a technique for differential staining of specific chromosome regions launched the field of sperm chromosome analysis. From this, the total aneuploidy rate was estimated as 38%, with a rate of 1.4% for sex chromosomes and an average of 2% for autosomes (10). These rates were considered excessively high and were attributed to the low chromosomal specificity of the technique. In 1978, a technique was developed to fuse sperm with hamster oocytes without zona pellucida (11). This was standardized in 1982 (12), and provided information on the full chromosome content of sperm. However, this technique was complex, laborious, and could only analyze fertilization-competent spermatozoa. In the mid-80s, *in situ* hybridization techniques using specific DNA probes labeled with radioisotopes were developed (13),

DOI: 10.1201/9781003024941-6

followed, in the 90s, by probes labeled with fluorophores, allowing visualization of sperm chromosomes by the fluorescence in situ hybridization (FISH) technique (14).

METHODOLOGY FOR CHROMOSOMAL ANALYSIS OF SPERM

Sperm aneuploidy tests use FISH with probes directed to specific DNA sequences in the interphase sperm nuclei. By visualizing the hybridization signals using fluorescence microscopy, numerical chromosome abnormalities can be identified in the nucleus of ejaculated, epididymal, and testicular sperm. The simultaneous use of multiple probes specific for different chromosomes allows a rapid and relatively simple evaluation of a large number of sperm, enabling the detection of structural and numerical abnormalities (15, 16). Before performing FISH, spermatozoa must be fixed and spread on glass slides, without overlapping, and disulfide bridges between protamines that condense the nuclear chromatin must be broken by reducing agents to decondense the nucleus and allow DNA probe access. Then, the sperm double-stranded DNA and FISH probes are denatured by incubation at high temperatures. Finally, the DNA probes and sperm nuclei are co-incubated to hybridize and form a duplex of complementary strands (see Figure 6.1).

FISH in sperm is commonly performed using centromeric, locus-specific, and sub-telomeric fluorescent DNA probes. Specific combinations of these three types of probes are designed to study segregation of specific structural rearrangements. However, in carriers of numerical sex chromosome abnormalities and in normal-karyotype infertile men, the most widely analyzed chromosomes are 13, 18, 21, X, and Y. Aneuploidies of these chromosomes are compatible with life (see Figure 6.2).

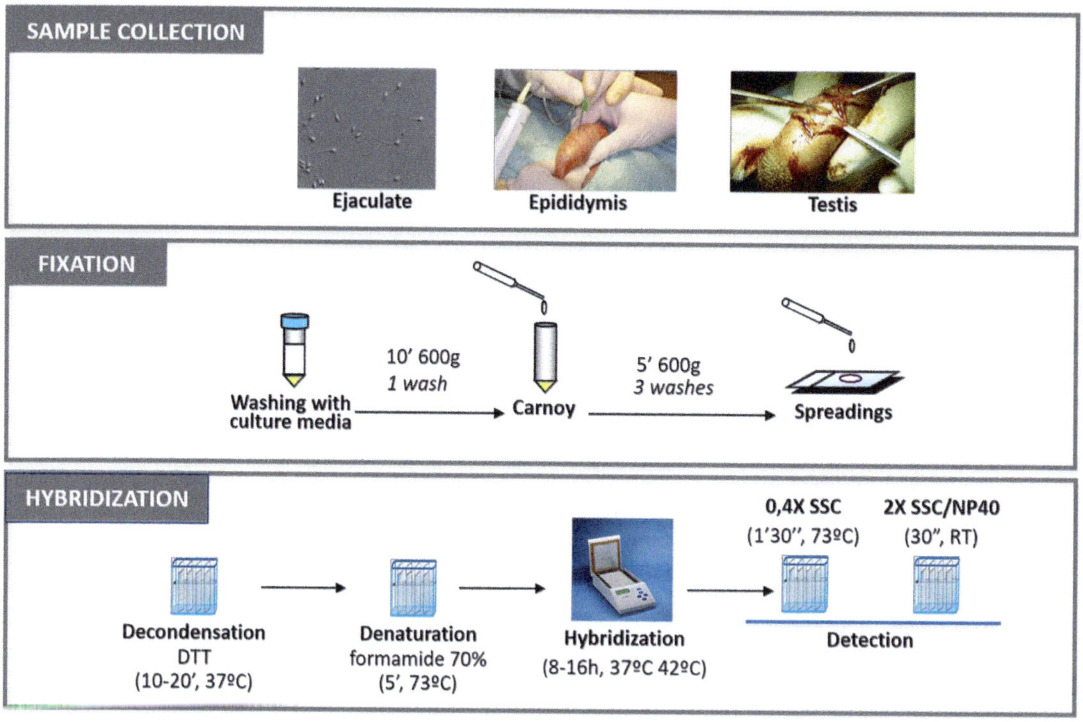

FIGURE 6.1 Sperm sample extraction and preparation for FISH analysis. After sample collection from the ejaculate, epididymis, or testis, spermatozoa are fixed and spread on glass slides. Hybridization is performed using specific fluorescently labeled probes.

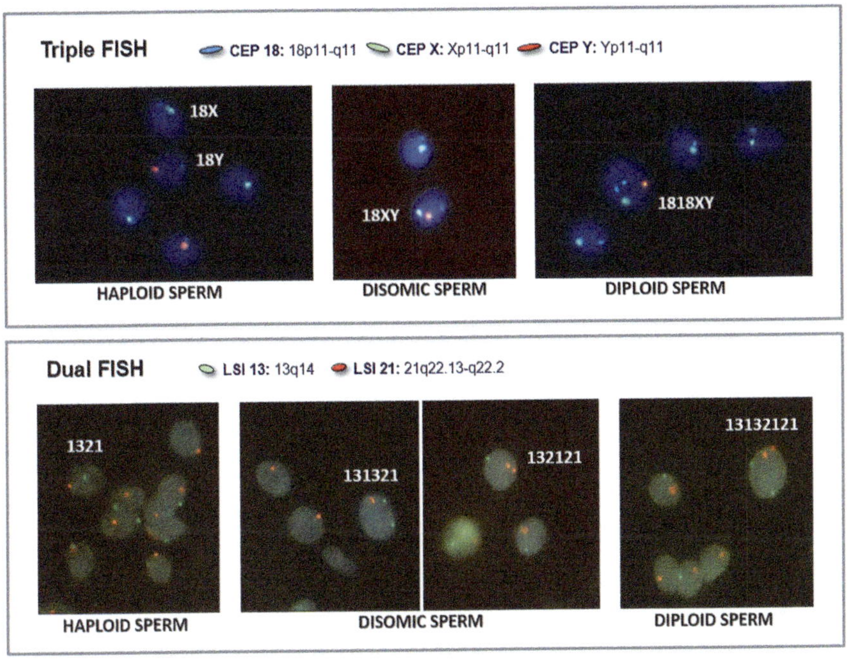

FIGURE 6.2 Evaluation of FISH signals using epifluorescence microscopy. Spermatozoa were hybridized using a triple FISH with centromeric probes (CEP) for chromosomes 18, X, and Y or a dual FISH with locus-specific probes (LSI) for chromosomes 13 and 21. Spermatozoa with one signal for each of the evaluated autosomes and one signal for the sex chromosomes (X or Y) are considered *normal haploid*; spermatozoa with two signals for a single chromosome and one signal for each remaining chromosome are considered *abnormal disomic*; and spermatozoa with two signals for all analyzed chromosomes are considered *abnormal diploid*.

In normozoospermic men with normal karyotypes, the total sperm aneuploidy rate is estimated at 6%, with a 0.31% and 0.12% disomy rate for the sex chromosomes and autosomes, respectively (16). Due to this low aneuploidy rate, at least 1,000 sperm per sample should be scored for clinical applications; however, this number may be smaller in cases of low sperm count, e.g., cryptozoospermia and azoospermia. An abnormal FISH result is considered when the sample shows a significant increase in sperm with numerical chromosome abnormalities (aneuploidies and/or diploidies) relative to the incidence of abnormal sperm seen in a control population of normozoospermic fertile males.

INDICATIONS

Currently, FISH analysis of sperm is used in diagnostic protocols for male infertility, allowing evaluation of the transmission risk of chromosomal defects to offspring. Infertile males with normal and abnormal karyotypes should be considered for evaluation of sperm aneuploidy in the following circumstances:

A) Infertile men with abnormal karyotype
   - *Carriers of numerical abnormalities for sex chromosomes.*

     Men with Klinefelter (47,XXY) or *47,XYY syndromes* are at risk of low sperm production with poor sperm quality and abnormal chromosome constitution (17–20). Blanco and coauthors described incidences of 1–20% sex chromosomal aneuploidies in sperm and 1% diploid sperm in these males (21).

- *Carriers of structural chromosomal abnormalities.*

  Carriers of *balanced chromosomal rearrangements*, e.g., Robertsonian or reciprocal translocations and inversions, can experience a variable range of alterations during gametogenesis and can be normozoospermic, oligozoospermic, or even azoospermic. After spermatogenesis, the spermatozoa can be chromosomally unbalanced in a variable range (22). The incidence of unbalanced sperm for the chromosomes of the rearrangement is 10–40% in Robertsonian translocations, 50–65% in reciprocal translocations, and 1–55% in inversion carriers (23).

B) Infertile men with normal karyotypes

- *Impaired meiosis in testicular analysis.*

  A low recombination frequency in meiotic pachytene cells is related to high aneuploidy frequency in sperm (24), mainly for the sex chromosomes (25). In fact, a significant correlation was described between cells with sexual vesicles without recombination sites and sex chromosome disomy in sperm (26). Additionally, Peinado and coauthors recently observed that 81.2% of non-obstructive azoospermic males with lower recombination levels than post-vasectomized obstructive azoospermic controls had an abnormal sperm FISH result, with a four-fold increase in disomy for all analyzed chromosomes (27).

- *Impaired sperm parameters.*

  Oligoasthenozoospermic males are reported to have higher incidences of aneuploid and diploid sperm than normozoospermic men (28–35). The sex chromosomes are the most affected, and the incidence of aneuploidy seems directly correlated with the severity of the oligozoospermia, being higher in patients with sperm concentrations < 5 million (36–46) (see Figure 6.3). The same correlation is observed in testicular sperm from azoospermic patients, mainly in those with *non-obstructive azoospermia* (37, 44, 47–51), with up to 42% having an abnormal FISH result (52).

  It is unclear whether abnormal sperm motility or morphology correlates with aneuploidy. No correlation was found between *isolated asthenozoospermia* and meiotic errors (42) even when FISH analysis was performed on motile and non-motile sperm from the same sample (53, 54). Nevertheless, isolated severe asthenozoospermia, particularly involving specific deformities of sperm flagella, strongly correlated with sperm meiotic defects (35, 55, 56). Also, an increased incidence of aneuploidy and polyploidy seems to associate with isolated teratozoospermia and specific morphological defects such as large-headed and multiple-tailed spermatozoa (57–62).

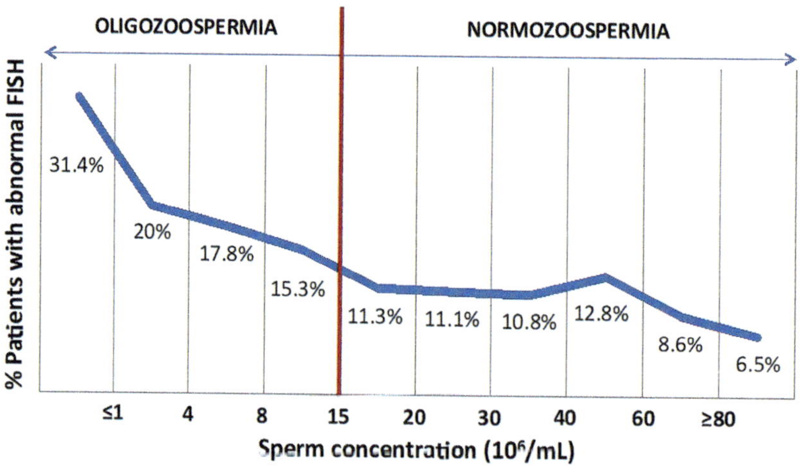

FIGURE 6.3 Effect of sperm concentration on FISH results. There is an inverse correlation between sperm concentration and the percentage of patients with increased sperm aneuploidy. Data obtained from Rodrigo et al. (46).

- *Chemotherapy and radiotherapy treatments.*

  Most *chemotherapy* or *radiotherapy* treatments have gonadal toxicity and affect spermatogenesis to a variable degree, depending on the type and duration of treatment. Reports indicate five-fold increases, relative to pretreatment levels, of diploid sperm and sperm with aneuploidies for autosomes and gonosomes after six months of treatment (63–65), and, in general, these rates decline to basal levels 18–24 months post-treatment (66). Combining sperm FISH protocols, a recent study found increased aneuploidies and chromosomal structural aberrations after cancer therapy in Hodgkin's lymphoma patients (67). Several studies have also described an association between Hodgkin's lymphoma and impaired spermatogenesis, some finding a significant increase in aneuploid sperm even before treatment (68, 69). These data suggest that the emergence of cancer itself induces problems in meiosis.
- *Clinical history of unknown recurrent pregnancy losses.*

  Approximately 66% of the abnormal karyotypes from miscarriages may have a male origin (70). Meiotic abnormalities (8, 49) and sperm aneuploidy are reported in recurrent pregnancy loss (RPL) patients (42, 71–76). Most reports describe increased incidences of sperm with sex chromosome disomy, and an increase in the diploidy rate was noted in a subset of patients with recurrent miscarriage after ovum donation (72). Moreover, the proportion of men with elevated aneuploid sperm is higher in couples with RPL (77). However, the proportion of patients with elevated levels of chromosomally abnormal sperm may decrease as the number of miscarriages increases, suggesting other etiologies at play when there is a clinical history of more than four miscarriages (46).
- *Clinical history of repetitive implantation failure.*

  Oocyte fertilization by a chromosomally abnormal sperm may cause implantation failure (30). A study of patients with three or more failed ICSI cycles reported that 31.6% had increased levels of spermatozoa with sex chromosome disomy (42). Also, later studies found an association between an abnormal FISH result in spermatozoa and a decrease in pregnancy and implantation rates in ICSI cycles (78–80).
- *Previous pregnancy with chromosomopathy.*

  Men with *chromosomally abnormal offspring of paternal origin*, e.g., Down syndrome (trisomy 21), Klinefelter syndrome (trisomy XXY), and Turner syndrome (monosomy X), exhibit a 1–20% incidence of sperm aneuploidy affecting the altered chromosome (81–84).

CLINICAL IMPACT OF SPERM CHROMOSOMAL ABNORMALITIES

As a clinical diagnostic tool, sperm FISH analysis should provide clinical prognostic value to inform the reproductive decisions of infertile couples. Therefore, an abnormal FISH result should be evaluated in terms of how much it affects clinical outcomes.

When considering *clinical outcome*, an increase in spermatozoa with chromosomal abnormalities is associated with decreased pregnancy rates and increased miscarriage risk for infertile couples undergoing ICSI cycles. Rubio et al. (2001) (42) compared the outcome of 108 ICSI cycles from patients with a normal sperm FISH result with 23 ICSI cycles from patients with an abnormal result. The groups had similar fertilization rates (71.5% vs. 74.5%), but the patients with normal FISH had a higher pregnancy rate (36.5% vs. 23.6%) and a lower miscarriage rate (54.8% vs. 80.0%). Similarly, Burrello et al. (2003) (79) studied 48 patients undergoing ICSI and observed lower pregnancy (34% vs. 75%) and implantation (13% vs.

34%) rates and a higher miscarriage rate (38.9% vs. 11.1%) in patients with higher incidences (> 1.55%) of aneuploid sperm. However, fertilization rate and embryo quality were similar in both cases. More recently, Rodrigo et al. (46) retrospectively analyzed the reproductive outcome of 1,527 male factor infertility couples performing conventional IVF/ICSI cycles. Couples with an abnormal sperm FISH result had significantly lower embryo transfer rates (73.2% vs. 82.6%), higher mean numbers of transferred embryos (2.3 ± 0.9 vs. 2.1 ± 0.7), lower pregnancy (18.1% vs. 36.3%) and implantation rates (8.9% vs. 22.8%), and higher miscarriage rates (62.5% vs. 34.9%) than patients with normal sperm FISH results. Nicopoullos et al. (80) also found that patients failing to achieve pregnancy after ICSI had a significantly higher sperm aneuploidy rate (2.37%) than patients that achieved pregnancy (1.18%). Additionally, Petit et al. (78) found that men whose partners failed to conceive after ≥ 4 ICSI cycles had higher rates of aneuploid and diploid sperm than men whose partners became pregnant after 1–3 ICSI cycles. Moreover, the probability of achieving a clinical pregnancy, of miscarrying, and of having a live birth decreases 2.6× (80), increases 3.6×, and decreases 0.4× (46), respectively, for every 1% increase in the sperm aneuploidy rate.

When considering *embryonic outcomes*, preimplantation genetic testing for aneuploidy (PGT-A) can be used to evaluate the impact of higher rates of sperm numerical chromosome abnormalities on the chromosomal constitution of embryos. Several reports described higher rates of abnormal embryos, noting a high incidence of aneuploidy for sex chromosomes and mosaicism (17, 85–90). The percentage of abnormal embryos from patients with oligozoospermia, azoospermia, and abnormal sperm FISH or impaired meiosis were reported to range from 43% to 78% (17, 86–94). Severe oligozoospermic or cryptozoospermic patients with Y-chromosome microdeletions also had a higher incidence of aneuploid embryos, principally monosomy X (95). Notably, different types of sperm aneuploidies appear to have distinct effects on embryo chromosome constitution: an increased percentage of spermatozoa with sex chromosome disomies associated with an increase in embryo aneuploidies compatible with life (Patau, Edwards, Down, Klinefelter, and Turner syndromes; and trisomies XXX and XYY), but an increase in diploid spermatozoa associated with an increase in triploid embryos, which mostly miscarry before delivery (17).

For translocation carriers, the percentage of abnormal gametes appears to correlate with the percentage of abnormal embryos (96). Also, for patients with normal karyotypes but isolated severe male factor infertility, a retrospective analysis of 249 PGS cycles showed a linear increase in the percentage of abnormal embryos with increasing percentages of total aneuploid sperm (97).

When considering *offspring*, fathers of children with Down, Klinefelter, or Turner syndromes may exhibit increases in sperm chromosomal abnormalities associated with the chromosomopathy observed in their children. Two fathers of children with Down syndrome of paternal origin were found to have sperm disomy 21 rates of 0.75% and 0.78% (81). Similar studies of couples with miscarriages or of children from carriers of sex chromosome abnormalities (Turner or Klinefelter syndrome) reported high incidences (0.20–24.7%) of sperm aneuploidy for sex chromosomes (82–84, 98–100).

CLINICAL REPRODUCTIVE COUNSELING

When an abnormal FISH result is found in sperm, genetic counseling should be offered to the couple. Additionally, depending on the severity of the abnormalities observed in the sperm, several clinical options can be proposed (see Figure 6.4):

1) If FISH indicates a significant increase in sperm aneuploidies, PGT-A may help improve the possibility of healthy pregnancies (86–88). In a retrospective study, patients with abnormal FISH in sperm that underwent PGT-A cycles analyzing a battery of nine chromosomes showed higher pregnancy (44.0% vs. 31.5%) and implantation rates (36.2% vs. 24.9%) than patients with normal sperm FISH results. Interestingly, patients with normal FISH in sperm had similar clinical results regardless of conventional IVF/ICSI or PGT-A; however, patients with abnormal

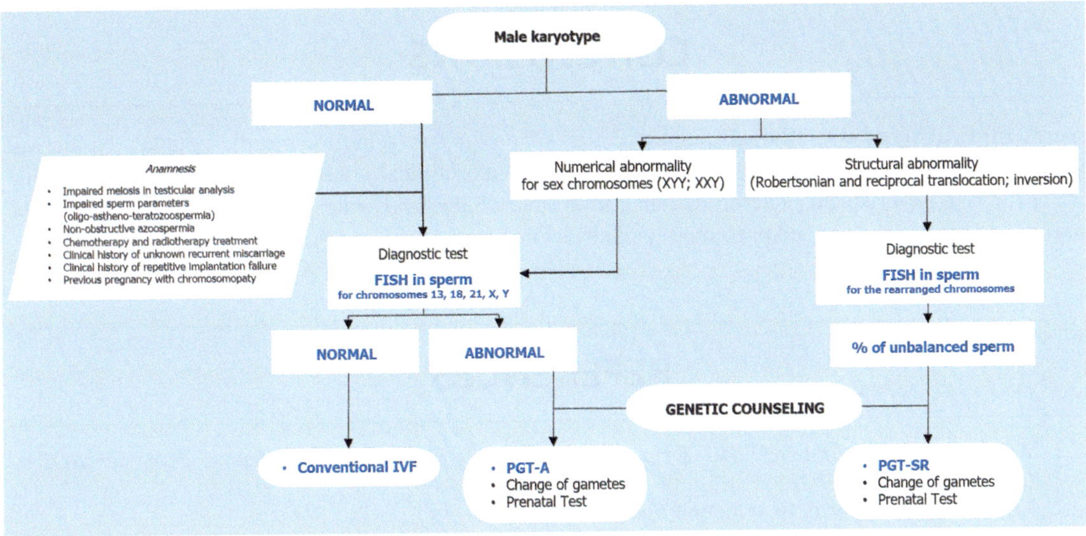

FIGURE 6.4 Algorithm of testing options for male factor infertility. FISH, fluorescence in situ hybridization; IVF, in vitro fertilization; PGT-SR, preimplantation genetic testing for structural rearrangements; PGT-A, preimplantation genetic testing for aneuploidies.

FISH in sperm showed better pregnancy and implantation rates after PGT-A (46). Moreover, aneuploidy screening of all 24 chromosomes in couples with male factor infertility provided even better clinical results, with 83.6% of cycles yielding at least one euploid embryo to transfer, resulting in a pregnancy rate per transfer of 62.9%, an implantation rate of 54.2%, and a take-home baby rate of 50.9% (94).

2) In cases where a severe meiotic impairment results in extremely high levels of abnormal sperm, donor sperm should be considered to provide better clinical results and better chances for healthy babies.

STRENGTHS AND LIMITATIONS

Sperm FISH is a non-invasive method that can easily and rapidly analyze thousands of spermatozoa, i.e., the products of male meiosis, for chromosomal defects. High-specificity DNA probes are commercially available, which, when applied to different samples from the same patient, allow any chromosome to be studied. However, FISH in sperm is not a comprehensive analysis of all 24 chromosomes, and its results only refer to the specific loci being tested. Thus, it cannot detect abnormalities outside the chromosomal segments specifically labeled by the DNA probes.

For clinical applications, a minimum of 1,000 sperm should be scored per chromosome analyzed; however, this number may not be obtained in ejaculated samples with low sperm count and in testicular samples from azoospermic patients. In addition to sampling constraints, FISH is subject to technical issues. For example, an absence of a signal, i.e., a nullisome, is often not considered in the analysis because it may represent a hybridization problem. Also, most of the technology used for FISH analysis is handmade and evaluation is often subjective. Standardizing and automating sperm aneuploidy assessment is necessary to reduce inter-laboratory variability and establish clinical references. In the absence of a reference, the use of internal controls for sperm aneuploidy comparisons is highly recommended.

CONCLUSIONS

Sperm FISH analysis is a clinical diagnostic tool of prognostic value to infertile couples. Numerical chromosome abnormalities in sperm can be translated into clinical consequences, including infertility or genetic risk for offspring. Therefore, when high sperm aneuploidy rates are found, couples should be offered genetic counseling and presented with clinical options, including PGT-A.

REFERENCES

1. World Health Organization. *WHO Technical Report Series 820. Recent Advances in Medically Assisted Conception*. Geneva; 1992.
2. Van Steirteghem A, et al. *Hum Reprod Update*. 2002;8:111–116.
3. Bonduelle M, et al. *Hum Reprod*. 2002;17:2600–2614.
4. Van Opstal D, et al. *Hum Reprod*. 1997;12:682–686.
5. Meschede D, et al. *Hum Reprod*. 1998;13(3):576–582.
6. In't Veld P, et al. *Lancet*. 1995;346:773.
7. Bonduelle M, et al. *Hum Reprod*. 1998;13:781–782.
8. Vendrell JM, et al. *Hum Reprod*. 1999;14:375–378.
9. Egozcue S, et al. *J Assist Reprod Genet*. 2000;17:307–309.
10. Pawlowitski IH,et al. *Humangenetik*. 1972;16:119–122.
11. Rudak E, et al. *Nature*. 1978;274:911–913.
12. Martin RH, et al. *Am J Hum Genet*. 1982;34:459–468.
13. Joseph AM, et al. *Hum Genet*. 1984;66:234–238.
14. Wyrobek AJ, et al. *Mol Reprod Dev*. 1990;27:200–208.
15. Downie SE, et al. *Mol Hum Reprod*. 1997;3:585–598.
16. Egozcue J, et al. *Hum Reprod Update*. 1997;3:441–452.
17. Rodrigo L, et al. *Fertil Steril*. 2010;94:1380–1386.
18. González-Merino E, et al. *Fertil Steril*. 2007;88:600–606.
19. García-Quevedo L, Blanco J, Serrate Z, et al. *Hum Reprod*. 2011;26:3486–3493.
20. Vialard F, et al. *J Androl*. 2012;33:1352–1359.
21. Blanco J, et al. *Hum Reprod*. 2001;16:887–892.
22. Rosenbusch B. *J Genet*. 2010;89:105–108.
23. Antón E, et al. *Hum Reprod*. 2007;22:2088–2092.
24. Ma S, et al. *Hum Reprod*. 2006;21:980–985.
25. Ferguson KA, et al. *Hum Mol Genet*. 2007;16:2870–2879.
26. Sun F, et al. *Mol Hum Reprod*. 2008;14:399–404.
27. Peinado Cervera V. *Estudio de anomalías meióticas y aneuploidías en pacientes con azoospermia secretora*. Universidad de Valencia. 2016; https://roderic.uv.es/handle/10550/53947.
28. Moosani N, et al. *Fertil Steril*. 1995;64:811–817.
29. Lähdetie J, et al. *Am J Med Genet*. 1997;71:115–121.
30. Pang MG, et al. *Hum Reprod*. 1999;14:1266-1273.
31. Pfeffer J, et al. *Fertil Steril*. 1999;72:472–478.
32. Gole LA, et al. *J Androl*. 2001;22:759–763.
33. Schmid TE, et al. *Hum Reprod*. 2003;18(7):1474–80.
34. Liu CH, et al. *J Assist Reprod Genet*. 2004;21(4):119–26.
35. Tempest HG, et al. *Reprod Biol Endocrinol*. 2004;2:82.
36. Bernardini L, et al. *Hum Reprod*. 1998;13:3406–3413.
37. Bernardini L, et al. *Hum Reprod*. 2000;15:2165–2172.
38. Rives N, et al. *Hum Genet*. 1999;105:266–272.
39. Vegetti W, et al. *Hum Reprod*. 2000;15:351–365.

40. Ushijima C, et al. *Hum Reprod.* 2000;15:1107–1111.
41. Nishikawa N, et al. *J Assist Reprod Genet.* 2000;17:97–102.
42. Rubio C, et al. *Hum Reprod.* 2001;16:2084–2092.
43. Calogero AE, et al. *Hum Reprod.* 2001;16:1433–1439.
44. Martin RH, et al. *Biol Reprod.* 2003;69:535–539.
45. Miharu N. *Cytogenet Genome Res.* 2005;111(3–4):347–51.
46. Rodrigo L, et al. *Biol Reprod.* 2019;101(6):1091–1101.
47. Levron J, et al. *Fertil Steril.* 2001;76:479–484.
48. Burrello N, et al. *Eur J Hum Genet.* 2002;10:362–366.
49. Mateizel I,et al. *Hum Reprod.* 2002;17:2249–2257.
50. Palermo GD, et al. *Hum Reprod.* 2002;17:570–575.
51. Rodrigo L, et al. *Hum Reprod.* 2004;19:118–123.
52. Rodrigo L, et al. *Fertil Steril.* 2011;95:1005–1012.
53. Samura O, et al. *Hum Reprod.* 1997;12(11):2437–42.
54. Zeyneloglu HB, et al. *Hum Reprod.* 2000;15(4):853–6.
55. Hristova R, et al. *Biol Reprod.* 2002;66(6):1781–3.
56. Rives NM. *Cytogenet Genome Res.* 2005;111(3–4):358–62.
57. In't Veld PA, et al. *Hum Reprod.* 1997;12(4):752–4.
58. Viville S, et al. *Hum Reprod.* 2000;15(12):2563–6.
59. Devillard F, et al. *Hum Reprod.* 2002;17(5):1292–8.
60. Vicari E, et al. *J Androl.* 2003;24(4):598–603.
61. Lewis-Jones I, et al. *Fertil Steril.* 2003;79(1):212–5.
62. Mateu E, et al. *J Androl.* 2006;27(1):6–10.
63. Robbins WA, et al. *Nat Genet.* 1997;16:74–78.
64. Martin RH, et al. *Cancer Genet Cytogenet.* 1999;108:133–136.
65. Frias S, et al. *Cancer Res.* 2003;63(1):44–51.
66. Tempest HG, et al. *Hum Reprod.* 2008;23(2):251–8.
67. Frias S, et al. *PLoS One.* 2020;15(12):e0242218.
68. Viviani S, et al. *Eur J Cancer.* 1991;27(11):1389–92.
69. Rueffer U, et al. *Ann Oncol.* 2001;12(9):1307–11.
70. Kim JW, et al. *BMC Med Genet.* 2010;3;11:153. doi: 10.1186/1471-2350-11-153.
71. Giorlandino C, et al. *Fertil Steril.* 1998;70(3):576–7.
72. Rubio C, et al. *J Assist Reprod Genet.* 1999;16(5):253–8.
73. Carrell DT, et al. *Obstet Gynecol.* 2003;101(6):1229–35.
74. Somprasit C, et al. *Reprod Biomed Online.* 2004;9(2):225–30.
75. Bernardini LM, et al. *Reprod Biomed Online.* 2004;9(3):312–20.
76. Al-Hassan S, et al. *Arch Androl.* 2005;51(1):69–76.
77. Ramasamy R, et al. *Fertil Steril.* 2015;103(4):906–909.e1. doi: 10.1016/j.fertnstert.2015.01.029.
78. Petit FM, et al. *Androl.* 2005;26:235–241.
79. Burrello N, et al. *Hum Reprod.* 2003;18:1371–1376.
80. Nicopoullos JD, et al. *Hum Reprod.* 2008;23:240–250.
81. Blanco J, et al. *Am J Hum Genet.* 1998;63:1067–72.
82. Martínez-Pasarell O, et al. *Hum Genet.* 1999;104:345–49.
83. Eskenazi B, et al. *Hum Reprod.* 2002;17:576–583.
84. Tang SS, et al. *Hum Reprod.* 2004;19:147–151.
85. Gianaroli L, et al. *J Endocrinol Invest.* 2000;23:711–716.
86. Silber S, et al. *Fertil Steril.* 2003;79:30–38.
87. Rubio C, et al. *Reprod Biomed Online.* 2005;11:497–506.
88. Gianaroli L, et al. *J Histochem Cytochem.* 2005;53:261–267.
89. Sánchez-Castro M, et al. *Hum Reprod.* 2009;24:1516–1521.
90. Al-Asmar N, et al. *Fertil Steril.* 2012;98:145–150.
91. Arán B, et al. *Reprod Biomed Online.* 2004;8:470–476.
92. Platteau P, et al. *Hum Reprod.* 2004;19:1570–1574.
93. Magli MC, et al. *Reprod Biomed Online.* 2009;18:536–542.
94. Rodrigo L, et al. *Biomed Res Int.* 2014;2014:517125.
95. Mateu E, et al. *Fertil Steril.* 2010;94:2874–2877.
96. Escudero T, et al. *Fertil Steril.* 2003;79:1528–34.

97. Rodrigo Vivó L. *Valor de los estudios cromosómicos en espermatozoides mediante la técnica de hibridación in situ fluorescente (FISH) en parejas estériles.* Universidad de Valencia; 2013; https://roderic.uv.es/handle/10550/30746.
98. Moosani N, et al. *Hum Reprod.* 1999;14:1137–1138.
99. Martínez-Pasarell O, et al. *Hum Reprod.* 1999;14:2735–2738.
100. Lowe X, et al. *Am J Hum Genet.* 2001;69:1046–1054.

7

Epigenetics in Sperm, Epigenetic Diagnostics, and Transgenerational Inheritance

Jennifer L. M. Thorson, Millissia Ben Maamar, and Michael K. Skinner

INTRODUCTION

Sperm are highly specialized cells transmitting the paternal genome to the oocyte and thus, to the next generation. For many years, it was believed that sperm was only contributing its DNA and nothing else. A growing body of evidence now supports the importance and relevance of the sperm epigenome for male fertility and offspring health. Sperm DNA methylation, sperm histones alterations, and non-coding RNA are all involved and can contribute to observable epigenetic alterations inherited by subsequent generations (1–4). Moreover, sperm has been shown to be particularly sensitive to different environmental exposures which can subsequently affect embryonic development and offspring (4). Several studies have observed that such environmental insults cause changes in the sperm epigenome which can be passed to subsequent generations transgenerationally (4–11). Recent data suggest that the different sperm epigenetic mutations (DNA methylation, ncRNA, histone modifications, and retention) offer predictive diagnostic tools in the assessment of male infertility and potential impacts on subsequent generations' disease etiology (8, 12–18). This chapter will focus on the different epigenetic factors, epigenetic diagnostics, concept of transgenerational inheritance, and the potential use of epigenetics in the context of male infertility.

DOI: 10.1201/9781003024941-7

EPIGENETIC MODIFICATIONS

Individuals typically respond to their environment through changes in gene expression and these changes are generally mediated by epigenetic processes. Additionally, the mechanisms which generate individual cell types allow one cell type to differentiate and develop into another cell type are primarily mediated by epigenetic mechanisms. Epigenetics is defined as the "molecular factors/processes around the DNA that regulate genome activity independent of DNA sequence, and that are mitotically stable" (19). Several distinct epigenetic factors or processes act to control genome activity in a cell and regulate gene expression. These include DNA methylation, histone modification/retention, chromatin remodeling, non-coding RNAs (ncRNAs), and RNA methylation (Figure 7.1).

DNA Methylation

DNA methylation was the first epimutation to be characterized. DNA methylation involves a small (methyl) chemical group that is enzymatically attached to DNA through DNA methyltransferase (DNMT), primarily at the cytosine base when adjacent to guanine in most species (20, 21) and the product is 5-methylcytosine (5mC) (Figure 7.1). In mammals, the addition of a methyl group to a CpG site results in the alteration (often inhibition) of a transcription factor binding to DNA, while the removal of a methyl group to a CpG site facilitates the recruiting of proteins involved in gene expression (22). Other chemical modifications of cytosine bases in DNA have since been described, but are infrequent and function not elucidated. The ten-eleven translocation (TET) family of enzymes can successively oxidize 5mC to 5-hydroxymethylcytosine (5hmC), which is the precursor to DNA methylation erasure (23). In broad terms, the presence of 5mC often represses DNA transcription, while 5hmC is permissive to transcription (24, 25). The functions of the other epigenetic modifications to cytosine are under investigation. N(6)-methyladenine is an epigenetic modification to the adenine base of DNA that was once thought to only be present in prokaryotic organisms, but has now been described in mammalian embryonic stem cells associated with RNA methylation (26).

Histone Modification and Retention

The histone proteins that DNA is wrapped around create the nucleosome and can be chemically modified to alter gene expression (Figure 7.1). Histone proteins are often subject to post-translational modifications

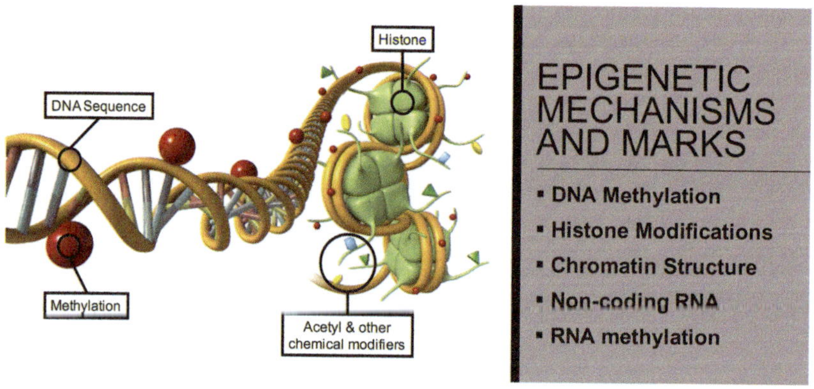

FIGURE 7.1 Epigenetic mechanisms and marks. Modified from (98).

which form a complex molecular mechanism that subsequently results in regulation of gene expression and downstream biological functions (27). Numerous different histone post-translational modifications interact and generate combinatorial patterns to influence gene expression. Among the known histone modifications are lysine acetylation, lysine and arginine methylation, arginine citrullination, lysine ubiquitination, lysine sumoylation, ADP-ribosylation, proline isomerization, and serine/threonine/tyrosine phosphorylation (28).

In many species, including humans, the sperm histones are removed from DNA following meiosis during spermiogenesis. The DNA is condensed with protamines resulting in highly compact nucleoprotamine complexes (29). In human sperm, roughly 5%–15% of the histones are retained in nucleosomes while the remainder is condensed in nucleoprotamine complexes (30, 31). Environmental exposures transgenerationally can modify the histone retention sites in sperm and dramatically increase the number of retention sites (32). The differential distribution of genes in protamine-associated versus histone-associated regions in the sperm yields crucial epigenetic information which is delivered to the oocyte and zygote (33, 34).

Chromatin Structure

The coiling, looping, and general structure of DNA, termed chromatin structure, is also an epigenetic factor (35) (Figure 7.1). The three-dimensional structure of DNA can make certain regions of the genome accessible to transcriptional machinery, or bring enhancer regions near gene promoters to affect gene expression. Many of the major events in the developmental progression of an organism, notably gametogenesis and fertilization, involve dramatic remodeling events of the germ cell chromatin (36). Chromatin modifications serve as crucial epigenetic modifications in development and reproduction. The condensation of DNA with protamines and removal of histones transcriptionally silences the sperm to allow effective delivery of the sperm DNA to the oocyte.

Non-Coding RNA (ncRNA)

Non-coding RNA molecules can act as epigenetic factors (37) (Figure 7.1). These are small RNA molecules that do not code for a protein, but rather function as RNA to regulate gene expression. The non-coding RNA molecules that act as epigenetic factors have secondary structure to facilitate DNA and protein interactions, but are not DNA sequence-dependent, so the majority do not depend on having a nucleotide sequence that is complimentary to a specific DNA or RNA region in order to function. Long non-coding RNAs (lncRNAs) (38) and small non-coding RNAs (sncRNAs) are the two major types. The sncRNA have many sub-families such as transfer RNA-derived small RNAs (tsRNAs) (39), which are examples of ncRNA classes that are present in sperm and can act as epigenetic factors that affect subsequent generations (39, 40).

RNA Methylation

Chemical modifications of RNAs can have dynamic regulatory roles similar to the epigenetic modifications of DNA and histone proteins (41) (Figure 7.1). The most prevalent known mammalian RNA modification is N6-methyladenosine (m6A), a reversible methylation of the messenger RNA (mRNA) (42). The methylation of RNA alters the structure of the RNA to change function and protein or DNA association. Methylation of numerous RNA species results in a diversity of functions on RNA including biophysical, biochemical, and metabolic stabilization of RNA and further crucial functional processes (43).

EPIGENETIC ALTERATIONS (EPIMUTATIONS) AND MALE INFERTILITY

Male infertility has increased at an alarming rate over the past 40 years and this crisis demands new approaches for prevention, diagnosis, and treatment (44). It is estimated as many as 1 in 20 men currently face reduced fertility (45). A myriad of environmental conditions ranging from increased exposure to toxicants to increased rates of obesity may explain this increase in male infertility (46). The molecular mechanisms driving this increase in male infertility likely include epigenetic alterations (47).

DNA Methylation

Idiopathic male infertility is associated with a reduction in the quality or quantity of sperm. An association between disrupted DNA methylation and abnormal human sperm was first described by Navarro-Costa (48). This group proposed a correlation between male gametogenic defects and incorrect epigenetics alterations (epimutations) in male germline genes. DNA isolated from poor quality sperm collected from an infertility clinic exhibits abnormal methylation at numerous sequences, implicating improper programming of the male germline (49). Poplinski (50) compared the DNA methylation of sperm among men exhibiting idiopathic infertility against that of control men. This research group found aberrant methylation patterns among the men with idiopathic infertility and suggest that these aberrant patterns may be passed onto children conceived through assistive-reproductive technologies (ART). Differential DNA methylation regions were found to be associated with male patients who are responsive to follicle-stimulating hormone (FSH) use, a therapeutic treatment for idiopathic infertility (51). Changes in the methylation of histones in sperm are suggested to have cumulative detrimental effects on fertility among men (52). Recent efforts have developed diagnostic tools that will identify aberrant DNA methylation and will provide a diagnostic tool in a clinical setting (18, 51).

Histone Modification and Retention

When the histones are replaced by protamines during spermatogenesis a strict ratio of protamine-1 (P1) to protamine-2 (P2) is regularly maintained. This ratio is strictly regulated and highly conserved among mammalian species (53). Aoki (54) found an association with a reduction in the P1/P2 ratio among a population of infertile men, indicating abnormal concentrations of either type of protamine are strongly associated with male infertility. In normal spermatogenesis, not all histones are replaced with protamines. There is a programmed pattern to which histones are retained in normal spermatogenesis. A population of infertile men exhibited random or abnormal histone retention as well as elevated levels of histone retention among their sperm (52, 55).

Acetylation of histones is a normal part of spermatogenesis, when histones are exchanged with protamines. Among infertile men, impaired spermatogenesis is associated with a decrease of histone acetylation in spermatids resulting in incorrect histone-to-protamine exchange (56). Aberrant histone acetylation of the promoters of developmentally important genes was found among subfertile men and suggested to adversely affect the transfer of epigenetic information to the oocyte (57).

Chromatin Modification

The structure of chromatin is known to affect gene expression and can be crucial for proper spermatogenesis. An increase in heterochromatic variations is strongly associated with male infertility (58). Any perturbation to the normal chromatin remodeling involved in precursor sperm cells has negative consequences on spermatogenesis and eventually on male fertility (59).

Non-Coding RNA (ncRNA)

Sperm RNAs represent a portion of epigenetic information that is delivered to the oocyte from the paternal germ cells. Sperm RNA elements (SRE) from males presenting idiopathic infertility and found around 30% presented an incomplete set of the required SREs (60, 61). Short non-coding RNAs play an essential role in spermatogenesis and have recently been implicated in male fertility disorders (reviewed in [62]). Long non-coding RNAs (lncRNAs) are shown to play a role in sperm motility and potentially fertility (63). Zhang (64) identified lncRNAs which are crucial for human spermatogenesis and sperm function and the resulting causes of male infertility.

RNA Methylation

Human sperm fertility has been associated with altered levels of N6-methyladenine (m6A) in human sperm (65). These authors suggest that increased levels of m6A may be a risk factor for decreased sperm motility. A depletion of spermatagonial stem cells (SSC) is also associated with a reduction of m6A resulting in abnormal spermatogenesis (66).

EPIGENETIC TRANSGENERATIONAL INHERITANCE

The environment has been shown to be one of the most critical factors to impact the biology of an organism. An exposure to one or multiple environmental factors (e.g., nutrition, toxicants, stress) can trigger changes in the transcriptome and impact the development of pathologies or phenotypic variation. As described above, epigenetic factors are the molecular mechanisms an organism uses to respond to an environmental change with modifications in gene expression. Most environmental factors and toxicants do not possess the capacity to alter DNA sequence or promote genetic mutations (67). However, the environment is able to dramatically influence epigenetic processes which then affect gene expression and development. In human and animal models, several studies have demonstrated that exposure to certain environmental toxicants at a specific window of development, especially when the epigenome is reprogramming, can affect the mechanisms involved in the establishment of the sperm epigenome. Since the sperm epigenome has been shown to be crucial for the fertility of the individuals, any variations could be related to male infertility (68). Some epimutations have also been shown to be transmitted via the sperm to the offspring and subsequent generations which has been defined as the concept of epigenetic transgenerational inheritance (4, 6, 69). Therefore, epigenetics provides a molecular mechanism for the environment to directly alter the biology of an organism (70). The presence of an altered epigenetic factor at a specific chromosomal location in response to an environmental factor is called an "epimutation" (71).

These environmentally induced epigenetic changes can be passed to subsequent generations and can result in changes in gene expression and phenotype in the offspring, even when there is no continued direct environmental exposure. This transgenerational inheritance is defined as: "Germline (sperm or egg) transmission of epigenetic information between generations in the absence of any continued direct exposures or genetic manipulations" (71). In 2005, Anway et al. studied the toxicant actions of the agricultural fungicide vinclozolin and showed its epigenetic transgenerational inheritance phenomenon (Figure 7.2) (6). F0 gestating female rats were exposed to an environmental toxicant, vinclozolin, during the fetal sex determination windows. The direct effects on the offspring from the F1 generation

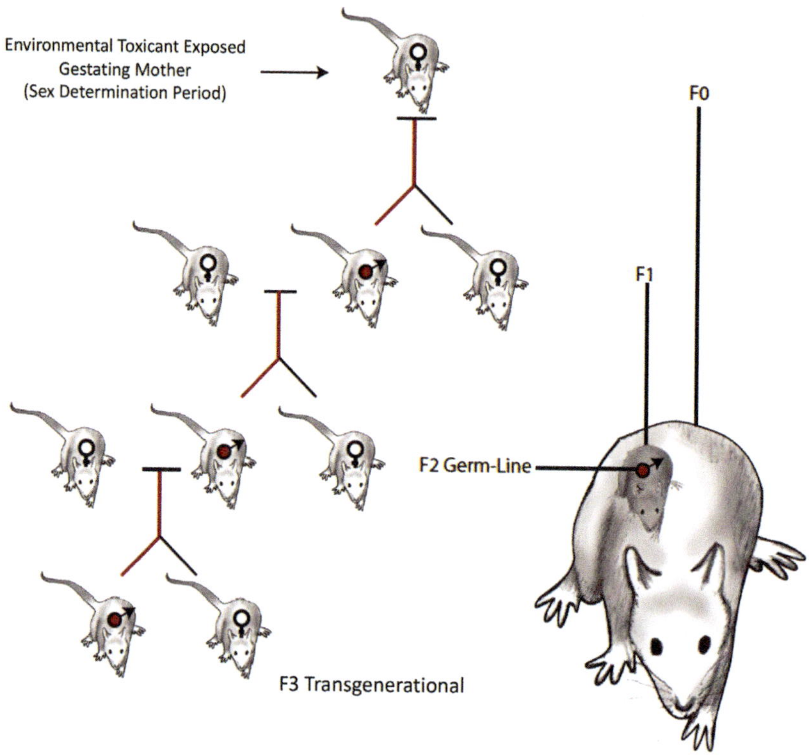

FIGURE 7.2 Epigenetic transgenerational inheritance. Summary of environmentally induced epigenetic reprogramming of primordial germ cells in the fetus that leads to the germline transmission of epimutations, resulting in all somatic cells having altered gene expression. This can result in changes in phenotype or increased disease susceptibility in the F1, F2, or transgenerational F3 generation. The F0, F1, and F2 generations have direct exposure so are multigenerational, while the F3 generation has no direct exposure so is transgenerational. Modified from (4).

were identified where a testis abnormality was observed. In the F2 generation grand offspring, the males demonstrated the same testis defect as their parents. When the F2 animals were bred to the transgenerational F3 generation, the testis disease appeared in over 90% of the male progeny (6). The abnormality incidence did not decrease through the generations but stayed high, suggesting a non-Mendelian phenomenon not relying on classic genetic processes. When the F3 generation vinclozolin males were outcrossed to wildtype females, the transgenerational phenotype was maintained at the same frequency. When the F3 generation vinclozolin females were outcrossed to wildtype males, the phenotype was lost (6). The authors concluded that the transgenerational phenomenon was transmitted through the male germline (sperm). This phenomenon has been shown to occur in females with different toxicants, such that the transgenerational phenotype was transmitted in a parent of origin allelic manner, similar to imprinted genes (72) (Figure 7.3).

It is critical to distinguish direct environmental exposure versus germline-mediated transgenerational events to understand the transgenerational inheritance phenomenon. When an F0 generation gestating female is exposed, the F1 generation fetus and its germ cells that will provide the F2 generation, are also directly exposed. Therefore, the effects observed in the F0, F1, and F2 generations are due to direct exposure toxicity as well as to environmentally induced epigenetic changes. The first generation displaying transgenerational epigenetic effects without any direct exposure toxicity is the F3 generation (Figure 7.2) (69). Because these transgenerational events require the sperm or the oocyte to be transmitted to the subsequent generations, they are mediated through the germ cells.

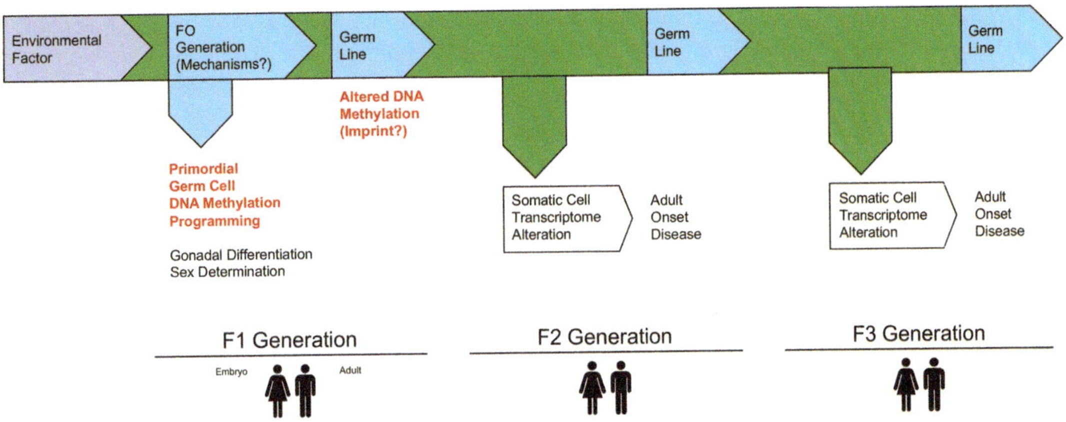

FIGURE 7.3 Role of germline in epigenetic transgenerational inheritance. Summary of environmentally induced epigenetic reprogramming of primordial germ cells that leads to the germline transmission of epimutations resulting in all somatic cells having an altered transcriptome that results in disease susceptibility. Modified from (4).

SPERM EPIGENETIC BIOMARKERS/ DIAGNOSTICS FOR PATHOLOGY

Sperm Epigenetic Diagnostics for Male Infertility

Molecular diagnostics are shifting the evaluation and treatment of human diseases. In reproductive health, male factors are involved in the couple's infertility in 50% of cases (73). The diagnosis to establish a male infertility is currently primarily through semen analysis, which evaluates sperm concentration, motility, and morphology using light microscopy (74). However, semen analysis is a poor predictor of a male fertility except in cases of oligospermia, azoospermia, or oligozoospermia (75, 76).

Several independent studies have shown that alterations in human sperm DNA methylation profiles were correlated with a decreased fecundity and an increased risk of abnormal embryo development (6, 77, 78). In 2015, Aston et al. showed that aberrant sperm DNA methylation patterns could be used to evaluate male fertility status and embryo quality (76). The Illumina Infinium array fertility test was developed as a way to assess male fertility potential and embryo development quality by measuring DNA methylation at a limited number of locations (CpG island sites) across the genome and comparing these methylation levels to those of the average fertile male (the average taken from 156 semen samples from men with normal semen parameters and a history of normal pregnancies). Abbesi et al. expanded this technique by identifying these aberrant DNA methylation patterns in sperm based on bisulfite sequencing combined with Illumina's Infinium technology for profiling human methylation levels which makes it suitable for clinical epigenetic work, as long as appropriate quality control procedures are established and followed (18). A recent genome-wide analysis for DNA methylation alterations in male infertility patients was found to generate a more robust and accurate diagnostic for male infertility (51). Future studies and diagnostics will need to use genome-wide analysis to develop the optimal diagnostics for disease.

A promising approach for the clinical therapy of male infertility is the use of endocrine therapeutics such as FSH, similar to what is currently used in the female (79). While this treatment is successful to stimulate oogenesis in the female, the response in stimulating spermatogenesis is much more variable within the infertile population (80). A recent study demonstrated a genome-wide analysis of DNA methylation identifying a male infertility signature of DMRs present in male infertility patients. Fertile patients were efficiently separated from the infertile population with minimal overlap showing the potential of this molecular biomarker (51). These results further support the potential use of an epigenetic biomarker diagnostic tool for patients with male infertility.

Sperm Epigenetic Diagnostic for Transgenerational Disease

Various environmental exposures such as stress, chemical exposures, and nutrition have been demonstrated to promote the epigenetic transgenerational inheritance of adult onset disease in a wide variety of organisms from plants to humans (4). These epigenetic changes could be used as potential biomarkers of exposure and disease (81). Several environmental exposures and toxicants have been shown to promote the epigenetic transgenerational inheritance of disease (6–11). Recent studies have also shown that ancestral environmental exposures could promote alterations in differential DNA methylated regions (DMRs), differential retention sites (DHRs), and non-coding RNA (2, 3). The agricultural fungicide vinclozolin (82, 83), the pesticide DDT (dichloro-diphenyl-trichloroethane) (13, 84), the herbicide atrazine (8), and herbicide glyphosate (85) have all been shown to promote the epigenetic transgenerational inheritance of disease. Moreover, unique epigenetic signatures of differential DNA methylation were associated with the pathologies observed in the transgenerational F3 generation males and females (8, 13, 82).

Studies have further investigated the individual animals and the specific pathologies, such as testis disease, observed in the transgenerational F3 generation were associated with specific DMRs for each disease and environmental exposure. Disease specific DMRs were identified for a number of these transgenerational pathologies which demonstrates that the establishment of an epigenetic biomarker for a specific disease and a specific exposure is possible (8, 13–17, 85). Transgenerational testis disease biomarkers/diagnostics were identified with most of these studies. Observations suggest male infertility and testis disease include transgenerational components which potentially can be detected with epigenetic diagnostics.

Genome-wide association studies (GWAS) have found specific genetic mutations associated with these human pathologies; however, these genetic mutations typically appear in less than 1% of the diseased population. In contrast, in a rodent model, the epigenetic alterations seem to have a higher frequency and appear in most individuals with the disease (8, 13–17, 85). Future studies in humans are now required to translate these animal studies. These studies have shed some light on potential new approaches to diagnose and prevent pathologies or diseases that may be transmitted through the transgenerational epigenetic inheritance phenomenon (Figure 7.3). Epigenetic biomarkers have a high frequency of association with pathologies, and their incorporation into medical diagnostics will facilitate preventative medicine not only for infertility diagnostics but for a number of different diseases.

ASSISTIVE REPRODUCTIVE TECHNOLOGIES, INFERTILITY, AND EPIGENETICS

Decreasing male infertility has been observed for the past 60 years (86–90). Several parameters of the male reproductive system have been affected such as a decrease in sperm counts and quality (87), an increase in the incidence of testicular cancer (88), an increase in hypospadias, and cryptorchidism cases (86). Most recent studies have linked this decline with environmental exposures rather than genetic factors due to the rapid pace of the decrease of semen quality and the increase in occurrence of male infertility (87, 91, 92). Abnormal methylation patterns and aberrant protamine insertions have been shown to influence male fertility. CpG islands at gene promoter regions which are usually hypomethylated are especially susceptible to aberrant methylation for specific genes such as DAZL and MTHFR or imprinted loci. Various forms of infertility have been linked and sperm defects in men have been associated with epigenetic abnormalities in these regions (93). Abnormal DNA methylation is detrimental in imprinted genes as they are known to be directly inherited from the parental germline and they are thought to be involved in part in the transgenerational effect phenomenon. Oligospermic patients (men with less than 10 million spermatozoa per 1 milliliter of semen) have been shown to contain a greater number of DNA methylation anomalies compared to men containing normal sperm counts (94). Interestingly, these imprinted loci are suspected to be inherited by the offspring during *in vitro* fertilization (IVF) treatments (95).

Epigenetic dysfunction can also occur during spermatogenesis and may be a contributing factor in male infertility. During the crucial step of histones replacement by protamines, many epigenetic regulators work together to facilitate the paternal genome reorganization and packaging through histone variation, specific histone modification, and their related chromatin remodelers. Any defects during this step are thought to be linked to male infertility (96). Besides, the differential DNA methylation patterns observed with male infertility are often accompanied by an altered protamine ratio and indicate the interaction between epigenetic phenotypes. Histone retention has been shown to occur sporadically throughout the genome, and loss of histone retention at imprinted loci is also associated with male infertility (52). The use of assisted reproductive technologies has increased in recent years. Understanding the epigenetic processes underlying male factor infertility will aid proper diagnosis, as well as help develop sperm selection techniques used for IVF (97).

SUMMARY

Our knowledge of epigenetics has considerably increased in the past two decades. Both DNA methylation and histone modifications are two well-studied epigenetic modifications in the germline. However, histone retention, ncRNA, and RNA methylation are also emerging as potentially important factors that could help understand the epigenetic mechanisms. The novel aspect brought by transgenerational inheritance involves a non-genetic form of inheritance, a non-genetic etiology of disease etiology, a molecular mechanism of how environmental factors (diet, stress, or chemical insults) can indirectly influence genome activity and disease, and the existence of the epigenetic transgenerational inheritance of disease and phenotypic variation. While many environmental conditions (such as toxicants or obesity) might explain the increase in male infertility, evidence suggests that the molecular mechanisms driving this increase involve epigenetic alterations. The constant advancement of molecular biology is anticipated to enhance the diagnosis and management of male infertility patients, as well as improve general therapeutic options and development. Many studies have shown that epigenetic diagnostics could be developed and applied to pathologies and disease. Expanded clinical trials are now needed to help validate and apply this novel technology to the management and treatment of male infertility.

REFERENCES

1. Jenkins TG, et al. *Syst Biol Reprod Med.* 2017;63(2):69–76.
2. Skinner MK, et al. *Epigenetics Chromatin.* 2018;11(1):8.
3. Ben Maamar M, et al. *Environ Epigenet.* 2018;4(2):1–19.
4. Nilsson E, et al. *Environ Epigenet.* 2018;4(2):1–13.
5. Nilsson E, et al. In *Transgenerational Epigenetics.* 2nd ed. (Tollefsbol T). Elsevier; 2019;13–24.
6. Anway MD, et al. *Science.* 2005;308(5727):1466–9.
7. Guerrero-Bosagna C, et al. *PloS One.* 2010;5(9):1–17.
8. McBirney M, et al. *PloS One.* 2017;12(9):1–37.
9. Anway MD,et al. *Prostate.* 2008;68(5):517–29.
10. Manikkam M, et al. *Reprod Toxicol.* 2012;34(4):708–19.

11. Manikkam M, et al. *PloS One*. 2013;8(1):1–18.
12. Ben Maamar M, et al. *Environ Res*. 2021;192:110279.
13. King SE, et al. *Environ Epigenet*. 2019;5(2):1–15.
14. Ben Maamar M, et al. *Reprod Toxicol*. 2020;S0890–6238(20):30198-2.
15. Thorson JLM, et al. *PloS One*. 2020;15(12):e0239380.
16. Thorson JLM, et al. *Environ Health*. 2020;19(1):109.
17. Nilsson E, et al. *Environ Epigenet*. 2020;6(1):dvaa020.
18. Abbasi M, et al. *Clin Epigenet*. 2018;10(1):119.
19. Skinner MK. *Epigenetics*. 2011;6(7):838–42.
20. Holliday R, et al. *Science*. 1975;187(4173):226–32.
21. Singer J, et al. *Science*. 1979;203(4384):1019–21.
22. Moore LD, et al. *Neuropsychopharmacology*. 2013;38(1):23–38.
23. Kriaucionis S, et al. *Cold Spring Harb Perspect Biol*. 2014;6(10):a018630.
24. An J, et al. *Exp Mol Med*. 2017;49(4):e323.
25. Mellen M, et al. *Proc Natl Acad Sci U S A*. 2017;114(37):E7812–E21.
26. Wu TP, et al. *Nature*. 2016;532(7599):329–33.
27. Su Z, et al. *ACS Chem Biol*. 2016;11(3):564–74.
28. Rothbart SB, et al. *Biochim Biophys Acta*. 2014;1839(8):627–43.
29. Oliva R, et al. *Prog Nucleic Acid Res Mol Biol*. 1991;40:25–94.
30. Gatewood JM,et al. *Science*. 1987;236(4804):962–4.
31. Oliva R. *Hum Reprod Update*. 2006;12(4):417–35.
32. Ben Maamar M, et al. *Sci Rep*. 2018(8):5308, 1–10.
33. Arpanahi A, et al. *Genome Res*. 2009;19(8):1338–49.
34. Hammoud SS, et al. *Nature*. 2009;460(7254):473–8.
35. Yaniv M. *Cancer Genet*. 2014;207(9):352–7.
36. Ooi SL, et al. *Curr Opin Cell Biol*. 2007;19(3):257–65.
37. Kornfeld JW, et al. *Front Genet*. 2014;5:57.
38. Wei JW, et al. *Oncol Rep*. 2017;37(1):3–9.
39. Chen Q, et al. *Nat Rev Genet*. 2016;17(12):733–43.
40. Schuster A, et al. *Environ Epigenet*. 2016;2(1):1–10.
41. Fu Y, et al. *Nat Rev Genet*. 2014;15(5):293–306.
42. Wu R, et al. *Mol Biotechnol*. 2016;58(7):450–9.
43. Motorin Y, et al. *Wiley Interdiscip Rev RNA*. 2011;2(5):611–31.
44. Ravitsky V, et al. *Biol Reprod*. 2019;101(5):872–4.
45. Levine H, et al. *Hum Reprod Update*. 2017;23(6):646–59.
46. Craig JR, et al. *Fertil Steril*. 2017;107(4):848–59.
47. Rajender S, et al. *Mutat Res*. 2011;727(3):62–71.
48. Navarro-Costa P, et al. *Hum Reprod*. 2010;25(10):2647–54.
49. Houshdaran S, et al. *PloS One*. 2007;2(12):e1289.
50. Poplinski A, et al. *Int J Androl*. 2010;33(4):642–9.
51. Luján S, et al. *Sci Rep*. 2019;9(1):16786.
52. Hammoud SS, et al. *Hum Reprod*. 2011;26(9):2558–69.
53. Corzett M, et al. *Mol Reprod Dev*. 2002;61(4):519–27.
54. Aoki VW, et al. *Hum Reprod*. 2005;20(5):1298–306.
55. Zhang X, *J Androl*. 2006;27(3):414–20.
56. Sonnack V, et al. *Andrologia*. 2002;34(6):384–90.
57. Vieweg M, et al. *Clin Epigenet*. 2015;7(1):31.
58. Minocherhomji S, et al. *Fertil Steril*. 2009;92(1):88–95.
59. Singh I, et al. *Epigenet Reprod Health*. 2021;21:47–67.
60. Jodar M, et al. *Sci Transl Med*. 2015;7(295):295re6.
61. Jodar M, et al. *Hum Reprod Update*. 2013;19:604–24.
62. McSwiggin HM, et al. *Reproduction*. 2018;156(2):R9–R21.
63. Liu Y, et al. *Sci Rep*. 2017;7(1):9055.
64. Zhang X, et al. *Biol Reprod*. 2019;100(4):982–93.
65. Yang Y, et al. *Sci Rep*. 2016;6:24345.
66. Lin Z, et al. *Cell Res*. 2017;27(10):1216–30.
67. McCarrey JR. *Mol Cell Endocrinol*. 2012;354(1–2):9–15.
68. Camprubi C, et al. *Reprod Biomed Online*. 2016;33(6):709–19.

69. Skinner MK. *Reprod Toxicol.* 2008;25(1):2–6.
70. Jirtle RL, et al. *Nat Rev Genet.* 2007;8(4):253–62.
71. Skinner MK, et al. *Trends Endocrinol Metab.* 2010;21(4):214–22.
72. Manikkam M, et al. *PloS One.* 2014;9(7):1–19.
73. Jungwirth A, et al. *Eur Urol.* 2012;62(2):324–32.
74. Wang C, et al. *Fertil Steril.* 2014;102(6):1502–7.
75. Nagler HM. *Nat Rev Urol.* 2011;8(1):16–7.
76. Aston KI, et al. *Fertil Steril.* 2015;104(6):1388–97.
77. Tunc O, et al. *J Assist Reprod Genet.* 2009;26(9–10):537–44.
78. Hammoud SS, et al. *Fertil Steril.* 2010;94(5):1728–33.
79. Duca Y, et al. *Expert Opin Pharmacother.* 2019;20(1):55–67.
80. Caroppo E, et al. *Fertil Steril.* 2003;80(6):1398–403.
81. Manikkam M, et al. *PloS One.* 2012;7(2):1–12.
82. Nilsson E, et al. *PloS One.* 2018;13(8):1–29.
83. Stouder C, et al. *Reproduction.* 2010;139(2):373–9.
84. Skinner MK, et al. *BMC Med.* 2013;11:228, 1–16.
85. Ben Maamar M, et al. *Environ Res.* 2020;192:110279.
86. Main KM, et al. *Best Pract Res Clin Endocrinol Metab.* 2010;24(2):279–89.
87. Sharpe RM. *Philos Trans Phys Sci Eng.* 2010;365(1546):1697–712.
88. Skakkebaek NE, et al. *Int J Androl.* 2007;30(4):198–204; discussion-5.
89. Andersson AM, et al. *Int J Androl.* 2008;31(2):74–80.
90. Skakkebaek NE, et al. *Int J Androl.* 2006;29(1):2–11.
91. Giwercman A, et al. *Best Pract Res Clin Endocrinol Metab.* 2011;25(2):391–402.
92. Nordkap L, et al.. *Mol Cell Endocrinol.* 2012;355(2):221–30.
93. Carrell DT. *Fertil Steril.* 2012;97(2):267–74.
94. Montjean D, et al. *Andrology.* 2015;3(2):235–40.
95. Kappil M, et al. *Curr Environ Health Rep.* 2015;2(2):155–62.
96. Bao J, et al. *Reproduction.* 2016;151(5):R55–70.
97. Murphy KE, et al. In *Transgenerational Epigenetics* (Tollefsbol, ed.). London: Elsevier Academic Press; 2014;43–55.
98. Skinner MK. *Sci Am.* 2014;311(2):44–51.

Aneuploidy in Human Oocytes and Preimplantation Embryos

8

Eva R. Hoffmann

INTRODUCTION

A Historical Context for Human Aneuploidy

After 30 years of widely held belief that humans had 48 chromosomes, the refinement of karyotyping techniques in the 1950s facilitated the discovery that humans have 46 chromosomes, including the XY sex chromosomes (1). Using improved cytogenetic techniques, Jacobs and Strong (1959) reported that Klinefelter Syndrome in males was caused by an extra X chromosome (2). In the same year, Jejeune, Gauthier, and Turpin (3, 4) and the Jacobs group (5) independently discovered that Down Syndrome was caused by an extra chromosome 21. Ford and colleagues (6) found that Turner Syndrome was caused by the loss of an X chromosome (45,X) in females and also reported the first mosaic individual (XXY/XX) (7). These studies reported in 1959 led to an explosion in the investigations into aneuploidy (8–12) and initiated epidemiological and extensive cohort studies of both spontaneous miscarriages as well as live births (Figure 8.1). Of spontaneous miscarriages, nearly 50% are chromosomally abnormal, mainly due to aneuploidy (one in three), but triploid conceptions are also common (13). To date, the most comprehensive cohort study is the US National Down Syndrome Project, which was initiated by Terry Hassold and Stephanie Sherman (14).

The population-based studies have revealed three important facets of human aneuploidies. First, the trisomic conceptions that reach clinical recognition predominantly derive from the mother and their incidence increases with maternal age. The chromosomes also display different aging profiles suggesting complex interactions between general aging features of oocytes and chromosome-specific susceptibilities (15). Second, besides maternal age, recombination has emerged as a critical factor that influences the chromosomal content of the egg (16). Chromosome-specific recombination patterns could thus contribute towards the different age-related aneuploidy curves displayed by specific chromosomes. Consistent with the importance of recombination, higher maternal (but not paternal) recombination rates are associated

DOI: 10.1201/9781003024941-8

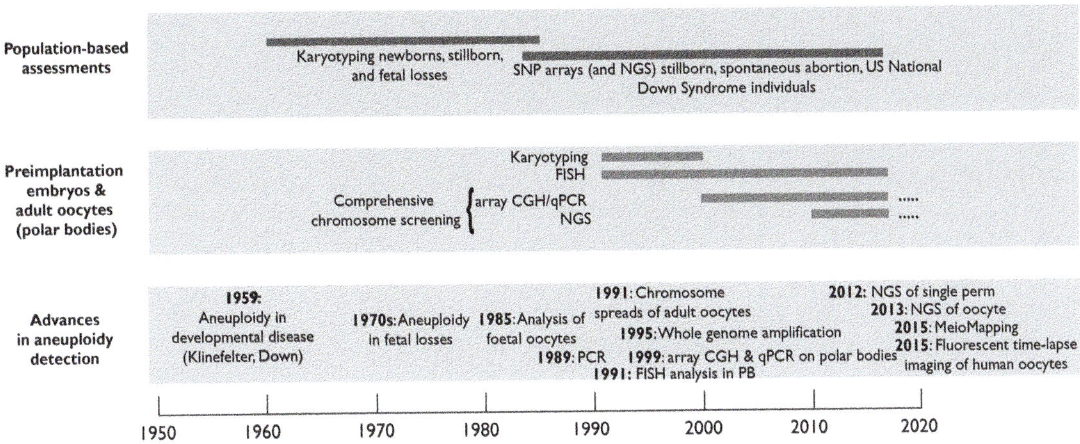

FIGURE 8.1 Historical overview of developments in the importance and detection of aneuploidy in human health. Abbreviations: SNP, single nucleotide polymorphisms; NGS, next generation sequencing; FISH, fluorescence *in situ* hybridization; array CGH, comparative genomic hybridization; qPCR, quantitative polymerase chain reaction; MeioMapping, recovery of both chromosome content and genetic variants (SNPs) from the same cell in oocytes and matched polar bodies.

with increased reproductive success (more children) (17). The implication is that recombination, which occurs during fetal development, affects chromosome segregation decades later in adult life, when oocytes ovulate. Third, the larger cohort studies have allowed other epidemiological factors to be uncovered, including the role of socioeconomic status (18). This raises important questions about how lifestyle and health issues might exert their influence on the quality of human eggs and embryonic development.

The U-Shaped Curve of Human Aneuploidies

Recent studies of human oocytes and preimplantation embryos has shown that their incidence follows a U-shape curve according to maternal age (Figure 8.2) (19, 20). Aneuploidy rates are initially high for teenagers, then decrease for 20–30 years, before increasing again as women enter their mid-30s. The U curve is also apparent in preimplantation embryos, where the magnitude of aneuploidy is much higher than in oocytes due to additional age-independent mitotic errors during preimplantation development (19, 21–26). In pregnancy losses (27) as well as live births of trisomy 21 (28), the U curve is also apparent consistent with chromosome errors during the meiotic divisions in oocytes shaping human fertility curves (27).

The incidence of aneuploidy is much higher in oocytes and preimplantation embryos compared to live births. This is due to selection against embryos with chromosome constellations that are incompatible with embryonic and early fetal development, such as monosomies of the autosomal chromosomes and complex aneuploidies, where multiple chromosomes are affected (29). Aneuploid embryos may be selected against due to their aberrant triggering of calcium signaling in decidual cells at implantation (30).

Population-based assessments are a vast underestimate of the chromosomal aneuploidies generated in gametes and embryos (Figure 8.3) (24, 26, 31–41). This wide range reflects the curvature of the aneuploidy rate according to the age distribution of women included in different studies. However, the use of different methodologies may also influence the variability in estimates of aneuploidy rates. For example, in some cases, fluorescence *in situ* hybridization (FISH)-based assessments have been plagued by suboptimal processing (42). Regardless of this, egg donation programs, where oocytes from younger women result in live births at rates equivalent to the donor's age, as opposed to the recipient's age, have firmly implicated chromosome errors in the egg as causative of infertility and pregnancy loss (43, 44).

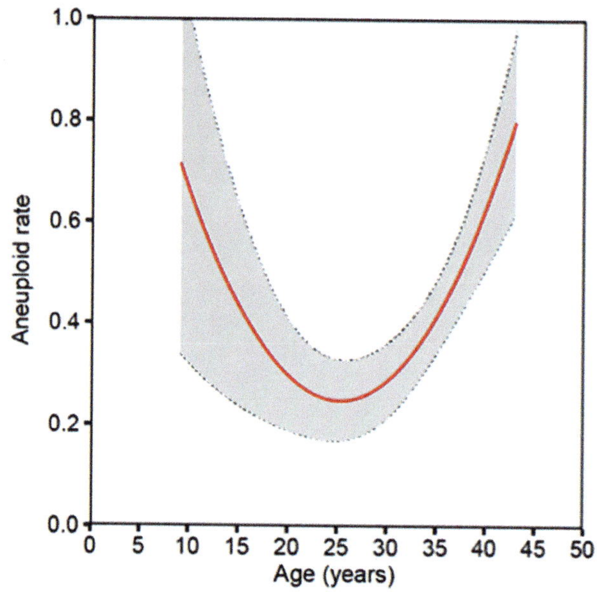

FIGURE 8.2 U curve of aneuploid conceptions. The aneuploidy incidence across female age in human oocytes. The red line shows the modeled curve and the gray area represents the 95% confidence interval. Data from (19).

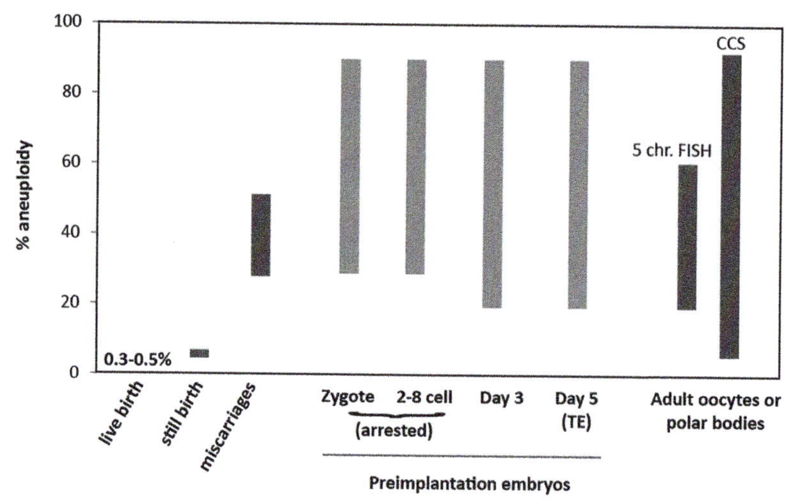

FIGURE 8.3 Variability in reported aneuploidy rates. Data from (24, 26, 31, 33–41, 151, 152).

Chromosome-Specific Effects and Unequal Contributions from Monosomies and Trisomies in Human Oocytes and Preimplantation Embryos

The genomic studies on oocytes and preimplantation embryos show that the U curve in oocytes and preimplantation embryos is best modeled by a quadratic equation; there is thus no specific inflection point at which aneuploidy in oocytes or preimplantation embryos becomes elevated. In contrast, in prenatal analyses, an exponential phase occurs around the mid-30s and that has given rise to the advanced maternal age classification. The differences in the curves reflect that all chromosomes are assessed in embryos (Figure 8.4),

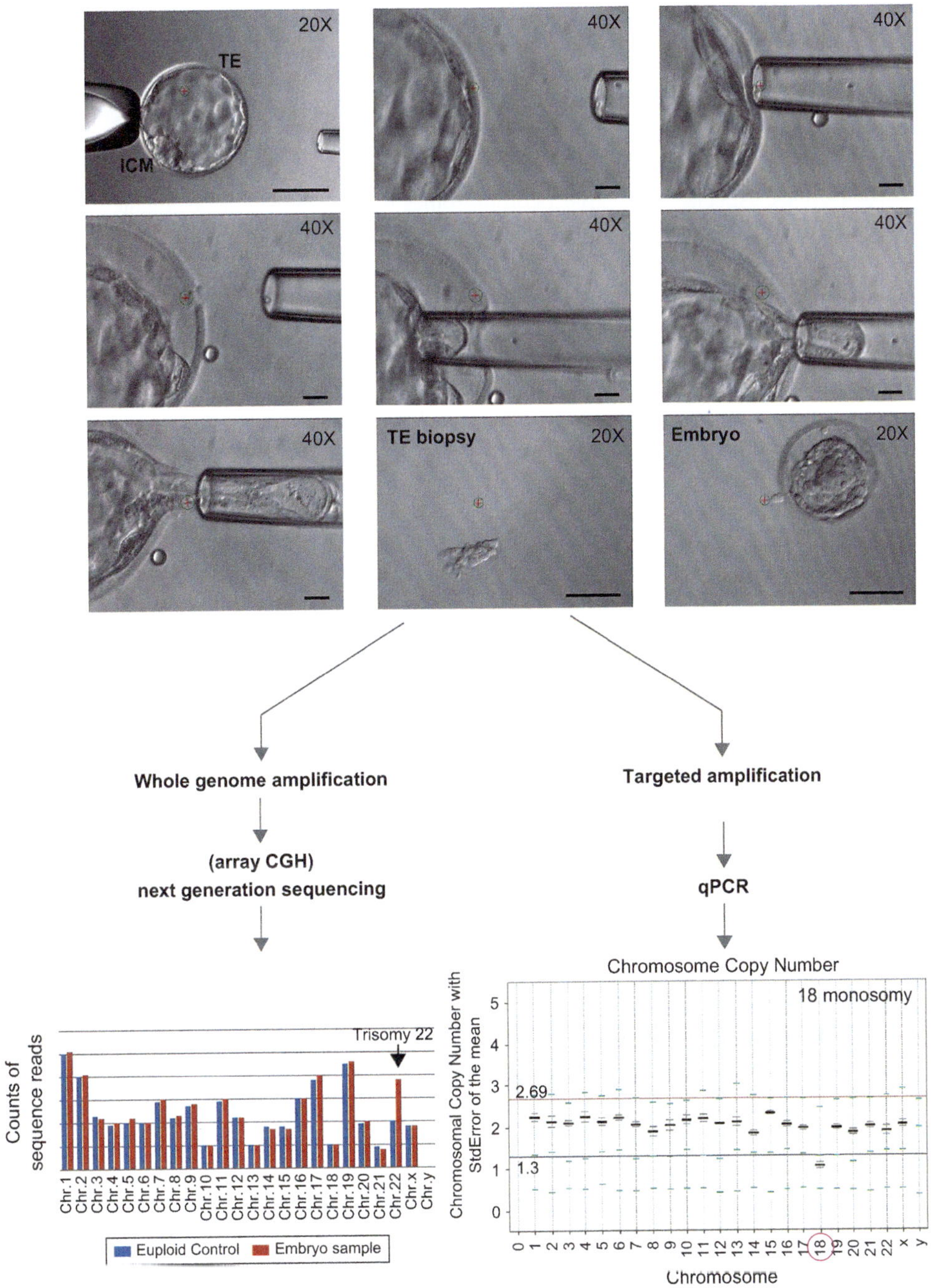

FIGURE 8.4 Detection of chromosome content in trophectoderm samples from blastocyst embryos. A sample of 5–10 cells is biopsied from the trophectoderm. One method uses whole-genome amplification coupled with next-generation sequencing (left), whereas the other uses a qPCR method (right). The whole-genome amplification method has been used to obtain sequence and chromosomal content on single cells, including oocytes and polar bodies. Data from (19, 50, 51).

whereas the exponential elevation in aneuploidies in human conceptions curve is highly dependent upon trisomies of chromosomes 13, 16, 18, and 21 that implant and develop into products of conceptions, including live births. Clinically, the different curves are important because the advanced maternal age classification does not apply to aneuploidies generated in eggs and preimplantation embryos.

In human oocytes, not all chromosomes are created equal with regards to their risk of aneuploidy and their maternal age-related curves (19, 24–26, 45). This includes the preponderance of aneuploidies for specific chromosomes, especially the largest chromosomes (1 to 5), which missegregate in young teenagers (19), whereas the two smallest chromosomes 21 and 22, but also 15 and 16, tend to be error-prone in women of advanced maternal age. There are also some chromosomes, such as chromosome 13, that initially show a high incidence in young teenagers, then improve, before showing elevated aneuploidy in women of advanced maternal age. However, in general, size correlates negatively with age-related aneuploidy (26).

The direct assessment of aneuploidies from oocytes and embryos has allowed new insight into their origins. Several important findings from ART settings include that aneuploidy rates for monosomies are at least, if not more, common than trisomies (23, 26, 31, 33, 46). In contrast, amongst natural conceptions, the only viable monosomy is 45; X0 (Turner Syndrome), suggesting selection or poor developmental potential of chromosomally abnormal conceptions at peri-implantation stages and during gestation (29, 47). The preponderance of monosomies in preimplantation embryos might be due to DNA damage that is poorly repaired, resulting in chromosome loss, since Cas9 double-strand breaks were recently shown to result in chromosome loss in nearly 50% of human embryos (48, 49).

Features Influencing the U Curve in Oocytes and Preimplantation Embryos

The origin of aneuploidies in human eggs due to meiotic errors can be accurately inferred when all three cells from a single meiosis are assessed together (MeioMapping) (50, 51). Such "trio" analyses have revealed that errors during meiosis I can be "corrected" by a compensating error at meiosis II, resulting in an euploid egg (reviewed in [20]). Approximately half of the segregation errors result in an aneuploid egg, whereas the polar bodies are affected in the remaining cases (50, 51). As a result, when analyzing embryos or fetuses without available polar body information, we capture only half of the segregation errors that occurred during meiosis I.

For preimplantation embryos, the shape of the U curve is heavily dependent upon the stage during development in which the embryo is sampled. High rates of aneuploidy due to mitotic errors (postzygotic) were originally identified using fluorescence *in situ* hybridization (e.g., [52]). Using genomic assessments, day 3 cleavage-stage embryos have been reported to have a high rate of aneuploidy, frequently characterized by chaotic karyotypes (26, 53, 54). In contrast, day 5 embryos have a lower rate of chaotic karyotypes, which has been proposed to be due to selection against embryos with high levels of chromosome instability at the day 3 cleavage stage (26). Consistent with this, a relatively high rate of preimplantation embryos arrest at day 3 (20–30%), although some of these are also euploid (54).

The causes of chromosome instability and embryonic arrest are currently not clear. A genome-wide association study suggested that common variants of *PLK4* may contribute to mitotic instability in early preimplantation embryos (55), but functional assessment and mechanistic studies are yet to be conducted on how specific *PLK4* alleles would alter the cell divisions. DNA damage during the embryonic divisions may also contribute towards the high arrest rate of cleavage-stage embryos (48, 49, 56). This is consistent with the transcriptional upregulation of cell cycle and DNA damage response genes in mature oocytes (57), which drive the first 2–3 mitotic divisions until the embryonic genome activation (EGA; Figure 8.5) (58).

Other factors may also contribute towards the highly variable rates of aneuploidies in preimplantation embryos reported in the literature (59). These include methodological differences such as whether the hormone stimulation used to treat women affects aneuploidy rates (60–63), the type of culture medium as well as the technology used to assess chromosomal content (reviewed in [64]). A substantial degree of variability in aneuploidy rates across different centers is also seen in oocyte donors (65), suggesting that the variability

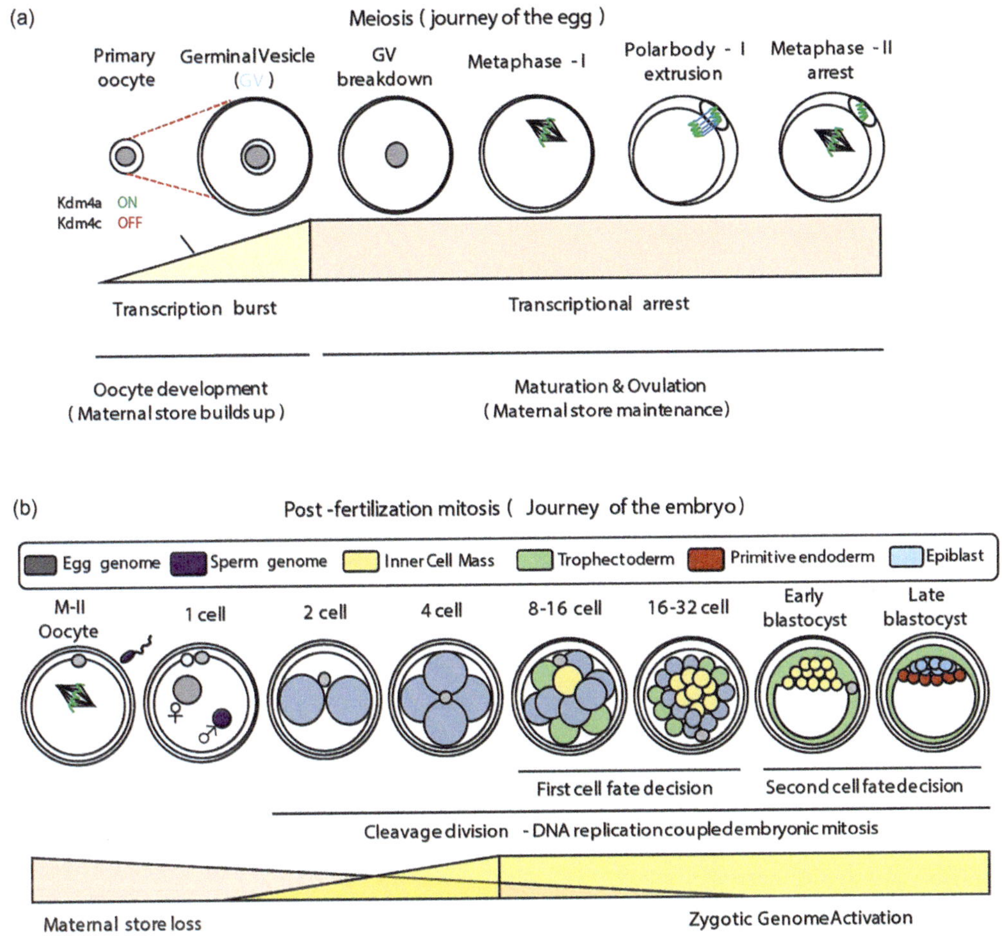

FIGURE 8.5 Human embryo development. A summary of the seminal events occurring during preimplantation development in human. Courtesy of Dr. Aditya Sankar.

in aneuploidy rates may not be specific to female patients undergoing fertility treatment. Variability in aneuploidy rates across different geographical settings is also seen for natural conceptions (66). Understanding whether such differences are biologically meaningful is important since they may reveal new insights into the origins of aneuploidies. As our methods to detect whole chromosome as well as "segmental" aneuploidies improve (Table 8.1), our focus will shift towards understanding the biological phenomena that drive chromosome instability in preimplantation embryos and their functional impact on the embryo (Table 8.1).

The Origins of Maternal Chromosome Errors

Oocytes and spermatocytes halve their chromosome numbers by a specialized division known as "meiosis." Although the chromosome content in sperm and activated oocytes are the same, their development to mature gametes is highly dimorphic. Primordial germ cells migrate to the genital ridges during fetal development and the gonadal environment determines whether the cells induce meiosis (oocyte) or remain arrested until puberty (males) (20). Mouse studies have shown that fetal oocytes initiate meiosis and differentiation due to the presence of retinoic acid in the fetal ovary (67, 68). The fetal oocytes replicate their DNA forming sister chromatids that are cohesed together. During meiotic prophase I, the homologous chromosomes align and recombine (69), which forms the bivalent chromosome structure (Figure 8.6).

TABLE 8.1 Estimated Incidence and Types of Aneuploidies at Different Stages of Human Reproduction

DEVELOPMENTAL STAGE	*INCIDENCE*	*MOST COMMON*	*REFERENCES*
Oocytes • Whole chromosome aneuploidies • *Trisomies* • Monosomies • Complex aneuploidies • Ploidy alteration • UPDs (uniparental disomy) • Pathogenic CNVs (copy number variation) • VOUS (variants of unknown significance) • Mosaic	20–90% *55.0% of aneuploidies* *45.0% of aneuploidies* 4–30% of aneuploid oocytes ND ND ND ND NA	± 15, ± 16, ± 18, ± 19, ± 21, ± 22 ND NA NA NA NA NA	(31, 35, 39, 50, 51, 137)
Preimplantation embryos • Whole chromosome aneuploidies • *Trisomies* • Monosomies • Complex aneuploidies • Ploidy alteration • UPDs • Pathogenic CNVs • VOUS • Mosaic	25–90% *50% of aneuploidies* *50% of aneuploidies* 5–30% of aneuploid embryos < 2% < 2% ND ND 4–5%	± 15, ± 16, ± 18, ± 19, ± 21, ± 22 ND Haploidy, Triploidy ND NA NA ND	(24, 38, 40, 45, 55, 138, 139)
POCs (miscarriages ≤ 12 gestational weeks) • Whole chromosome aneuploidies • *Trisomies* • Monosomies • Complex aneuploidies • Ploidy alteration • UPDs • Pathogenic CNVs • VOUS • Mosaic	40–70% *85% of aneuploidies* *15% of aneuploidies* Absent < 2% < 2% 2–5% 2–3% < 2%	+ 13, + 14, + 15, + 16, + 17, + 18, + 21, + 22, 45X NA Triploidy ND del 7q11.23, del/dup 8p23.1, del 15q11.2q13, del/dup 17p11.2, del 17q21.31, del 22q11.2 dup 4q35.2, dup 8p22, del 15q11.2, dup 15q13.3, del/dup 16p13.11, del 16p11.2, dup Xp22.11 2, 8, 16, sex, structural (7, 13, 18, 20, 21)	(140–144)
Ongoing pregnancies (> 12 gestational weeks) • Whole chromosome aneuploidies • *Trisomies* • Monosomies • Complex aneuploidies • Ploidy alteration • UPDs • Pathogenic CNVs • VOUS • Mosaic	< 5% *95% of aneuploidies* *5% of aneuploidies* Absent Absent-0.01% 0.01% ≈0.5% 2–3% < 0.5%	+ 13, + 18, + 21, 45X, 47XXX, 47XXY NA Triploidy ND del 7q11.23, del/dup 8p23.1, del 15q11.2q13, del/dup 17p11.2, del 17q21.31, del 22q11.2 dup 4q35.2, dup 8p22, del 15q11.2, dup 15q13.3, del/dup 16p13.11, del 16p11.2, dup Xp22.11 16, sex, structural (7, 13, 18, 20, 21)	(144–147)

(*Continued*)

TABLE 8.1 (CONTINUED) Estimated Incidence and Types of Aneuploidies at Different Stages of Human Reproduction

DEVELOPMENTAL STAGE	*INCIDENCE*	*MOST COMMON*	*REFERENCES*
Newborns	0.1–4%	+ 13, + 18, + 21, 45X, 47XXX, 47XXY	(146–150)
• Whole chromosome aneuploidies	*95% of aneuploidies*	NA NA	
• *Trisomies*	*5% of aneuploidies*	ND	
• Monosomies		del 7q11.23, del/dup 8p23.1, del 15q11.2q13, del/dup 17p11.2, del 17q21.31, del 22q11.2	
• Complex aneuploidies	Absent		
• Ploidy alteration	Absent		
• UPDs	< 0.01%	dup 4q35.2, dup 8p22, del 15q11.2, dup 15q13.3, del/dup 16p13.11, del 16p11.2, dup Xp22.11	
• Pathogenic CNVs	≈0.5%		
• VOUS	2–3%		
• Mosaic	< 0.5%	Sex	

Abbreviations: CNVs, Copy Number Variations; NA, Not Applicable; ND, Not Determined; POCs, Product Of Conceptions; UPDs, UniParental Disomies; VOUS, Variants Of Unknown Significance

The oocytes then enter a prolonged arrest (dictyate) until menarche, when a single follicle matures and the egg completes the first meiotic division. Oocyte numbers within the ovary are depleted from fetal development until the onset of menopause. Only 450 of the 5–7 million fetal oocytes complete the first meiotic division, and even fewer go on to segregate sister chromatids after fertilization, which induces the completion of the second meiotic division (70).

During meiosis I, the sister kinetochores are co-oriented such that they are pulled to the same side of the spindle. At the same time, cohesion between sister chromatids along chromosome arms is released, allowing the homologous chromosome to separate. In contrast, cohesion near the centromeres is protected until meiosis II, when it is released to allow sister chromatids to separate (Figure 8.6).

Aneuploidies originating from female meiosis have traditionally been classified as meiosis I or meiosis II errors, depending upon whether the extra chromosome contained genetic information from both or only one of the maternal chromosomes, respectively (chromosome fingerprint; Figure 8.7). Meiosis I errors are more frequent compared to meiosis II (50, 51). Direct assessment of human oocytes cytologically (71–73) or via genomic characterization has revealed three error types in meiosis I: meiosis I nondisjunction (73), precocious sister chromosome segregation or predivision (74, 75), meiosis II nondisjunction, and more recently, reverse segregation (19, 51, 76).

Error Type in Meiosis I of Oocytes Depends on Female Age

One of the most surprising findings from studies of oocytes and polar bodies is that not only do chromosomes that are affected depend upon female age, but also the type of errors. Meiosis I nondisjunction is prevalent in young teenagers and this incidence decreases with age. Precocious separation of sister chromatids (PSSC), also known as predivision, appears to be two- to ten-fold more frequent than classical meiosis I non-disjunction in women above 20 years and increases linearly with maternal age. In women in the advanced maternal age group, reverse segregation is the most prevalent error type (19).

Reverse Segregation: A Novel Segregation Pattern

Chromosome fingerprinting of all three meiotic cells has firmly established that homologous chromosomes segregate at meiosis I and sister chromatids at meiosis II (classical meiosis). Fingerprinting relies on

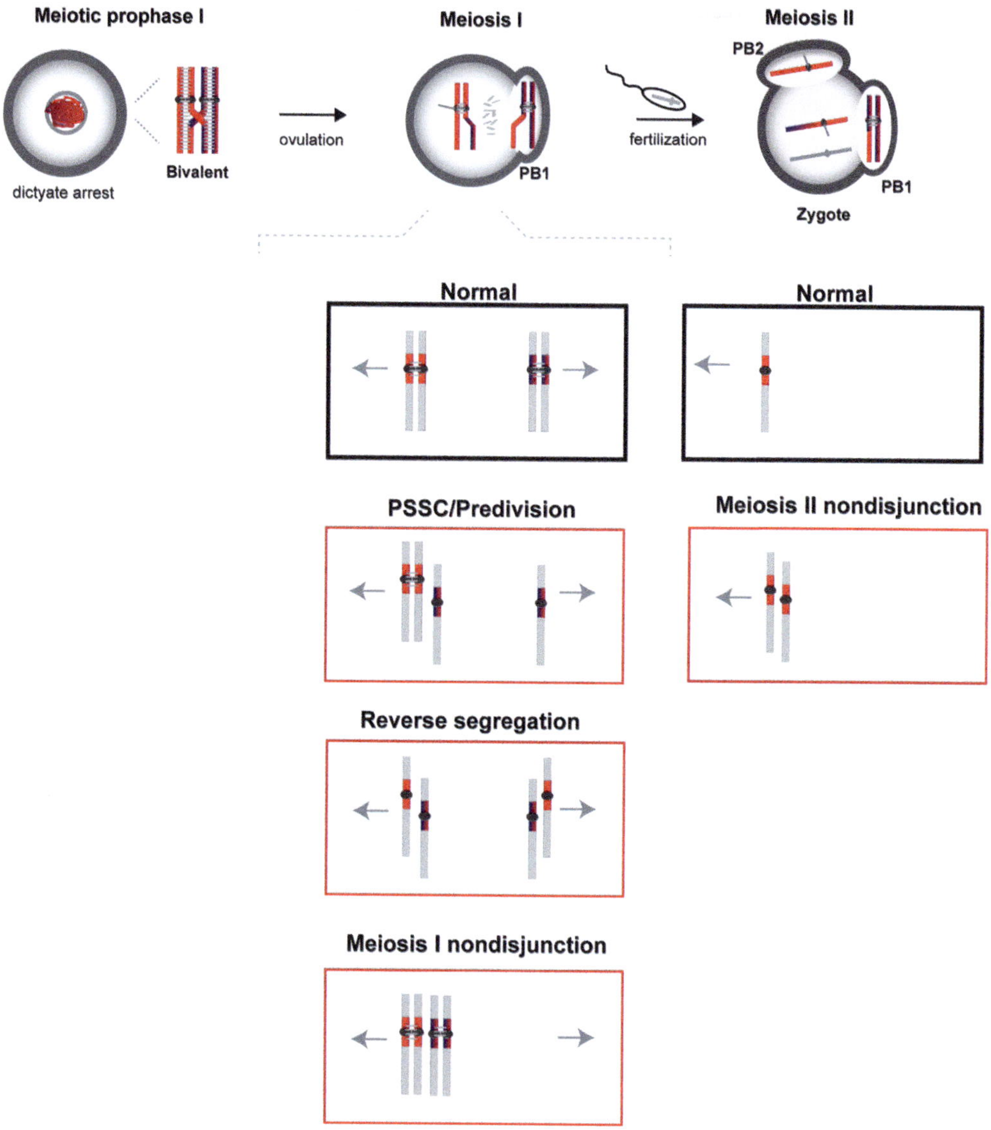

FIGURE 8.6 Chromosome segregation errors in human female meiosis. The two chromosomes are shown in red and blue, respectively. Recombination occurs in fetal oocytes, which arrest at the dictyate stage, once bivalent chromosomes have formed. Their formation depends upon recombination as well as cohesion between the two sister chromatids. Different chromosome segregation patterns can be discerned when the genetic variants (SNPs) are available. Data from (50, 51).

the assumption that crossing over (or recombination) is suppressed near centromeres, such that two sister chromatids share the same genetic markers such as single nucleotide polymorphism. After meiosis I, the oocyte should contain two sister chromatids (one chromosome) and the first polar body should contain two sister chromatids of the second homolog. However, Ottolini, et al. (2015) found that frequently the first polar body contained two non-sister chromatids, i.e., chromosome fingerprints from both maternal chromosomes. The frequency of this event was more than 100× increased compared to the incidence expected from two independent predivision events, which suggests a common origin. Moreover, it was also reported that there was a weak preference for the two non-sister chromatids to segregate in a balanced fashion at meiosis II, such that the oocyte and polar body contained one chromatid each. Thus, overall,

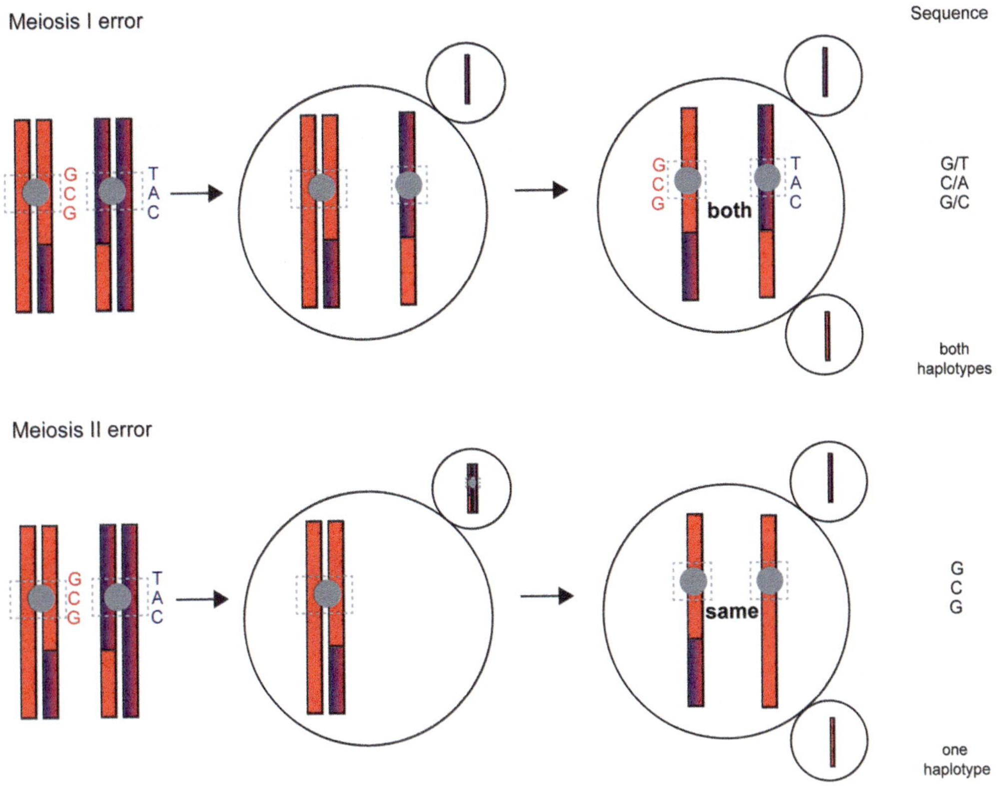

FIGURE 8.7 Chromosomal fingerprinting allows inference about the origin of extra chromosomes. Recombination between homologous chromosomes causes shuffling of the genetic material. However, because recombination is suppressed near centromeres, the origin of the chromosome is preserved (gray box, red, and blue). The genetic markers such as single nucleotide polymorphisms around the centromere can be used as a chromosomal fingerprint. Meiosis I errors lead to two maternal chromosomes with information from both blue and red chromosomes (upper panel). In contrast, meiosis II errors lead to the same chromosomal fingerprint of both maternal chromosomes.

the entire meiosis was chromosomally balanced. However, the pattern of segregation was reversed: sisters segregated at meiosis I, followed by homologs in meiosis II (Figure 8.8). The report also showed that only a single or few of the chromosomes followed a reverse segregation pattern within oocytes (19, 51). This suggests chromosome-specific vulnerability to aneuploidies amidst general aging effects of the oocytes as a whole.

Time-Lapse Imaging Reveals Chromosomal "Aging" Defects Precede Errors in Segregation

One of the major challenges is to determine what affects the error types that chromosomes make in meiosis I. Analyses of fixed chromosomes have revealed that the bivalent chromosome configurations that should be established in fetal oocytes are either not established properly (69, 77–79) and frequently deteriorate in adult oocytes. In particular univalents and the loss of cohesion between sister chromatids have been reported as "vulnerable" crossover configurations (Figure 8.9). Time-lapse imaging in mouse oocytes revealed that univalents are at risk of meiosis I nondisjunction as well as PSSC/predivision (80). In humans, the recent development of time-lapse imaging where oocytes are injected with mRNA of

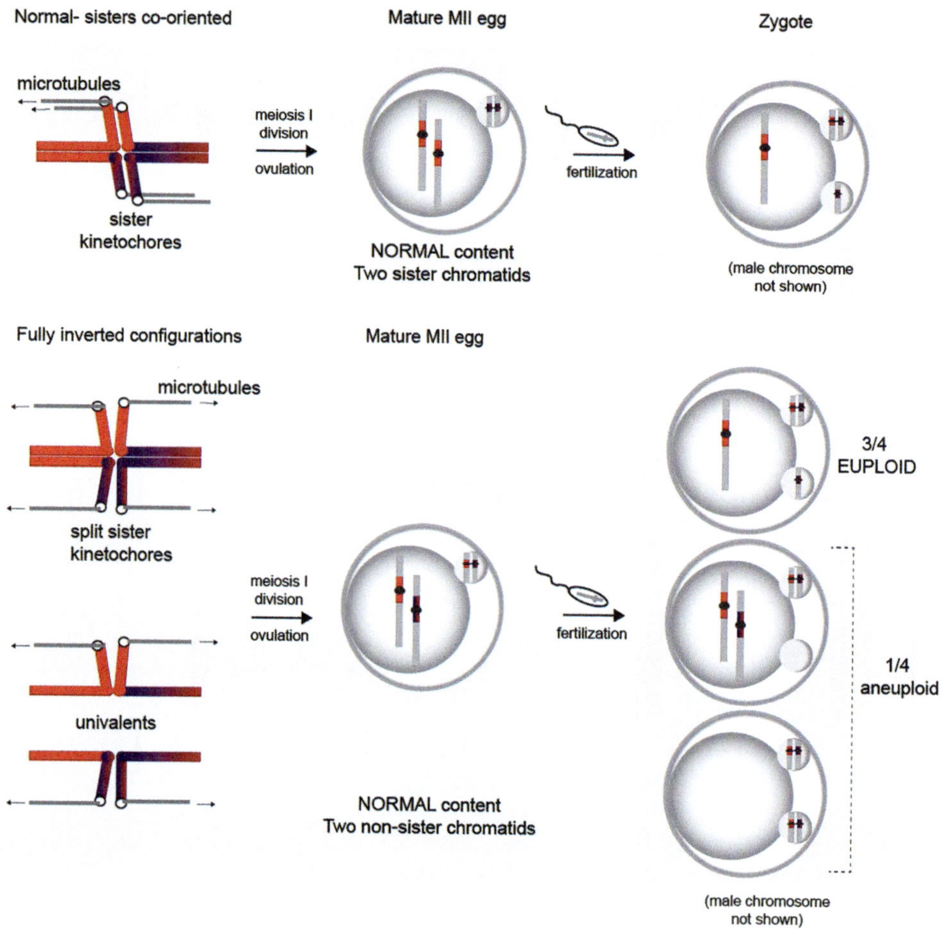

FIGURE 8.8 Reverse segregation—a new segregation pattern in human meiosis. Genomic analyses of oocytes and their matching polar bodies revealed that reverse segregation is the most common non-canonical segregation pattern in human oocytes. Rather than having a single chromosome and genotype in the first polar body, two non-sister chromatids are often present (purple and red). Time-lapse imaging revealed that sister kinetochores are often separated on both chromosomes, termed "fully inverted" and that this configuration precedes reverse segregation.

fluorescently labeled kinetochores and tubulin showed that sister kinetochores often "invert" at meiosis I. Thus, rather than being co-oriented, sister kinetochores are attached to microtubules from opposite spindle poles. Fully inverted configurations occur when both chromosomes display the behavior. Half-inverted configurations occur when only one of the homologs is affected. In several instances, univalent formation is associated with bi-orientation of sister kinetochores (Figure 8.8). Moreover, univalent formation precedes bi-orientation indicating that the inverted kinetochores may be a response to bivalent deterioration (19, 81, 82).

The loss of cohesion and structural integrity of bivalent chromosomes are age-dependent and is one manifestation of "chromosomal aging" in human oocytes (83–85). However, the molecular mechanism underlying loss of cohesion is currently not known. Studies from rodents suggest that the loading of cohesin complexes may be restricted to fetal development (86–88) and subsequently affected during the extended dictyate arrest (89, 90). Intriguingly, SMC1β, a meiosis-specific component of meiotic cohesin complexes is haploinsufficient for maintaining the bivalent configuration in mouse oocytes (91, 92). Whether depletion of meiotic cohesin complexes underlies cohesion loss in human oocytes is not clear, since loss of cohesin staining does not appear to preferentially affect the chromosomes that have lost their bivalent structure (93). Since only a proportion of cohesin complexes are thought to mediate sister chromatid cohesion in mitotic cells (94, 95), it is possible that the cohesive function of the cohesin complexes, an acetylated form of SMC3, may be affected during the extended dictyate arrest. Common genetic variants in cohesin genes have been linked to trisomy 21 risk in a recent genome-wide association study (96).

Recombination Affects Chromosome Segregation in Human Females

The findings that univalent formation precedes errors in segregation in both mouse and human oocytes can explain the importance of recombination in aneuploidy (19, 81, 82, 97). In population-based studies, altered recombination patterns and maternal age remain the two major factors that influence aneuploidy in human conceptions (16). Although defects in female meiosis are more pronounced, altered recombination also influences chromosome segregation in sperm (98). As chromosome fingerprinting, single nucleotide polymorphism (SNP) arrays and next-generation sequencing have become available, recombination patterns can also be studied. Recombination, which occurs during fetal development, has been inferred to influence chromosome segregation in complex and chromosome-specific ways. Studies on miscarriages and in trisomic individuals revealed three patterns associated with increased risk of aneuploidy: failure to recombine (non-exchange or achiasmate); recombination points near chromosomal ends (telomeric); and recombination in the vicinity of centromeres.

The recombination patterns associated with meiotic errors and aneuploidy are highly chromosome-specific. All three types of recombination patterns were associated with errors in the segregation of chromosome 21, whereas chromosome 18 was predominantly due to the non-exchange pattern (16). Since recombination together with sister chromatid cohesion physically links homologous chromosomes prior to their segregation at anaphase I, the non-exchange or achiasmate pairs were suggested to predispose to classical meiosis I non-disjunction (Figure 8.9). Achiasmate pairs would generate univalents and analyses of fetal oocytes have suggested that up to 8% of fetal oocytes lack an MLH1 focus on one or more chromosomes, with specific vulnerability for the smallest chromosomes (21 and 22) (99).

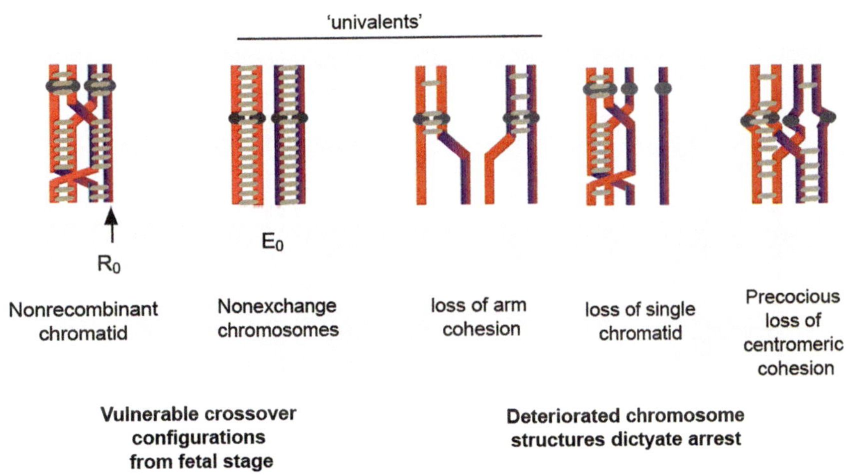

FIGURE 8.9 Vulnerable chromosome configurations in meiosis.

Recombination Rates Are Established during Fetal Life and Affect Chromosome Segregation in Adult Oocytes

Recent studies on single oocytes have allowed inferences about recombination and chromosome segregation at the DNA sequence level (50, 51). Oocytes with higher recombination rates are more likely to be euploid. Thus, selection against aneuploidy conceptions during the peri-implantation stages can explain why children born to older mothers tend to have higher recombination rates compared to their siblings born whilst the mother was young (17)—aneuploid eggs tend to have lower recombination rates and therefore do not develop to healthy babies. One model is that as age-related depletion of cohesion occurs, chromosome pairs with multiple crossovers may be more likely to retain their bivalent configuration as oocytes age (Figure 8.10).

MeioMapping also revealed some unexpected features of recombination and its effect on chromosome segregation. Normally, crossing over has been considered on a per bivalent basis such that if a crossover occurred between the two homologous chromosomes, chromosome segregation would proceed with high fidelity. This was referred to as the "obligate" crossover (e.g., [100]). However, Ottolini, et al. (2015) found that what happens to individual chromatids also influences their segregation. Non-recombinant chromatids that did not engage in recombination reactions (although its sister chromatid did) are vulnerable and at increased risk of undergoing PSSC/predivision (Figure 8.9). These findings open new areas of investigation as access to oocytes and new single-cell technologies improve.

Paradoxically, fetal oocytes have more vulnerable configurations compared to spermatocytes despite higher overall global recombination rates (69). However, oocytes display extreme heterogeneity in crossover rates (77) and the two smallest acrocentric chromosomes 21 and 22 are frequently without an MLH1

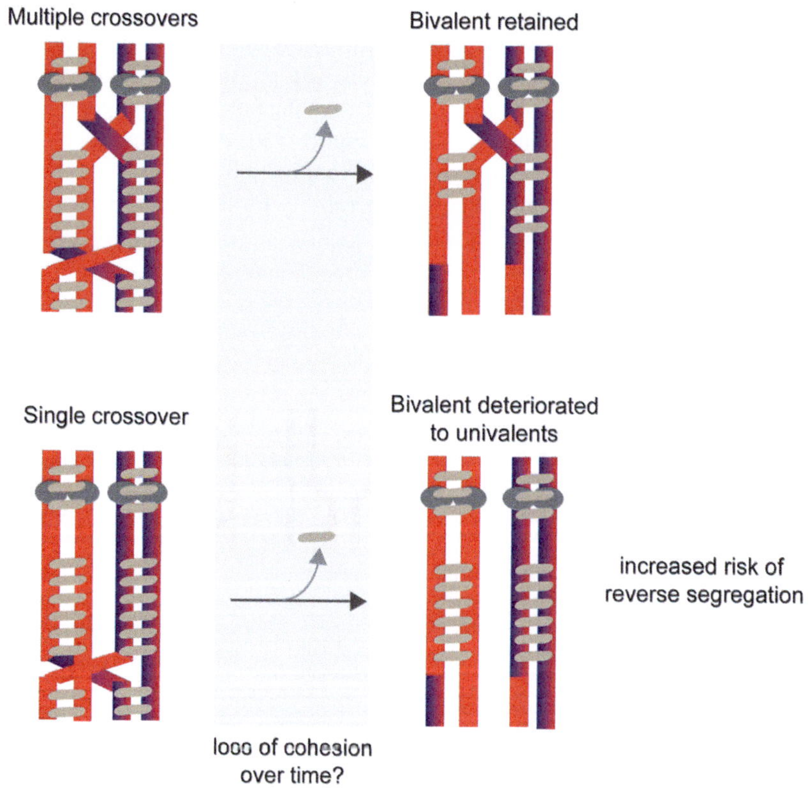

FIGURE 8.10 A hypothetical model for how recombination may be protective of bivalent structures as oocytes age.

focus (79, 99, 101). This pattern of non-exchange is virtually never seen in spermatocytes (79, 101). Thus, despite the higher rate of recombination in female germ cells, their distribution amongst chromosome pairs appears to be less regulated. The defect in patterning of crossovers has been proposed to be due to inefficiency in the maturation of crossovers (78), which would generate vulnerable crossover configurations at increased risk of chromosome errors, decades later in adult life.

Meiotic Spindles Are Inherently Error-Prone in Human Female Meiosis

In mitotic cells, spindle formation is driven largely from centrosomes, microtubule organizing centers (MTOC) from which microtubules emanate and capture chromosomes by their kinetochores. However, in many species, oogenesis is acentrosomal, although MTOCs are formed (102). In human oocytes MTOCs are not apparent and the spindle formation is driven by the chromosomes. The first metaphase I takes an astonishing 12–15 hours (83, 103) and the spindles are highly unstable (103). Although spindle instability correlates with missegregation, this facet of human oogenesis does not appear to be age-related (103).

Microtubules are composed of alpha and beta tubulin subunits. Recently, mutations in *TUBB8*, the major β-tubulin expressed only in oocytes and preimplantation embryos, were shown to be associated with maturation arrest (104) as well as defective divisions in both oocytes and embryos (104–106). *TUBB8* is particularly interesting because it evolved in the primate clade and could thus contribute to the substantial differences in spindle dynamics in mouse and human oocytes.

General Aging Features of Human Oocytes

Approaches to studying aging features of human oocytes that may predispose to aneuploidy have largely derived from transcriptional comparisons or immunocytological staining of factors identified in mouse oocytes or from human cell lines (57). The spindle assembly checkpoint (SAC) is essential in mouse oocytes to facilitate accurate chromosome segregation (107) and has been suggested to be affected by aging in human oocytes (108). Similarly, the removal of acetylation marks on histone H4K12 or H3K9 is important for chromosome compaction in human oocytes and is associated with aberrant meiosis (109, 110). This higher level of histone H4 acetylation in oocytes from women of advanced maternal age (109) indicates a general decrease in the capacity of aged oocytes to remove histone acetylation marks. Collectively, several cellular and chromosomal factors conspire to cause the high levels of aneuploidy in human eggs. Understanding aneuploidy requires sophisticated approaches. A recent study of aneuploidy risk in human oocytes identified potential genetic variants in Aurora kinases that are important for correct chromosome attachment to the meiotic spindle (111). Although such mutations may be rare, precision medicine for women at risk of aneuploid conceptions may become a possibility (112).

Mitotic Chromosome Errors in Preimplantation Embryos

Mitotic errors in chromosome segregation lead to mosaic embryos. The incidence of mosaicism in human embryos is currently being debated (64, 113), but in contrast to our growing knowledge of basic biological mechanisms that lead to aneuploidy in human oocytes, analysis of cell divisions and chromosome segregation errors in preimplantation embryos is less developed. Many models are inferred from knowledge obtained in mitosis in cell lines (e.g., reviewed by [114]). However, ethical regulations and the technical challenges of linage tracing and single-cell genomics limit our ability for direct observations in embryos. In particular, classical loss-of-function and gain-of-function analyses that are used to infer gene function are challenging in embryos. This area of research will likely evolve rapidly in the coming years since

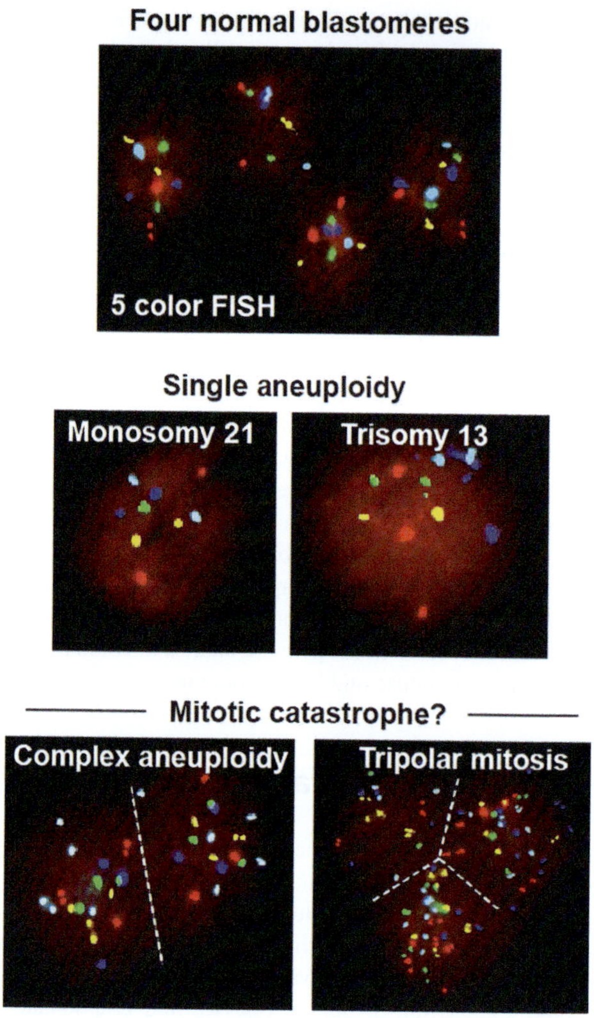

FIGURE 8.11 Tripolar mitosis and "chaotic" karyotypes in a human preimplantation embryo. FISH used against five chromosomes. The chromosomes have segregated in highly chaotic fashion into the three daughter cells.

ethical permission has been granted in several countries to use CRISPR-Cas9 for gene editing. Despite the limitations, major insights into chromosomal aneuploidy in embryos have been obtained from the use of preimplantation genetic screening and diagnosis (PGD/PGS) to prevent aneuploid conceptions and the inheritance of monogenic diseases.

The current emerging picture is that DNA damage, cellular depletion (56), and aneuploidy (115) cause developmental arrest in 20–30% of preimplantation embryos. Individual blastomeres can display highly chaotic karyotypes (mosaicism, Figure 8.11), especially during the cleavage stage (26, 53, 116, 117). Confocal microscopy has revealed that normally fertilized zygotes can undergo tripolar divisions, which would give rise to highly chaotic karyotypes (118).

Expanded blastocyst embryos are less mosaic than cleavage-stage embryos and contain fewer chaotic karyotypes (26). Nevertheless, meiotic aneuploidies including "complex" events that affect several chromosomes can persist to the blastocyst stage (23, 26, 51, 119). The relatively low rate of whole aneuploidy mosaicism in day 5 blastocyst embryos is concordant with the low level (less than 1%) in chorionic villi biopsies (120). Thus, aneuploidies *per se* do not appear to influence preimplantation development. The emerging picture, however, is complex, since clinics report highly variable aneuploidy rates not only for patients but also oocyte donors (59, 65). It is currently unclear what factors cause the variability, but understanding its causes is critical, especially for preimplantation genetic diagnosis (121).

Genome-Wide Association Study to Map Variants That Affect Chaotic Mitotic Aneuploidies

The first genome-wide association study (GWAS) to identify common variants that affect aneuploidy in preimplantation embryos has been conducted. Using aneuploidy data from around 46,000 embryos (a mix of day 3 and day 5) from 4700 individuals, McCoy and colleagues found no association with putative maternal meiotic origin and maternal genotypes at the statistical significance threshold used. However, they found that a 600 Kb region (quantitative trait loci, QTL) of low recombination on chromosome 4 that was associated with multiple, complex aneuploidies of mitotic origin (they followed paternal chromosomes, since the meiotic error rate is very low in sperm). The QTL contained common variants of *PLK4*, a polo-like kinase that regulates centrosome numbers and whose dysregulation can lead to large-scale chromosome missegregation (122, 123). Importantly, it was the maternal *PLK4* variants that influenced mitotic chromosome segregation, consistent with the model that maternal factors drive the initial mitotic divisions in human preimplantation embryos (58, 124, 125). Although the QTL contains seven other gene variants, *PLK4* is an attractive candidate since it regulates centriole duplication, a critical part of the centrosome cycle, and also mediates spindle formation during the initial cell divisions in mouse and bovine embryos (122, 123). Thus, *PLK4* variants may cause tripolar spindle formation that results in chaotic karyotypes (Figure 8.11). Importantly, tripolar spindles originating from normally fertilized human embryos have been observed (118).

The first follow-up clinical study of the rs2305957 minor variant of *PLK4* in the Chinese population revealed significant association with blastocyst formation in women with the AA genotype undergoing IVF as well as in a cohort of women suffering from early, recurrent pregnancy loss (126). It is intriguing, however, that although the efficiency of blastocyst formation was decreased in the women with the AA genotype undergoing IVF, implantation, early miscarriage rates, and live-birth rates of the blastocyst that were transferred was not affected by the minor variant. One possibility is that the rs230597 is a risk factor that results predominantly in aberrant cell divisions resulting in embryonic arrest, thus embryos that developed to day 5 were genomically stable. Although this cannot be ruled out, such a model cannot explain why women carrying the rs2305957 allele would be at elevated risk of early pregnancy loss. Further studies on the chromosomal constitution will be required to determine possible links between *PLK4* variants and aneuploidies that sustain fetal development. Another possibility is that the other variants within the QTL contribute to the genome instability or embryonic loss, as several of the genes have been implicated in cell cycle regulation or embryogenesis. As functional studies become feasible, assessing the function of the minor variant allele and determining whether alterations in *PLK4* actually occur in human oocytes or during early divisions are critical next steps in assessing cause and effect.

Aneuploidy and Embryonic Arrest

The association of complex and often chaotic aneuploidies with embryonic arrest in early stage preimplantation embryos suggests that grossly aneuploid blastomeres may have poor viability. Indeed, aberrant cell divisions leading to gross cellular and genomic defects may cause arrest (56, 76). In such cases, aneuploidy is a result of other cellular defects. In other cases, aneuploidy may drive embryonic arrest or affect embryo development. The effect of aneuploidy on embryo development and the assignation of aneuploidy cells to specific tissues are studied by the generation of chimeras (127–134). Regarding cells carrying a single chromosomal aneuploidy, there is no or little evidence for the preferential assignment of aneuploidy cells to the placental precursor line (trophectoderm) or the fetus. Even when chaotic karyotypes are produced by inhibiting components of the spindle assembly checkpoint, there is no evidence for the active allocation of aneuploid cells to specific lineages (135). However, in mouse embryos with chaotic chromosome constitutions, blastocysts display a depletion in cell numbers and the mechanism may depend upon lineage. Aneuploidy in the fetal lineage leads to apoptosis whereas senescence limits aneuploidy within the placental lineage (135). It would be interesting to elucidate whether human embryos display similar mechanisms and in particular, what causes apoptosis versus senescence in the two different lineages.

Future Directions

In this chapter, we have reviewed current findings and emerging themes in the human aneuploidy field in the context of our knowledge of population genetics. We envisage that technology-driven investigations into aneuploidy and its intersection with other cellular features, such as epigenetics, will shape our increasing appreciation of our genetic inheritance. Early embryonic divisions are largely governed by maternal factors, until the activation of the embryonic genome (124, 136). It follows that the robustness in the developmental program of mature oocytes as well as early embryos is important for genome stability in the germline. With current advances in embryology, stem cells, and gene editing, it may soon be possible to use some of these tools to gain fundamental insight into the mechanisms and cellular causes of aneuploidy, a condition that afflicts such a large proportion of human conceptions and limits reproductive lifespan in women.

ACKNOWLEDGMENTS

I am grateful to my colleagues, who have provided their expertise to ensure the accuracy of the content of this chapter. Alan Handyside and Antonio Capalbo were co-authors and contributed to the first edition from which this chapter is updated. Aditya Sankar contributed Figure 8.5. We apologize to colleagues whose work we could not cite due to restrictions in length. ERH is funded by a Novo Nordisk Young Investigator Award.

REFERENCES

1. Tjio JH,, et al. *Hereditas*. 1956;42(1–2):U1–6.
2. Jacobs PA,, et al. *Nature*. 1959;183(4657):302–3.
3. Gautier M,, et al. *Hum Genet*. 2009;126(2):317–24.
4. Lejeune J, et al. *C R Hebd Seances Acad Sci*. 1959;248(11):1721–2.
5. Jacobs PA, et al. *Lancet*. 1959;1(7075):710.
6. Ford CE, et al. *Lancet*. 1959;1(7075):711–3.
7. Ford CE, et al. *Nature*. 1959;183(4667):1030–2.
8. Gottlieb MI, et al. *Am J Med*. 1962;33:763–73.
9. Gibson DA, et al. *Med Biol Illus*. 1963;13:80–8.
10. Gorlin RJ. *J Dent Res*. 1963;42:SUPPL1297–306.
11. Bowen P. *Clin Orthop Relat Res*. 1964;33:40–58.
12. Miller OJ. *Am J Obstet Gynecol*. 1964;90:SUPPL:1078–139.
13. Jacobs PA. *Annu Rev Genom Hum G*. 2014;15:29–46.
14. Freeman SB, et al. *Public Health Rep*. 2007;122(1):62–72.
15. Hassold TJ, et al. *Annu Rev Genet*. 1984;18:69–97.
16. Nagaoka SI, et al. *Nat Rev Genet*. 2012;13(7):493–504.

17. Kong A, et al. *Nat Genet.* 2004;36(11):1203–6.
18. Christianson RE, et al. *Genet Med.* 2004;6(6):487–94.
19. Gruhn JR, et al. *Science.* 2019;365(6460):1466–9.
20. Shukla V, et al. *Semin Cell Dev Biol.* 2021;113:132–47.
21. Van Blerkom J, et al. *Hum Reprod.* 1992;7(3):379–90.
22. Pellestor F, et al. *Eur J Med Genet.* 2006;49(2):103–16.
23. Handyside AH, et al. *Eur J Hum Genet.* 2012;20(7):742–7.
24. Franasiak JM, et al. *Fertil Steril.* 2014;101(3):656–63 e1.
25. Vera-Rodriguez M, et al. *Nat Commun.* 2015;6:7601.
26. McCoy RC, et al. *PLoS Genet.* 2015;11(10):e1005601.
27. Hassold TJ. *Basic Life Sci.* 1985;36:103–15.
28. Erickson JD. *Ann Hum Genet.* 1978;41(3):289–98.
29. Hassold T, et al. *Nat Rev Genet.* 2001;2(4):280–91.
30. Brosens JJ, et al. *Sci Rep.* 2014;4:3894.
31. Pellestor F, et al. *Hum Reprod.* 2002;17(8):2134–45.
32. Angell R. *Am J Hum Genet.* 1997;61(1):23–32.
33. Kuliev A, et al. *Cytogenet Genome Res.* 2005;111(3–4):193–8.
34. Anahory T, et al. *Mol Hum Reprod.* 2003;9(10):577–85.
35. Kuliev A, et al. *Reprod Biomed Online.* 2011;22(1):2–8.
36. Fragouli E, et al. *Cytogenet Genome Res.* 2006;114(1):30–8.
37. Fragouli E, et al. *Reprod Biomed Online.* 2009;19(2):228–37.
38. Munne S, et al. *Reprod Biomed Online.* 2007;14(5):628–34.
39. Gabriel AS, et al. *J Med Genet.* 2011;48(7):433–7.
40. Handyside AH, et al. *Eur J Hum Genet.* 2012;20(7):742–7.
41. Magli MC, et al. *Reprod Biomed Online.* 2012;24(3):331–8.
42. Munne S, et al. *Fertil Steril.* 2010;94(2):408–30.
43. *clinics SfARTCsraSm.* 2007.
44. Check JH, et al. *J Reprod Med.* 2011;56(3–4):149–52.
45. Franasiak JM, et al. *J Assist Reprod Genet.* 2014;31(11):1501–9.
46. Pellestor F, et al. *Hum Genet.* 2003;112(2):195–203.
47. Capalbo A, et al. *Hum Reprod.* 2014; 29(6):1173–81.
48. Zuccaro MV, et al. *Cell.* 2020;183(6):1650–64 e15.
49. Hoffmann ER, et al. *Cell.* 2020;183(6):1464–6.
50. Hou Y, et al. *Cell.* 2013;155(7):1492–506.
51. Ottolini CS, et al. *Nat Genet.* 2015;47(7):727–35.
52. Munne S, et al.. *Hum Reprod.* 1993;8(12):2185–91.
53. Vanneste E, et al. *Nat Med.* 2009;15(5):577–83.
54. Vega M, et al. *Fertil Steril.* 2014;102(2):394–8.
55. McCoy RC, et al. *Science.* 2015;348(6231):235–8.
56. Kort DH, et al. *Hum Reprod.* 2016;31(2):312–23.
57. Grondahl ML, et al.. *Hum Reprod.* 2010;25(4):957–68.
58. Braude P, et al. *Nature.* 1988;332(6163):459–61.
59. Munne S, et al. *Fertil Steril.* 2012;97(4):935–42.
60. Labarta E, et al. *J Clin Endocrinol Metab.* 2012;97(10):E1987–94.
61. Massie JA, et al. *Fertil Steril.* 2011;95(3):970–2.
62. Plachot M. *Mol Cell Endocrinol.* 2001;183 Suppl 1:S59–63.
63. Rubio C, et al. *Hum Reprod.* 2010;25(9):2290–7.
64. Capalbo A, et al. *Hum Reprod.* 2016;31(1):199–208.
65. Munne S, et al. *Hum Reprod.* 2017;32(4):743–9.
66. Loane M, et al. *Eur J Hum Genet.* 2013;21(1):27–33.
67. Koubova J, et al. *Proc Natl Acad Sci U S A.* 2006;103(8):2474–9.
68. Bowles J, et al. *Science.* 2006;312(5773):596–600.
69. Tease C, et al. *Am J Hum Genet.* 2002;70(6):1469–79.
70. Hartshorne GM, et al. *Mol Hum Reprod.* 2009;15(12):805–19.
71. Jagiello GM, et al. In *Chromosome Today.* Eds. PL. Pearson, KR Lewis. 1975:43–63.
72. Michelmann HW, et al. *Fertil Steril.* 1985;43(2):320–2.
73. Martin RH, et al. *J Reprod Fertil.* 1986;78(2):673–8.
74. Angell RR. *Hum Genet.* 1991;86(4):383–7.

75. Angell RR, et al. *Hum Reprod.* 1991;6(4):568–73.
76. Ottolini CS, et al. *Sci Rep.* 2017;7(1):9744.
77. Lenzi ML, et al. *Am J Hum Genet.* 2005;76(1):112–27.
78. Wang S, et al. *Cell.* 2017;168(6):977–89 e17.
79. Gruhn JR, et al. *PLoS One.* 2013;8(12):e85075.
80. Kouznetsova A, et al. *Nat Genet.* 2007;39(8):966–8.
81. Zielinska AP, et al. *Elife.* 2015;4:e11389.
82. Zielinska AP, et al. *Curr Biol.* 2019;29(22):3749–65 e7.
83. Angell RR. *Cytogenet Cell Genet.* 1995;69(3–4):266–72.
84. Garcia-Cruz R, et al. *Hum Reprod.* 2010;25(9):2316–27.
85. Zielinska AP, et al. *Elife.* 2015;4:e11389.
86. Revenkova E, et al. *Curr Biol.* 2010;20(17):1529–33.
87. Tachibana-Konwalski K, et al. *Genes Dev.* 2010;24(22):2505–16.
88. Burkhardt S, et al. *Curr Biol.* 2016;26(5):678–85.
89. Lister LM, et al. *Curr Biol.* 2010;20(17):1511–21.
90. Chiang T, et al. *Curr Biol.* 2010;20(17):1522–8.
91. Hodges CA, et al. *Nat Genet.* 2005;37(12):1351–5.
92. Murdoch B, et al. *PLoS Genet.* 2013;9(2):e1003241.
93. Garcia-Cruz R, et al. *Hum Reprod.* 2010;25(1):179–91.
94. Zhang J, et al. *Mol Cell.* 2008;31(1):143–51.
95. Nishiyama T, et al. *Cell.* 2010;143(5):737–49.
96. Chernus JM, et al. *PLoS Genet.* 2019;15(12):e1008414.
97. Sakakibara Y, et al. *Nat Commun.* 2015;6:7550.
98. Gonsalves J, et al. *Hum Mol Genet.* 2004;13(22):2875–83.
99. Hassold T, et al. *Am J Hum Genet.* 2021;108(1):16–24.
100. Housworth EA, et al. *Am J Hum Genet.* 2003;73(1):188–97.
101. Cheng EY, et al. *PLoS Genet.* 2009;5(9):e1000661.
102. Schuh M, et al. *Cell.* 2007;130(3):484–98.
103. Holubcova Z, et al. *Science.* 2015;348(6239):1143–7.
104. Feng R, et al. *N Engl J Med.* 2016;374(3):223–32.
105. Feng R, et al. *J Med Genet.* 2016;53(10):662–71.
106. Chen B, et al. *Hum Reprod.* 2017;32(2):457–64.
107. Touati SA, et al. *Nat Commun.* 2015;6:6946.
108. Eichenlaub-Ritter U. *Int J Dev Biol.* 2012;56(10–12):841–52.
109. van den Berg IM, et al. *Hum Reprod.* 2011;26(5):1181–90.
110. Huang J, et al. *Fertil Steril.* 2012;97(1):178–84 e3.
111. Nguyen AL, et al. *Mol Hum Reprod.* 2017;23(6):406–16.
112. Cimadomo D, et al. *Hum Reprod Update.* 2020;26(4):453–73.
113. Vera-Rodriguez M, et al. *Fertil Steril.* 2017;107(5):1107–12.
114. Ambartsumyan G, et al. *Hum Mol Genet.* 2008;17(R1):R10–5.
115. Plachot M, et al. *Ann Genet.* 1987;30(1):22–32.
116. Kola I, et al. *Biol Reprod.* 1987;37(2):395–401.
117. Bongso A, et al. *Fertil Steril.* 1991;56(1):66–70.
118. Chatzimeletiou K, et al. *Hum Reprod.* 2005;20(3):672–82.
119. Capalbo A, et al. *Hum Reprod.* 2013;28(2):509–18.
120. Malvestiti F, et al. *Prenat Diagn.* 2015;35(11):1117–27.
121. Sermon K, et al. *Mol Hum Reprod.* 2016;22(8):845–57.
122. Coelho PA, et al. *Dev Cell.* 2013;27(5):586–97.
123. Liang S, et al. *Theriogenology.* 2016;85(4):754–61 e1.
124. Niakan KK, et al. *Development.* 2012;139:829–41.
125. Tadros W, et al. *Development.* 2009;136(18):3033–42.
126. Zhang Q, et al. *Fertil Steril.* 2017;107(4):1034–40 e5.
127. Bogart MH, et al. *Anat Embryol (Berl).* 1990;181(2):137–47.
128. Everett CA, et al. *Reproduction.* 2007;134(6):799–809.
129. Lue YH, et al. *Pediatr Endocrinol Rev.* 2010;8 Suppl 1:140–4.
130. Buganim Y, et al. *Cell Stem Cell.* 2014;15(3):295–309.
131. Fundele R, et al. *Roux Arch Dev Biol.* 1985;194(3):178–80.
132. Cox DR, et al. *Dev Biol.* 1984;101(2):416–24.

133. Magnuson T, et al. *J Embryol Exp Morphol.* 1982;69:223–36.
134. Epstein CJ, et al. *Proc Natl Acad Sci U S A.* 1982;79(14):4376–80.
135. Bolton H, et al. *Nat Commun.* 2016;7:11165.
136. Lee MT, et al. *Annu Rev Cell Dev Biol.* 2014;30:581–613.
137. Fragouli E, et al. *Cytogenet Genome Res.* 2011;133(2–4):107–18.
138. Alfarawati S, et al. *Fertil Steril.* 2011;95(2):520–4.
139. Gueye NA, et al. *Fertil Steril.* 2014;101(1):232–6.
140. Fritz B, et al. *Eur J Hum Genet.* 2001;9(12):910–6.
141. Levy B, et al. *Obstet Gynecol.* 2014;124(2 Pt 1):202–9.
142. Rodriguez-Purata J, et al. *Fertil Steril.* 2015;104(6):1460–66 e1–12.
143. Wang Y, et al. *Clin Genet.* 2017;91(6):849–58.
144. Wapner RJ, et al. *New Engl J Med.* 2012;367(23):2175–84.
145. Forabosco A, et al. *Eur J Hum Genet.* 2009;17(7):897–903.
146. Martin CL, et al. *Clin Perinatol.* 2015;42(2):227–42, vii.
147. Sudmant PH, et al. *Science.* 2015;349(6253):aab3761.
148. Hamerton JL, et al. *Clin Genet.* 1975;8(4):223–43.
149. Hassold T, et al. *Environ Mol Mutagen.* 1996;28(3):167–75.
150. Wellesley D, et al. *Eur J Hum Genet.* 2012;20(5):521–6.
151. Lim AS, et al. *Fertil Steril.* 1997;68(2):265–71.
152. Girardi L, et al. *Am J Hum Genet.* 2020;106(4):525–34.

Epigenetics from Oocytes to Embryos

9

Dagnė Daškevičiūtė, Marta Sanchez-Delgado, and David Monk

INTRODUCTION

Gene expression drastically changes during gametogenesis in mammals, halting completely before these cells are fully mature. In mice, expression resumes shortly after fertilization, in a process termed zygotic genome activation (ZGA) (note, here we are using the strictest definition of the zygotic, the one-cell stage preceding the first cleavage division), in which developmentally important genes are expressed from the late one-cell stage. Fertilization represents the first step in the creation of an embryo and requires the successive completion of multiple complex processes before the formation of a totipotent embryo during a period termed the maternal-to-zygotic transition (MZT). Syngamy, the fusion of the two terminally differentiated haploid gametes to form a zygote, is followed by epigenetic reprogramming, genome activation, and the depletion of maternal-derived transcripts to give rise to unspecialized cells of the cleavage-stage embryo. Totipotency of the blastomeres is gradually lost as cleavages progress allowing for initial specification and the development of the blastocyst, resulting in the formation of the inner cell mass (ICM), which will develop into the embryo proper, and the trophectoderm (TE), the forbearer of the extraembryonic tissues and placenta. Recent studies describing these mechanisms have benefited from advances and innovations in low-input and single-cell technologies revealing that, despite subtle differencing in timing (Table 9.1), these events are largely conserved between mammalian species. In this chapter we will describe the recent progress in understanding epigenetic regulation during mammalian pre-implantation development including global DNA demethylation, paternal protamine exchange, redistribution of histone modifications, and gradual formation of higher-order architecture, as well as the defining steps in regulating ZGA and embryonic genome activation (EGA). Furthermore, we discuss their implications for embryo selection during assisted reproductive cycles and whether faults in these coordinated processes could contribute to human embryo arrest observed during assisted reproductive cycles.

EPIGENETIC MECHANISMS

DNA methylation and histone modifications ultimately control chromatin accessibility and packaging, allowing for gene expression or repression. DNA methylation is the covalent modification of a methyl

 DOI: 10.1201/9781003024941-9

TABLE 9.1 Comparison of Timings for Key MZT Events during Human and Mouse Pre-Implantation Development

MZT FEATURE	*HUMAN*	*MOUSE*
First Cleavage division	27–30 hours post fertilization.	16–20 hours post fertilization.
Stage of implantation	6–8 days post fertilization.	4–4.5 days post fertilization.
Genome activation	4–8 cells.	2 cells.
DNA demethylation	Late two-cell stage.[13]	One-cell stage.[14]
Histone remodeling	Depends on modification. No noncanonical H3K4me3 in oocyte or zygote.[43]	Depends on modification. Oocyte-derived noncanonical H3K4me3 distribution until two-cell stage.[44,45]
Chromatin accessibility	Increases during cleavage stages and promoter accessibility correlates with initial genome activation.[56,57]	
Higher-order chromatin architecture	No compartmentation in sperm. Gradual establishment of TADs during cleavage stages.[61]	Sperm compartmentation, but not in oocytes. Gradual establishment of TADs during cleavage stages.[58–60]
Imprint ICR maintenance	Some ICRs requires ZFP57.[34]	ZFP57 essential for all ICRs.[31]

group to the carbon-5 atom of cytosine (5mC) residues in DNA. This reaction is catalyzed by the DNA methyltransferases (DNMTs).[1,2] In the mammalian genome, 5mC mainly occurs in the context of CpG dinucleotides, forming a symmetrical pattern on both DNA strands. Non-CpG methylation is rare and restricted to exceptional cell types, including oocytes and neurones.[3,4] Overall, 5mC distribution is bimodal, with the majority of the genome being heavily methylated, whereas CpG islands are usually devoid of methylation. CpG islands are discrete intervals enriched for CpG dinucleotides (200-2000 bp, GC content > 50%, observed-to-expected CpG ratio > 60%) which often overlap transcriptional start sites. Overall, 5mC is considered to be a repressive mark, especially when observed at heterochromatin, functional elements such as centromeric and pericentromeric repeats, as well as gene promoters and transposable elements. DNA methylation is essential for mammalian development being involved in lineage inappropriate gene silencing, X chromosome inactivation, imprinting, and safeguarding overall genome integrity (reviewed in [5]).

In order to be contained within the nucleus, the two meters of DNA in each cell need to be efficiently packaged. To achieve this, DNA in somatic cells and oocytes is wrapped around nucleosomes comprised of an octamer of histone proteins (two subunits of each type: H2a, H2b, H3, and H4). In sperm, approximately 70–80% of the genome is organized by protamines; however, CpG-rich sequences retain histones.[6] Fascinatingly, each histone has a protruding N-terminal tail that can be modified by a plethora of post-translational modifications, including acetylation and methylation.[7–9] These histone modifications have distinct functions and are located at specific sites throughout the genome.

DNA METHYLATION TRANSITORY STATES AND REPROGRAMMING

As a stable epigenetic mark that is heritable through cell division, 5mC is not only an important component of maintaining cell identity but is essential for genome stability. However, the epigenome, including global DNA methylation, needs to be reprogrammed to allow for the progeny to develop. There are two waves of demethylation and subsequent remethylation that occur during mammalian development (Figure 9.1a). The first occurs during germ cell development and the other immediately after fertilization to ensure a pluripotent state. Both sperm and oocytes are considered to be the most terminally

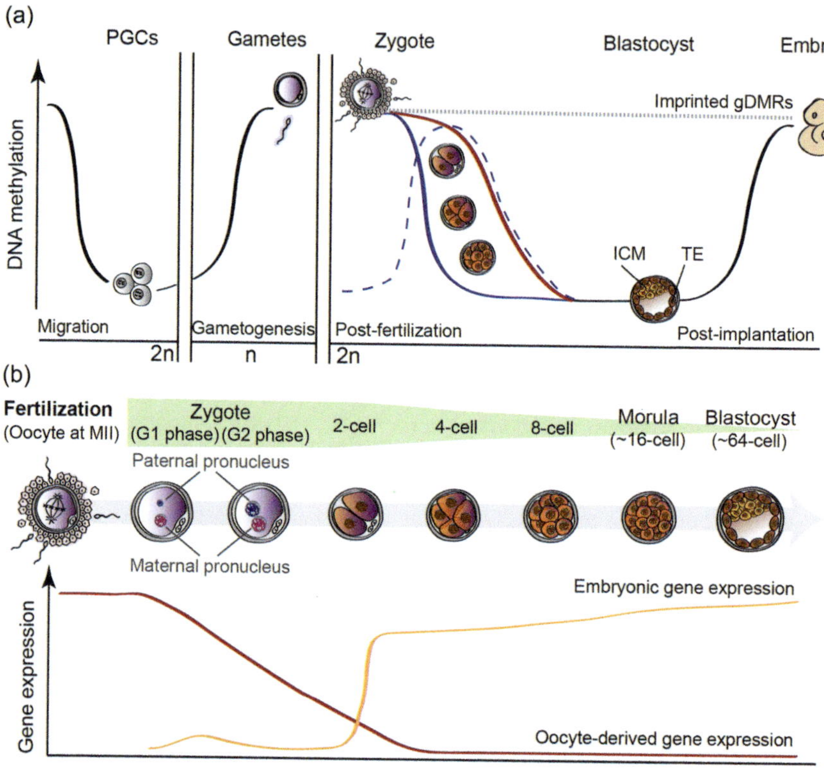

FIGURE 9.1 Methylation dynamics during development in a mouse. (a) The left panel of the figure represents erasure of methylation that occurs during migration of primordial germ cells to the gonadal ridge during embryonic development, with the subsequent sex-specific methylation establishment in the gametes shown in the middle panel. Following fertilization, the embryonic genome methylation is erased methylation asynchronously in the maternal and paternal pronuclei. The sperm-derived paternal genome is demethylated by a combination of 5mC oxidation to 5hmC and diluted during replication, whilst the oocyte-derived maternal genome is passively lost during replication. Following implantation, the embryonic genome is remethylated. The dotted grey line represents maintenance of 5mC and imprinted ICRs. Black (general), blue (sperm-derived), and red (oocyte-derived) solid lines represent the global 5mC patterns, while dashed lines denote 5hmC. The X-axis depicts the mammalian developmental stages; the Y-axis signifies general DNA methylation quantity. (b) The lower panels show declining totipotency with subsequent cleavage cell divisions (green line) concomitant with oocyte-derived transcript degradation and embryonic genome activation.

differentiated cell types of the body, and the associated epigenetic signatures of these haploid cells need to be erased to gain totipotency.[10] Key events for embryonic reprogramming include the erasure of most 5mC marks inherited from the gametes, with the exception of imprinted loci and some retrotransposon sequences,[11] and the establishment of embryonic methylation patterns during the processes of blastocyst implantation.[12]

In the fertilized mouse oocytes, now termed "zygote," the two parental genomes are spatially separated within the maternal and paternal pronuclei. Even before the completion of the first cell division, the maternally and paternally derived DNA have asymmetric DNA methylation profiles, with the paternal genome having undergone active demethylation independent of replication in an enzymatic-mediated reaction performed by the ten-eleven translocation (TET) protein family.[13,14] The methylation dynamics are similar in human embryos, with the notable exception that the most drastic demethylation occurs from fertilization until the two-cell stage.[15,16] There are three members of the TET family: TET1, TET2 and TET3, all capable of oxidizing 5mC to intermediate states that include 5-hydroxymethylcytosine (5hmC), 5-formylcytosine (5fC) and 5-carboxylcytosine (5caC) (Figure 9.2a).[17] Paternal genome demethylation is carried out by TET3 which is abundantly expressed in the oocyte and present in the zygote at high levels.[18–20] 5fC and 5caC are committed intermediates of the oxidative demethylation process and are magnitudes less abundant than 5hmC, and unlike 5hmC, are not enriched in somatic tissues. Despite both 5fC and 5caC having the potential for active removal by thymine DNA glycosylases (TDG) through base excision repair,[21] they actually show a gradual decline in the paternal pronucleus, therefore a replication-coupled demethylation model is favored, since oxidized 5mC derivatives are poorly recognized by DNMT1.[22]

Following fertilization, both the maternally and paternally derived genomes are exposed to the same cellular environment, yet the two pronuclei show unequal methylation dynamics with the maternal genome being protected from TET3-mediated demethylation by the oocyte-derived factor DPPA3 (Figure 9.2b).[18,23] DPPA3 specifically interacts with H3K9me2 enriched in the maternal, but not paternal pronucleus[19,23] inhibiting the recruitment of TET3.[24] Ultimately, maternal demethylation is achieved by replication-dependent dilution,[25] largely due to the nuclear exclusion of DNMT1 and its co-factor UHRF1.[26] Therefore, 5mC levels on both parental genomes reach their lowest by the blastocyst stage, which is followed by widespread *de novo* remethylation during post-implantation development. These dynamics are largely conserved in mammalian development, with the human showing subtly slower kinetics.[27]

While we have discussed that the paternal genome undergoes widespread 5mC loss before the completion of the first S-phase following fertilization, recent studies have revealed that maternally derived DNMT3A is also responsible for low level *de novo* methylation of the zygotic paternal genome.[27] This ongoing process is thought to be required to maintain methylation at paternally methylated imprinted regions.[28]

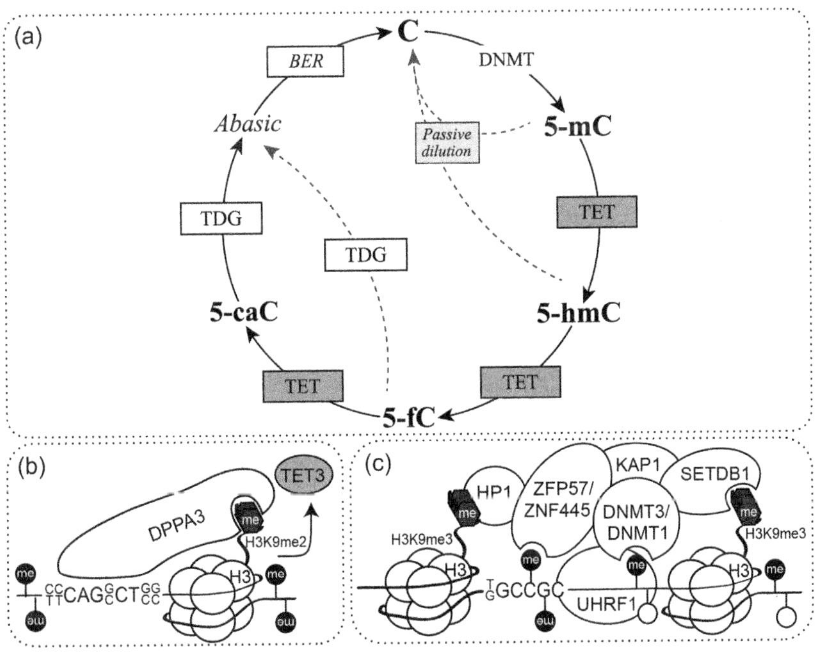

FIGURE 9.2 Protein complexes that protect imprinted ICRs from demethylation during embryonic reprogramming. (a) The dynamic pathway to demethylation, highlighting the oxidative intermediates. (b) Protection of ICR sequences from TET3-mediated demethylation by the factor DPPA3. (c) Imprint maintenance by the KAP1/TRIM28 co-repressor complex recruited to ICRs by the zinc finger proteins ZFP57 and ZNF445. In both examples, the consensus-binding sequences are shown.

THE ENIGMA OF IMPRINTED GENES

Despite the extensive global hypomethylation that results following epigenetic reprogramming in the pre-implantation embryo, the methylation marks at imprinting control regions (ICRs) are spared. This unique set of genes, characterized by their parent-of-origin monoallelic expression, results from differential allele methylation inherited from one of the gametes.[29] Recent studies have shown that there is extensive opposing methylation between sperm and oocytes;[30] however, the vast majority is successfully demethylated, except at imprinted ICRs. In some cases, these ICRs dictate classical imprinting in all subsequent somatic tissues, while a significant number of oocyte-derived, maternally methylated intervals only survive in the placenta.[10,30]

Even though nuclear exclusion of DNMT1 is essential for passive demethylation in cleavage-stage embryos, genetic evidence implies it is still required for maintenance of ICR methylation, suggesting that it is not exclusively retained in the cytoplasm. The limited amounts of DNMT1, as well as DNMT3A/B, are targeted to ICR via the TRIM28/KAP1 co-repressive complex that exhibits sequence-specificity due to the zinc-finger proteins (ZFPs) ZFP57 and ZNF445[31–34] (Figure 9.2c). Classically, ZFP-KAP1 complexes include the TRIM28/KAP1 scaffold protein, the histone methyltransferase SETDB1, the nucleosome remodeling histone deacetylation complex (NuRD), heterochromatin protein 1 (HP1), and the DNA methylation machinery, all of which are responsible for silencing of transposable elements, with underlying sequence-specificity attributed to unique combinations of ZFPs that target different retrotransposon families.[35] Once recruited to a sequence, transcriptional repression is mediated by H3K9me3 and DNA methylation.[36] ZFPs have been shown not only to recognize transposable elements, but an increasing number of these proteins also bind to single-copy regions in the genome as highlighted by ZFP57 and ZNF445.[33,34]

In addition to the ZFP-KAP1 co-repressive complex mediated maintenance of methylation marks at ICRs, mechanisms that preclude them from being erased are important, especially for the limited number of paternally methylated ICRs inheriting their methylation from sperm. The aforementioned DPPA3 protects methylation at maternally methylated ICRs and the *H19* and *Rasgrf1* ICRs, presumably by binding to the H3K9me2 and being refractory to TET3.[23,37,38]

HISTONE MODIFICATION ASYMMETRY

Histone modifications are key regulatory signals throughout pre-implantation embryo development that influence transcription factor (TF) interactions with the underlying chromatin. Until relatively recently, our knowledge of histone modification reprogramming was restricted to observations from immunofluorescent staining with no precise genome localization known.[39] However, the development of low-input and high-resolution techniques for mapping histone modifications, such as ChIP-seq, CUT&Run, and CUT&Tag[40–42] have allowed for histone modification demarcation of distinct genomic features in pre-implantation embryos[43–46] (Figure 9.1). For example, H3K4me3 is a hallmark of active promoters; H3K27me3 is a repressive histone mark that preferentially occupies promoters of developmental genes, whilst H3K27ac is widely used to differentiate active enhancers (reviewed in [7]). Interestingly, in mice several histone modifications show noncanonical distributions and functions in oocytes and early pre-implantation embryos, including H3K4me3 and H3K27me3 which display widespread domains rather than distinct peaks of enrichment, which are briefly inherited and regulate zygotic expression.[40,45,46]

H3K4me3 is a histone modification largely restricted to regions of DNA with unmethylated gene promoters that are transcriptionally active in embryos, similar to somatic tissues.[13,46] Unlike histone acetylation, histone methylation does not change chromatin-DNA interactions by altering charge, but promotes binding of transcriptional activating TFs such as CHD1 and CFP1,[47] whilst blocking transcriptional repressors including the NuRD complex.[48] After fertilization in the mouse, H3K4me3 in the paternal genome is rapidly depleted while the noncanonical form covering promoter and distal sequences in the maternal genome persists until ZGA, upon which it is re-established with a canonical profile by the two-cell stage.[40,44] Since H3K4me3 remodeling is rapid, this implies that yet to be identified oocyte-derived histone demethylase must be responsible for the remodeling. Unexpectedly, mapping H3K4me3 in human oocytes identified defined peaks in promoters, unlike the broad distribution defining the noncanonical profile, revealing that such patterns are not observed in humans.[43]

H3K27me3 is a polycomb-based histone modification associated with gene repression[49] and can be inherited from the oocyte regulating enhancer function and lineage-specific genes.[50] In the mouse zygote, H3K27me3 located in promoter regions in both the maternal and paternal genomes is lost, followed by rapid redistributions during the cleavage stages.[50] The rapid erasure throughout the paternal genome is accompanied by selective depletion at maternal promoters. Recently, oocyte-derived, maternal H3K27me3-mediated imprinting has been described in mouse pre-implantation embryos for several genes, including *Gabl*, *Phf17*, *Sfmbts*, and *Slc38a4*.[50] This occurs at DNA unmethylated promoter regions, but is transient, being lost by implantation, reflecting the complementary roles of H3K27me3 and 5mC in regulating imprinting. During human pre-implantation development, the H3K27me3 oocyte and embryonic profiles differ from mice. By the eight-cell cleavage stage, human embryos exhibit complete erasure on both parental genomes,[43] consistent with the largely non-existent H3K27me3-mediated, germline 5mC-independent temporal imprinting.[51,52] H3K27me3 also appears in intergenic regions, where it is widespread and may be involved in transcriptional repression, compensating for the lack of 5mC and H3K9me3 during reprogramming. This is especially important at repeat-rich sequences such as long terminal repeat (LTR) retrotransposons, which must be properly regulated to avoid recombination, self-duplication, and genome instability. Since 5mC is largely demethylated shortly after fertilization, silencing of LTRs requires a switch from 5mC to repressive histone modifications, including H3K27me3 and H3K9me3.[53] Promoters demarked by H3K9me3 are erased at fertilization and re-established at post-implantation stages, indicating that unique sequences and LTRs have different regulatory mechanisms, likely involving ZFP-KAP1 complexes.

PROTAMINE TO HISTONE EXCHANGE POST-FERTILIZATION

The chromatin of mature sperm differs markedly from oocyte and somatic chromatin as the majority of DNA is not associated with histones or arranged in nucleosomes, but is complexed with protamines, resulting in approximately 6–20-fold greater compaction than nucleosome-based structures.[54] Such compaction is not only required to facilitate motility, but also to help prevent DNA damage since mature sperm lack DNA repair machinery. During spermatogenesis, protamines replace approximately 70–80% of histones, which results in a void in epigenetic marks, somewhat restricting paternal transmission of these epigenetic signals. At fertilization, the highly compacted and transcriptional quiescent sperm chromatin needs to be remodeled into a decondensed state to facilitate in zygotic development and fusion of parental pronuclei. Despite being an essential event, the timing and regulation of protamine removal are poorly understood. Following fertilization, sperm DNA is decondensed and expands to approximately three times the size of the mature sperm nucleus, resulting in the formation of the paternal pronucleus. These events are thought to coincide with the complete removal of protamine from the paternal chromatin and exchange with canonical and variant histones that must be completed prior to replication and the initiation of transcription. The oocyte plays a pivotal role in the removal of protamines from the paternal chromatin by supplying not only glutathione, which reduces disulfide bonds between proteins, but also by providing the pool of histones and chaperones. Subsequently, the paternal pronucleus has a relatively "naïve" chromatin structure containing minimal repressive histone marks that ultimately influence higher-order chromatin architecture at the early cleavage stage and correlates with the observation that the bulk of transcription in mouse zygotes occurs within the male pronucleus.[55]

POST-FERTILIZATION CHROMATIN REMODELING

Chromatin reorganization is critical for the reprogramming of the sperm- and oocyte-derived genomes into a pluripotent state. Open chromatin is tightly linked to transcriptional activation and is in part associated with repositioned or depleted nucleosomes and TF binding. Recently, studies using ATAC-seq (assay for transposase-accessible chromatin using sequencing) have revealed the landscape of chromatin accessibility during mouse development.[56] In contrast to the asymmetries reported for 5mC and histone modifications, the two parental genomes seem to have a relatively synchronized chromatin profile, except at imprinted loci, with open chromatin located at CpG island promoters, which largely precedes EGA[57] (Figure 9.1a, b).

ESTABLISHMENT OF HIGHER-ORDER CHROMATIN INTERACTIONS

The nucleus of an interphase cell contains chromatin organized into a hierarchical structure that is key for most biological events, including transcription. DNA can form loops bringing distal regulatory elements into close proximity with promoters. These looping events are constrained within self-interacting genomic regions known as topologically associated domains (TADs). The recent development of low-input Hi-C based on the 3C (chromosome conformation capture) method has revealed the dynamic nature of the genome architecture during pre-implantation development[58,59] (Figure 9.3a). The meiotic chromatin of MII mouse oocytes generally lacks TAD structures, whereas sperm present with both, TADs and A/B compartments, which largely correspond to gene-rich and repressive chromatin intervals respectively, and likely reflect protamine packaging.[58] Following fertilization, the higher-order structures are gradually created, with the establishment of maternal genome being weaker than the paternal, with TAD formation dependent upon DNA replication rather than EGA[59,60] (Figure 9.3b). In humans, sperm lack three-dimensional genome architecture and TADs, which is largely due to the absence of CTCF, which only becomes expressed following EGA, coinciding with TADs and A/B compartmentation.[61] Unlike in a mouse, TAD establishment in pre-implantation embryos appears to be EGA-dependent, suggesting that the mechanisms of establishing chromatin structures differ between mammalian species.[59–61]

ACTIVATING THE EMBRYONIC GENOME

The first detection of transcription following fertilization occurs in the mouse zygote and is referred to as ZGA. This is unique among mammals as it occurs before the first cleavage division, with genome activation generally initiating in four- to eight-cell embryos and termed EGA (Figure 9.1b). Genome activation in the mouse is divided into the minor ZGA, occurring during the one-cell stage, and the major wave during the two-cell stage. The transcripts generated during the minor wave are rather promiscuous, low-level,

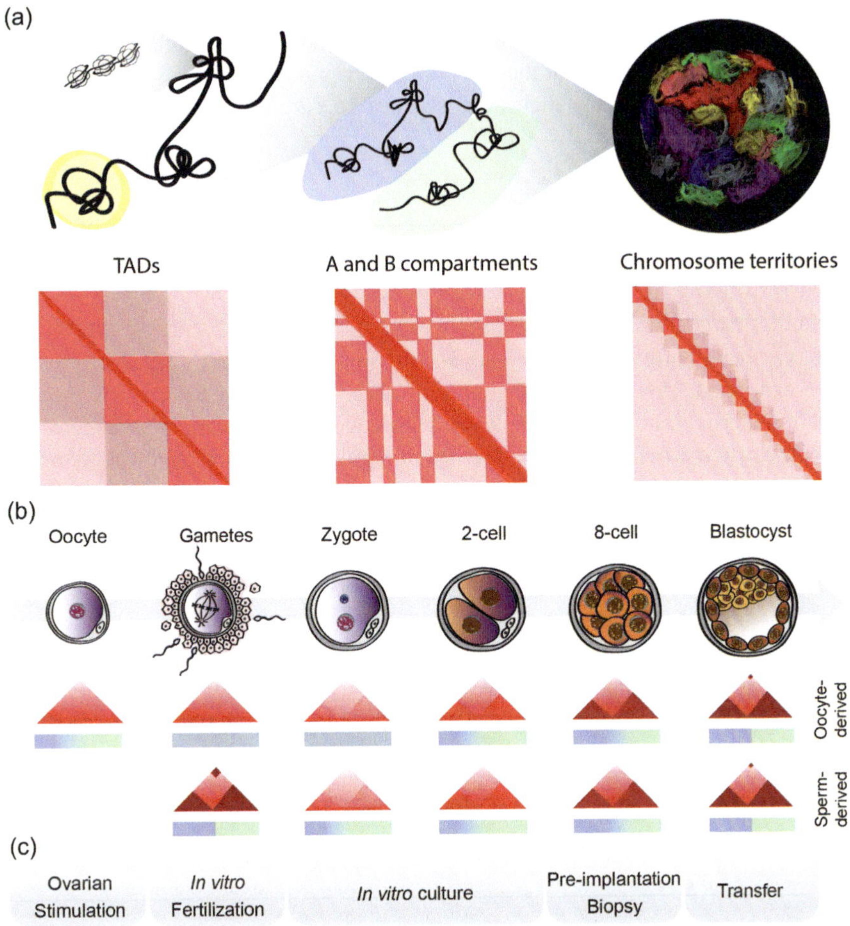

FIGURE 9.3 Higher-order genome organization during mouse maternal-to-embryo transition. (a, from left to right) Chromatin forms long-range interactions between enhancers and promoters. Topologically associated domains (TADs) comprise regions of the genome that frequently organize into self-interacting domains. TADs with similar epigenetic profiles defined by euchromatin and heterochromatin organize into A compartments associated with active transcription, whilst B compartments are enriched with repression regions. Each chromosome occupies distinct territories within the nucleus as revealed by individual chromosome paints. For each progressive structure, representative Hi-C contact maps are shown. (b) Chromatin dynamics throughout mouse pre-implantation development. (c) Comparison to related stages of human development important for assisted reproduction.

and genome-wide, often resulting in transcripts that are inefficiently spliced and polyadenylated.[62] Although the significance of the minor ZGA in the mouse remains unclear, and may reflect opportunistic transcription produced as the genomes are epigenetically reset, some significantly functional transcripts are produced. The D4Z4 repeats are located in the subtelomeric region of chr4q35 with each repeat unit containing the open reading frame for the DUX4 double-homeobox TF that acts as a transcriptional activator in both mice and humans.[63] The D4Z4 repeats are normally highly methylated and enriched with H3K9me3, but during pre-implantation reprogramming, these marks are temporally depleted allowing for a burst of *DUX4* expression.[64,65] Chromatin immunoprecipitation experiments (ChIP-Seq) indicate that DUX preferentially binds to genes and endogenous retroviral elements (ERVs) that are specific for early stage embryos, with binding profiles overlapping intervals with chromatin accessibility.[64] However, it is currently unknown if DUX acts as a pioneer factor, binding to heterochromatin and subsequently making it accessible, possibly through nucleosome remodeling and deposition of acetylation on histones.[66] The

DUX4 TF is not solely responsible for genome activation and other yet-to-be-identified pioneer factors/ TFs must be involved. These may be maternally derived mRNAs that might be translationally upregulated, whereas others may be regulated at the level of transcription. In all cases, these early activators are not present at fertilization, suggesting their activity must be carefully controlled to prevent premature transcriptional activation.

DEPLETION OF OOCYTE-DERIVE TRANSCRIPTS

As discussed above, for an embryo to develop beyond the initial reprogramming phase, the newly formed genome must be expressed. However, concurrent with ZGA/EGA is another equally important process, the degradation of maternal products (Figure 9.1b). These two events are not independent; rather, there is a complex interplay of maternal and zygotic factors in regulating various mechanistic aspects of MZT. In a similar fashion to transcriptional activation, the clearance of the maternally derived transcripts is a coordinated and gradual process during the initial stages of development (reviewed in [67]). Some transcripts are eliminated soon after fertilization, whereas others are degraded only after the major onset of transcription during cleavage development. Maternal mRNA processing is controlled by a variety of different RNA-binding protein complexes, which recognize sequences in the maternal RNAs to promote their degradation via cleavage, deadenylation, and decapping. Recently, novel regulatory mechanisms, such as N^6-methyladenosine RNA modifications,[68] have been implicated as important determinants of RNA stability during the MZT.

FUTURE CLINICAL APPLICATIONS

Major epigenetic changes occur throughout the epigenome during gametogenesis and early embryonic development, and it will be no surprise that occasionally these complex processes go awry for genetic or stochastic environmental reasons. Furthermore, it is entirely possible that some assisted reproductive techniques (ART) may inadvertently alter the gamete and/or embryonal epigenome and potentially generate adverse medical consequences in the offspring. Epigenetic mechanisms in the gametes and their resulting profiles remain largely hidden from view, hampering gamete preselection. It has been shown that individuals with fertility issues can have gametes with aberrant methylation profiles, including at imprinted loci, that may be evident in the offspring.[69] Similarly, it is well documented that specific genomic intervals are refractory to protamine exchange during male meiosis and retain histones. Given that genes escaping protamination are likely to be expressed in early embryonic development,[6] it is entirely possible that aberrant retention may directly influence genome activation and embryo viability. Recent studies have shown that decreased expression of factors involved in maternal RNA clearance, resulting in the abnormal accumulation of maternal transcripts, is observed in the developmentally arrested cleavage embryos of patients seeking ART.[70] These issues could be addressed in the future by the utilizing of low-input and single-cell technologies, allowing for embryonic screening following blastomere or TE biopsies and the selective transfer of embryos showing appropriate genome activation, lineage commitment, and a chromosome complement.

The development of proof-of-concept techniques has allowed for the simultaneous analysis of DNA and polyadenylated-captured RNA from single cells. By utilizing genome and transcriptome sequencing (G&T-seq),[71] both embryonic expression profiling and concurrent genome sequencing could be employed to track appropriate genome activation, underlying mutation status, and copy number variations. The G&T-seq protocol has recently been adapted so the isolated genomic DNA can be subject to bisulphite conversion and used for methylation profiling (single-cell methylation and transcriptome sequencing; scM&T-seq).[72] While these methodologies are in their infancy, future applications may allow for monitoring of EGA and the completion of epigenetic reprogramming. Furthermore, by exploiting additional single-cell multi-omics technologies, such as parallel nucleosome, methylation, and transcriptome sequencing (scNMT-seq),[73] the fundamental processes of genome activation, DNA methylation, and chromatin accessibility can be monitored ensuring the completion of MZT. Currently, this type of methodology is not commercially available and is restricted to a few laboratories worldwide, but the use of such state-of-the-art methodologies could assist in prenatal diagnosis and ART. Ultimately, the future testing of oocytes, via polar body biopsies, or embryonic cells by single-cell sequencing could result in the selection of healthy embryos for transfer and reduce incidence of congenital genetic diseases.

ACKNOWLEDGMENTS

The Monk lab is funded by the H2020 HUTER project from the European Union's Horizon 2020 research and innovation program under grant agreement No 874867. Dagnė Daškevičiūtė is a recipient of a BBSRC NRP DTP-PhD studentship and Marta Sanchez-Delgado received an EMBO Postdoctoral Fellowship (ALTF-934-2019).

REFERENCES

1. Okano M, et al. *Cell*. 1999;99(3):247–57.
2. Kaneda M, et al. *Nature*. 2004;429(6994):900–3.
3. Lister R, et al. *Nature*. 2009;462(7271):315–22.
4. Shirane K, et al. *PLoS Genet*. 2013;9(4):e1003439.
5. Monk D. *Int J Biochem Cell Biol*. 2015;67:128–38.
6. Hammoud SS, et al. *Nature*. 2009;460(7254):473–8.
7. Kouzarides T. *Cell*. 2007;128(4):693–705.
8. Nicetto D, et al. *Curr Opin Genet Dev*. 2019;55:1–10.
9. Mikkelsen TS, et al. *Nature*. 2007;448(7153):553–60.
10. Sanchez-Delgado M, et al. *PLoS Genet*. 2016;12(11):e1006427.
11. Smith ZD, et al. *Nature*. 2012;484(7394):339–44.
12. Argelaguet R, et al. *Nature*. 2019;576(7787):487–91.
13. Guo H, et al. *Nature*. 2014;511(7511):606–10.
14. Mayer W, et al. *Nature*. 2000;403(6769):501–2.
15. Okae H, et al. *PLoS Genet*. 2014;10(12):e1004868.
16. Ito S, et al. *Nature*. 2010;466(7310):1129–33.
17. Song C-X, et al. *Trends Biochem Sci*. 2013;38(10):480–4.
18. Wossidlo M, et al. *Nat Commun*. 2011;2:241.
19. Gu T-P, et al. *Nature*. 2011;477(7366):606–10.
20. Inoue A, et al. *Science*. 2011;334(6053):194.
21. Shen L, et al. *Cell*. 2013;153(3):692–706.
22. Nakatani T, et al. *EMBO Rep*. 2015;16(5):582–9.
23. Nakamura T, et al. *Nature*. 2012;486(7403):415–9.
24. Wang Q-Q, et al. *Reprod Fertil Dev*. 2019;31(2):347–56.
25. Howell CY, et al. *Cell*. 2001;104(6):829–38.
26. Li Y, et al. *Nature*. 2018;564(7734):136–40.
27. Matsuzaki H, et al. *Development*. 2015;142(22):3833–44.
28. Monk D, et al. *Nat Rev Genet*. 2019;20(4):235–48.

29. Smallwood SA, et al. *Nat Genet.* 2011;43(8):811–4.
30. Hanna CW, et al. *Genome Res.* 2016;26(6):756–67.
31. Li X, et al. *Dev Cell.* 2008;15(4):547–57.
32. Quenneville S, et al. *Mol Cell.* 2011;44(3):361–72.
33. Takahashi N, et al. *Genes Dev.* 2019;33(1–2):49–54.
34. Monteagudo-Sánchez A, et al. *Nucleic Acids Res.* 2020;48(20):11394–407.
35. Imbeault M, et al. *Nature.* 2017;543(7646):550–4.
36. Schultz DC, et al. *Genes Dev.* 2002;16(8):919–32.
37. Takahashi N, et al. *Cold Spring Harb Symp Quant Biol.* 2015;80:177–87.
38. Nakamura T, et al. *Nat Cell Biol.* 2007;9(1):64–71.
39. Santos F, et al. *Reproduction.* 2004;127(6):643–51.
40. Zhang B, et al. *Nature.* 2016;537(7621):553–7.
41. Meers MP, et al. *Elife.* 2019;8:e46314.
42. Kaya-Okur HS, et al. *Nat Commun.* 2019;10(1):1930.
43. Xia W, et al. *Science.* 2019;365(6451):353–60.
44. Dahl JA, et al. *Nature.* 2016;537(7621):548–52.
45. Liu X, et al. *Nature.* 2016;537(7621):558–62.
46. Shen J, et al. *Cell Res.* 2015;25(1):143–7.
47. Flanagan JF, et al. *J Mol Biol.* 2007;369(2):334–42.
48. Nishioka K, et al. *Genes Dev.* 2002;16(4):479–89.
49. Bernstein BE, et al. *Cell.* 2006;125(2):315–26.
50. Inoue A, et al. *Nature.* 2017;547(7664):419–24.
51. Okae H, et al. *Hum Mol Genet.* 2012;21(3):548–58.
52. Zhang W, et al. *Genes Dev.* 2019;33(7–8):382–7.
53. Wang C, et al. *Nat Cell Biol.* 2018;20(5):620–31.
54. Ward WS, et al. *Biol Reprod.* 1991;44(4):569–74.
55. Svoboda P. *Semin Cell Dev Biol.* 2018;84:118–26.
56. Liu L, et al. *Nat Commun.* 2019;10(1):364.
57. Wu J, et al. *Nature.* 2016;534(7609):652–7.
58. Du Z, et al. *Nature.* 2017;547(7662):232–5.
59. Ke Y, et al. *Cell.* 2017;170(2):367–381.e20.
60. Flyamer IM, et al. *Nature.* 2017;544(7648):110–4.
61. Chen X, et al. *Nature.* 2019;576(7786):306–10.
62. Yamamoto R, et al. *J Reprod Dev.* 2016;62(1):87–92.
63. Hewitt JE, et al. *Hum Mol Genet.* 1994;3(8):1287–95.
64. Hendrickson PG, et al. *Nat Genet.* 2017;49(6):925–34.
65. De Iaco A, et al. *Nat Genet.* 2017;49(6):941–5.
66. Choi SH, et al. *Nucleic Acids Res.* 2016;44(11):5161–73.
67. Walser CB, et al. *Curr Opin Genet Dev.* 2011;21(4):431–43.
68. Ivanova I, et al. *The RNA. Mol Cell.* 2017;67(6):1059–1067.e4.
69. Choux C, et al. *Epigenetics.* 2020;15(12):1386–95.
70. Sha Q-Q, et al. *Nat Commun.* 2020;11(1):4917.
71. Macaulay IC, et al. *Nat Methods.* 2015;12(6):519–22.
72. Angermueller C, et al. *Nat Methods.* 2016;13(3):229–32.
73. Clark SJ, et al. *Nat Commun.* 2018;9(1):781.

Embryo Kinetics and Aneuploidy

10

Fernando Meseguer, Noelia Ramírez, and Marcos Meseguer

INTRODUCTION

In this chapter we are going to summarize the most important studies related to the importance of morphokinetics in the *in vitro* fertilization laboratory. First, we are going to look at the evolution of embryo selection techniques with their advantages and disadvantages. Second, we will describe the importance of morphokinetic parameters and predictive algorithms in the improvement of reproductive outcomes. Next, we will define aneuploidy and explain the techniques available to detect aneuploidies and the importance of diagnosis. Subsequently, we will debate the most relevant morphokinetic parameters and algorithms capable of predicting the embryo's ploidy. Finally, we describe the limitations that prevent patients from accessing the technique and the future perspectives in the field of morphokinetics.

MORPHOLOGY VS. MORPHOKINETICS

Louise Brown's birth in 1978 was the result of years of hard work and research in the field of reproductive medicine and obstetrics (1). Since then, improvements in the field have resulted in an increase in the number of babies born by assisted reproduction techniques (ART) (2). However, despite these improvements, high rates of failed treatments keep happening, making clear the need to continue improving assisted reproduction techniques (3). These failures rates can be reduced by transferring more than one embryo, but increasing the probability of having multiple pregnancies with all the risks that this entails (4). Due to the improvements in evaluation and embryo selection, the policy of single embryo transfer (SET) has been achieved, without seeing any decrease in clinical results (5).

Traditionally, embryo selection in the vast majority of clinics is carried out taking into account only morphological parameters, despite lacking reliable criteria and being a rather subjective methodology (6). This selection assigns a quality grade to each embryo based on the microscopic evaluation of its morphology. To evaluate embryos on day 3 the following parameters are observed: the number of cells, fragmentation, cell symmetry, multinucleation, and the presence of other alterations such as an abnormal zona pellucida, vacuoles, or any other type of cytoplasmic abnormality. In the evaluation of the blastocyst

DOI: 10.1201/9781003024941-10

on day 5, the parameters are the degree of expansion, the number, cohesion, and homogeneity of the trophoectoderm (TE) cells, as well as the cohesion and area of the inner cell mass (ICM). The embryos of higher quality are those that present an expanded blastocyst, high number of cohesive and homogeneous TE cells, and compact ICM with an area between 1990 and 3800 μm^2 (7, 8).

The current trend of IVF laboratories is to transfer as few embryos as possible to minimize the risk of having a multiple pregnancy; for this reason it is crucial to select the optimal embryo (9). Although transfer on day 5 based solely on the blastocyst morphology brings certain improvements compared to embryo transfer on day 3, it remains a subjective methodology. It has been shown that there is no strong correlation between morphology and embryo competence due to the high degree of inter- and intra-observer variability. This is because there are no sufficiently detailed definitions of the different degrees of blastocysts and the controversy about establishing exact times for their observations (2). Furthermore, the embryos are exposed to uncontrolled atmospheric conditions during microscopic observation (10). In an attempt to reduce the above limitations, time-lapse (TL) was born as a technique capable of carrying out a dynamic and objective study of embryonic development, by the analysis of morphokinetic parameters (11). The first time-lapse system was created in 1997; since then, many companies have added new time-lapse systems to the market (12). Nowadays, there are different types of time-lapse systems and each one has different technical characteristics. However, the most widely used time-lapse systems are EmbryoScope and Geri with their respective plus versions (Table 10.1).

The first step in introducing TL systems into the clinic is to validate it as a safe incubator. There are many studies confirming the safety of TL systems, in that no differences have been found in the development of embryos cultured with this technology (14–19).

One of the great advantages of using time-lapse as an incubator is the ability to observe morphological phenomena in a dynamic way. These dynamic phenomena cannot be detected by static observation, while with conventional incubators we would miss too much information about embryo development such as irregular divisions, collapse and re-expansion of the blastocyst, pronuclei fading, and the time of starting blastulation (20). In addition, this technology has significantly increased knowledge about assisted reproduction (11).

Since the introduction of time-lapse technology in the clinic, a series of algorithms based on morphological and morphokinetic parameters have been developed. Over the years, these algorithms have evolved from the prediction of blastocyst arrival, essential for embryo selection and transfer on day 3, to the prediction of live birth (LB) and embryo ploidy. Different authors have described a series of algorithms with the ability to predict implantation based on morphology and embryo division times. The quantitative parameters with the greatest predictive capacity are the duration of the second cell cycle (cc2) or the division time at three cells (t3)—the division time at 2 cells (t2) (cc2 = t3-t2); and the division time to a 5-cell or fourth division (t5) (21–26); followed by the second synchrony (S2) or the 4-cell division time—the 3-cell division time (S2 = t4 – t3) (21, 22, 24); and the time of pronuclei fading (tPNf) (25–29). However, the predictive parameters considered by these algorithms are focused only on the morphokinetic variables of the early stages of embryo development.

Other authors have focused on creating algorithms capable of predicting LB, adding morphokinetic parameters of late stages of embryonic development. In this context, Fishel et al. carried out a remarkable study with a cohort of 1810 transferred blastocysts. They demonstrated an increase in LB with the use of an embryonic selection algorithm instead of the conventional morphological selection. This algorithm takes into account morphokinetic parameters of early stages of development such as t2, cc2, and t2 / t5-t2, and late-stage parameters such as the time from insemination to start of blastulation (tSB) and the duration of blastulation (dB = tB [time to blastulation]—tSB) (30–32). Other predictive morphokinetic parameters of the live birth are the disappearance time of two cells and the time point when the horizontal diameter of the expanded blastocyst reached 160 μm (texpB2) (33). To these parameters Rienzi et al. added the time of arrival at morula (tM) and the quality of the trophectoderm (34).

Notwithstanding the capacity of the aforementioned algorithms to predict clinical outcomes, they do not take into account the risk of embryo aneuploidies.

TABLE 10.1 Technical Characteristics of the Time-Lapse Systems Available on the Market (13)

TECHNICAL CHARACTERISTICS	*PRIMO VISION VO+*	*MIRI TL*	*EEVA*	*EMBRYOSCOPE D*	*EMBRYOSCOPE +*	*GERI / GERI +*
	Primo Vision VO– 	**MIRI TL** 	**EEVA** 	**Embryoscope D** 	**Embryoscope +** 	**Geri / Geri +** 
Optics	Bright Field	Bright Field	Dark Field	Bright Field	Bright Field	Bright Field / Dark Field
Integrated Incubator	No	Yes / Benchtop	No	Yes hybrid	Yes	Yes / Benchtop
Focal Planes	3 – 11	4	1	7	11	11
Embryo Imaging	Pictures by user (10′ – 60′)	Picture every 5′	Picture every 5′	Picture every 10′	Picture every 10′	Picture every 10′
Capacity Patients	1 x Microscope	6 x System	1 x Camera	6 x System	15 x system	6 x System
Embryos / PT & Culture	9–16 / Individual Culture	14 / Individual Culture	12 / Group Culture	12 / Individual Culture	16 / Individual Culture	16 / Individual Culture
Data Analyses	Manual	Semi-Automated	Real-Time Automated	Manual / Automated	Manual / Automated	Manual / Semi-Automated

ANEUPLOIDIES AND DIAGNOSIS: PREIMPLANTATIONAL GENETIC TESTING (PGT-A)

Aneuploidy is defined as an abnormal number of chromosomes in a cell, and is the most common chromosome abnormality in humans (5% of all clinically recognized pregnancies) (35, 36). In the preimplantation embryo, aneuploidies originate in meiotic and mitotic chromosome segregation errors, but the underlying cellular mechanisms still remain unclear (37, 38). These genetic aberrations are linked to reduced fertility, since they compromise embryonic development, reduce the implantation rate, and increase the risk of pregnancy loss (39).

There are many risk factors that contribute to embryo aneuploidies, but the most established is advanced maternal age (AMA) (40, 41). Specifically, aneuploidy rate rises from a 30% baseline in 30-year-old women up to 90% in mothers older than 40, prior to the menopause (42). Studies relate this dramatic increase with the defects of molecular and cellular processes that aged oocytes experience, like mitochondrial dysfunction, shortening of telomeres, cohesin dysfunction, and meiotic spindle abnormalities (43).

In addition, it is important to consider that aneuploidies are the main genetic cause of congenital birth defects (40, 44). The majority of embryos with aneuploidies are not viable, but some aneuploidies are compatible with live birth, causing chromosomal disorders. The most frequently detected aneuploidies involve the copy number change of an entire chromosome, e.g., trisomy 13, trisomy 18, trisomy 21, sex-chromosome trisomies (47XXX, 47XXY, 47XYY), and X monosomy (44–46). These chromosomal aneuploidies cause syndromes that affect seriously the children's health and their quality of life (MedGen UIDs: 56261, 56262, 4385, 113140, 4403, 473794, 21734). To avoid transfer of embryos with aneuploidies and their consequences, preimplantation genetic testing for aneuploidy (PGT-A) has been proposed.

PGT was introduced at the beginning of the 1990s; since then, the techniques to conduct genetic analysis have been constantly evolving (47–49). At first, this test began with the polymerase chain reaction (PCR) method, which was replaced by interphase fluorescence *in situ* hybridization (FISH) a few years later. At present, genome-wide technologies have emerged as new candidates, mainly in PGT-A where next-generation sequencing (NGS) is replacing the gold standard methods of FISH and PCR. Continuing with PGT-A, the guidelines recommend the following indications (50):

- Advanced maternal age (AMA).
- Recurrent implantation failure (RIF).
- Severe male factor (SMF).
- Recurrent miscarriage (RM): a recent evidence-based guideline (51) describes that couples with a history of RM have a high chance of conceiving naturally; for this reason they do not recommend PGT-A for RM without a genetic cause.

In those patients the aims of the PGT-A are to increase pregnancy rates per embryo transfer and decrease miscarriage rates (52, 53), as well as to get a healthy child into the home without chromosomal disorders. Moreover, this test also purports to increase elective single embryo transfer and reduce time to pregnancy (54).

These days, PGT-A has been established worldwide as a clinical service. For patients with more than one embryo, IVF with PGT-A shortens the overall time spent in treatment, and reduces failed embryo transfer and clinical miscarriage when compared to IVF alone (55–57). Over the past decade, there have been substantial advancements in technology which have overcome the main limitations of PGT-A. Next-generation sequencing conducts a 24-chromosome testing with high hybridization efficiency and improves the detection of mosaicism when combined with TE biopsy (58). PGT-A is still an invasive test whose cost-effectiveness is questioned, and the number of biopsied embryos depends on the ART cycle. To overcome these limitations and enhance embryo selection, studies evaluate the combination of multiple approaches. Morphokinetic parameters together with PGT-A provide a promising strategy to improve embryo screening (59).

Finally, there are many papers describing good practice recommendations for PGT based on the experience and the scientific evidence (54, 55, 60–67). The main goal of these consortia is to standardize the different technical aspects of embryo biopsy and PGT, including recommendations on the organization of PGT and the basic requirements of an IVF/PGT center. However, PGT requests should be compliant with national legislation and local ethics, so these recommendations could vary. There are several ethical issues concerning the decision to undergo this invasive technique (68–70). Religion also plays a decisive role, because some religious convictions consider PGT as an antenatal option (71, 72). Furthermore, in some countries PGT is expressly forbidden by local laws. This is the case of Germany, where the Embryo Protection Law of 1990 mentions some points that might interfere with PGT (73). This legislation notes that PGT cycles do not transfer all the embryos that are generated *in vitro*, and considers each biopsied cell as a complete embryo capable of developing into a new individual. In these cases, where PGT is not allowed for legal, ethical, or religious reasons, the utilization of time-lapse imaging to predict aneuploidies could be an interesting choice.

MORPHOKINETICS AND ANEUPLOIDY

Time-lapse systems can overcome the disadvantages of PGT-A using morphokinetic parameters to predict the embryo ploidy. One of the first studies focused on the relationship between embryo ploidy and embryonic development times was carried out by Chávez's research group (74). In their study, they observed that euploid embryos had precise division times from the first cell division up to the 4-cell stage, while only 30% of the aneuploid embryos had division times within the normal time windows. It has been shown that most embryos with abnormal cell division times show chromosomal abnormalities (21,75–77). The current literature details morphokinetic parameters in all stages of embryonic development and their relation with embryo ploidy.

Pronuclei

The first embryonic development event that occurs and can be evaluated in IVF treatments is the appearance of the pronuclei. A human oocyte successfully fertilized contains two pronuclei and two polar bodies (78). It is widely accepted that mono- and tri-pronuclear embryos can also continue development to the blastocyst stage, but they are likely to be aneuploid. The time of pronuclei fading (tPNf) is an early morphokinetic parameter which can also predict clinical outcomes. While some authors did not find direct association between tPNf and aneuploidies assessed by TE biopsy and 24 chromosome-microarrays (79), others showed that the mean duration of pronuclei fading (tPNf) for normal embryos was significantly shorter than in abnormal ones (80–82). Other studies showed that the morphokinetic parameters most related to ploidy were the pronuclei duration (dPN = tPNf – tPNa) and the onset of the first cytokinesis, which were longer in aneuploid embryos compared to euploid embryos (83, 84). Nonetheless, time of pronuclei appearance and time for second polar body extrusion were not significantly different between normal and abnormal embryos (80).

The Morphokinetic Parameters of Early Stages of Embryonic Development

Initially, the early stages of embryonic development were used to detect the most relevant morphokinetic parameters. This is because exact timing of later divisional events becomes increasingly difficult to ascertain both manually and through image analysis, since it is difficult to keep track of the cells that have already been counted. Additionally, cells tend to cover each other and it is therefore necessary to keep track of the cells in several focal planes. Finally, the more objects (cells and fragmentation) there are in the image, the harder it is to count the cells and to determine if a division occurred. For these reasons, divisions to more than six cells become increasingly difficult to detect (21).

Some authors did not observe a correlation between early morphokinetic parameters and aneuploidy rates (82, 85–87), while others, using the EmbryoScope and Geri systems, showed that the time between the first and second mitosis, the prolonged duration of the first cytokinesis, and the third shortest synchrony (S3 = t8-t5) were correlated with aneuploidy (74, 88, 89). Moreover, in another study, the times of the early cleavage stage, t5-t2 and cc3, were correlated with embryo ploidy (90). Subsequently, Minasi and colleagues observed that the four-cell stage was reached significantly earlier in euploid embryos compared to aneuploid ones. In addition, the time of the second synchrony (s2) was faster in euploid compared to aneuploid embryos (91) (Table 10.2).

Swain in 2013 conducted a study to review the most relevant researches examining the relationship between embryonic morphokinetics and aneuploidy. Swain concluded that morphokinetic parameters together with morphology can aid in the selection of euploid embryos; however, the accuracy of this approach is far from ideal. Additionally, controversial results were obtained in the literature (79).

The Morphokinetic Parameters of Later Stages of Embryonic Development

As we have seen, early embryo kinetics were able to predict blastocyst development and reproductive outcomes. However, these early morphokinetic parameters did not have the capacity to predict ploidy status with good accuracy and the results were controversial. Nevertheless, later morphologic events and morphokinetic parameters may provide further insight (79).

TABLE 10.2 Early Morphokinetic Parameters Analyzed in Euploid and Aneuploid Human Blastocysts

STAGE	*NORMAL EMBRYOS (N = 197) MEAN DURATION (HOURS) (MEAN ± SD)*	*ABNORMAL EMBRYOS (N = 263) MEAN DURATION (HOURS) (MEAN ± SD)*	*P-VALUE*
tPB2	4.0 ± 2.2	4.5 ± 1.8	P > 0.05
tPNa	10.6 ± 3.4	10.8 ± 3.2	P > 0.05
tPNf	24.5 ± 4.3	25.8 ± 5.6	P < 0.05*
t2	28.3 ± 7.2	30.6 ± 9.7	P < 0.05*
t3	38.7 ± 7.0	39.7 ± 8.5	P > 0.05
t4	40.5 ± 7.2	41.5 ± 8.2	P > 0.05
t5	52.3 ± 8.6	50.1 ± 9.6	P < 0.05*
cc2	10.5 ± 4.2	9.1 ± 4.9	P < 0.05*
cc3	13.5 ± 5.6	10.8 ± 7.0	P < 0.05*
S2	1.8 ± 3.3	2.1 ± 4.0	P > 0.05
t5-t2	23.9 ± 7.4	20.1 ± 8.6	P < 0.05*

Notes: Chawla et al. made a retrospective cohort study (n = 460 biopsied embryos and analyzed by CGH microarrays). They showed that the mean duration of t2 for normal embryos was significantly shorter than that for abnormal embryos. Furthermore, in their study they observed how the mean duration of t5, cc2, cc3, t5-t2 for normal embryos was significantly higher than that of the abnormal embryos (80).

Abbreviations: *Statistically significant; tPB2: appearance of second polar body; tPNa: appearance of pronuclei; tPNf: both pronuclei faded; t2: division to 2 cells; t3: division to 3 cells; t4: division to 4 cells; t5: division to 5 cells; CC2: duration of second cell cycle—from 2 to 3 cells; CC3: duration of the third cell cycle CC3 (t5-t3); S2: second synchrony; t5-t2: difference between t5 and t2.

TABLE 10.3 Morphokinetic Parameters Analyzed in Euploid and Aneuploid Human Blastocysts (91)

KINETIC PARAMETER	*EUPLOID BLASTOCYSTS (N = 294)*			*ANEUPLOID BLASTOCYSTS (N = 634)*			*P-VALUE*
	N	*MEAN*	*95%CI*	*N*	*MEAN*	*95%CI*	
tSB	266	103.4	102.2–104.6	598	105.0	104.0–106.0	0.007
tB	263	110.2	108.8–111.5	593	112.8	111.7–113.9	< 0.001
tEB	222	118.7	117.0–120.5	493	122.1	120.7–123.4	< 0.001
tHB	174	133.2	131.2–135.2	328	137.4	135.7–139.1	< 0.001

Abbreviations: tSB: the time to start blastulation; tB: the time to reach the full blastocyst; tEB: the time to expanded blastocyst; tHB: the time to blastocyst hatching.

Despite this, conflicting results have been published regarding the ability of blastocyst formation times to predict embryonic ploidy. While some studies observed that morphokinetic parameters of the later stages of embryonic development were not able to predict aneuploidy (87, 92, 93), other studies show otherwise. Melzer observed that compaction was faster in aneuploid embryos compared to euploid ones (94). In contrast, these results do not agree with those of Campbell; her research group proved that aneuploid embryos initiated compaction (tSC) later than euploids (82). Another study that relates the morphokinetics of late stages and embryonic ploidy was carried out by Huang's research group. They demonstrated that blastocyst expansion time was directly associated with embryo ploidy status. The averaged euploid blastocyst was significantly faster than in aneuploid blastocysts (95). Furthermore, late time-lapse parameters such as the time to start blastulation (tSB), the time to expanded blastocyst (tB), the time to expanded blastocyst (tEB), and the time to blastocyst hatching (tHB) were different in euploid and aneuploidy embryos but did not improve clinical outcomes (91) (Table 10.3).

In a parallel effort, similar results were observed in the Desai study where they demonstrated that chromosomal status was correlated significantly with tSB, tEB, and the interval tEB-tSB (96). Campbell came to the same conclusions: late morphokinetic parameters (tSC, tSB and tB) increase the probability of selecting euploid embryos (82) (Table 10.4).

TABLE 10.4 Euploidy Rate for Biopsied Embryos in Each Kinetic Group Was Assessed. Chi-Square Analysis Was Performed to Compare Differences (96)

KINETIC PARAMETER	*EUPLOID*	*95% CI*	*P-VALUE*
tSB < 96.2 h	48.1 %	42 – 54	< 0.05
tSB > 96.2 h	37.6 %	33 – 42	
tEB < 116 h	46.8 %	42 – 51	< 0.001
tEB > 116 h	30 %	25 – 37	
tEB – tSB < 13 h	47.7 %	43 – 52	< 0.001
tEB – tSB > 13 h	31.5 %	26 – 37	

Abbreviations: tSB: the time to start blastulation; tB the time to reach the full blastocyst; tEB: the time to expanded blastocyst; tHB: the time to blastocyst hatching.

MODEL RISK AND ALGORITHMS

An algorithm is an ordered and finite set of operations through which we can find the solution to a problem. In our field, the problems are related to infertility such as ART, anatomic dimorphisms, morphokinetics, etc. There are some methodologies to design an algorithm. In the field of embryology, the vast majority of algorithms are based on hierarchical models whose dates are organized into a tree-like structure. In these models, a classification is made by clustering the embryos according to their morphological and morphokinetic parameters; it is then observed if there is a correlation between these parameters and the clinical outcomes (21). However, Desai used another model, called an "addition algorithm," to predict clinical outcomes. The overall morphokinetic score was determined by adding or subtracting points (maximum of 4 and minimum of –2.0):

- Negative points: cc2 = < 5 hours (–1), presence of multinucleation (–0.5), presence of irregular division (–0.5).
- Positive points: t5 = 45.8 – 57.0 Hours Post-Insemination (HPI) (+ 1), s2 = 0.0 – 0.1 hours (+ 1), s3 = 1.4 – 7.0 hours (+ 1), tSB = < 100 HPI (+1).

Initially, in the Desai algorithm embryos were screened by morphology, and then embryos with the highest morphokinetic score were selected for transfer (29).

Generally, the design of the algorithm consists of two phases: the first is the generation of the algorithm, and this phase uses cases in which the embryo selection is based on standard morphological assessment to build a hierarchical or addition classification model. The second phase consists of a standard morphological evaluation plus the relevant morphokinetic criteria for embryo selection. In this phase, the predictive capacity of the model is evaluated (23).

Two hierarchical algorithms based on risk models capable of predicting embryonic ploidy stand out in the bibliography. While one is based on the early stages of embryonic development, the other focuses on the later stages. On one hand, Basile and colleagues elaborated an algorithm which takes into account the interval of t5-t2 and cc3 to divide embryos into four grades (A-D). If the value of t5-t2 was greater than 20.5 hours, the embryo was categorized as A or B. Conversely, if the value of t5-t2 was less than 20.5 hours, the embryo was categorized as C or D. In parallel, if the value of cc3 was between 11 or 18 hours, the embryo was categorized as A or C depending on t5-t2. Likewise, if the value of cc3 was not between 11 or 18 hours, the embryo was categorized as B or D depending on t5-t2 (90) (Figure 10.1).

Therefore, embryos could be found in one group or another depending on their morphokinetics. They observed a significant decrease in the percentage of normal chromosomal status between the different categories (A = 35.90%, B = 26.40%, C = 12.10%, and D = 9.80%) (90) (Figure 10.2).

On the other hand, in Campbell's work, a correlation was observed between the morphokinetics of later embryonic stages (tSB and tB) and their chromosomal status. Considering these parameters, they used 97 biopsied embryos to develop a risk classification model. The model classified embryos into three classes of aneuploidy risk: low, medium, or high risk. First, if the value of tB was less than 122.9 hours, the embryo was categorized as low or medium risk. Contrariwise, if the value of tB was greater than 122.9 hours, the embryo was categorized as high risk. Second, if the value of tSB was less than 96.2 hours, the embryo was categorized as low risk. Hence, if the value of tSB was greater than 96.2 hours, the embryo was categorized as medium risk (82).

On the basis of Table 10.5, we can develop a hierarchical algorithm to classify embryos into four groups depending on their morphokinetic parameters (Figure 10.3).

Shortly thereafter, the same research group used their aneuploidy risk model in order to evaluate the effectiveness and potential impact of this model for unselected IVF patients without biopsy and PGT-A. The clinical results, fetal heartbeat (FHB), and live birth (LB), were calculated for each risk class. In a cohort of 88 embryos, a significant increase in the percentage of FHB and LB was observed in the low-risk class (72.7% and 61.1%) compared to the medium-risk class (25.5 and 19.2%), respectively. In the

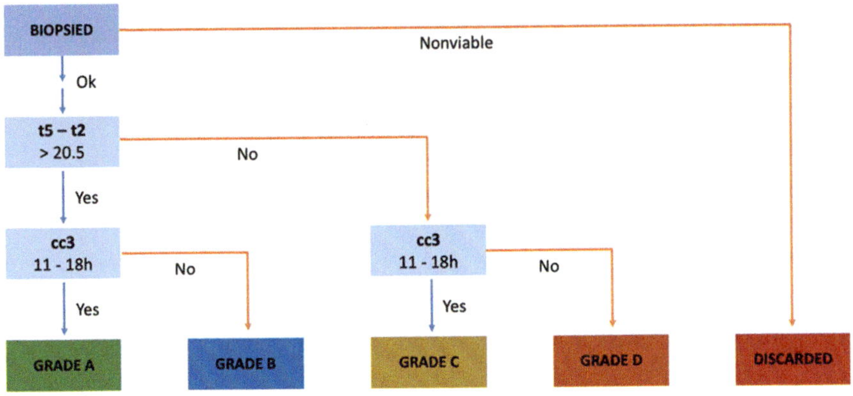

FIGURE 10.1 Hierarchical classification of embryos based on the embryos available for biopsy, interval t5-t2, and the duration of the third cell cycle (cc3). The algorithm classified embryos into four categories based on the expected percentage of chromosomally normal embryos (90).

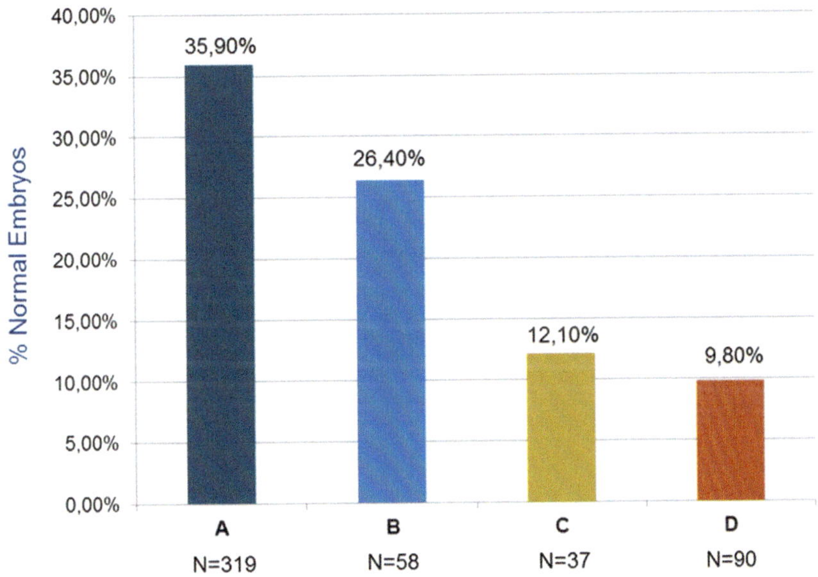

FIGURE 10.2 The correlation between the percentage of normal embryos and the grades of Basile's algorithm.

TABLE 10.5 Aneuploidy Risk Classification with Associated Incidence Rates and Probabilities of Aneuploidy

RISK CLASS	*DEFINITION*	*N*	*INCIDENCE*	*PROBABILITY*
Low	tB < 122.9 and tSB < 96.2	36	0.36	0.37
Medium	tB < 122.9 and tSB > 96.2	49	0.69	0.69
High	tB > 122.9	12	1.00	0.97
All		97	0.61	0.61

Notes: Corrected Akaike information criterion is 296; area under the receiver operating characteristic curve is 0.72; imputes = 1; n = number of embryos; incidence = incidence rate of aneuploidy embryos; probability = probability estimated using recursive partitioning-derived risk classification of an embryo being aneuploid.

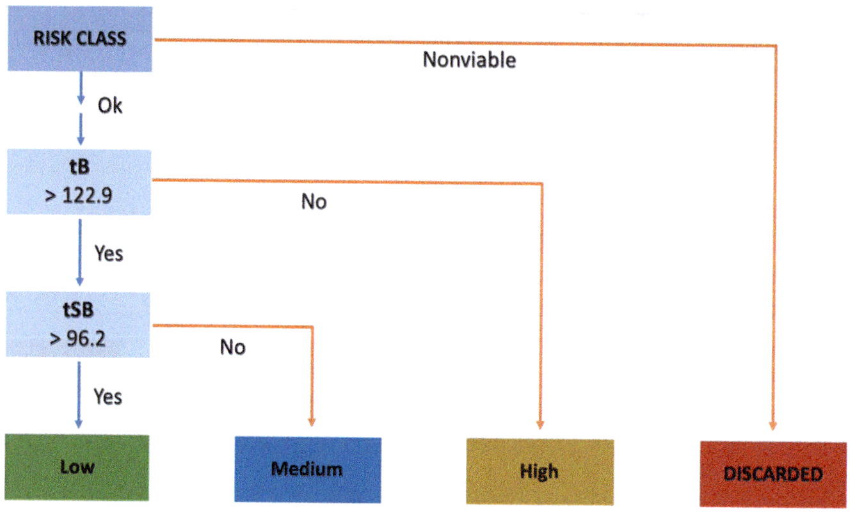

FIGURE 10.3 Hierarchical classification of embryos based on the available embryos, the time to blastulation (tB), and the time to start blastulation (tSB). The algorithm classifies embryos into four categories based on the expected percentage of chromosomally normal embryos (82).

TABLE 10.6 Comparison of the Fetal Heartbeat and Live Birth Rate for Each Aneuploidy Risk Class. All Embryos Used Were KID

	FHB KID		*LB KID*	
RISK CLASS	*Nº OF EMBRYOS*	*FHB KID RATE*	*Nº OF EMBRYOS*	*LB KID RATE*
All	88	42.0	46	39.1
Low	33	72.7**	18	61.1*
Medium	51	25.5**	26	19.2*
High	4	0	2	0
AUC	0.75		0.74	

Notes: LB KID data were calculated only from treatments where the information could have been obtained (over 10 months from time of embryo transfer).

Abbreviations: FHB = fetal heartbeat; KID = known implantation data; LB = live birth; AUC = area under the ROC curve. ** $p < 0.0001$ * $p < 0.01$.

same way, a significant increase was also shown between the medium- and high-risk classes (0% FHB and 0% LB) (81) (Table 10.6).

These studies show how clinical outcomes improve when ploidy is predicted by morphokinetic parameters and these values are taken into account in the embryo selection algorithms (81). Additionally, according to Basile, these morphokinetic parameters could be considered as a specific marker for the validation of an embryo selection algorithm (90).

BARRIERS TO ACCESS

Nevertheless, two subsequent articles observed that time-lapse parameters cannot be used to select euploid blastocyst. Kramer in 2014 and Rienzi in 2015 proved that the Campbell and Basile models failed (92, 93). This is because the database used in both algorithms was obtained from their own clinics. It was amply demonstrated that many variables may be affecting embryo morphokinetics such as type of stimulation, age of the patient, temperature, pH, medium, and culture conditions (97). This fact explains the necessity to modify and validate all previous algorithms using the embryos of your own clinic, since there are no universal algorithms.

However, the company Vitrolife has developed an algorithm called KIDScore D5; it was proposed as a possible universal algorithm (98), because it was developed using a database from 24 different clinics (99). The mathematical details of the algorithm are protected by copyright and have not yet been published (100). Additionally, this algorithm is capable of making annotations automatically, reducing the time and the volume of work, as well as inter- and intra-observer subjectivity because the algorithm evaluates all embryos under the same criteria (2). The KIDScore D5 integrates morphokinetic parameters of all stages of embryonic development (PN, t2, t3, t4, t5, tB, ICM, and TE), being able to find a correlation between its score and the embryo chromosome status (98) (Table 10.7).

This algorithm gives a score between 1 and 10, the optimal embryos being the ones with the highest score. Hence, when the score of the algorithm increases, the probability of being euploid also increases (98) (Figure 10.4).

Although the KIDScore D5 algorithm and the aneuploidy risk models may help to select euploid embryos, the accuracy of this approach at the moment is far from ideal (79). Furthermore, as detailed above, there is great controversy regarding the use of morphokinetic parameters to predict embryonic ploidy (84). For these reasons, morphokinetics do not have the rigor to replace molecular techniques used in PGT-A for identifying euploid embryos. This is partly due to the different laboratory conditions and protocols between clinics, the significant overlap in kinetics between euploid and aneuploid embryos, and the only moderate predictive ability (96).

TABLE 10.7 KIDScore D5 Mean and Standard Deviation in Euploid and Aneuploid Embryos

	KID SCORE DAY 5 VERSION 3		
PLOIDY	*MEAN*	*STANDARD DEVIATION*	*N*
Euploid embryos	**5.25**	**1.87**	**879**
Aneuploid embryos	**4.59**	**1.80**	**1058**

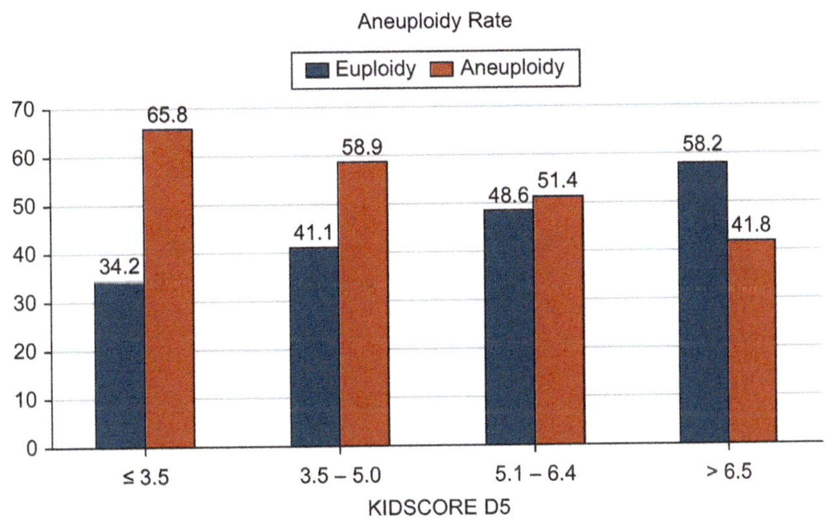

FIGURE 10.4 There is an inverse correlation between the aneuploidy rate and the score of the algorithm KIDScore D5.

Nonetheless, embryo biopsy is an essential procedure for PGT-A which is still invasive and requires highly qualified personnel (80). Hence, morphokinetics could be a useful tool to avoid a biopsy and PGT-A in younger patients or to reduce the number of embryos biopsied (79). Moreover, this is a unique opportunity for patients who cannot access PGT-A (101). Even where legislation allows embryo biopsy, the application of time-lapse algorithms could be a more economical alternative. In this approach all the embryos would be biopsied, but it would only diagnose the embryos in the aneuploid low-risk group (102).

CONCLUSION

Since the introduction of time-lapse systems in the clinic, numerous authors have developed a series of algorithms which take into account morphokinetic parameters to predict clinical outcomes successfully. Part of this chapter has focused on the prediction of embryonic aneuploidies using morphokinetic parameters. While some authors have shown a correlation between morphokinetic parameters and aneuploidy rates, others have not observed this correlation. Campbell and Basile's groups made two aneuploidy model risks based on morphokinetics which could predict aneuploidy. However, when both algorithms were analyzed by Kramer and Rienzi, they concluded that the algorithms were not capable of predicting embryonic ploidy. This is because the databases used by Campbell and Basile for developing their algorithms came from their own clinics; the laboratory conditions could also be affecting the times of embryonic development. Vitrolife's algorithm, KIDScore D5, can overcome this drawback because it is developed using a database from 24 different clinics. Additionally, this algorithm can predict aneuploidy rates. Currently, time-lapse algorithms and morphokinetics do not have enough accuracy to replace PGT-A. Nevertheless, this is an interesting option for people who will not be able to access PGT-A due to ethical or religious convictions and legal or economic restrictions.

FUTURE PERSPECTIVES

Future prospects in the field of embryo selection are directed towards automation. Consequently, the time and volume of laboratory work would be reduced, as well as subjective assessment and inter-intra-observer variability. The most recent projects on embryo selection introduce new tools to the sector such as artificial intelligence. The newest applications of artificial intelligence are machine learning and deep learning. The latter allows the recognition of a single piece of data from an immense set, making small changes in its parameters to get closer to the real or true data (103). While machine learning models use a series of rules to learn, deep learning is capable of learning using the "trial and error" model, where the repetition of a circumstance is responsible for learning. Deep learning is a model inspired by the biological nervous system, processing information through interconnected neurons (104). Hence, an algorithm using deep learning can analyze raw video from a TL system, without having annotated morphokinetic parameters (105, 106). The proposed artificial intelligence methodologies go beyond the study of individual variables and progress towards the study of multiple parameters hidden in the huge data set (big-data) provided by TL technology (107). Artificial intelligence may improve the prediction of embryonic ploidy, and together with time-lapse systems, they constitute a non-invasive methodology capable of replacing/complementing PGT-A, thus overcoming all the problems that this entails.

REFERENCES

1. Steptoe PC, et al. *Lancet.* 1978;312(8085):366.
2. Adolfsson E, et al. *J Bras Reprod Assist.* 2018;22(3):228–37.
3. Aparicio B, et al. *Reprod Biomed Online.* 2013;27(6):654–63.
4. Van Voorhis B, et al. 2017;108:757.
5. Cutting R. *Best Pract Res: Clin Obstet Gynaecol.* 2018;53:30–7.
6. Arav A, et al. *Reprod Biomed Online.* 2008;17(5):669–75.
7. Gardner DK, et al. *Fertil Steril.* 1998;69(1):84–8.
8. ASEBIR. *Criterios ASEBIR de Valoración Morfológica de Oocitos, Embriones Tempranos y Blastocistos Humanos.* 3º. Góbalo, editor. Madrid; 2015.
9. Milewski R, et al. *Reproduction.* 2017;154(2):R37–53.
10. Coticchio G, et al. *Hum Reprod.* 2018;33(1):23–31.
11. Mandawala AA, et al. *Anim Reprod Sci.* 2016;174:2–10.
12. Payne D, et al. *Hum Reprod.* 1997;12(3):532–41.
13. Campbell A, et al. *Atlas of Time Lapse Embryology.* 1st ed. New York: CRC Press; 2015:65–68.
14. Nakahara T, et al. *J Assist Reprod Genet.* 2010;27(2–3):93–6.
15. Cruz M, et al. *J Assist Reprod Genet.* 2011;28(7):569–73.
16. Meseguer M, et al. *Fertil Steril.* 2012;98(6):1481–1489.e10.
17. Kirkegaard K, et al. *Hum Reprod.* 2013;28(10):2643–51.
18. Park H, et al. *Hum Reprod.* 2015;30(2):268–75.
19. Pribenszky C, et al. *Reprod Biomed Online.* 2017;35(5):511–20.
20. Zaninovic N, et al. *Fertil Steril.* 2017;108:722–9.
21. Meseguer M, et al. *Hum Reprod.* 2011;26(10):2658–71.
22. Vermilyea MD, et al. *Reprod Biomed Online.* 2014;29(6):729–36.
23. Basile N, et al. *Hum Reprod.* 2015;30(2):276–83.
24. Goodman LR, et al. *Fertil Steril.* 2016;105(2):275–285.e10.
25. Liu Y, et al. *Fertil Steril.* 2016;105(3):656–662.e1.
26. Petersen BM, et al. *Hum Reprod.* 2016;31(10):2231–44.
27. Azzarello A, et al. *Hum Reprod.* 2012;27(9):2649–57.
28. Chamayou S, et al. *J Assist Reprod Genet.* 2013;30(5):703–10.
29. Desai N, et al. *Fertil Steril.* 2016;106(6):1370–8.
30. Fishel S, et al. *Reprod Biomed Online.* 2017;35(4):407–16.
31. Fishel S, et al. *Reprod Biomed Online.* 2018;37(3):304–13.
32. Fishel S, et al. *Reprod Biomed Online.* 2020;40(1):61–70.
33. Bodri D, et al. *Reprod Biol.* 2018;18(4):355–60.
34. Rienzi L, et al. *Fertil Steril.* 2019;112(6):1080–1093.e1.
35. Zegers-Hochschild F, et al. *Fertil Steril.* 2017;108(3):393–406.
36. Hassold T, et al. *Nat Rev Genet.* 2001;2:280–91.
37. Vázquez-Diez C, et al. *Reproduction.* 2018;155:R63–76.
38. Kubicek D, et al. *Reprod Biomed Online.* 2019;38(3):330–9.
39. Maxwell SM, et al. *Fertil Steril.* 2016;106(6):1414–1419.e5.
40. Hassold T, et al. *Hum Mol Genet.* 2007;16:R203–8.
41. Ubaldi FM, et al. *Front Endocrinol.* 2019;10:94.
42. Franasiak JM, et al. *Fertil Steril.* 2014;101(3).
43. Cimadomo D, et al. *Front Endocrinol (Lausanne).* 2018;9:1.
44. Chen Y, et al. *Hum Mutat.* 2017;38(6):669–77.
45. Griffiths AJ, et al. *An Introduction to Genetic Analysis.* New York: W. H. Freeman; 2000
46. Wellesley D, et al. *Eur J Hum Genet.* 2012;20(5):521–6.
47. Capalbo A, et al. *J Assist Reprod Genet.* 2016;1279–86.
48. Handyside AH. *Reproduction.* 2018;156:F75–9.
49. Niederberger C, et al. *Fertil Steril.* 2018;110:185–324.e5.
50. Harton GL, et al. *Hum Reprod.* 2011;26(1):33–40.
51. Bender Atik R, et al. *Hum Reprod.* 2018;2018(2).
52. Blockeel C, et al. *Reprod Biomed Online.* 2008;17(6):848–54.

53. Coll L, et al. *Zygote*. 2018;26(3):191–8.
54. Carvalho F, et al. *Hum Reprod Open*. 2020;2020(3):1–12.
55. De Rycke M, et al. *Hum Reprod*. 2017;32(10):1974–94.
56. Neal SA, et al. *Fertil Steril*. 2018;110:896–904.
57. Shao YH, et al. *Eur J Obstet Gynecol Reprod Biol*. 2019;235:1–5.
58. García-Pascual CM, et al. *Genes (Basel)*. 2020;11(7):1–10.
59. Penzias A, et al. *Fertil Steril*. 2018;109(3):429–36.
60. Claustres M, et al. *Eur J Hum Genet*. 2014;22(2):160–70.
61. De Los Santos MJ, et al. *Hum Reprod*. 2016;31(4):685–6.
62. Coonen E, et al. *Hum Reprod Open*. 2020;2020(3):1–20.
63. Harton GL, et al. *Hum Reprod*. 2011;26(1):41–6.
64. Practice Committee of the Society for Assisted Reproductive Technology. *Fertil Steril*. 2008;90(5 SUPPL.).
65. Kuliev A. *Reprod Biomed Online*. 2008;16(1):134–47.
66. The Preimplantation Genetic Diagnosis International Society (PGDIS). *Reprod Biomed Online*. 2004;9(4):430–4.
67. Thornhill AR, et al. *Hum Reprod*. 2005;20(1):35–48.
68. Zuckerman S, et al. *J Assist Reprod Genet*. 2020;37(8):1903–12.
69. Harper J, et al. *Hum Reprod*. 2014;29(8):1603–9.
70. Klitzman R. *J Genet Couns*. 2018;27(4):909–19.
71. Winkelman WD, et al. *J Assist Reprod Genet*. 2015;32(5):665–75.
72. David BE, et al. *J Genet Couns*. 2012;21(5):625–30.
73. Ludwig M. *Trend Genet*. 2001;17:473–4.
74. Chavez SL, et al. *Nat Commun*. 2012;3:1–12.
75. Cruz M, et al. *Reprod Biomed Online*. 2012;25(4):371–81.
76. Del Carmen Nogales M, et al. *Fertil Steril*. 2017;107(1):229–235.e2.
77. Barrie A, et al. *Reprod Biomed Online*. 2017;34(5):455–62.
78. Feenan K, et al. *Hum Fertil*. 2006;9(3):157–69.
79. Swain JE. *J Assist Reprod Genet*. 2013;30(8):1081–90.
80. Chawla M, et al. *J Assist Reprod Genet*. 2015;32(1):69–75.
81. Campbell A, et al. *Reprod Biomed Online*. 2013;27(2):140–6.
82. Campbell A, et al. *Reprod Biomed Online*. 2013;26(5):477–85.
83. Vera-Rodriguez M, et al. *Nat Commun*. 2015;6(May):1–14.
84. Pennetta F, et al. *Curr Opin Obstet Gynecol*. 2018;30(3):185–96.
85. Semeniuk L, et al. *Fertil Steril*. 2013;99(3):S6.
86. Basile N, et al. *Fertil Steril*. 2012;98(3):S142.
87. Yang Z, et al. *BMC Med Genomics*. 2014;7(1):1–13.
88. Friedman BE, et al. *Fertil Steril*. 2012;98(3):S38.
89. Bayram A, et al. *Hum Reprod*. 2012;27(suppl 2):103–5.
90. Basile N, et al. *Fertil Steril*. 2014;101(3):699–704.e1.
91. Minasi MG, et al. *Hum Reprod*. 2016;31(10):1–7.
92. Kramer YG, et al. *J Assist Reprod Genet*. 2014;31(9):1231–42.
93. Rienzi L, et al. *Reprod Biomed Online*. 2015;30(1):57–66.
94. Melzer KE, et al. *Fertil Steril*. 2012;98(3):S136.
95. Huang TT, et al. *Reprod Biomed Online*. 2019;39(1):27–39.
96. Desai N, et al. *Fertil Steril*. 2018;109(4):665–74.
97. Zaninovic N, et al. *Fertil Steril*. 2017;108(5):722–9.
98. Bori L, et al. *ESHRE Oral Communications*. 2020.
99. Gazzo E, et al. *J Bras Reprod Assist*. 2020;24(1):55–60.
100. Reignier A, et al. *J Assist Reprod Genet*. 2019;36(11):2279–85.
101. Montag M. *Reprod BioMedicine Online*. 2013;26:528–30.
102. Montag M. *Reprod Biomed Online*. 2013;26(6):528–30.
103. Lecun Y. *Nature*. 2015;521:436–44.
104. Kim KG. *Healthc Inform Res*. 2016;22(4):351.
105. Khosravi P, et al. *NPJ Digit Med*. 2019,2(1):21.
106. Tran D, et al. *Hum Reprod*. 2019;34(6):1011–8.
107. Curchoe CL, et al. *J Assist Reprod Genet*. 2019;36:591–600.

Preimplantation Genetic Testing of Aneuploidies (PGT-A)

11

Daniela N. Bakalova, Darren K. Griffin, Maria E. Póo, and Alan R. Thornhill

INTRODUCTION

Preimplantation genetic testing for aneuploidy (PGT-A) aims to avoid genetic disease in newborns, miscarriage, and increase live-birth rates (LBR) per embryo transfer/treatment cycle. Aneuploidy refers to a phenomenon whereby the total number of chromosomes in an individual deviates from the usual 23 pairs (46 chromosomes). Broadly this can manifest as an additional chromosome (trisomy), a missing chromosome (monosomy), or combinations of the above (e.g., double trisomy, complex aneuploidy). Aneuploidy is frequently observed in human IVF embryos and is the leading cause of early pregnancy loss and IVF failure [1]. Indeed, large datasets from chromosomal screening of human embryos consistently demonstrate that approximately half of the embryos generated during *in vitro* fertilization (IVF) are aneuploid [2]. PGT-A thus can identify and allow selective transfer of embryos with an apparently normal chromosome complement (euploid) in patients undergoing IVF treatment.

The development of PGT-A has been controversial. Initial concerns were primarily due to technological limitations using fluorescence *in situ* hybridization (FISH), where aneuploidy results could be obtained for chromosomes 13, 18, 21, X, and Y (and later 16 and 22) only. Uncertainties were raised about the inherent issue with analyzing a fluorescence probe panel of restricted scope on a single cell from a cleavage-stage embryo or polar body. Despite the expansion of the FISH screening protocol to more chromosomes, detection rates were sub-optimal for routine clinical use. A shift in diagnostic technologies towards array comparative genomic hybridization (aCGH) and next-generation sequencing (NGS) saw an improvement in the data generated by PGT-A trials. These new methodologies offer greater resolution, reliability, and chromosome coverage to provide a more comprehensive analysis that expands beyond traditional whole-chromosome aneuploidy detection in single cells, to include the identification of segmental (or partial) aneuploidies and mosaicism on arguably more representative multi-cellular trophectoderm samples from blastocyst stage embryos. A second concern was the inherent damage to the developmental potential of the embryo as a result of cleavage-stage biopsy—a problem later circumvented by improvements in blastocyst development rates facilitating trophectoderm biopsy.

DOI: 10.1201/9781003024941-11

Whilst PGT-A remains a controversial topic, the fact that aneuploidy is the leading cause of pregnancy loss and IVF failure in humans should at least make it a treatment option worthy of consideration and discussion. PGT-A also has the potential to increase the proportion of single embryo transfers, helping to avoid multiple gestations and their associated clinical challenges [1].

In this chapter, we evaluate the available data for past, present, and future PGT-A methodologies. We review the benefits and limitations of clinical application of this treatment option and consider some of the ethical issues surrounding it.

PGT-A: HISTORY AND PRESENT DAY

Over two decades prior to the first clinical use of PGT in patients undergoing IVF, Gardner and Edwards performed the first sexing of rabbit blastocysts by applying the principle of embryo biopsy set out by Seidel in the early 1950s [3, 4]. They successfully detected the presence of Barr bodies—inactive sex chromatin—and used this to determine the sex of embryos prior to implantation [5].

The first clinical case of PGT was performed in 1990, when identification of embryos diagnosed as female (and hence not at risk of having sex-linked disorders) was performed using polymerase chain reaction (PCR) [6]. Here, Handyside et al. amplified a Y chromosome-specific repeat sequence, to detect (or not) the presence of a Y chromosome in embryos from couples at risk of transmitting adrenoleukodystrophy and X-linked mental retardation. Embryos assumed to be female by virtue of absence of a band on an agarose gel were selectively transferred in an attempt to eliminate transmission of the X-linked conditions. In the first treatment cycles, it led to two confirmed pregnancies with normal female twins. In 1992, Handyside followed up with the first PGT for a monogenic (autosomal recessive [AR]) disorder—cystic fibrosis (again with a PCR approach)—reporting the first healthy birth following PGT-M [7].

Despite the initial success of PGT for sexing using PCR, it was inherently prone to false positive or negative results [8]. Following the first misdiagnosis for embryo sexing resulting from a failed amplification of the Y-specific sequence, the technique was superseded by fluorescence *in situ* hybridization (FISH) [9]. FISH uses fluorescent nucleic acid probes complementary to chromosomal DNA to visualize specific regions of interest using fluorescent microscopy. Using FISH for X and Y chromosome specific probes, it thereby displayed distinct diagnostic patterns for male and female embryos. A year after its first clinical application for sex determination, it was used for the detection of aneuploidy [10].

Only a subset of the total number of chromosomes could be screened simultaneously using FISH due to the small number of available fluorochromes. The method was therefore targeted to chromosomes related to the most common aneuploidies compatible with life: 13, 18, 21, X, and Y [11–13] and later for two chromosomes associated with early pregnancy loss, i.e., 16 and 22. PGT-A using this probe combination reportedly led to a reduction in the incidence of aneuploid syndrome-affected live births; however, whether it resulted in significant improvements in implantation rates has been the topic of discussion and intense debate ever since [13].

While aneuploidy may be more commonly seen in specific chromosomes at miscarriage, at the embryo stage it can affect any chromosome and thus a broader platform is required if PGT-A is to be clinically effective, accepted, and implemented widely. The development of novel technologies in the late 2000s facilitated the transition from FISH-based analysis to comprehensive chromosomal screening (CCS) methodologies. CCS is a catch-all term that includes array comparative genomic hybridization (aCGH), real-time quantitative PCR (RT-qPCR), and next-generation sequencing (NGS). NGS, at the time of writing, is the preferred platform in terms of current usage. In addition to identifying chromosome copy number differences across all chromosomes, these techniques also offer superior resolution and additional karyotype information, such as detection of segmental aneuploidy and embryonic mosaicism [14].

THE EVOLUTION OF EMBRYO BIOPSY STRATEGIES

To date, all PGT procedures require access to DNA from preimplantation embryos or eggs harvested during IVF procedures; an invasive biopsy process was required to remove either polar body (PB), blastomeres, or trophectoderm cells (TE) depending upon the developmental stage and the desired approach relevant for particular conditions. Figure 11.1 summarizes the advantages and limitations of each strategy.

First and Second Polar Body Biopsy

Polar bodies are by-products of meiosis, and their testing represents an indirect approach for genetic analysis of oocyte status. Removal of polar body I (PBI), polar body II (PBII), or both should pose little compromise to subsequent development as they are not required for further embryonic development. Despite the benefit of not removing embryonic cells, the applicability of this strategy is controversial, due to criticisms of high false positive and negative error rates when using this method [15]. Polar bodies are usually retrieved on the day of fertilization from MII oocytes and/or zygotes through mechanical or laser-assisted breach of the zona pellucida [16]. Several studies have stated that this breaching does not have negative effects upon embryo development and quality parameters, such as oocyte lysis and/or activation [16–19]. Several groups have applied the PB methodology for aneuploidy screening using FISH and aCGH, with neonatal outcomes similar to those obtained using cleavage-stage biopsy, suggesting its effective clinical application [20–23], despite disagreement from at least one group [24]. Nevertheless, the clinical application of PGT by PB biopsy is steadily decreasing, perhaps due to its time-consuming nature

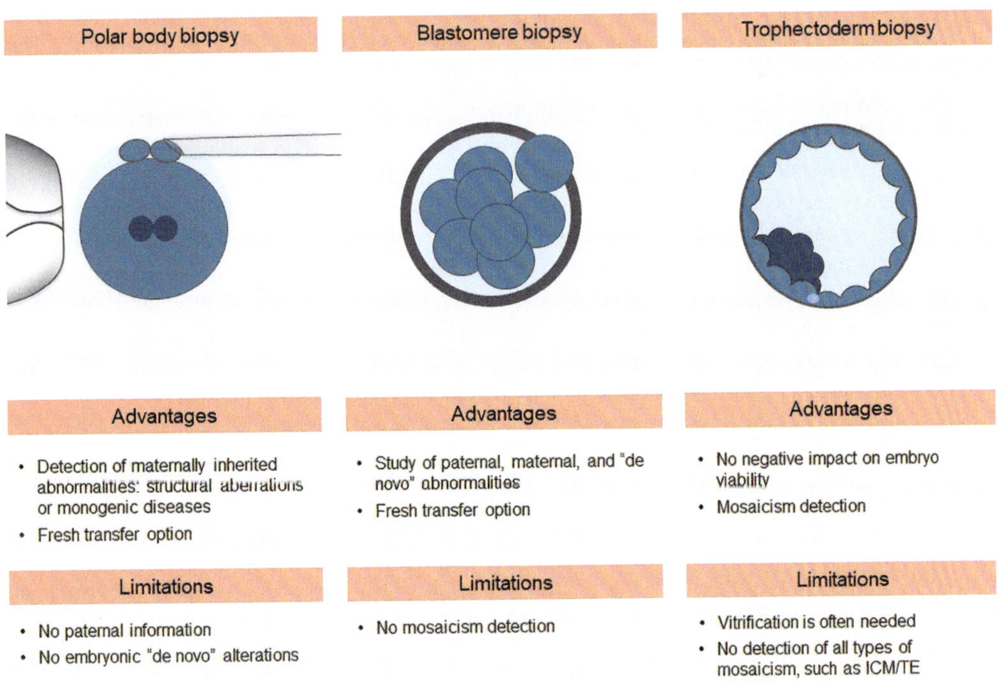

FIGURE 11.1 Overview of biopsy methods and the associated advantages and limitations. (By courtesy of Igenomix.)

(as both PBI and PBII are required for analysis) and the limited nature of the technique diagnostically. The fact remains that since analysis is performed on waste products of female meiosis, aneuploidies of paternal and post-zygotic mitotic origin cannot be detected; this may lead to incorrectly categorized embryos and associated clinical, ethical, and financial consequences. Moreover, the ESTEEM trial highlighted additional shortcomings of the technique, including reduced diagnostic accuracy as a result of frequently degraded DNA and lower predictive value as the technique cannot identify mosaicism in the embryo [24].

Blastomere: Cleavage-Stage Biopsy

Cleavage-stage biopsy involves the aspiration of one blastomere from a day 3 embryo, which has a minimum of six cells (usually six to eight cells) and a low fragmentation level (< 20%). This process is usually facilitated by Ca^{++}/Mg^{++} free media to loosen cell-cell adhesion reversibly [25]. The biopsy can be carried out via laser-assisted, mechanical, or acid Tyrode's drilling of the zona pellucida, with some data suggesting that methods do not negatively impact clinical outcomes [26–29]. Laser-assisted biopsy is most prevalent, perhaps due to standardization of the aperture generated as compared to the variable nature of using acid Tyrode's. However, this is a controversial issue as refuting evidence demonstrates a decrease in sustained implantation rates following blastomere biopsy (from 50% to 30%). This means that a proportion of competent embryos will be sufficiently harmed during the biopsy process to prevent them from progressing into a pregnancy [30]. It is speculated that this may be related to the Ca^{++}/Mg^{++} free media, with several studies in mice demonstrating that Ca^{++} depletion caused cellular cytoskeleton remodeling which in turn had a negative impact of cell compaction [31]. Collectively, these data indicate that the cleavage-stage biopsy procedure may cause significant harm to the embryo.

The blastomere biopsy approach allows identification of aneuploidies of both maternal and paternal origin: an advantage over PB methodologies. Additionally, blastomere biopsy is the preferred approach for fresh embryo transfer (FET) cycles. Day 3 blastomeres enter an arrest phase at the eight-cell stage, during which cell growth, DNA duplication, and protein synthesis take place [32]. This state is maintained for a period long enough to allow preimplantation analysis before progression to the blastocyst stage (day 5) and subsequent embryo transfer.

Blastomere biopsy is, however, affected by limitations related to the single cell nature of analysis, such as a high rate of amplification failure and concerns about incorrectly (non-representative) called results. There is strong evidence that embryos at the cleavage stage carry a higher frequency of chromosome abnormality [33]. Mosaicism, a phenomenon where there are two or more chromosomally different cell lines present in the embryo, is primarily caused by mitotic errors post-fertilization [34] and the earlier an error occurs, the higher proportion of the embryo will have an abnormal karyotype. Therefore, should only one cell be recovered at the cleavage stage, there is a risk that it will not be representative of the entire embryo—this could result in a false positive or negative result. To overcome this, two cells could be retrieved at the day 3 stage; however, this could result in a 25–33% loss of cell content (removal of two cells from an eight-cell or six-cell blastomere, respectively), which could negatively impact developmental potential and clinical outcomes [31].

Some groups and two RCTs have demonstrated positive results following good laboratory practice, including improved live-birth rates as compared to untested day 5 blastocysts [36]. However, as summarized in a meta-analysis comparing IVF with and without PGT-A using blastomere biopsies, data from several groups concluded no overall benefit of PGT-A on live-birth rates [37]. It should be noted that the studies of the meta-analysis employed an outdated FISH methodology, as compared to a more comprehensive aCGH approach used in one of the RCT studies. Again, this supports the theory that the success of blastomere biopsies is governed by good lab techniques.

Trophectoderm: Blastocyst-Stage Biopsy

The first successful clinical use of trophectoderm (TE) biopsy in the clinic was by de Boer et al. in 2004, shortly followed by the first live births in 2005, as published by two separate groups [38, 39].

Since then, TE biopsy has become the most widely used approach for PGT-A, largely replacing both PB and blastomere biopsies. This notable rise is a result of improved culture conditions leading to higher blastulation rates, as well as advances in vitrification and cryopreservation techniques, overcoming reservations due to a narrow window of implantation [25, 38]. Trophectoderm biopsy is performed on day 5, 6, and occasionally day 7 embryos. At this stage, embryos have entered the blastocyst phase and undergone the first cell differentiation into two cell lineages: the inner cell mass (ICM) and the trophectoderm (TE). The ICM will develop into the fetus, whereas the TE will form the placenta, thus collecting cells from the TE should have no direct detrimental effect on embryo development, if performed correctly [40].

TE biopsy is beneficial in several ways. Firstly, blastocysts suitable for biopsy usually contain between 60–100 cells, therefore the removal of 2–10 cells should not pose an adverse effect on embryo viability. Indeed, an RCT conducted by Scott et al. suggested that, unlike blastomere biopsy, TE biopsy does not hinder implantation potential or progressive embryo development [40]. Additionally, the increased number of analyzed cells overcomes technical challenges with regards to amplification failure, allele drop-out (ADO, a phenomenon involving selective allelic amplification), and misdiagnosis as compared to its cleavage-stage counterpart [41]. As previously noted, chromosome abnormality is found at higher levels in the earlier stages of embryo development and is noted less frequently during the blastocyst stage, which should lead to a higher frequency of euploid embryos detected. Moreover, the sample size of cells is larger reducing the likelihood of non-representative results.

In addition, a strict and effective cryopreservation program is a central pillar for PGT-A involving blastocyst biopsy; this can be a limiting factor to this strategy. Some reassurance lies in reports that show consistent live-birth rates and neonatal outcomes between previously vitrified embryos and fresh cycles. For this reason, various studies and RCTs demonstrate the benefits of using PGT-A with trophectoderm samples, with an increase in implantation and delivery rates, as well as a decrease in multiple gestations due to the successful identification of euploid embryos and consequent elective single embryo transfers (eSET) [42, 43].

The prospect of non-invasive PGT-A (niPGT-A) is an exciting opportunity to negate the requirement of sourcing genetic material directly from the embryo. Following recent progress and modifications in both the testing and clinical embryology laboratories, a non-invasive PGT-A approach is fast becoming a very real prospect for routine clinical use and is dealt with in more detail elsewhere in this book.

TESTING METHODOLOGIES

An ideal PGT-A test requires genome-wide and non-targeted screening techniques that are able to detect all chromosomes accurately, reliably, and reproducibly at a reasonable cost in a realistic timeframe. A number of methods allow for the detection of aneuploidies in human embryos—from FISH, which provides limited information for a small number of chromosomes—to aCGH-, qPCR-, NGS-, and SNP-based technologies. The evolution of aneuploidy screening technologies has provided detailed insight into the genetic status of human embryos. Together, these enable the selection of embryos more likely to be euploid for transfer during PGT-A treatment cycles.

Fluorescence *In-Situ* Hybridization (FISH)

FISH is a cytogenetic test that exploits the ability of double-stranded DNA, once denatured, to anneal to complementary sequences by hybridization [44]. Labeled probes targeting a specific chromosomal region are hybridized to the slide-fixed sample and are visualized using fluorescence microscopy. The application of FISH to PGT-A involves retrieving one or more biopsied cells from the embryo. The presence of a fluorescent signal indicates the presence of a chromosome, and a lack of signal is indicative of a missing chromosome but clearly confidence in the accuracy of the approach may be called into question when making a diagnosis on a single cell rather than hundreds (or thousands) of cells for which the technique was developed.

FISH on preimplantation embryos was clinically introduced in the early 90s using X and Y chromosome probes to identify families at risk of transmitting sex-linked disorders, followed by early PGT-A protocols screening for sporadic aneuploidy of common chromosomes associated with live-birth defects. Soon after, chromosomes 16 and 22 were also included, both of which are commonly seen in first trimester pregnancy loss. Preliminary data obtained from these more comprehensive protocols demonstrated a doubling in implantation rates, albeit showing no notable increase in pregnancy rates [45]. However, these must be applied over multiple rounds of hybridization, and assume that the morphology and integrity of the single nucleus sample remain sufficient for accurate and consistent chromosomal analysis. Subsequent hybridization to allow for expanded screening of all 23 pairs of chromosomes proved especially challenging due to overlapping signals, probe failures, and nuclear organization [8]. The combination of these limitations (as well as study-specific deleterious effects of embryo biopsy) provides a convincing argument why randomized control trials (RCTs) conducted in the early 2000s did not demonstrate an improvement in live-birth rates using cleavage-stage biopsy and FISH, and thus led to more general and lasting concerns surrounding the clinical application of PGT-A [44, 45].

Nevertheless, FISH is still used by some laboratories today for the identification of segmental aneuploidies resulting from balanced translocations, specifically those involving small segments near the telomeres of chromosomes, which cannot be detected using other screening methodologies [46].

First-Generation Whole-Genome Amplification Technologies

The development of new technologies has facilitated the transition from FISH analysis of a limited number of chromosomes to the simultaneous screening of all 23 chromosome pairs. The starting material for PGT-A is very limited, approximately 6pg of DNA per cell. This minute quantity is not sufficient for more than a single PCR amplification, thus comprehensive chromosomal screening (CCS) applications require whole-genome amplification (WGA) to increase genome-wide representative levels of DNA suitable for analysis. Commonly used protocols for WGA fit into three categories: PCR-based, multiple displacement amplification (MDA)-based, and hybrid. PCR-based methods include degenerate oligonucleotide primed PCR (DOP-PCR). This method combines DOP primer annealing, expansion, and amplification through two cycling stages to expand DNA content from picogram quantities to micrograms. MDA approaches use high-fidelity bacteriophage DNA polymerase that denatures the double-stranded DNA and amplifies single-stranded templates. An example of a hybrid method is multiple annealing and looping-based amplification cycles (MALBAC); although more labor-intensive, this offers reasonable uniformity across the genome [1].

Comparative Genomic Hybridization: CGH

Among WGA-dependent technologies is comparative genomic hybridization (CGH). This technique uses fluorescently tagged genomic test and reference DNA—each labeled with a different fluorochrome—co-hybridized to slide-fixed metaphase chromosomes. This allows visualization of chromosome copy number across the entire karyotype [22]. CGH was first applied in an IVF setting in 1999, when single blastomeres were successfully analyzed following WGA [47].

Despite the expanded genome coverage compared to FISH, so-called metaphase CGH (mCGH) faces challenges of its own. First, research shows that the technique has limited resolution due to background fluorescence; this leads to the omission of specific chromosomal regions. Second, balanced translocations and other structural rearrangements, such as inversions, cannot be detected using this method since these changes are not associated with whole chromosome gains or losses [48]. Moreover, mCGH is time-consuming and labor-intensive [21]. The demanding nature of this technique means that IVF centers have to cryopreserve the tested embryos in order to allow for the analysis period, thus fresh IVF cycles are not compatible with PGT-A analysis using mCGH. However, most clinics now cryopreserve embryos as standard practice in PGT cycles; therefore this limitation would only be applicable for a limited number of cases. In the early 2000s when mCGH first became clinically applicable, this proved to be a concern for many as freezing protocols were not sufficiently advanced at the time.

Moreover, mCGH is unable to detect translocations (event where a section of a chromosome breaks and reattaches to a different chromosome) and uniparental disomy (a phenomenon where both copies of a chromosome are inherited from the same parent) [1] as the ratio of the test DNA would remain consistent with that of the reference.

Despite these limitations, mCGH provided the first insight into whole chromosome analysis, negating the need for unreliable sequential hybridization using FISH, providing a stepping stone for the development of more advanced CCS methodologies.

Array-CGH

Array CGH (aCGH) technology, as with conventional mCGH, involves the comparative analysis of copy number variations between a fluorescently labeled test (clinical) sample (red) and euploid reference (green) DNAs, following cycles of WGA [1]. Unlike slide-based CGH methodologies, in aCGH, hybridization takes place on bacterial artificial chromosome (BAC) or oligonucleotide microarrays, with chromosome coverage of around 1Mb (megabase: a unit of measure of DNA which is equal to 1 million nucleotides). The results are then visualized and interpreted using software that analyzes the red:green fluorescence ratios at specific chromosomal loci. As with mCGH, chromosomal status is determined based on the relative intensity of the fluorescent signals. If a specific position presents with a higher ratio of red to green, this indicates a gain of chromosomal content in the test (sample) DNA, which is indicative of a trisomy. A higher ratio in favor of the green signal would represent a loss in chromosomal content from the test DNA, indicating a monosomy. Chromosomal status calling is automatically generated by the analysis software by calculating standard deviation ratios of the raw data. In addition to identifying whole chromosomal loss and/or gain, this method can also detect sub-chromosomal structural imbalances and segmental aneuploidies.

Unsurprisingly, aCGH has the same limitations as mCGH regarding translocations, uniparental disomy, and other anomalies where the net loss or gain of chromosomal regions is zero and detection is not possible. Additionally, visualization is binary without a scale to indicate additional gain or loss, meaning that aCGH cannot differentiate between the presence or absence of one or several copies of a chromosome thus limiting its ability to detect triploidy. Despite these limitations, aCGH has been a game-changing technology for the analysis of chromosomes in IVF with hundreds of groups worldwide successfully applying the technique to samples retrieved via polar body, blastomere, and blastocyst biopsies [49]. Removing the need to fix and locate individual nuclei onto microscope slides combined with the introduction of objective, fully automated calling software following automated scanning (rather than microscopy) strongly supports this solution-based approach.

Single Nucleotide Polymorphism Microarrays: SNPs

Single nucleotide polymorphisms (SNPs) are variations in DNA sequence found in various regions of the genome that are highly variable within the human population. As for aCGH, SNP microarray methodology requires DNA hybridization and fluorescence microscopy. However, SNP arrays interrogate specific polymorphisms of the 300 million present across the human genome (with microarray densities typically ranging from 100K to 1M); this provides a genotype for the test sample which is then compared to known maternal and paternal SNP patterns to identify a ploidy status based on predicted heterozygous patterns at specific loci [50]. When compared to classic microarray technologies for PGT-A analysis that can only detect gains and losses at specific chromosomal regions, SNP arrays can simultaneously identify the ploidy status of samples, the parental origin of chromosomal abnormalities, uniparental disomy (UPD), and specific artefactual errors such as allele dropout (ADO).

Drawbacks of the technique are the limit of resolution and the requirement for parental reference samples. There are many possible abnormalities associated with miscarriage and implantation failure that could arise from *de novo* mutations, which are below the resolution of SNP arrays typically used; this could lead to missed diagnoses and adverse pregnancy outcomes. Additionally, it is inherent to the principle of SNP-based analysis that maternal and paternal DNAs are required that can serve as a reference for the tested embryos; this makes the process more time-consuming and expensive. However, SNP arrays have been successfully applied for aneuploidy screening by several groups [51–53] and the technique allows simultaneous PGT-A and PGT-M (preimplantation genetic testing for monogenic disorders) detection.

PGT-A Upgraded: Next-Generation Sequencing

The latest and most frequently applied approach for aneuploidy screening is next-generation sequencing (NGS). This is an umbrella term for recently developed platforms that can generate aneuploidy data from a small number of cells (e.g., from TE biopsies) and are rapid, reliable, accurate, scalable, and thus cost-effective. Most NGS protocols, analogous to aCGH, begin with WGA. Next, a barcoding step is performed with each sample being labeled with a unique identification sequence. Barcoded DNA samples from individual embryos can be pooled together, and processed simultaneously, saving both time and cost. Following sequencing, each sample is aligned to a known reference human genome. Aneuploidies and large deletions and duplications can be identified, and an automated "call" is generated regarding ploidy status at specific chromosomal regions.

The two dominant, commercially available NGS solutions for PGT-A are VeriSeq PGS (SBS – Illumina) and Reproseq (Ion Torrent – Thermo Fisher), each using a different method for base calling according to the core sequencing technology. During the sequencing stage, the Illumina "sequencing by synthesis" (SBS) method platform enriches the DNA template with fluorescently labeled nucleotides whose presence/absence is determined by optical detection using specialized cameras. Alternatively, the Ion Torrent platform utilizes a pH-based approach: nucleotide incorporation to the template DNA causes a change in pH of the surrounding solution via the release of protons, and the relative change is measured against the number of incorporated nucleotides to determine a call. Despite the different sequencing approaches the output data is remarkably similar.

NGS has shown an improvement in clinical outcomes compared to aCGH and has also been validated for the detection of mosaicism at lower levels (~10% limit of detection) than previous technologies [54]. Indeed, a recent study validated the use of a semi-automated NGS protocol for the detection of mosaic aneuploidies at different levels of mosaicism (0%, 40%, 60%, 100%), with 100% informativity (18/18) and 94.4% concordance (17/18) with manipulated cell line models that mimicked various degrees of mosaicism [55]. The same group also validated the protocol for uniform whole aneuploidies with 100% informativity (96/96) and 100% concordance (96/96).

There are several lines of evidence that indicate reduced miscarriage rate, time to pregnancy, and improved implantation rate per transfer when applying an NGS-based approach for PGT-A using TE biopsies [56]. Moreover, NGS can simultaneously detect monogenic disorders and be applied to the analysis of embryonic cell free DNA (cfDNA) in spent culture media samples [57, 58]. To date, NGS is considered to be the preferred PGT-A technology with the most positive IVF outcomes.

INDICATIONS FOR PGT-A: WHAT DO THE DATA SAY?

Advanced Maternal Age

Advanced maternal age (AMA) remains the most common referral category for PGT-A [59]. According to SART (Society for Assisted Reproductive Technology) data, the proportion of PGT cycles as a total of ART treatments increased annually in the USA from 13% to 27% between 2014 and 2016, with the majority of these performed for PGT-A [60]. In contrast, HFEA (Human Fertilisation & Embryology Authority) data for the same time period shows that 2% of all IVF cycles in the UK were PGT-A treatment cycles, with 64% of these being in women of AMA [60]. The prevalence of aneuploidy and miscarriage rates increases with maternal age, and female patients are considered to be of AMA once they reach 35. Aneuploidy significantly increases in women above 35 years, reaching an incidence greater than 70% in patients over the age of 43 (Figure 11.2).

The usefulness of PGT-A is controversial and under constant scrutiny owing to conflicting evidence regarding the benefits of the process. A meta-analysis review of five RCTs that utilized FISH for PGT-A, where AMA was the indication for testing, showed no overall benefit in the PGT-A group (537 women) compared to the control (525 women). The live birth and ongoing pregnancy rates were significantly lower in the PGT-A group (18% vs. 26%, respectively), and the difference in miscarriage rate between the two groups was negligible (11% vs. 12%) [37]. Moreover, both groups presented with cases of intrauterine death and live born trisomic offspring. However, the studies included in this meta-analysis have been widely criticized because of flaws in the inclusion criteria, embryo biopsy procedures, and the way testing platforms were used that, together, could have collectively impaired the outcome data. Indeed, some research supports the clinical utility of PGT-A in patients with AMA. First, one group who used the FISH platform published a notable increase in live-birth rates in women aged 41–44 years between the PGT-A and the conventional blastocyst transfer groups (32.3% vs. 15.5%, respectively). Due to the limited technical scope of FISH, the same group later investigated their hypothesis as an RCT, using a more comprehensive chromosome screening methodology: aCGH. The group confirmed their initial findings of higher LBR in the PGT-A arm when compared to morphological embryo

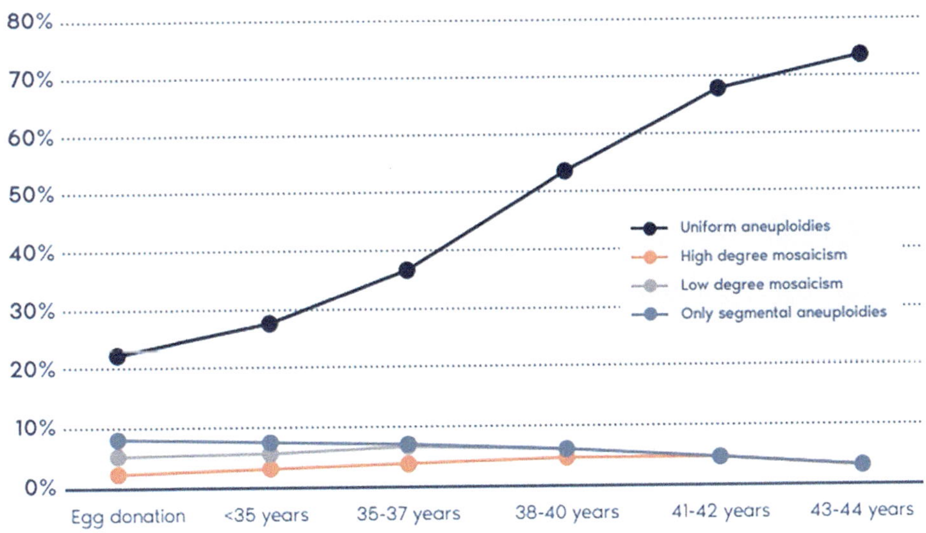

FIGURE 11.2 Overview of the incidence of aneuploidy according to maternal age. There is a clear trend showing an increase in the percentage of screened embryos identified to be aneuploid as maternal age also increases. (By courtesy of Igenomix: internal data of trophectoderm biopsies analyzed in 2019.)

selection and reported 52.9% vs. 24.2% LBR upon first transfer, and 36.0% vs. 21.9% per patient. Additionally, they demonstrated a significant decrease in the miscarriage rate between the PGT-A and control groups (2.7% vs. 39.0%) [61]. The STAR trial, the largest multicenter RCT to date investigating the benefits of NGS for PGT-A, showed no overall benefit across all ages, but a significant increase in ongoing pregnancy rate (OPR) per embryo transfer in the 35–40 years subgroup versus the control arm (51% vs. 37%) [62]. A recently published meta-analysis of 11 RCTs reached the same conclusion, whereby PGT-A improved the LBR in patients over 35 years but seemed ineffective in younger women [62]. Additionally, another group has published a more than doubling of live-birth rates between PGT-A and non-PGT-A blastocyst transfer groups (32.3% vs. 15.5%, respectively) [62].

Despite limited RCT evidence, HFEA "real world data" reports that LBR (per embryo transfer) for fresh and frozen cycles was 26% and 34%, respectively [60]. These findings highlight the benefits of clinically applying PGT-A for patients of AMA, who have a high risk of embryonic aneuploidy. Without the ability to screen for aneuploidy, a large number of patients may experience recurrent miscarriages, leading to several rounds of unsuccessful IVF treatment. This is not only financially challenging but also brings emotional burden and the possibility of medical complications. Nevertheless, the use of PGT-A for patients with AMA indication may provide a more cost- and time-effective route to pregnancy.

Recurrent Pregnancy Loss

Recurrent pregnancy loss (RPL) has varying definitions but can be considered as a diagnosis after two or more consecutive miscarriages in the first 24 weeks of gestation [63]. It has been estimated that the incidence of spontaneous pregnancy loss in the general population is approximately 15% [63]. There are several diagnostic explanations underpinning the etiology of RPL, with a parental balanced structural rearrangement contributing to approximately 2–4% of miscarriages [63]. There is also evidence to demonstrate that the incidence of chromosomal abnormalities in embryos from patients with previous aneuploid conceptions that have resulted in a miscarriage, is much higher than control groups (65.8% vs. 34.0%) [64].

Bianco et al. investigated the association between spontaneous miscarriage and aneuploidy in subsequent pregnancies. Of the 46,939 women included in the study, the risk of aneuploidy increased significantly from 1.39% in women with no history of previous pregnancy loss to 2.18% in the group of women who had experienced three or more miscarriages [65]. Rubio et al. demonstrated improved miscarriage rates following PGT-A testing in couples with a history of aneuploid miscarriages, when compared to natural conception (NC) (9% PGT-A, 28% NC) [23]. A systemic review investigating pregnancy outcomes following PGT-A in couples with unexplained recurrent miscarriages found that implantation rates and ongoing pregnancy rates in the PGT-A group are considerably higher when compared to the NC control (52.63% vs. 19.15% and 61.54% vs. 32.49%, respectively) [66].

Conversely, a more recent study compared the results between PGT-A and non-PGT-A IVF treatment groups. No significant differences in the LBR per patient were reported (PGT-A 26.8%; IVF alone 21.1%), and interestingly the difference in miscarriage rate per clinical pregnancy between the two groups was also negligible (PGT-A 14.3%, IVF alone 20.0%). However, PGT-A improved the LBR per embryo transfer (52.4% vs. 21.6%, respectively) and was also shown to reduce biochemical pregnancy loss per biochemical pregnancy (12.5% vs. 45.0%) [66].

The significant increase in clinical pregnancy and live-birth rates with the use of PGT-A in women with RPL, when compared to no-PGT-A embryo transfer cycles, support the use of PGT-A in patients with a history of repeated pregnancy loss in the eyes of some [67].

Recurrent Implantation Failure

Recurrent implantation failure (RIF) has no clear definition nor generic criteria to fit every couple experiencing failed ART cycles. RIF is generally "diagnosed" in patients who have experienced three or more failed IVF treatments following multiple transfers of ten or more embryos [67]. The causes of RIF are yet to be fully identified; however, it is thought that the parental karyotype is not the only factor, and the synergy between the embryo and endometrium may have an important role in defining implantation success [68].

An observational study revealed a significant increase in clinical pregnancies and implantation rates in RIF patients who underwent PGT-A using CGH when compared to the no PGT-A group (68.3% vs. 22.0% and 68.3% vs. 21.2%, respectively) [69]. An RCT by Rubio et al. showed higher, but not significant, live-birth rates per patient in the PGT-A arm versus conventional blastocyst transfer when investigating the clinical benefits of PGT-A using FISH for patients with RIF indication (47.9% vs. 27.9%, respectively) [35]. Moreover, other studies have reported a significantly reduced incidence of pregnancy loss and increase of clinical and sustained implantation rates in PGT-A treated RIF patients [70, 71].

Male Factor Infertility

The male factor also plays a critical role in the success of IVF treatment, implantation, and pregnancy rates. Studies have shown that there is an increased incidence of chromosomal abnormalities in sperm with poor parameters such as oligo- and tetrazoospermia [72]. Although this may be in part expected, it is clear that chromosomal abnormalities can be elevated in some infertile men despite their normal karyotypes [73].

A large-scale review showed that couples where the male partner presented with abnormal sperm FISH results had improved clinical outcomes where embryos were selected by PGT-A using aCGH and NGS platforms [74]. Furthermore, a retrospective study that included 266 severe male factor (SMF) infertility couples demonstrated a trend towards improved LBR, implantation, and clinical pregnancy rates in the PGT-A versus non-PGT-A groups (55.6% vs. 51.1%, 65.6% vs. 64.2%, and 62.2% vs. 58.0%, respectively) [75]. Perhaps the highest quality evidence for the benefit of PGT-A in SMF patients is a recent RCT performed on 101 cycles showing significantly lower miscarriage rates (6.6% vs. 26.3%) and significantly higher ongoing pregnancy rates (65.8% vs. 29.8%) per transfer, between the PGT-A and control group (< 2 million sperm/ml), respectively [56].

These results support findings that blastocyst biopsies from embryos generated using sperm with abnormal parameters have a higher incidence of sex chromosomal abnormalities, compared to those derived from semen with normal parameters [56].

Implantation and Pregnancy Rates in Good Prognosis Patients

Data consistently shows that aneuploidy increases with age—though it is notable that even in younger patients (maternal age 30–35), embryo aneuploidy rates ranging from 44.9% to 47.0% are detected [2, 49]. These abnormalities are the primary cause of pregnancy loss, as most ploidy aberrations are not compatible with life.

For patients with a good prognosis, PGT-A following TE biopsy and aCGH or NGS diagnosis has the potential to increase overall pregnancy rates. One RCT showed a significantly higher clinical pregnancy rate per embryo transfer in the CCS group (70.9% vs. 45.8%) as compared to the conventional morphology approach in patients without an increased risk of aneuploidy. This study revealed additional advantages of aneuploidy screening: the aCGH group presented with lower rate (2.6% vs. 9.1%) [49].

In a second RCT, good prognosis patients (≤ 42 years of age, AMH levels of ≥ 1.2 ng/ml) were randomized to one of two groups: euploid blastocyst single embryo transfer (SET) following PGT-A using qPCR, or untested double embryo transfer (DET). A similar ongoing pregnancy rate per patient was noted; however, implantation rates were more favorable in the SET group [76]. In another RCT, performed by Scott et al., infertile couples were randomized to a PGT-A group (using qPCR) or an untreated group. All female patients had AMH levels of ≥ 1.2 ng/ml and presented with no more than one failed IVF attempt. The results showed an overall live-birth rate of 57% (403/707) and a miscarriage rate of 15.25% (72/475). Collectively, these well-founded RCTs demonstrate that some benefits are to be found when using PGT-A over morphology-based embryo selection for good prognosis patients, regardless of the analysis platform.

STRENGTHS OF PGT-A: WHAT THE RESULTS SAY

Clinical Outcomes

Some critics of PGT-A argue that there is limited RCT-based or high-quality evidence that the technique improves IVF patients' chances of having a live birth [76]. This is true to an extent, as RCTs often focus on secondary outcomes (miscarriage and implantation rates) [42, 49, 61] or do not show favorable results since they investigate cumulative LBR. However, PGT-A proponents often state that measurement of a procedure's effectiveness in assisting IVF outcomes should also include parameters such as miscarriage rates and time to pregnancy. The selectivity of which evidence is deemed suitable for assessment by some regulating bodies, including "RCT-only" approaches, may lead to meaningful, non-RCT data being ignored. Despite the long-standing skepticism from the UK's fertility regulator, their own validated outcome data shows a significantly higher LBR (per embryo transfer [Figure 11.3] and transfer cycle [Figure 11.4]) at all age groups for PGT-A cycles compared to non-PGT-A, demonstrating clear benefits for the use of aneuploidy screening [62, 77, 78].

Time to Pregnancy

Couples undergoing IVF treatment are often faced with a multitude of financial and emotional burdens. The impact of age is most apparent in female patients, as ovarian reserve begins to diminish over the age of 35 and, with it, aneuploidy in eggs (and subsequently embryos). By selectively transferring euploid embryos identified using PGT-A, the time to pregnancy can be significantly reduced. A recent retrospective cohort

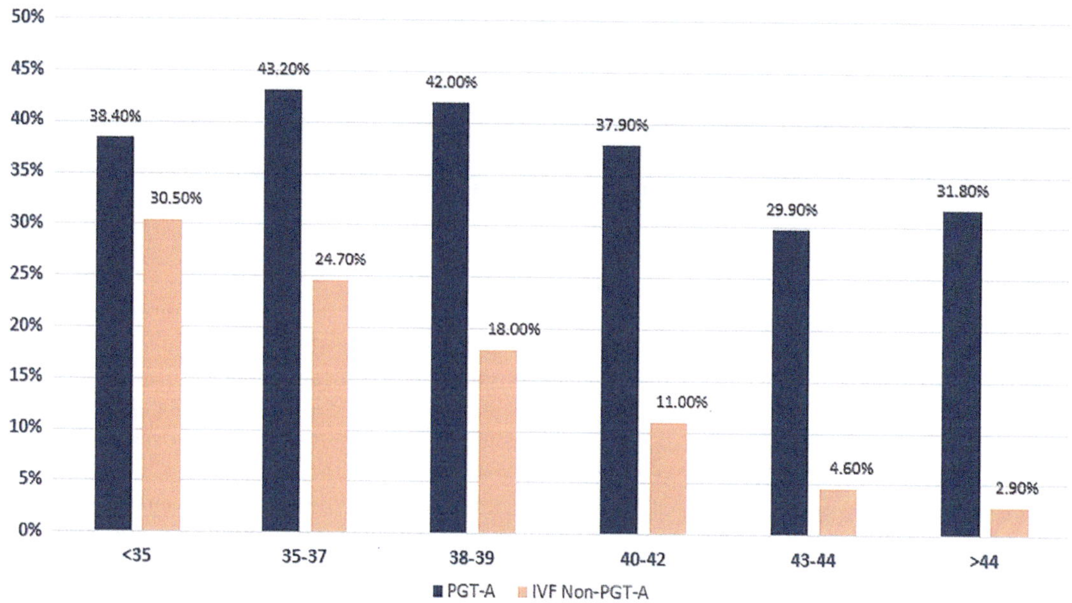

FIGURE 11.3 Summary table of the LBR per embryo transferred. These data show the highest LBR per embryo transferred was 43.2%, for the 35–37 yr maternal age group. This was followed by a decreasing trend, reaching a lowest LBR of 29.9% in the 43–44 yr group. The > 44 yr age group reported a slightly higher LBR of 31.8%; however, it should be noted that only 22 embryos were transferred in this age group. (From ref 82, with permission.)

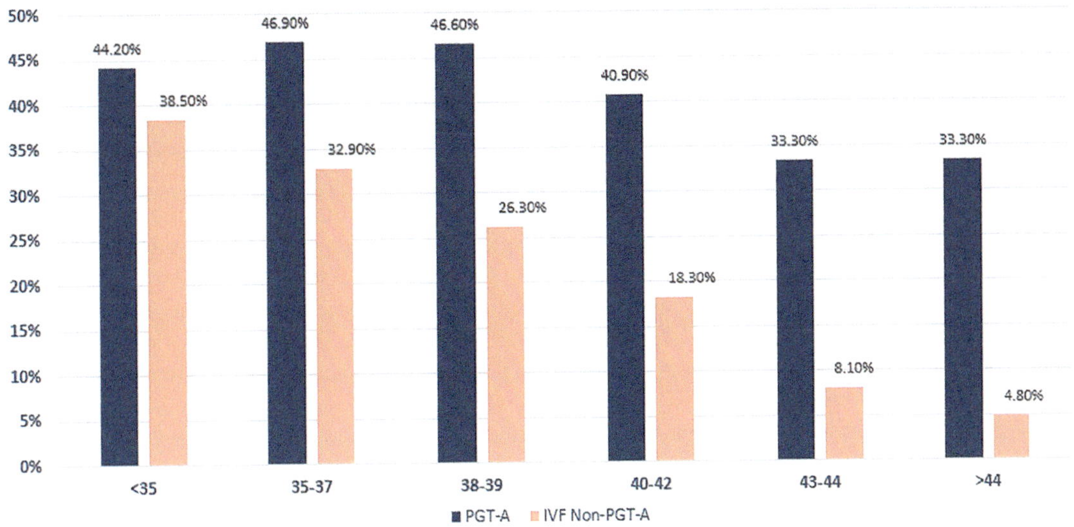

FIGURE 11.4 Summary table of the LBR per treatment cycle. This data shows that across all age groups the increase between LBR per treatment cycle vs. LBR per embryo transferred was greater in the non-PGT-A group. However, the LBR per treatment cycle is consistently higher in the PGT-A groups vs. the non-PGT-A group. (From ref 82, with permission.)

study involving women of AMA showed that clinical pregnancy leading to a live birth was achieved in a shorter time in the PGT-A group as compared to a control (104.8 days vs. 140.6 days) [79]. This is further supported in an RCT performed by Rubio and colleagues, where time to pregnancy was nominally reduced in patients undergoing PGT-A treatment [61]. Prioritization of normal embryos for transfer may lead to a couple conceiving a healthy baby in a shorter time; this may be especially advantageous for AMA patients. It is also notable that a shorter time to pregnancy is associated with a reduced number of cycles and transfers, which may reduce the costs and emotional burden of ART treatment.

Single Embryo Transfer and Cryopreservation

Simultaneous multiple pregnancies are associated with adverse outcomes, such as premature birth, intrauterine growth restriction, perinatal death, and maternal complications [80]. Research highlights a lack of evidence for multiple embryo transfers improving pregnancy rates [80]. In an effort to reduce the clinical (and financial) risks associated with multiple gestations, the UK guidelines advise that wherever possible only a single embryo should be transferred. For this purpose, PGT-A provides the opportunity to select the most appropriate embryo for transfer based on ploidy status; this is referred to as "elective SET" (eSET). Most importantly there is no evidence to suggest that eSET has a negative impact on cumulative LBRs [81]. In fact, several studies demonstrate consistently good clinical pregnancy rates per transfer coupled with a significant reduction in multiple pregnancies [41, 43, 82], meaning that blastocyst selection combined with PGT-A could be an effective approach for improving IVF outcomes using eSET. Moreover, Figure 11.4 shows the clinical outcomes using PGT-A with deferred transfer cycles, compared to regular IVF/ICSI.

Additionally, several euploid embryos may be identified using PGT-A in a single cycle, and the cryopreservation of unused good quality embryos could provide patients with the opportunity for better family planning. The reassurance that euploid embryos are available for transfer at a later date may have a positive impact on the prospective parents' well-being.

Cost Effectiveness of PGT-A

Despite increasing utilization and supporting evidence for the value of PGT-A, the question still remains: is PGT-A an expensive adjunct to IVF treatment or a cost-effective solution? Simultaneously performing PGT-A in an IVF cycle can increase the total cost by 8% compared with a regular blastocyst transfer [61], so it is important to consider whether this extra expenditure confers benefits [82].

One group demonstrated that PGT-A has no cost benefit to patients that were under the age of 40 [83]. However, a study by Neal et al. demonstrates that, overall, PGT-A is cost-saving per live birth; results in fewer miscarriages and failed transfers; and shortens the duration of treatment [84]. Additionally, an RCT that considered the clinical results when assessing the cost-effectiveness deduced that the price for PGT-A could decrease the cost per baby by 12% in Europe and by 10% in the United States, compared with a non-PGT-A cycle [36]. Some evidence suggests that PGT-A is a cost-effective solution for specific patient populations only. For perhaps the most frequent indication, AMA, research shows that older women that have more than one embryo for analysis benefit the most from aneuploidy screening [85]. That the financial benefit is most marked in AMA patients seems to agree with the argument that PGT-A is most effective in this group, who have the highest aneuploidy rates. Therefore, investing in treatment that could identify embryos unfit for transfer or storage and expedite the time to pregnancy may eliminate unnecessary costs for storage or futile treatment, and reduce spontaneous miscarriage and aneuploid pregnancy, which may otherwise result in adverse maternal outcomes.

The costs associated with management of miscarriages, multiple pregnancies, and the emotional burden of repeated pregnancy loss should also be considered. For some patients, the emotional cost-effectiveness of PGT-A outweighs any financial burdens.

LIMITATIONS OF PGT-A AND OTHER CONSIDERATIONS

Mosaicism: Diagnostic Accuracy and Clinical Applications

Chromosomal mosaicism refers to the presence of two or more chromosomally different cell lines in the embryo; this usually occurs as a result of post-zygotic segregation errors. Embryonic mosaicism in TE biopsies can be identified in PGT-A by some testing platforms.

The true incidence of mosaicism is difficult to determine due to technical limitations with the testing platforms and because TE biopsies are taken from a single, random sample of the tissue and are not necessarily representative of the whole blastocyst. It is possible that the mosaicism is compartmentalized to the TE or ICM, and therefore the whole collection of cells in the biopsy may have a different chromosomal make up to the rest of the embryo. This is the main diagnostic uncertainty associated with PGT-A, since aneuploidy lines present in only the ICM could be missed using a TE biopsy, or vice versa. This may lead to transfer of embryos incompatible with life, or the discarding of those with reproductive potential, respectively. Nevertheless, patients undergoing PGT-A have better outcomes, including higher pregnancy per transfer and implantation rates, as well as reduced miscarriages compared to no treatment groups [40, 49, 86].

Recently, some studies have proposed the possibility of transferring mosaic embryos in the absence of euploid options. One group found that embryos with low level mosaicism (< 50%) and retrieved at younger maternal age were more successful than those with high-level mosaicism (≥ 50%) (LBR: low vs. high: 44.5% vs. 36%; MR: low vs. high: 5.1% vs. 30.7%) [87]. Moreover, a double-blinded trial comparing LBR after transfer of uniformly euploid, low-degree, and moderate-degree mosaic embryos showed no significant difference between outcomes (43.4%, 42.9%, and 42%, respectively) [88], concluding that low mosaics are associated with a good prognosis and should therefore not be discarded.

Finally, it is important to consider that, for some results, the detection of mosaicism may be artefactual, owing to inconsistencies and background noise within testing platforms. This can result in the discard of healthy embryos and potentially increase the time to pregnancy until more embryos are available for transfer in a later IVF cycle. The topic of mosaicism in human embryos is complex and is dealt with in greater detail elsewhere in this book (see Chapter 13).

The Invasive Biopsy

Currently, PGT-A requires an invasive biopsy which is a complex process, requiring highly trained embryologists, and when performed by sub-competent practitioners there is an increased risk of harm to the embryo. For example, the STAR trial demonstrates how the variations in laboratory protocols led to discrepancies between centers in the proportion of embryos identified as euploid [62]. It is speculated that this variability contributed to the lack of significant results overall obtained in this study [62].

While embryo hatching (lysis of the zona pellucida facilitating escape of the embryo) is a natural process, to facilitate herniation of cells for biopsy, assisted hatching of a small region of the zona pellucida is usually performed via thermal laser ablation. Studies show that this procedure could impact embryonic development, resulting in delayed compaction and smaller blastocysts with thicker zona pellucida [31]. Moreover, it has been estimated that the use of lasers could elevate culture media temperatures up to 80°C; however, in experienced hands, it has been suggested that this process has no adverse effects on embryo development and clinical outcomes [31].

A paired RCT reported a reduction in implantation rate using embryos that had undergone cleavage-stage biopsies, while confirming that blastocyst stage biopsy does not affect embryo viability or reduce implantation potential. This could be explained by the smaller proportion of the blastocyst being removed (~5 out of 100 cells) when compared to cleavage-stage (1–2 out of 6–8 cells) [31].

ETHICAL CONCERNS

Moral Status of the Embryo

Concerns regarding the discard of aneuploid or high mosaic embryos that are considered unsuitable for transfer remain widespread. In some cultures, embryos hold full moral status equal to that of a human [89], thus their discarding would be comparable to taking the life of a living person; this ethical position restricts the use of PGT. There have been some suggestions to use PGT-A for embryo prioritization instead of selection. Through this approach, embryos are classified into a preferential order for transfer. Prioritization gives every embryo the potential for transfer and should markedly reduce the discard of embryos thereby making the process more ethically acceptable.

In reality, whether PGT-A is applied or not, many potentially viable embryos are discarded as part of routine IVF practice, often as a result of subjective quality criteria (e.g., "poor morphology" embryos being discarded rather than cryopreserved for future clinical use).

Reproductive Autonomy

Prospective parents should be able to make independent and fully informed reproductive decisions, whether that would be to choose aneuploidy screening or proceed with untested embryos. In some situations, clinic and governmental regulators limit the use of PGT-A, such as in the Nordic countries where despite improved technological advances in PGT-A, legislation remains skeptical and often disputes potential benefits. Only Iceland and Finland offer clinical testing for chromosomal abnormalities, whereas Denmark and Sweden offer it within a research protocol approved by local ethics authorities, and Norway currently prohibits its use completely. This has meant that accessibility to aneuploidy testing is significantly restricted when compared with other countries where this treatment is available [90]. While this limits patients' choice, regulators argue that the benefits of PGT-A are not yet fully evidenced.

Additional societal concerns relate to the suggestion that PGT-A implies that individuals affected by aneuploidy syndromes (e.g., Down syndrome) are undesirable. Some groups argue that reducing the number of aneuploidy syndrome births will lead to reduced research funding and care standards for affected individuals [91].

Nevertheless, one must recognize that ultimately the aim of PGT-A is to aid couples with having a successful pregnancy, which, without testing, could otherwise lead to pregnancy loss and significant emotional and financial burdens.

Psychological Aspects of Miscarriage and Pregnancy Loss

While it is reasonable to consider live birth as a measure of success when applying adjuvant treatments to IVF, the possibility of reducing miscarriage and repeated pregnancy loss should not be dismissed. Ongoing pregnancy loss can result in persistent feelings of anxiety, unresolved grief, and guilt in hopeful parents. This could have an impact on attachment and engagement in subsequent pregnancies [92]. In a non-PGT-A cycle, on the basis of chance alone, a euploid embryo may not be transferred until the third or even subsequent cycles, whereas PGT-A would aim to selectively transfer a euploid embryo in the first cycle. Offering PGT-A to couples who have experienced fertility challenges may, alongside the other potential benefits of PGT-A, spare them the emotional burden of further adverse outcomes.

Conversely, the impact of informing patients that they have no viable embryos for transfer—e.g., if all embryos are found to be aneuploid with PGT-A—should also be considered. This may be particularly difficult for AMA patients or those who do not have the financial means to undergo further IVF cycles. Instances where the lack of viable embryos is a result of mosaicism may pose further challenges since the true incidence and clinical meaning of a mosaic result are not yet fully understood. For some patients, the emotional burden of not knowing whether a mosaic embryo could have been successfully transferred may have a more significant impact than a poor pregnancy outcome. Genetic counseling, informed consent, and case-by-case assessments help identify for which patients PGT-A is most appropriate.

CONCLUSIONS

Despite the ongoing controversy regarding the field of PGT-A, recent developments in IVF practice (improved embryo culture and cryopreservation techniques) and testing laboratories (development of higher resolution platforms and improved molecular/analytical techniques) have increased the reliability and efficacy of PGT-A. Numerous recent studies have shown the benefits of testing embryos for chromosomal abnormalities, including increased implantation and pregnancy rates per transfer, decreased miscarriage rates per patient, and shorter time to pregnancy when compared to conventional embryo scoring by morphology alone. Additionally, with an increase in couples choosing to start families later in life, PGT-A will prove a valuable tool, as advanced maternal age is a primary cause of chromosomal abnormalities in embryos. While PGT-A cannot guarantee a healthy baby, it offers a critical, objective assessment of an embryo's chance of creating a live birth. The combined approach of preferential selection of a euploid embryo and elective single embryo transfer avoids the transfer of embryos whose ploidy statuses are strongly associated with negative outcomes while virtually eliminating the chance of multiple pregnancy—the biggest single risk factor following IVF treatment.

The Future: The Prospects of Non-Invasive PGT-A

Current dogma dictates that an invasive biopsy is required to obtain sufficient material to identify chromosomal abnormalities in embryos accurately and reliably. While TE biopsies have been shown to be safe for the embryo, the biopsy process is complex and if incorrectly performed can be detrimental to the embryo. This may reduce an embryo's chances of successful development and the chance of pregnancy. Invasive embryo biopsies also require highly trained embryologists and specialized equipment, meaning it is a time- and cost-consuming procedure. Moreover, embryos not meeting a subjective, lab-specific quality threshold are not biopsied and are frequently discarded. The recent identification of embryonic cell-free DNA (cfDNA) in spent blastocyst media (SBM) has opened a new avenue of possibilities for non-invasive embryo aneuploidy testing. Some have proposed that this DNA originates from cells discarded by embryos as a correction mechanism for aneuploidies. The application of non-invasive PGT-A is fast becoming a clinical reality and is dealt with in more detail elsewhere in this book.

REFERENCES

1. D. Griffin, et al. *Preimplantation Genetic Testing: Recent Advances in Reproductive Medicine*, CRC Press, 2020.
2. E. Fragouli, et al. *Human Genetics*, vol. 132, pp. 1001–1013, 2013.
3. R. Edwards. *Nature*, vol. 214, pp. 576–577, 1967.
4. A. Tarkowski, et al. *Journal of Embryology*, vol. 18, pp. 155–180, 1967.
5. K. Sanders, et al. *Journal of Fetal Medicine*, vol. 4, pp. 51–56, 2017.
6. A. Handyside, et al. *Nature*, vol. 344, pp. 768–770, 1990.
7. A. Handyside, et al. *New England Journal of Medicine*, vol. 327, pp. 905–909, 1992.
8. S. Munne, et al. *Molecular Human Reproduction*, vol. 4, pp. 863–870, 1998.
9. E. Kontogianni, et al. In Verlinsky, Y. and Strom, C. (eds), *Preimplantation Genetics*. Plenum Press, 1991, pp. 139–14.
10. J. Delhanty, et al. *Human Molecular Genetics*, vol. 2, no. 8, p. 1183, 1993.
11. B. Schrurs, et al. *Human Reproduction*, vol. 8, no. 2, pp. 296–301, 1993.
12. S. Munne, et al. *Human Reproduction*, vol. 8, no. 12, pp. 2185–2191, 1993.
13. L. Gianaroli, et al. *Human Reproduction*, vol. 12, no. 8, pp. 1762–1767, 1997.
14. E. Fragouli. *Fertility and Sterility*, vol. 109, no. 4, pp. 606–607, 2018.
15. A. Capalbo, et al. *Human Reproduction*, vol. 25, no. 2, pp. 509–518, 2013.
16. M. Montag, et al. *Reproductive BioMedicine Online*, vol. 18 Supplement 1, pp. 6–11, 2009.
17. M. Magli, et al. *Human Reproduction*, vol. 26, no. 11, pp. 3181–3185, 2011.
18. M. Montag, et al. *Fertility and Sterility*, vol. 69, no. 3, pp. 539–542, 1998.
19. I. Hammoud, et al. *Journal of Assisted Reproduction and Genetics*, vol. 27, no. 7, pp. 423–427, 2010.
20. Y. Verlinsky, et al. *Journal of Assisted Reproductive Genetics*, vol. 15, no. 5, pp. 285–289, 1998.
21. D. Wells, et al. *Fertility and Sterility*, vol. 78, no. 3, pp. 543–549, 2002.
22. J. Geraedts, et al. *Human Reproduction*, vol. 26, no. 11, pp. 3173–3180, 2011.
23. S. Mastenbroek, et al. *Human Reproduction*, vol. 17, no. 4, pp. 454–466, 2011.
24. W. Verpoest, et al. *Human Reproduction*, vol. 33, no. 9, pp. 1767–1776, 2018.
25. D. Griffin, et al. *Reproduction*, vol. 156, pp. 29–50, 2018.
26. T. Eldar-Geva, et al. *Fertility and Sterility*, vol. 102, no. 4, pp. 1016–1021, 2014.
27. A. De Vos, et al. *Prenatal Diagnosis*, vol. 21, no. 9, pp. 767–780, 2011.
28. S. Geber, et al. *Reproductive Biology and Endocrinology*, vol. 28, no. 9, p. 58, 2011.
29. A. Jones, et al. *Fertility and Sterility*, vol. 85, no. 2, pp. 481–491, 2006.
30. K. Scott, et al. *Fertility and Sterility*, vol. 100, no. 3, 2013.

31. D. Cimadomo, et al. *BioMed Research International*, 2016;7193075.
32. Y. Kalma, et al. *Human Reproduction*, vol. 33, pp. 32–38, 2018.
33. A. Capalbo, et al. *Human Reproduction*, vol. 32, no. 3, pp. 492–498, 2016.
34. F. Grati, et al. *Reproductive BioMedicine Online*, vol. 36, no. 4, pp. 442–449, 2018.
35. C. Rubio, et al. *Fertility and Sterility*, vol. 99, no. 5, pp. 1400–1407, 2013.
36. S. Mastenbroek, et al. *Human Reproduction Update*, vol. 17, no. 4, pp. 454–466, 2011.
37. S. McArthur, et al. *Fertility and Sterility*, vol. 84, no. 6, pp. 1628–1636, 2005.
38. G. Kokkali, et al. *Human Reproduction*, vol. 20, no. 7, pp. 1855–1859, 2005.
39. R. Scott, et al. *Fertility and Sterility*, vol. 100, no. 3, pp. 624–630, 2013.
40. E. Forman, et al. *Fertility and Sterility*, vol. 96, no. 3, p. 390, 2011.
41. E. Forman, et al. *Fertility and Sterility*, vol. 100, no. 1, pp. 100–107, 2013.
42. R. Scott, et al. *Fertility and Sterility*, vol. 100, no. 3, pp. 687–703, 2013.
43. R. Bishop. *Bio Horizons*, vol. 3, pp. 85–95, 2010.
44. E. Baart, et al. *Prenatal Diagnostics*, vol. 27, pp. 55–63, 2007.
45. T. Hassold, et al. *Nature Reviews Genetics*, vol. 2, pp. 280–291, 2001.
46. D. Wells, et al. *Nucleic Acids Research*, vol. 27, pp. 1214–1218, 1999.
47. Z. Yang, et al. *Molecular Cytogenetics*, vol. 5, no. 24, 2012.
48. D. Johnson, et al. *Human Reproduction*, vol. 24, no. 4, pp. 1066–1075, 2010.
49. N. Treff, et al. *Fertility and Sterility*, vol. 95, no. 5, pp. 1606–1612, 2011.
50. A. Handyside. *Reproductive Biomedicine Online*, vol. 31, no. 3, pp. 307–308, 2015.
51. M. Rabinowitz, et al. *Fertility and Sterility*, pp. 395–401, 2012.
52. C. Rubio, et al. *Biology of Reproduction*, vol. 101, no. 6, pp. 1083–1090, 2019.
53. C. García-Pascual, et al. *Genes*, vol. 11, 2020.
54. C. Rubio. *Reproductive BioMedicine Online*, vol. 38, no. 1, E8, 2019.
55. N. Treff, et al. *European Journal of Medical Genetics*, vol. 62, no. 8, p. 103647, 2019.
56. C. Rubio, et al. *Fertility and Sterility*, vol. 112, no. 3, pp. 510–519, 2019.
57. C. Simon, et al. *Handbook of New Genetic Diagnostic Technologies in Reproductive Medicine: Improving Patient Success Rates and Infant Health*, CRC Press, 2018.
58. R. Theobald, et al. *Human Reproduction*, vol. 35, no. 4, pp. 986–998, 2020.
59. C. Rubio, et al. *Fertility and Sterility*, vol. 107, no. 5, pp. 30254–30256, 2017.
60. S. Munne, et al. *Fertility and Sterility*, vol. 112, no. 6, pp. 1071–1079, 2019.
61. M. Simopoulou, et al. *Journal of Assisted Reproduction and Genetics*, 2021.
62. H. Ford, et al. *Reviews in Obstetrics & Gynecology*, vol. 2, no. 2, pp. 76–83, 2009.
63. K. Bianco, et al. *Obstetrics and Gynecology*, vol. 107, pp. 1098–1102, 2006.
64. T. Sato, et al. *Human Reproduction*, vol. 34, no. 12, pp. 2340–2348, 2019.
65. S. Bhatt, et al. *Human Reproduction*, pp. 1–6, 2021.
66. A. Thornhill, et al. *Human Reproduction*, vol. 20, no. 1, pp. 35–48, 2005.
67. E. Greco, et al. *BioMed Research International*, vol. 2014, no. 457913, 2014.
68. M. Cozzolino. *Archives of Gynecology and Obstetrics*, vol. 308, pp. 849–850, 2020.
69. E. Mateu, et al. *Journal of Andrology*, vol. 27, no. 1, pp. 6–10, 2006.
70. C. Rubio, et al. *Human Reproduction*, vol. 16, no. 10, pp. 2084–2092, 2001.
71. L. Rodrigo, et al. *Biology of Reproduction*, vol. 101, no. 6, pp. 1091–1101, 2019.
72. M. Asoglu, et al. *Reproductive BioMedicine Online*, vol. 41, no. 4, pp. 595–603, 2020.
73. E. Forman, et al. *Fertility and Sterility*, vol. 100, no. 1, pp. 100–107, 2013.
74. Human Fertilisation & Embryology Authority, "Pre-implantation genetic testing for aneuploidy (PGT-A)," [Online]. [Accessed May 2021].
75. K. Sanders, et al. *J Assist Reprod Genet*, vol. 38, pp. 3277–3285, 2021.
76. D. Griffin, *How Valuable Are Randomised Controlled Trials in Reproductive Medicine?*, Open Acces Government, 2021.
77. E. Lee, et al. *Australia and New Zealand Journal of Obstetrics and Gynaecology*, vol. 58, no. 5, pp. 525–532, 2018.
78. M. Dahan, et al. *Middle East Fertility Society Journal*, vol. 25, no. 32, 2020.
79. Z. Pandian, et al. *Cochrane Database System Review*, vol. 2013, no. 7, p. CD003416, 2013.
80. S. Gordts, et al. *Reproductive Biomedicine Online*, vol. 10, no. 4, pp. 436–441, 2005.
81. P. Scriven. *Journal of Assisted Reproductive Genetics*, vol. 37, no. 3, pp. 637–676, 2020.
82. S. Neal, et al. *Fertility and Sterility*, vol. 110, no. 5, pp. 896–904, 2018.
83. R. Robinson, et al. *Fertility and Sterility*, vol. 111, no. 6, pp. 1115–1116, 2019.
84. P. Lin, et al. *Journal of Clinical Medicine*, 2020.

85. A. Capalbo, et al. *The American Journal of Human Genetics*, vol. 108, pp. 1–10, 2021.
86. M. Brown. *The Journal of Medicine and Philosophy*, vol. 43, no. 2, pp. 132–158, 2018.
87. J. Hreinsson, et al. *Acta Obstetricia et Gynecologica Scandinavica*, 2020.
88. K. Morris. *The Lancet*, vol. 372, no. 9641, pp. 791–792, 2008.
89. J. O'Leary, et al. *Journal of Prenatal & Perinatal Psychology and Health*, vol. 25, no. 3, pp. 173–193, 2011.
90. E. Hammond, et al. *Fertility and Sterility*, vol. 107, pp. 220–228, 2017.
91. C. Rubio, et al. *Human Reproduction*, vol. 24, no. 8, pp. 2045–2047, 2009.
92. M. Stephenson, et al. *Human Reproduction*, vol. 17, pp. 446–451, 2002.

Preimplantation Genetic Testing for Structural Rearrangements

12

Inmaculada Campos-Galindo and Vanessa Peinado

INTRODUCTION

Infertility is a multifactorial condition of increasing prevalence [1, 2]. Chromosome abnormalities appear to play important roles in its etiology [3]—these aberrations are 2–3 times more common in infertile individuals than in the general population [4, 5] and are responsible for gestational pregnancy losses, implantation failures, and congenital malformations. Indeed, the prevalence of infertility in individuals with chromosome abnormalities is up to 25 times higher than in the general population [6–8]. Consequently, chromosome analysis is an essential part of the infertility evaluation (both members of the couple) to maximize the probability of pregnancy and minimize the transmission of chromosomal/genetic anomalies to the offspring.

Cytogenetics is the study of chromosome structure and properties that capitalizes on the constant morphology and size of normal chromosomes to detect abnormalities [9]. Karyotyping is a useful cytogenetic tool that uses *in vitro* culture and banding techniques to perform structural and numerical chromosomal analyses in metaphases to describe normal or abnormal chromosome complement of an individual, tissue, or cell line [10]. Currently, karyotyping of peripheral blood samples remains the first-line method for detecting chromosomal abnormalities in infertile couples.

Constitutional chromosomal rearrangements can be inherited from a carrier parent or occur *de novo* in the gametes that result in zygotes [11]. Structural chromosomal rearrangements indicate changes in the native structure of the chromosomes. There are two general types of structural chromosomal rearrangements: balanced with no loss or gain of genetic material, and unbalanced with loss or gain of genetic material (Table 12.1). Carriers of structural chromosome rearrangements have an increased risk of fertility problems, recurrent miscarriages, and offspring with congenital abnormalities and mental retardation. Thus, preimplantation genetic testing for structural rearrangements (PGT-SR) is widely applied when an abnormal karyotype is identified in the infertility work-up of couples.

DOI: 10.1201/9781003024941-12

TABLE 12.1 Balanced and Unbalanced Structural Chromosome Rearrangements

		REARRANGEMENT		*ISCN NOMENCLATURE*	*EXAMPLES*
Structural chromosome rearrangements	**Balanced**	**Translocation**	**Reciprocal translocation**	**t**	46,XY,**t**(3;21)(p25;q22.1)
			Robertsonian translocation	**rob/der**	45,XX,**der**(13;14)(q10;q10) 45,XX,**rob**(13;14)(q10;q10)
		Inversion	**Paracentric inversion**	**inv**	46,XY,**inv**(1)(q25q42)
			Pericentric inversion		46,XX,**inv**(4)(p12q13)
	Unbalanced	**Deletion**		**del**	46,X,**del**(X)(q27)
		Duplication		**dup**	46,XX,**dup**(16)(q12.1)
		Marker chromosome		**mar**	47,XY,+**mar**
		Isochromosome		**i**	46,XY,**i**(18q)

STRUCTURAL CHROMOSOME REARRANGEMENTS AND RISK FOR THE CARRIER

Carriers of Balanced Structural Rearrangements

Balanced structural chromosome rearrangements occur when there is heterozygous rearrangement with no loss or gain of genetic material. They are the most frequent chromosome abnormalities in the general population, with a prevalence of 0.4% in prenatal samples and 0.2% in newborns [12, 13]. The incidence of chromosome rearrangements is approximately 5% higher in couples with infertility that have experienced recurrent *in vitro* fertilization (IVF), repeated implantation failure (RIF), or recurrent pregnancy loss (RPL) [6, 7]. In addition, about 2% of infertile males are carriers of structural chromosome abnormalities, which is six times higher than that observed in the general population [14].

Most balanced carriers of chromosome rearrangements do not display an obvious phenotype: if there are no genetic material gains or losses or no truncation of a gene in the breakpoints, all of the necessary genetic information is present. However, these carriers may present with reduced fertility, repeated miscarriage, or offspring with an abnormal phenotype. In particular, heterozygotic carriers of balanced chromosome rearrangements are predisposed to produce unbalanced gametes during meiosis, through unequal meiotic segregation in translocation carriers or recombination events in inversion carriers during gametogenesis [15, 16]. Thus, their condition can remain undetected until they attempt to reproduce [17].

Carriers of Translocations

Translocations are interchromosomal rearrangements that involve two different chromosomes, and may occur between non-homologous chromosomes; these are the most common type of chromosome rearrangement. Translocations occur after a break in two or more chromosomes and interchange of chromosomal fragments between two or more chromosomes. Inherited translocations are commonly innocuous for the carrier, although the carrier usually has high rates of unbalanced gametes following meiotic segregation and unbalanced embryos [18, 19]. The risk of unbalanced gametes depends on the chromosome involved, size of the region involved, position of the breakpoints, and the maternal or paternal origin. Consequently, translocation carriers have a greater chance of being infertile and/or having chromosomally abnormal pregnancies that lead to recurrent spontaneous abortions or children with congenital anomalies and mental retardation [16]. The risk of having abnormal offspring is specific for each translocation event.

Reciprocal translocations are rearrangements produced by a single breakage and exchange of distal segments between two non-homologous chromosomes, creating two abnormal derivatives (Figure 12.1a). All chromosomes participate in reciprocal translocations [11], in which chromosomes form a quadrivalent at the pachytene stage of meiosis (Figure 12.1b). Five defined modes of segregation are possible, resulting in 16 possible gametes which, once fertilized, can lead to the formation of 32 possible embryo outcomes [20]. Only one mode of segregation can result in a normal/balanced embryo and genetically healthy ongoing pregnancy. The gametes produced are frequently imbalanced with partial trisomy (duplication) or partial monosomy (deletion), or rather frequently, a combination of partial trisomy and partial monosomy [21], leading to the generation of chromosomally unbalanced embryos.

Robertsonian translocations involve two acrocentric chromosomes (in humans, chromosomes 13, 14, 15, 21, and 22 are acrocentric chromosomes) joined at the centromere with loss of their short arms (p arm); this reduces the chromosome number to 45 (instead of 46) with no phenotypic effect (Figure 12.2a). In these translocations, chromosomes form a trivalent at the pachytene stage of meiosis (Figure 12.2b). Then the chromosomes can segregate by four modes, resulting in eight different embryo outcomes [20]. Only one mode can result in a genetically healthy ongoing pregnancy. The remaining modes of segregation

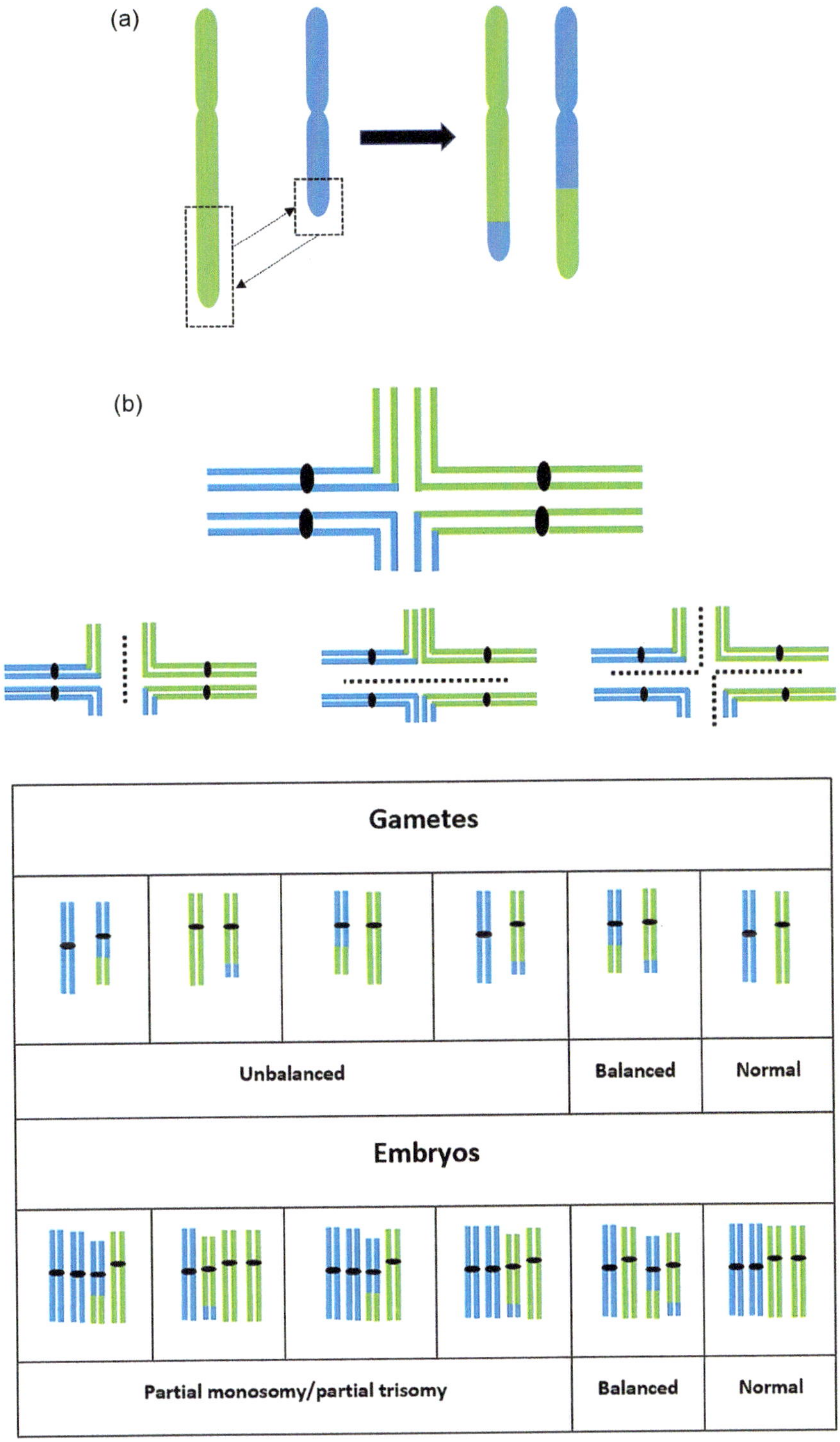

FIGURE 12.1 Segregations, gametes, and embryos in reciprocal translocation carriers: (a) reciprocal translocation; (b) segregations in gametes and embryos in reciprocal translocation carriers.

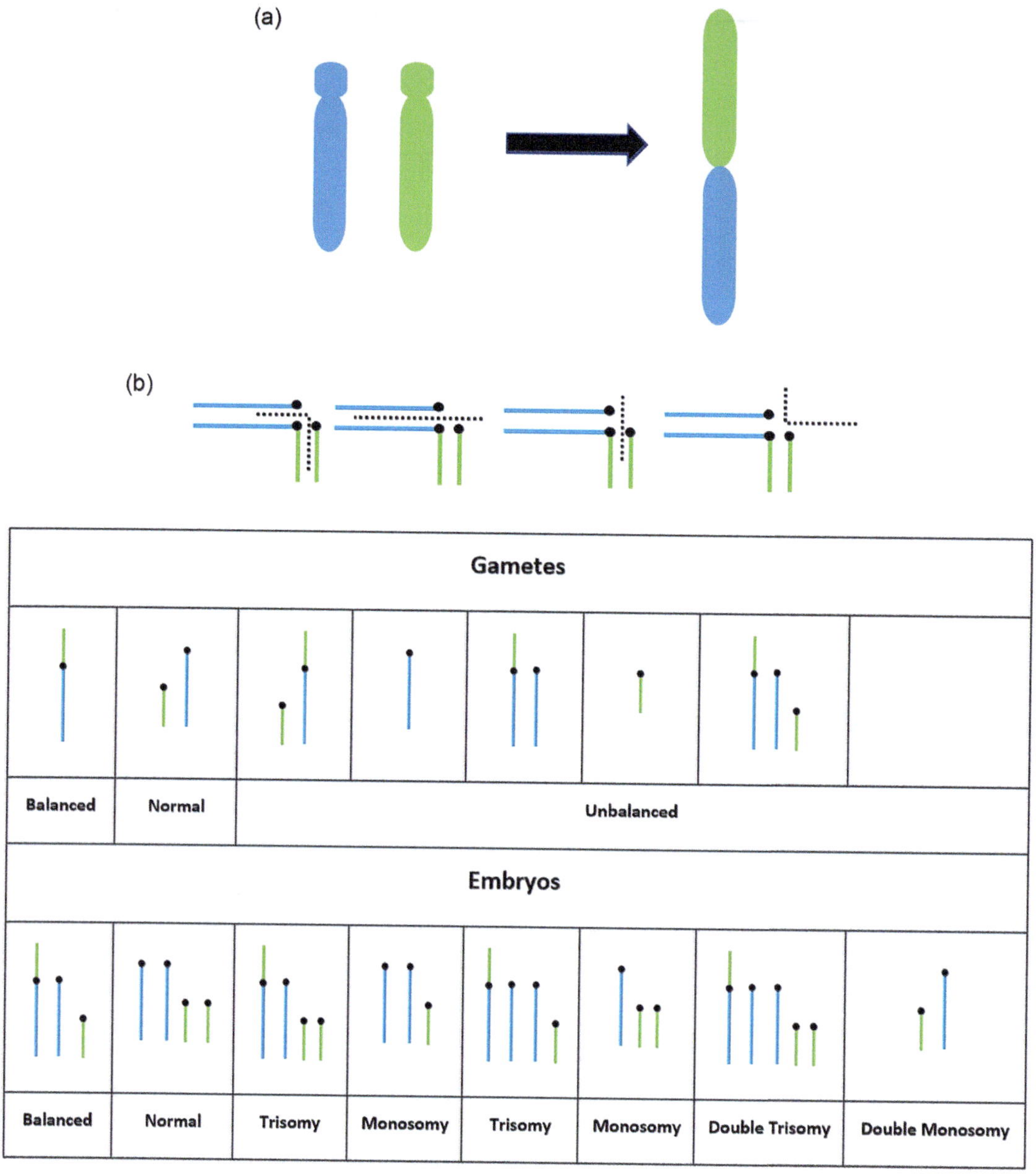

FIGURE 12.2 Segregations, gametes, and embryos in reciprocal translocation carriers: (a) Robertsonian translocation; (b) Segregations in gametes and embryos in Robertsonian translocation carriers.

lead to nullisomic or disomic gametes, generating chromosomally unbalanced embryos (monosomic or trisomic embryos) [22].

Complex chromosomal rearrangements (CCRs) are structural rearrangements involving more than two breakpoints and often more than two chromosomes. CCRs are extremely rare events in the entire population [23], with an incidence of approximately 0.5% in newborns [24]. Familial transmission mainly occurs through female carriers [25]. Nevertheless, the majority of CCRs (~70–75%) are *de novo* [26]; this complicates determination of risk [25]. In inherited cases, reproductive risk can be estimated by specific family history [27], and CCRs resulting in viable offspring are rare [26]. Theoretically, meiotic segregation of a CCR in heterozygous balanced carriers can lead to numerous unbalanced chromosomal gamete configurations with a high risk of spontaneous miscarriage or chromosomally abnormal offspring [26].

There is a link between infertility and chromosomal translocations, which occur at a rate of 0.6% among infertile couples compared with an incidence of 0.2% in the general population [8, 28]. The prevalence of balanced translocation is higher in females than males, and higher still if there is a family history of a stillborn or abnormal liveborn [29] or RPL [30]. A parental carrier of a structural chromosome rearrangement, most commonly a reciprocal or Robertsonian translocation, is associated with a history of RPL in approximately 3.5% of couples [31]. In infertile men, that chromosomal translocation may adversely affect spermatogenesis, leading to azoospermia, oligozoospermia, and infertility [32].

Carriers of Inversions

Inversions are rearrangements that occur after a double intrachromosomal break, 180° rotation of the fragment located between two breakpoints, and subsequent insertion of the fragment into the chromosome. During normal meiosis, crossing over must occur to ensure orderly disjunction. In inversions, this process must occur within the inverted segment [21]. Carriers of inversions are at risk of having offspring with unbalanced karyotypes, and this risk depends on the implicated chromosome and length of the inverted segment [33]. The chance of recombination may increase when the size of the inverted fragment accounts for more than 30% of the whole chromosome, and there may be a high risk of producing aneuploid embryos when the size of the inverted fragment is at least 100 Mb or over 50% [34, 35].

Pericentric inversions involve a break in each arm of the chromosome and inversion of the segment containing the centromere [36] (Figure 12.3a). Pericentric inversions during meiosis may produce gametes with partial trisomy or partial monosomy (duplication in the end of short arm (p-arm) and a deletion in the end of the long arm (q-arm) or vice versa), resulting in non-viable embryos, miscarriages, or livebirths with anomalies. There is a 5–10% overall risk that a carrier of pericentric inversion will have a child with unbalanced chromosome rearrangement [37], specifically 10% for female carriers and 5% for male carriers [38]. The risk of unbalanced segregation is dependent on the chromosome involved, breakpoints and size of the inverted fragment, and sex of the carrier [39].

Paracentric inversions occur after a two-break event in the same chromosome arm, and inversion and rejoining of the segment between two breakpoints (Figure 12.3b) [40]. Paracentric inversions produce gametes that have either no centromere (acentric) or two centromeres (dicentric) and are not viable. Therefore, most paracentric inversions are considered harmless. However, a possible meiotic crossover in the inverted segment may generate chromosomally unbalanced gametes. Most zygotes resulting from chromosomally unbalanced gametes are lost very early, even before implantation, but there is still a risk of having a phenotypically abnormal child [40, 41]. Spontaneous abortion, infertility, and mental retardation and/or congenital malformation of offspring have also been reported among paracentric inversion carriers [41], and the pooled risk of unbalanced viable offspring has been reported in 4% [22, 42].

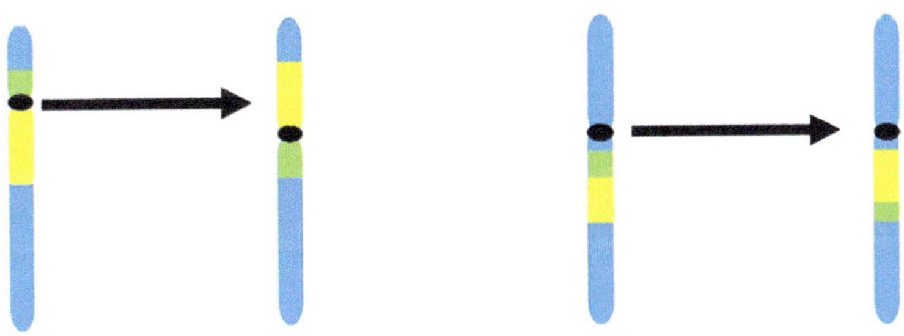

FIGURE 12.3 Pericentric and paracentric inversions: (a) pericentric inversion; (b) paracentric inversion.

Carriers of Unbalanced Structural Rearrangements

In unbalanced structural chromosome rearrangements/unbalanced rearrangements, genetic material is gained or lost. Unbalanced rearrangements encompass interstitial and terminal deletions, interstitial duplications, and marker chromosomes and isochromosomes, among others.

Deletions are structural abnormality consisting of loss of interstitial or terminal chromosomal fragments that produces partial chromosomal monosomies. Cytogenetically, visible deletions occur in 1 in ~7,000 (0.014%) live births [43]; 4.7% of all reported chromosome abnormalities are deletions, including microdeletions, with a prevalence of 1.99 per 10,000 (0.02%) births [44]. Deletion carriers can generate unbalanced gametes with deletion, producing deleted embryo carriers.

Duplications are repetitions of a segment of an arm. Duplications are even less common than deletions with a prevalence of 0.7 per 10,000 births, representing 1.6% of all reported chromosome abnormalities [44]. Some gametes inherit the duplication, producing unbalanced embryos.

Small supernumerary marker chromosomes (sSMCs) are centric chromosome fragments that can originate from any of the 24 chromosomes (most cases involve acrocentric chromosomes), and their frequency in the general population is estimated to be approximately 0.04% [45]. The incidence is three-fold higher in infertile couples, and they are more commonly found in the male partner [46]. Approximately one-third of sSMC carriers exhibit an identifiable phenotype, ranging from subfertility such as azoospermia or oligospermia in men, to birth defects and/or intellectual disability [47]. Heterochromatic sSMC can be inherited in embryos with the same karyotype as the parent carrier. In sSMC with euchromatin, the carrier can be balanced or unbalanced and the presence or absence of sSMC in gametes can lead to unbalanced embryos.

INTERCHROMOSAL EFFECT

In 1963, Lejeune observed an increased rate of carriers of balanced reciprocal translocations among the parents of children with trisomy 21 [48]. This phenomenon of interference during meiosis is known as an interchromosomal effect (ICE). During meiosis, structural chromosome rearrangements can influence the disjunction and segregation of other chromosome pairs (structurally normal) leading to full chromosomal aneuploidy in the resulting gametes and offspring [48, 49].

It has long been debated whether the chromosomes involved in rearrangement affect segregation of the structurally normal chromosomes. Some studies in PGT-SR embryos from carriers of chromosome rearrangements support the existence of an ICE [50–53] that is dependent on the rearranged fragment size, patient, and chromosome [54]; however, other studies suggest that an ICE is either entirely absent or negligible in these patients [51, 55]. Furthermore, evidence for an ICE is absent in some studies [49, 56, 57], or the increased aneuploidy rate may result from factors such as oligoasthenoteratozoospermia (OAT) syndrome, which is frequently observed in these patients [53, 58, 59].

Robertsonian translocations appear to be the most prone to producing such interference, closely followed by reciprocal translocations [60, 61]. The presence of complex rearrangements seems to be an important factor for promoting ICE: 71.4% of these carriers have increased rate of gametes with numerical abnormalities [60].

PGT-SR

Historically, couples carrying a chromosomal rearrangement have had no therapeutic options for reducing the risk of pregnancy loss or fetal anomalies beyond invasive prenatal testing. PGT-SR provides an opportunity to distinguish between euploid balanced embryos (potentially resulting in a successful ongoing pregnancy) before implantation occurs, thereby minimizing the risk of implantation failure, miscarriage, or birth of a chromosomally unbalanced child [20]. Prenatal follow-up is recommended in cases of pregnancy after PGT-SR.

Another potential benefit to be considered is that PGT-SR allows the transfer of euploid balanced embryos, reducing the time needed to achieve a viable pregnancy from 4–6 years to < 4 months and decreasing the incidence of miscarriage from > 90% to < 15% [62, 18, 19] compared with spontaneous conceptions [18, 62–64].

Methodology for PGT-SR

PGT-SR was initially conducted using strategies such as fluorescence *in situ* hybridization (FISH), and later on with array comparative genomic hybridization (aCGH), and next-generation sequencing (NGS). FISH is still useful on embryo biopsies taken at the cleavage stage (day 3 post-insemination), while aCGH and NGS are applied on trophectoderm biopsies from day 5/6 blastocysts. Depending upon the technology, the testing detects only the chromosomes likely involved in rearrangement, or all 23 pairs of chromosomes (Table 12.2). PGT-SR requires a personalized review of parental karyotypes during the preclinical

TABLE 12.2 Comparisons of PGT-SR Technical Approaches

	CHROMOSOMES INVOLVED IN REARRANGEMENT	*ANEUPLOIDY EVALUATION FOR ALL 24 CHROMOSOMES*	*PRECLINICAL WORK-UP*	*BIOPSY DAY*	*FRAGMENT SIZES*
FISH-based PGT-SR	Yes	No[a]	Yes	Cleavage stage	Rearrangements involving small fragments or subtelomeric regions of chromosomes that are difficult or impossible to detect using other methods. Cleavage-stage biopsies.
Arrays based PGT-SR	Yes	Yes	No[b]	Cleavage-stage Trophectoderm	Feasible if three of four segments for two-way reciprocal translocations are above the resolution of the platform.
NGS-based PGT-SR	Yes	Yes	No[b]	Cleavage-stage Trophectoderm	Feasible if three of four segments for two-way reciprocal translocations are above the resolution of the platform.

[a] Combined with screening aneuploidies of only a limited number of chromosomes.
[b] Preclinical work-up is not required unless the carrier has an unbalanced karyotype.

work-up. Before treatment initiation, providers must perform an in-depth analysis of all potential segregation results from different recombination modes, to confirm that all potential unbalances can be detected with the technology applied. To this end, the European Society of Human Reproduction and Embryology (ESHRE) PGT Consortium has good practice recommendations on the technical aspects of aCGH, NGS, and FISH for PGT-SR including laboratory issues, work practice controls, pre-examination validation, preclinical work-up, risk assessment, and limitations [65]. Periodically, the ESHRE PGT Consortium updates the guidelines with new techniques used in IVF and genetic laboratories.

FISH-Based PGT-SR Analyses

Early studies performed FISH using chromosome probes applied to polar bodies to diagnose translocations of maternal origin [66, 67]. A common FISH strategy used for the detection of unbalanced segregation in translocations involves the use of centromeric (CEN), locus-specific (LSI), and subtelomere (TEL) probes in polar bodies and interphase nuclei at cleavage-stage biopsy [63, 67, 68]. This strategy is acceptable for rearrangements involving small fragments or subtelomeric regions of chromosomes that are difficult or impossible to detect using other methods [65]. FISH has several technical limitations such as hybridization failure, cross-hybridization, poor probe hybridization, signal overlap, signal splitting, and problems related to quality fixation and loss of micronuclei or chromosomes during fixation [69–71]. PGT-SR by FISH focuses on chromosomes involved in rearrangement, and combined with screening aneuploidies of only a limited number of chromosomes (i.e., all 23 pairs of chromosomes are not evaluated for the presence of aneuploidy), might result in the transfer of abnormal embryos with aneuploidy for chromosomes not analyzed. The error rates of FISH protocols for translocation reportedly in range from 0% to 10%, with an average error rate of 6% [18, 71, 72].

As noted in the methodology section, for PGT-SR cases in which FISH is the technology of choice, case-specific preclinical work-up is recommended to assess PGT-SR feasibility, identify informative probes (hybridization efficiency, false-positive rate, and false-negative rate of the probe set), and work on a clinical testing strategy (Figure 12.4) [65]. Based on karyotype reports, the laboratory evaluates the theoretical segregation of chromosome/s implicated in structural rearrangements, to ensure that PGT-SR by FISH allows for the detection of all expected genotypes in the embryos using specific probes. The blood sample is fixed on a slide and FISH assays are performed. Working on clinical testing strategies is accepted on sperm cells in cases of male translocation carriers [65].

Array- and NGS-Based PGT-SR Analyses

PGT with molecular-based tests has been developed to solve several of the limitations of FISH with improvements in the comprehensive analysis of the 23 pairs of chromosomes, automation, turnaround time, cost effectiveness, sensitivity, and reliability. Such approaches involve whole-genome amplification (WGA) followed by microarray single-nucleotide polymorphisms (SNPs) [73–77], microarray of comparative genome hybridization (aCGH) [78–81], or next-generation sequencing (NGS) [82, 83].

PGT-SR protocols based on arrays involve competitive fluorescence hybridization of differentially labeled samples and reference DNA on a slide with fixed DNA probes. The DNA probes correspond to specific chromosomal regions. The evaluation of fluorescence ratios allows the identification of chromosomal losses or gains. Two types of array platforms are used for PGT-SR:

- aCGH platforms based on oligonucleotides, providing a resolution of 5 to 10 Mb.
- Array platforms based on SNPs, providing a resolution of 2.4 to 5 Mb. Karyotype is inferred from the haplotypes detected in DNA from the embryo biopsy. Based on haplotype information, embryos carrying the balanced form of the rearrangement can be distinguished from normal diploid non-carrier embryos [65].

NGS allows direct reading of sequenced DNA fragments and their quantification based on sequence read numbers [65]. Compared with arrays, NGS may offer advantages such as reduced DNA sequencing cost,

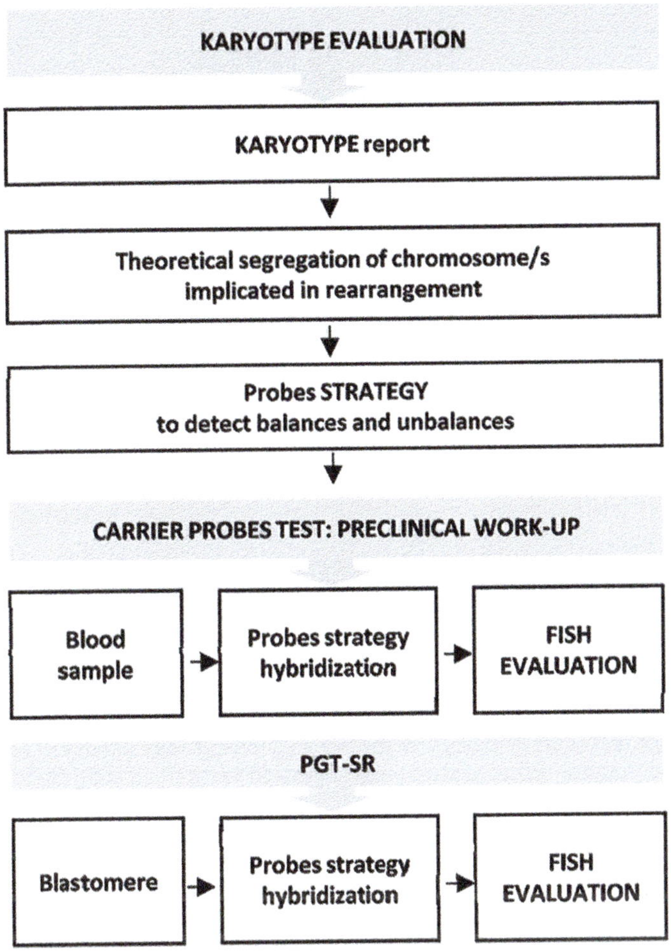

FIGURE 12.4 Case-specific preclinical work-up in FISH-based PGT-SR.

larger number of samples simultaneously sequenced, detection of chromosomal mosaicism in trophectoderm biopsies, and potential automation of the protocol (the limitations are indicated in Table 12.3).

Case-specific preclinical work-up is not required when performing arrays or NGS for balanced carriers of structural rearrangements (Figure 12.5). The size of the translocated fragment is inferred from the karyotype formula. In cases of reciprocal translocations, it is acceptable to perform PGT-SR by arrays or NGS, if three of four segments for two-way reciprocal translocations are above the threshold of resolution of the platform used (Figure 12.5; [65]). Exceptionally, preclinical work-up is required when performing arrays or NGS for unbalanced carriers of structural rearrangements (Figure 12.5). Preclinical work-up allows evaluation if the platform identifies the expected unbalances in the blood sample of the carrier (Figure 12.5; [65]).

PGT-SR RESULTS

Incidence of Chromosomal Abnormalities

The incidence of aneuploid unbalanced embryos in carriers of reciprocal translocations is significantly higher than that observed in Robertsonian translocation carriers in both cleavage-stage biopsies and at the

TABLE 12.3 Limitations of PGT-SR Technical Approaches as Summarized in Reference

TEST	*LIMITATIONS*
FISH-based PGT-SR	Does not allow for a distinction between embryos with a normal or balanced karyotype. Does not allow for the detection of uniparental disomy (UPD). Can only assess the copy number of the chromosomes targeted by the DNA probes used. Number of chromosomes that can be simultaneously detected is also restricted; therefore, sequential rounds of FISH may be required. Commercial probes are available for only a limited number of loci. Impossible to detect mosaicism if FISH is performed in a single-cell biopsy.
Array-based PGT-SR	Does not allow for a distinction between embryos with a normal or balanced karyotype. Does not allow for the detection of UPD. Detection of translocation segments is limited by the resolution of the platform. If the size of more than one of the four translocated segments is below this resolution limit, aCGH-based PGT is not possible. Detection of unbalanced segregations that have breakpoints near the telomere or in the subtelomere region is not always possible since the probe coverage in these regions is low. For each aCGH-based PGT-SR case, limitations should be investigated during preclinical work-up. Array-based PGT-SR analysis is less sensitive to detect mosaicism than NGS.
NGS	Does not allow for a distinction between embryos with a normal or a balanced karyotype Does not allow for the detection of UPD. Cannot detect whole ploidy changes. Cannot discriminate balanced from normal results. Cannot detect low level chromosomal mosaicism. Cannot detect abnormalities below the predefined resolution. If the size of more than one of the four translocated segments is below this resolution limit, NGS for PGT-SR is not possible.

Source: Based on data from ref 65.

blastocyst stage. Reciprocal translocation carriers also have a higher percentage of aneuploid unbalanced embryos in cleavage-stage biopsies compared to blastocyst-stage biopsies. The percentage of aneuploid embryos with abnormalities only for chromosomes involved in the rearrangement (without ICE) is also significantly higher for reciprocal translocations compared to Robertsonian translocations in the cleavage stage and blastocyst stage, and is higher than in inversions. By contrast, the incidence of embryos with aneuploidies affecting only chromosomes not involved in the rearrangement (pure ICE) is significantly lower in reciprocal translocations compared to Robertsonian translocations, in both cleavage- and blastocyst-stage biopsies and compared to inversions (Table 12.4; [61, 84, 85]).

Therefore, in reciprocal translocation carriers, the main reason for non-transferable embryos is the presence of aneuploidies affecting only the chromosomes involved in the rearrangement. However, in Robertsonian translocations and inversion carriers, ICE seems to be the main cause of non-transferable embryos. Female translocation carriers have a higher incidence of abnormal embryos than male carriers, but this gender effect is not observed in inversion carriers. In addition, the adverse effect of the presence of a structural chromosomal rearrangement is increased in advanced maternal age and in female rearrangement carriers [61].

Clinical Outcome of PGT-SR Cycles

To date, no randomized controlled trial has been performed to validate the benefits of PGT-SR, and most studies have been retrospective. In retrospective studies, PGT-SR produced a significant decrease in spontaneous miscarriage [18, 64]. Yet, other studies did not find significant differences in live birth

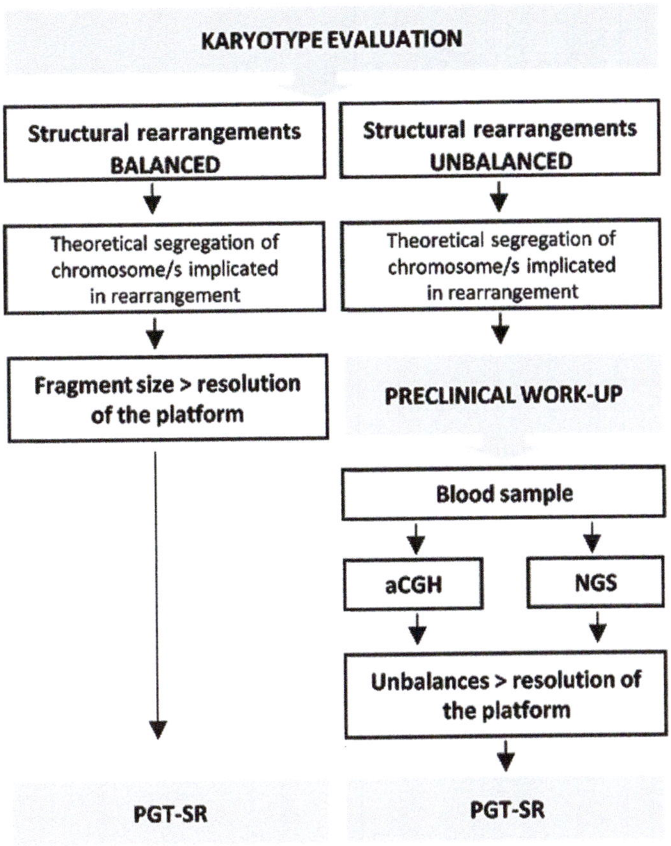

FIGURE 12.5 Case-specific preclinical work-up in array- and NGS-based PGT-SR.

rates between PGT-SR and natural conception groups [86, 87]. As expected, higher pregnancy rates were observed in couples with a higher percentage of euploid balanced embryos (> 50%; [18]).

The last publication from the ESHRE PGT Consortium collected data from 3459 PGT-SR cycles performed from 2013 to 2015 (25.4% in Robertsonian translocation carriers, 60.4% in reciprocal translocation carriers, 6.2% in inversion carriers, and 2.5% in deletions carriers). Of the embryos successfully biopsied, one in four was genetically transferable, which was the lowest percentage of transferable embryos found in the reciprocal translocation group (on average, 23% for male and 20% for female carriers). Of all transferable embryos, between 54% and 65% were transferred in a fresh or a frozen cycle. Human chorionic gonadotropin was detected in 23% of the cases, with a positive heartbeat in 19% per oocyte retrieval (OR) and 34% per embryo transfer. Historically, the clinical pregnancy rate is stable at about 18% per OR and 28% per embryo transfer procedure [88].

STRENGTHS AND LIMITATIONS OF PGT-SR

PGT-SR may be valuable to reduce spontaneous miscarriage rates [18], but there have been mixed results regarding the most clinically relevant outcomes of live birth rate. Two studies that compared PGT-SR and natural conception did not find significant differences in live birth rates (66.6% vs. 55.8% and 67.5% vs. 65.4%) between groups [89], although miscarriage rates were lower in the PGT-SR groups in both

TABLE 12.4 Incidence of Chromosomal Abnormalities in Reciprocal Translocations, Robertsonian Translocations, Pericentric Inversion, and Inversion (Paracentric and Pericentric Inversion)

	RECIPROCAL TRANSLOCATION				*ROBERTSONIAN TRANSLOCATION*				*PERICENTRIC INVERSION*	*INVERSION (PERICENTRIC AND PARACENTRIC)*
	MATEU-BRULL, 2019	*MATEU-BRULL, 2019*	*CAMPOS-GALINDO, 2019*	*BOYNUKALIN, 2021*	*MATEU-BRULL, 2019*	*MATEU-BRULL, 2019*	*CAMPOS-GALINDO, 2019*	*BOYNUKALIN, 2021*	*MATEU-BRULL, 2019*	*CAMPOS-GALINDO, 2019*
	D+3	*D+5/6*	*D+5/6*	*D+5/6*	*D+3*	*D+5/6*	*D+5/6*	*D+5/6*	*D+3*	*D+5/6*
Aneuploid unbalanced embryos	89.3	80.3	69.49	70.2	73.8	61.4	51.32	63.5	68.1	52.30
Euploid unbalanced embryos (without ICE)	25.9	34.8	35.86	34.8	13.2	15.9	20.28	19.2	6.5	7.42
Aneuploid balanced embryos (pure ICE)	33.4	21.3	18.01	14.1	33.4	36.4	25.35	30.8	36.2	37.46

studies [86, 87]. More recently, PGT-SR was shown to be associated with increased live birth rates and decreased rates of miscarriage and birth defects in reciprocal translocation carriers with two prior adverse pregnancy outcomes [90].

CONCLUSIONS

Unequal meiotic segregation in structural chromosomal rearrangement carriers produces unbalanced gametes generating unbalanced and/or abnormal embryos, thus increasing the risk of fertility problems, recurrent miscarriages, and offspring with congenital abnormalities and mental retardations. PGT-SR can distinguish between euploid balanced embryos and exclude aneuploid and/or unbalanced embryos, improving reproductive efficiency and reducing the risk of pregnancy loss or fetal anomalies.

REFERENCES

1. Mascarenhas MN, et al. *PLoS Med.* 2012;9(12):e1001356.
2. Izzo CR, et al. *Rev Assoc Med Bras (1992).* 2015;61(6):557–9.
3. Shah K, et al. *Reproduction.* 2003;126(1):13–25.
4. Jesus AR, et al. *J Assist Reprod Genet.* 2019;36(7):1471–1479.
5. Clementini E, et al. *Hum Reprod.* 2005;20(2):437–42.
6. Campana M, et al. *Am J Med Genet.* 1986;24(2):341–56.
7. Fryns JP, et al. *Eur J Obstet Gynecol Reprod Biol.* 1998;81(2):171–6.
8. Stern C, et al. *Hum Reprod.* 1999;14(8):2097–101.
9. Hare WCD, et al. *Cytogenetics in Animal Reproduction.* Slough: Commonwealth Agricultural Bureaux, UK, 1979.
10. McGowan-Jordan J, et al., editors. *ISCN 2020—An International System for Human Cytogenomic Nomenclature (2020).* Karger, 2020.
11. Shaffer LG, et al. *Curr Opin Genet Dev.* 2012;22(3):264–71.
12. Jacobs PA, et al. *Ann Hum Genet.* 1974;37(4):359–76.
13. Van Dyke DL, et al. *Am J Hum Genet.* 1983;35(2):301–8.
14. Mau-Holzmann UA. *Cytogenet Genome Res.* 2005;111(3–4):317–36.
15. Neri G, et al. *Am J Med Genet.* 1983;16(4):535–61.
16. Scriven PN, et al. *Prenat Diagn.* 1998;18(13):1437–49.
17. Alfarawati S, et al. *PLoS Genet.* 2012;8(10):e1003025.
18. Munné S, et al. *Fertil Steril.* 2000;73(6):1209–18.
19. Verlinsky Y, et al. *Reprod Biomed Online.* 2005;11(2):219–25.
20. Beyer CE, et al. *J Assist Reprod Genet.* 2017;34(11):1483–1492.
21. Gardner RJM, et al. *Chromosome Abnormalities and Genetic Counseling*, Fourth Edition. OUP USA, 2011.
22. Morin SJ, et al. *Fertil Steril.* 2017;107(1):19–26.
23. Liao Y, et al. *Mol Cytogenet.* 2014;7:39.
24. Durmaz B, et al. *Genet Couns.* 2016;27(4):461–470.
25. Batista DA, et al. *Am J Med Genet.* 1994;53(3):255–63.
26. Pellestor F, et al. *Hum Reprod Update.* 2011;17(4):476–94.
27. Gorski JL, et al. *Am J Med Genet.* 1988;29(2):247–61.
28. Hook EB, et al. In: Hook EB, Porter IH (eds) *Population Cytogenetics--Studies in Humans.* Academic Press, 1977:63–79.
29. Simpson JL, et al. *Fertil Steril.* 2019;112(4):622–630.
30. Schmid W. *Cytogenetics.* 1962;1:199–209.
31. Stephenson MD. *Fertil Steril.* 1996;66(1):24–9.
32. Dong Y, et al. *J Int Med Res.* 2012;40(6):2274–83.
33. Bocian E, et al. *J Appl Genet.* 2005;46(1):109–14.

34. Morel F, et al. *Hum Reprod.* 2007;22(1):136–41.
35. Anton E, et al. *Cytogenet Genome Res.* 2005;111(3–4):297–304.
36. Kaiser P. *Hum Genet.* 1984;68(1):1–47.
37. Nussbaum RL, et al. *Thompson & Thompson Genetics in Medicine.* 7th ed. Saunders/Elsevier, 2007.
38. Sutherland GR, et al. *Clin Genet.* 1976;10(1):54–9.
39. Merrion K, et al. *J Assist Reprod Genet.* 2019;36(12):2557–2561.
40. Madan K. *Hum Genet.* 1995;96(5):503–15.
41. Pettenati MJ, et al. *Am J Med Genet.* 1995;55(2):171–87.
42. Anton E, et al. *Fertil Steril.* 2006;85(3):661–6.
43. Jacobs PA, et al. *J Med Genet.* 1992;29(2):103–8.
44. Wellesley D, et al. *Eur J Hum Genet.* 2012;20(5):521–6.
45. Warburton D. *Am J Hum Genet.* 1991;49(5):995–1013.
46. Liehr T, et al. *Int J Mol Med.* 2007;19(5):719–31.
47. Bartsch O, et al. *Eur J Hum Genet.* 2005;13(11):1192–204.
48. Lejeune J. *Pediatrics.* 1963;32:326–37.
49. Estop AM, et al. *Hum Genet.* 2000;106(5):517–24. Erratum in: *Hum Genet.* 2000;107(1):95.
50. Gianaroli L, et al. *Hum Reprod.* 2002;17(12):3201–7.
51. Vozdova M, et al. *Fertil Steril.* 2011;96(1):e66–70.
52. Blanco J, et al. *Cytogenet Cell Genet.* 1998;83(3–4):275–80.
53. Pang MG, et al. *Hum Reprod.* 1999;14(5):1266–73.
54. Boerke A, et al. *Theriogenology.* 2007;68 Suppl 1:S147–55.
55. Munne S, et al. *Reprod Biomed Online.* 2005;10:363–9.
56. Guttenbach M, et al. *Hum Reprod.* 1997;12(3):468–73.
57. Honda H, et al. *Hum Genet.* 1999;105(5):428–36.
58. Moosani N, et al. *Fertil Steril.* 1995;64(4):811–7.
59. Rives N, et al. *Hum Genet.* 1999;105:266–272.
60. Anton E, et al. *Syst Biol Reprod Med.* 2011;57(6):268–78.
61. Mateu-Brull E, et al. *J Assist Reprod Genet.* 2019;36(12):2547–2555.
62. Fischer J, et al. *Fertil Steril.* 2010;94(1):283–9.
63. Munné S, et al. *J Assist Reprod Genet.* 1998;15(5):290–6.
64. Otani T, et al. *Reprod Biomed Online.* 2006;13(6):869–74.
65. ESHRE PGT-SR/PGT-A Working Group, et al. *Hum Reprod Open.* 2020;2020(3):hoaa017.
66. Munne S, et al. *Prenat Diagn.* 1998;18:1459–1466
67. Verlinsky Y, et al. *Mol Hum Reprod.* 1999;5(2):89–95.
68. Conn CM, et al. *Hum Genet.* 1998;102(1):117–23.
69. DeUgarte CM, et al. *Fertil Steril.* 2008;90(4):1049–54.
70. Munne´ S. *Reprod Biomed Online.* 2002;4:183–196.
71. Velilla E, et al. *Reprod Biomed Online.* 2002;4:210–217.
72. Li M, et al. *Fertil Steril.* 2005;84:1395–1400.
73. Lukaszuk K, et al. *AJP Rep.* 2015;5:e172–e175.
74. Tan YQ, et al. *Hum Reprod.* 2013;28:2581–2592.
75. Treff NR, et al. *Fertil Steril.* 2011;96:e58–e65.
76. van Uum CM, et al. *Eur J Hum Genet.* 2012;20:938–944.
77. Yanxin X, et al. *J Assist Reprod Genet.* 2017;35(1):177–186.
78. Alfarawati S, et al. *Hum Reprod.* 2011;26(6):1560–74.
79. Fiorentino F, et al. *Hum Reprod.* 2011;26:1925–1935.
80. Li G, et al. *Syst Biol Reprod Med.* 2014;60:119–124.
81. Tobler KJ, et al. *J Assist Reprod Genet.* 2014;31:843–850.
82. Bono S, et al. *Prenat Diagn.* 2015;35:938–944.
83. Wang L, et al. *Biol Reprod.* 2014;91:37.
84. Campos-Galindo I, et al. *Efecto de los reordenamientos cromosómicos equilibrados sobre la tasa de aneuploidías y desequilibrios cromosómicos en biopsia de trofoectodermo.* Poster P-087, ASEBIR, Cáceres, Spain, 2019.
85. Boynukalin FK, et al. *J Assist Reprod Genet.* 2021;38(2):387–396.
86. Franssen MT, et al. *BMJ.* 2006;332(7544):759–63.Erratum in: *BMJ.* 2006;332(7548):1012.
87. Bedaiwy MA, et al. *Fertil Steril.* 2016;106(3 Suppl):e343.
88. Coonen E, et al. *Hum Reprod Open.* 2020;2020:hoaa043.
89. Harris BS, et al. *FS Rev.* 2020;2:43–56.
90. Huang C, et al. *J Assist Reprod Genet.* 2019;36(11):2325–2331.

Mosaicism in Preimplantation Embryos

13

Maurizio Poli and Antonio Capalbo

INTRODUCTION

Mosaicism is defined as the presence of two or more genotypically different cell lines in a given organism, embryo, or cell line. Chromosomal imbalances present in all cells of an organism are defined as uniform aneuploidies and mostly derive from aberrant meiotic processes occurring during the final stages of oocyte maturation (1–3). Differently, embryo mosaicism originates from mitotic segregation errors occurring in post-zygotic developmental stages when progenitor cells sequentially divide into daughter cells. These defective events can occur in either euploid or aneuploid conceptuses (4).

Although less than 2% of prenatal specimens (i.e., amniocentesis or chorionic villus sampling [CVS]) have shown evidence of chromosomal mosaicism, some studies have detected it in up to 73% of cleavage-stage preimplantation embryos (5). Because of the potential impact the presence of multiple cell lines in an embryo could have on both clinical outcomes and preimplantation diagnostic efficacy, mosaicism has recently attracted special attention from scientists working in human embryology and reproductive genetics. The latest application of NGS-based technologies in preimplantation genetic testing has the potential to provide technically higher sensitivity in detecting intermediate chromosome copy number variations and, consequently, improved tools for identifying mosaicism in trophectoderm (TE) biopsies. In this chapter, we discuss the fundamental concepts on embryo mosaicism formation and provide data from the latest studies available on this topic. Finally, we contextualize these findings in the clinical landscape, providing a logic framework for the interpretation and management of embryo mosaicism in IVF treatments.

ORIGINS OF MOSAICISM

The specific cytogenetic mechanisms involved in the formation of chromosomal mosaicism in embryos have not yet been fully characterized. Current hypotheses are based on theoretical models that still lack experimental evidence and confirmation in epidemiological studies. Defective mitotic processes suspected to have a causative role in chromosomal mosaicism include chromatid non-disjunction, anaphase

DOI: 10.1201/9781003024941-13

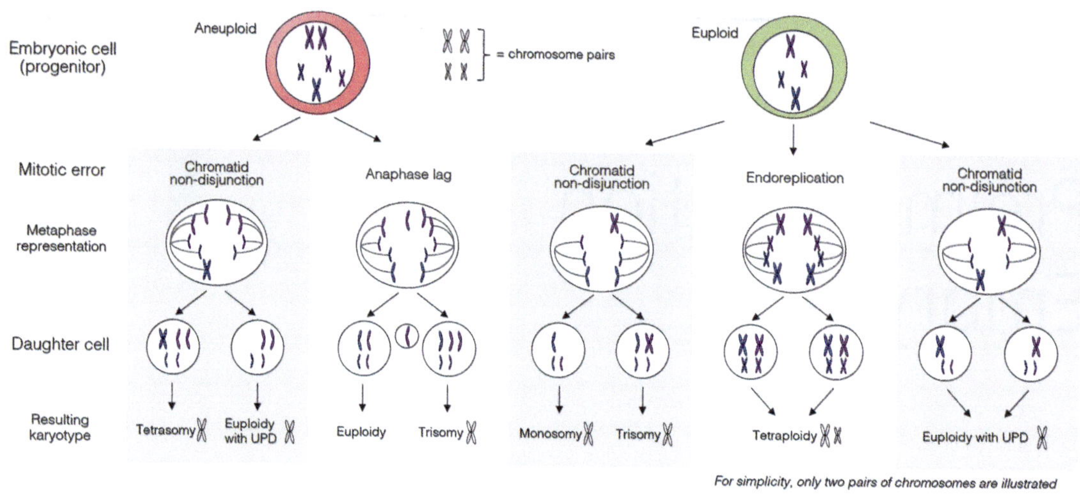

FIGURE 13.1 Most common mitotic errors that could lead to chromosomal mosaicism. Examples from both an aneuploid and a euploid progenitor cell are followed through their typical metaphase plate and the embryo's resulting karyotype.

lag, endoreplication, and trisomy rescue (which may lead to UPD) (6–8). UPD consists of the presence of chromosome copies (two or more) originated from a single parent and is correlated with pathological conditions determined by aberrant genetic imprinting (i.e., Prader–Willi and Angelman syndromes). A visual representation of these defective mitotic processes is shown in Figure 13.1.

Defective mitotic events appear to be mainly caused by faulty cell cycle control, centrosome and mitotic spindle functional deficiency, and inefficiency in the chromosome cohesion apparatus (9). Although ovarian stimulation regimens were also suggested to impact the incidence of mitotic errors (10), this hypothesis has yet to be validated through more comprehensive, sensitive, and up-to-date technologies and properly designed studies.

Depending on whether the defective event occurs in a euploid or aneuploid embryo, the ensuing mosaicism can be classified in two main categories:

Aneuploid—aneuploid: if the original zygote carried an aberrant set of chromosomes and the defective process gives origin to secondary aneuploid cell lines. When multiple defective cell lines accrue as a result of consecutive aberrant mitotic processes, the embryo is defined as a "chaotic mosaic."

Euploid—aneuploid: generally deriving from a euploid zygote undergoing defective mitosis thus originating aneuploid cell lines within the embryo. More unlikely, this conformation could derive from an aneuploid zygote where the chromosomal defect is rescued through an unbalanced chromosomal segregation. In this case, UPD can ensue if homologous chromosomes with the same parental origin are left in the "corrected" euploid cell. When, typically due to defective cytokinesis, different ploidy statuses ($2n$, $4n$) are present in the same embryo, its configuration is described as "ploidy mosaic."

THE MOSAIC EMBRYO

The developmental stage at which the mitotic error occurs has a direct impact on the proportion of embryonic cells showing the secondary karyotype (6) (Figure 13.2). Typically, the developmental stage of the

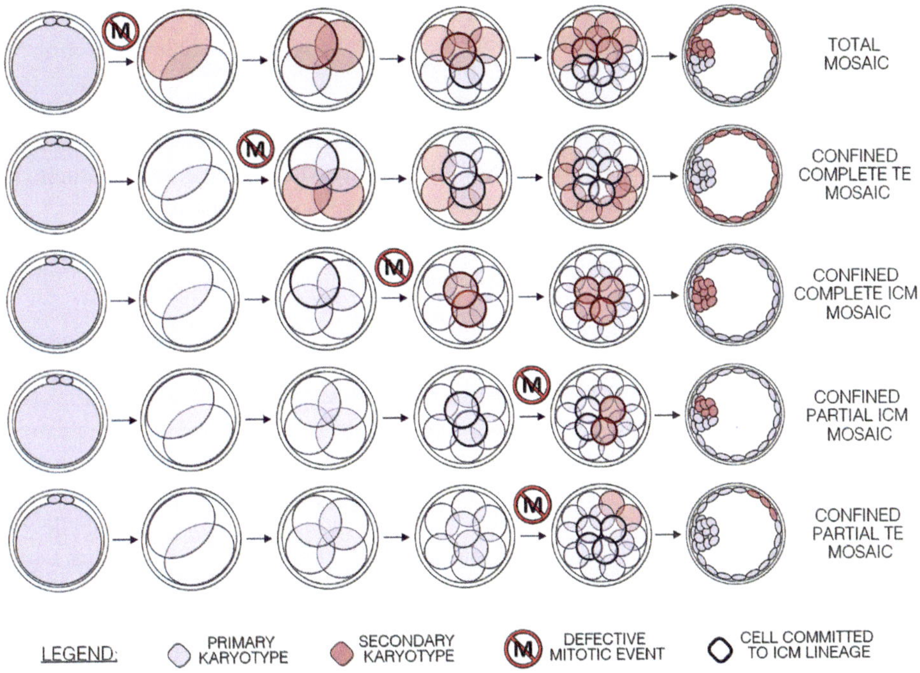

FIGURE 13.2 Theoretical model of propagation of a mitotic error through embryo development. The development stage at which the error occurs, and cellular lineage commitment of the cells affected determines the type of mosaicism derived. Note that differential cellular fitness or possible clonal depletion are not considered in this model.

embryo and the differentiation pathway undertaken by the cell affected by the mitotic error can determine the type of mosaicism and its extent in the embryo. In theory, the earlier the error occurs, the higher the portion of the embryo showing the secondary karyotype. Chromosomal abnormalities occurring at an early stage may have a deleterious effect on embryo development and arrest its progression to later stages. The evidence that chromosomal mosaicism is detected at higher frequencies in cleavage-stage embryos than in blastocyst-stage ones may be attributed to this process (11). In cases where the mitotic error affects a smaller portion of the embryo, chromosomal imbalances (especially if complex) may cause cell cycle arrest, resulting in the reduction or complete clonal depletion of the specific cell line within the ensuing embryo. This model would explain why mosaicism is rarely detected at later reproductive stages (i.e., prenatal diagnosis or live birth) (11). Additionally, the cell lineage commitment of the cells affected by the mitotic error determines whether mosaicism is disseminated throughout the embryo or restricted to a specific tissue (e.g., inner cell mass [ICM], TE) (12). As a consequence, the earlier the defective mitotic event occurs in development, the higher the chance that mosaicism will affect both embryonic and extra-embryonic tissues (Figure 13.2).

MOSAICISM DETECTION TOOLS

Mosaicism was first detected in human embryos using fluorescent *in situ* hybridization (FISH) on blastomeres collected from cleavage-stage embryos (13). At that time, embryos were cultured on for three days in the laboratory and only direct assessment of chromosomes (e.g., FISH) was possible. Since then, several advancements have been made in both IVF and genetics laboratories. Nowadays, multicellular trophectoderm biopsy is the gold standard for preimplantation genetic testing and the technology employed for indirect cytogenetic testing has gone through several updates from FISH to comprehensive technologies (i.e., array comparative genomic hybridization [aCGH], array single nucleotide polymorphism [aSNP], quantitative polymerase chain reaction [qPCR], and eventually culminating in next-generation sequencing [NGS]). With the introduction of NGS, analytical resolution has significantly increased, providing a higher detection capability of subtle chromosomal variations. The ability of NGS to discriminate intermediate chromosomal values was validated using known mixtures of euploid and aneuploid cell lines at different ratios.

The 2019 PGDIS guidelines on the clinical use of putative mosaic embryos recommended that i) embryos with intermediate chromosomal values can be considered for transfer after uniformly euploid, ii) embryos with lower mosaicism grade should be prioritized over those with higher grades, and that iii) embryos with mosaicism involving chromosomes associated with uniparental disomy or syndromes compatible with life should be used as a last resource. Moreover, patients should be fully informed of the limitations and accuracy of PGT-A testing in detecting mosaicism.

In fact, it has been shown that parallel assessment of biological specimen and cell line mixture models leads to suboptimal sensitivity and specificity in mosaicism calling (14), suggesting limited applicability of intermediate chromosomal value identification in trophectoderm biopsies. This finding was further supported by Victor and colleagues by showing that the analysis of subsequent biopsies from alleged mosaic embryos led to inconsistent confirmatory results (uniformly euploid profiles, mosaic profiles with reciprocal aneuploidy, or intermediate profiles albeit at different degrees [i.e., low vs. high]) (15). Further, the same group showed that the transfer of 100 embryos with single mosaic aneuploidy led to comparable clinical outcomes to fully euploid embryos, irrespective of their level of mosaicism. This study suggested that intermediate chromosomal levels don't provide sufficient accuracy and robustness for mosaicism diagnosis.

Popovic and colleagues re-analyzed the karyotype of outgrowths of previously biopsied human blastocysts, showing that if mosaicism was not considered during analysis, the concordance between TE biopsy and outgrowth karyotypes was 100%, with 0% false negative and positive results, compared with an accuracy of 80%, 0% false negative, and 18.5% false positive results when mosaicism was considered (16). Apart from technical variations intrinsic to the analytical methodology employed, these findings could be explained by the ability of an embryo to facilitate targeted apoptosis in the abnormal cells, as well as differential fitness between euploid and aneuploid cells, eventually resulting in karyotype rescue in subsequent developmental stages. Alternatively, profile plots suggesting the presence of mosaicism may be determined by biological phenomena alternative to mosaicism that are occurring in part or the whole embryo (i.e., intermediate S-phase cell analysis, monosomy/trisomy in polyploidy embryo). In fact, the positive predictive value of putative mosaicism analysis performed in TE biopsies has been consistently shown to be extremely poor, not only at the embryo level, but also at subsequent developmental stages (i.e., day 12 outgrowth) (16). For all these reasons, reporting a diagnosis of mosaicism in embryos can be considered conceptually biased and flawed (17).

LIMITATIONS IN DIAGNOSING MOSAICISM

Cleavage-stage human embryos have been associated with an incidence of mosaicism as high as 90% (18–21). However, these extremely high rates may be overestimated due to low sensitivity of the methodology employed in these studies (e.g., FISH). Moreover, one abnormal cell was generally sufficient to determine presence of mosaicism, providing a weak criterium for mosaicism detection (6). In fact, it has been shown that FISH has a conservative positive error rate of 10% (22), which inherently results in high detection of abnormal results even in uniformly euploid embryos. Parallel FISH and comprehensive chromosome testing (CCT) analyses of cells deriving from the same embryo showed that the technique employed for analysis has a significant impact on the rate of mosaicism detected (100% mosaicism diagnosis in FISH vs. 31% in SNP array) (22).

Moreover, it has been shown that CCT analysis of multifocal TE biopsies leads to a lower incidence of mosaicism compared to cleavage-stage embryos' blastomere analysis (5% vs. 31%, respectively) (23). This difference in mosaicism incidence may be due to multicellular samples providing a more robust and representative analytical template compared to single blastomere analysis. However, a selective depletive mechanism of aneuploid/mosaic cells occurring between cleavage and blastocyst stage may also play a role.

Nonetheless, it should be noted that the combination of TE biopsy and NGS-based copy number analysis is not a bias-free strategy for mosaicism detection. Firstly, TE specimens may have higher heterogeneity in terms of quantity and quality compared to precise mixtures derived from cell lines, increasing variability in analytical data signal. Moreover, sampling is prone to technical and experimental variability that results in a data distribution outside the assigned thresholds. In practice, when low limits are employed both the incidence of mosaicism detection and the overall diagnostic uncertainty increase. This means that when the thresholds used for detection of intermediate chromosome values are placed as low as 20%, we identify an increasing number of off-range datapoints which are interpreted as chromosomal mosaicism, although the actual chromosomal status of the embryo may be different. As a consequence, a minority of euploid samples produce datapoints outside the ranges used to define normal profiles, being automatically reported as abnormal (or mosaic in this case), or vice versa. Nonetheless, a diagnosis of embryo mosaicism generates a course of action that often impacts IVF treatment management and overall clinical outcomes. A recent web-based survey showed that around half of the IVF centers performing PGT consider an embryo as mosaic when abnormal cells are detected in 20% of the tested sample (24), resulting in reduced embryo availability for transfer.

For these reasons, the manipulation of thresholds used for ploidy status calling should be employed with prudence and driven by clinical studies aimed at determining their positive and negative predictive values and clinical utility.

Finally, the extent to which a randomly sampled population of cells collected during biopsy is representative of the rest of the blastocyst is the main diagnostic uncertainty associated with PGT-A. This aspect becomes even more prominent when intermediate CNV are detected. This diagnostic uncertainty should therefore be acknowledged by denominating these instances as "putative mosaicism."

IMPACT OF MOSAICISM ON TREATMENT OUTCOME

The lack of unequivocal evidence on the actual chromosomal status and reproductive potential of mosaic embryos generates uncertainty on their employability and safety for use in clinical treatments. As discussed, by widening the thresholds for identification of mosaicism, the number of embryos falling into this category significantly increases, further impacting the number of embryos considered healthy and available for transfer. In turn, this approach results in lower efficacy of the IVF treatment, especially for those patients with reduced ovarian reserve that produce a limited number of embryos. However, this diagnostic impasse raises questions on whether embryos generating profiles with intermediate chromosome copy values actually pose a health threat to receiving patients or if their abandonment is one of the biggest wastages ever recorded in reproductive medicine.

A diagnosis of chromosomal mosaicism is usually followed by genetic counseling sessions aimed at explaining not only what mosaicism consists of, but also what scientific and clinical evidence is available both in terms of diagnostic accuracy and prognostic values. Recently, it has been shown how a diagnosis of chromosomal mosaicism is usually followed by the abandonment of the putatively affected embryo (25). In this study by Munné and colleagues, only 143 out of 6,368 (2.2%) embryos diagnosed with some types of mosaicism were transferred to the patients, whilst the vast majority were discarded or left in the cryotank. However, after the transfer of thousands of putative mosaic embryos worldwide, only a single case of mosaic pregnancy has been reported so far (26), suggesting that the detection of intermediate CNVs at the PGT stage is hardly related to adverse clinical outcomes. Therefore, it is likely that a large portion of the abandoned embryos may actually be fully euploid and with high reproductive potential.

Despite the diagnostic uncertainty and unknown significance of intermediate chromosome copy number findings, an increasing number of laboratories started to include in their diagnostic reports the presence of putative mosaicism, leaving the decision on how to manage the employment of such embryos for treatments to IVF clinics. However, only a few studies had been conducted to assess the reproductive potential of putative mosaic embryos. In the previously mentioned study by Munné and colleagues, embryo transfer outcomes and pregnancies derived from euploid and mosaic embryos (including complex, single, double, monosomic, trisomic, and segmental at both low mosaicism of 20–40%, and high mosaicism > 40%) were followed up. Cumulative results showed that uniformly euploid and putative mosaic embryos produced significantly different results in terms of implantation and miscarriage rates (25). These findings were in contrast with another study by Fragouli and colleagues, where no statistical difference was observed between euploid embryos and embryos displaying several types of mosaic-consistent diagnosis (implantation rate, $P = 0.1$; miscarriage rate, $P = 0.462$) (27). A subsequent study by Spinella and colleagues showed that the grade of mosaicism detected significantly impacts embryos' reproductive potential with low-grade mosaics and uniformly euploid embryos producing similar implantation and miscarriage rates (28). However, high-grade mosaic embryos showed significantly reduced implantation and clinical pregnancy compared to low-grade mosaic embryos. Interestingly, combining the results deriving from low- and high-grade mosaic embryos, a decrease in clinical performance of the group was observed in terms of implantation, but not in miscarriage rates (28). The contrasting results on the reproductive competence of mosaic embryos provided by these studies, combined with experimental limitations due to their retrospective design, didn't help the decision-making process in the case of putative mosaic embryos. In fact, these studies showed crucial biases in terms of population selection as patients receiving mosaic embryos were not matched with those receiving uniformly euploid ones, thus crucially limiting the strength of their findings.

For instance, patients with good prognosis were likely to have several embryos to choose from for transfer, thus reducing the number of mosaic embryos replaced in this group of patients. Similarly, poor prognosis patients were more likely to have fewer embryos to select from, increasing their chance to receive mosaic embryos. This way, uniformly euploid embryos were systematically transferred to patients with better prognosis and vice versa. Moreover, aneuploid embryos may have been erroneously classified as highly mosaic embryos, biasing the basic characteristics of the group and the corresponding outcomes. Finally, part of the mosaic embryos employed for transfer may have been patient's last resource of embryos after a few failed transfers of uniformly euploid embryos. There is evidence that success rates decrease as subsequent transfers are attempted, even in cycles not involving PGT analysis. All these limitations in experimental design are likely to have critically biased the results obtained from these studies, underestimating the true reproductive potential of putative mosaic embryos.

NON-SELECTION TRIAL

In order to address the experimental limitations that affected previous studies, we designed and registered a non-selection trial to generate strong evidence on the reproductive potential of putative mosaic embryos. For this purpose, we enrolled a total of 1,335 TE biopsy cases which led to 897 single embryo transfer cycles (490 cases resulted in all aneuploid embryos). When the biopsies were analyzed, embryos showing intermediate values below 50% for any of the chromosomes were automatically reported to the requesting IVF clinic as uniformly euploid. Clinics then prioritized the embryo for transfer based on its morphology. Following the transfer, raw data about chromosomes profiles were unblinded and intermediate chromosome copy numbers recorded, leading to three comparative groups: uniformly euploid (n = 484), low mosaic (20%–30%, n = 282), and moderate mosaic (30%–50%, n = 131). Clinical outcomes were followed up to compare positive pregnancy, biochemical pregnancy loss, implantation, miscarriage, and live birth rates across all groups (Table 13.1).

TABLE 13.1 Clinical Outcomes of the Only Prospective Non-Selection Trial on Mosaic Embryo Transfer Available to Date

	EUPLOID	*LOW PUTATIVE MOSAIC (20–30% VARIATION)*	*MODERATE PUTATIVE MOSAIC (30–50% VARIATION)*
Number of SET n	484	282	131
Confirmed pregnancy rate n. (%) (95%: range)	241/484 (49.8%) (95%CI = 45.2%–54.3%)	136/282 (48.2%) (95%CI = 42.3%–54.2%)	63/131 (48.1%) (95%CI = 39.3%–57.0%)
Miscarriage rate n. (%) (95%C: range)	29/241 (12.0%) (95%CI = 8.2%–16.8%)	15/136 (11.0%) (95%CI = 6.3%–17.5%)	8/63 (12.7%) (95%CI = 5.6%–23.5%)
Live birth rate n. (%) (95%C: range)	210/484 (43.4%) (95%CI = 38.9%–47.9%)	121/282(42,9%) (95%CI = 37.1%–48.9%)	55/131 (42.0%) (95%CI = 33.4%–50.9%)

Statistical analysis showed no significant difference in any of the outcomes measured, suggesting that the detection of low-moderate mosaicism doesn't provide any clinically useful predictive value (29).

These findings suggest that the abandonment of putative mosaic embryos leads to a decrease in cost-efficacy of the IVF treatment due to potential wastage of reproductively competent embryos, as well as unnecessary counseling sessions, additional embryo screening tests, and extra IVF treatment cycles.

CLINICAL CONSEQUENCES OF MOSAICISM

We have seen that not only embryo mosaicism may be overcalled from a biological and a diagnostic standpoint, but also that, if present at low and moderate grades, no effects on clinical outcomes or reproductive performance could be detected when assessed through a non-selection trial. Arguably, most negative effects of a putative embryo mosaicism diagnosis could be minimized by weighing the importance of the details reported to the patient, ensuring that their relevance is explained based on unbiased and comprehensive data.

As discussed, a diagnosis of mosaicism can have severe effects on the fate and employment of embryos for clinical use. In turn, the abandonment of putative mosaic embryos can drastically reduce the efficacy of an IVF treatment. However, until conclusive substantiation on the clinical impact of mosaicism on embryo biology and IVF treatment are gathered, the decisions following a mosaicism diagnosis should be carefully weighed against current evidence on associated consequences at the clinical, psychological, and financial levels. All these aspects should be covered during a thorough reproductive and genetic counseling session following the diagnostic report.

CONFIRMATORY PRENATAL DIAGNOSIS FOR MOSAIC EMBRYOS

Prenatal diagnosis is always advised in IVF treatments involving PGT. Importantly, PND is even more strongly recommended when potentially mosaic embryos categorized as high risk are transferred (i.e., trisomy 13, 18, 21, and 22). In these circumstances, amniocentesis should be preferred to other prenatal investigation strategies. Because certain embryos could be affected by mosaicism in the extra-embryonic cell lineage, providing a prenatal diagnosis based on chorionic villi sampling (CVS) may not be representative of the fetus. Similarly, non-invasive prenatal screening (NIPT) is based on DNA fragments derived from cytotrophoblast cells undergoing apoptosis, hence not representing the actual fetal karyotype (30, 31). Therefore, both CVS and NIPT would not be advantageous in determining the true chromosomal status of the fetus. On the other hand, due to the fact that a high proportion of fetal cells are contained in the amniotic fluid, amniocentesis is the diagnostic method of choice in cases of suspected fetal mosaicism (32). Nonetheless, invasive PND is associated with a well-defined procedural risk for the fetus and the pregnancy which may not be justified in all cases. Hopefully, the collection of larger evidence on the clinical risks associated with the transfer of putative mosaic embryos will further define for which cases PND is recommended and for which the risk is greater than the potential benefits.

CONCLUSIONS

In this chapter, we described how an altered chromosomal copy number can originate from both technical variation and biological phenomena other than mosaicism. Inherent inaccuracy in the methods employed for chromosome copy number estimation, as well as the inconsistency of the criteria used to define mosaicism detection, has led to an overestimation of its occurrence in IVF embryos. Since technical and biological explanations, alternative to embryo chromosomal mosaicism, are possible, these altered chromosomal profiles would be better reported as "profiles consistent with mosaicism" or "putative mosaic profile," rather than as a definite diagnosis. For this reason, mosaicism should be acknowledged in the PGT consent form as a limitation of the testing methodology as it is commonly referred to during PND (i.e., CVS and amniocentesis). Also, it should be stated that this shortcoming should not impose risks higher than IVF without PGT, considering that it will not have a major impact on clinical outcomes (as demonstrated by the low incidence observed in human blastocysts and pregnancies) and by the positive outcomes of published randomized clinical trials (RCTs) on PGT-clinical effectiveness (33–35). Also, it should be considered that, in the light of most recent evidence, avoiding mosaicism reporting might be preferable as it improves diagnostic accuracy, drastically reducing the occurrence of false positive calls and associated clinical consequences (15, 16).

The only prospective non-selection trial on the subject demonstrated that putative mosaic embryos produce similar clinical outcomes to uniformly euploid embryos (29). Further gathering of additional clinical data in similarly designed studies is of crucial importance for the production of uncontroversial evidence on the predictive value of mosaicism findings. This body of evidence will eventually improve decision-making procedures, fine-tuning of PGT-A cycles, and relieve patients and professionals of uncertain diagnostic outcomes.

REFERENCES

1. Hassold T, et al. *Nat Rev Genet.* 2001;2(4):280–91.
2. Ottolini CS, et al. *Nat Genet.* 2015;47(7):727–35.
3. Capalbo A, et al. *Hum Reprod Update.* 2017;23(6):706–22.
4. Grati FR, et al. *Reprod Biomed Online.* 2018;36(4):442–9.
5. van Echten-Arends J, et al. *Hum Reprod Update.* 2011;17(5):620–7.
6. Capalbo A, et al. *Fertil Steril.* 2017;108(6):1007–1015.e3.

7. Taylor TH, et al. *Hum Reprod Update.* 2014;20(4):571–81.
8. Gueye N-A, et al. *Fertil Steril.* 2014;101(1):232–6.
9. McCoy RC. *Trends Genet.* 2017;33(7):448–63.
10. Munne S, et al. *Hum Reprod.* 1997;12(4):780–4.
11. McCoy RC. *Trends Genet.* 2017;33(7):448–63.
12. Liu J, et al. *Biol Reprod.* 2012;87(6):148.
13. Delhanty JDA, et al. *Hum Mol Genet.* 1993;2(8):1183–5.
14. Goodrich D, et al. *J Assist Reprod Genet.* 2016;33(11):1473–80.
15. Victor AR, et al. *Fertil Steril.* 2019;111(2):280–93.
16. Popovic M, et al. *Hum Reprod.* 2019;34(4):758–69.
17. Capalbo A, et al. *Fertil Steril.* 2017;107(5):1098–106.
18. Daphnis DD, et al. *Hum Reprod.* 2005;20(1):129–37.
19. Gonzalez-Merino E, et al. *Genet Test.* 2003;7(2):85–95.
20. Santos MA, et al. *Hum Reprod.* 2010;25(8):1916–26.
21. Baart EB, et al. *Hum Reprod.* 2006;21(1):223–33.
22. Treff NR, et al. *Mol Hum Reprod.* 2010;16(8):583–9.
23. Capalbo A, et al. *Hum Reprod.* 2016;32(3):492–8.
24. Weissman A, et al. *Fertil Steril.* 2017;107(5):1092–7.
25. Munné S, et al. *Fertil Steril.* 2017;108(1):62–71.e8.
26. Kahraman S, et al. *Hum Reprod.* 2020;35(3):727–33.
27. Fragouli E, et al. *Hum Genet.* 2017;136(7):805–19.
28. Spinella F, et al. *Fertil Steril.* 2018;109(1):77–83.
29. Capalbo A, et al. *Fertil Steril.* 2020;114(3):e414.
30. Mardy A, et al. *Am J Med Genet Part C Semin Med Genet.* 2016;172(2):118–22.
31. Van Opstal D, et al. *Expert Rev Mol Diagn.* 2016;16(5):513–20.
32. Grati FR, et al. *Best Pract Res Clin Obstet Gynaecol.* 2017;42:39–52.
33. Forman EJ, et al. *Fertil Steril.* 2013;100(1):100–107.e1.
34. Yang Z, et al. *Mol Cytogenet.* 2012;5(1):24.
35. Scott RT, et al. *Fertil Steril.* 2012;97(4):870–5.

14 Embryo Cell-Free DNA in the Culture Medium and Its Potential for Non-Invasive Aneuploidy Testing

Carmen Rubio, Luis Navarro-Sánchez, and Carmen M. García-Pascual

INTRODUCTION

Aneuploidy rates are high in human embryos and increase with advancing female age [1, 2]. Trisomic and monosomic embryos account for at least 10% of human pregnancies, and that incidence may exceed 50% for women nearing the end of their reproductive lifespan [3, 4]. Age-related changes result in higher aneuploidy rates and increased spontaneous abortions, reducing ongoing implantation rates. Aneuploidy may also contribute to infertility in other contexts, such as in cases of recurrent miscarriages [5], repetitive implantation failure [6], or male infertility [7, 8].

Aneuploidy testing is central to the practice of assisted reproduction, where the ultimate goal is to achieve a healthy birth, ideally from a single transferred embryo. Consequently, many methods have arisen to enable embryo aneuploidy testing. Preimplantation genetic testing for aneuploidy (PGT-A) was originally incorporated into *in vitro* fertilization (IVF) programs to improve pregnancy rates per transfer and to decrease miscarriage rates. Improvements in the time to conceive, the number of transfers required, and the associated cost are important additional benefits of PGT-A compared to conventional IVF [9, 10]. PGT-A currently requires embryo manipulation, specifically a trophectoderm (TE) biopsy, to retrieve four to eight cells that then undergo analysis of chromosome copy number by next-generation sequencing (NGS). Because embryo biopsies require dedicated equipment and highly trained personnel, they are costly and time-intensive. Interest in a non-invasive approach to assess chromosome number is therefore increasing. Such an approach could offer the benefits of invasive PGT-A, like improved live birth at the first embryo transfer, reduction of miscarriage rates and multiple pregnancies, and shortened time to pregnancy, without the limitations of embryo manipulation and biopsy.

DOI: 10.1201/9781003024941-14

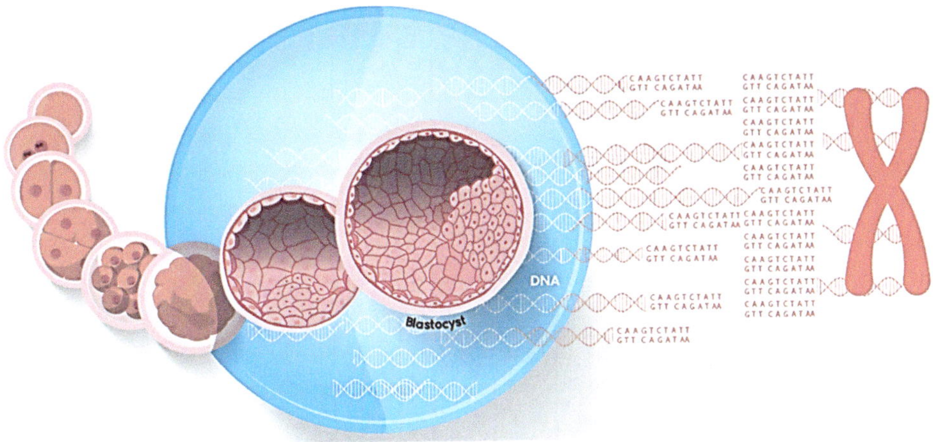

FIGURE 14.1 Cell-free DNA is secreted by preimplantation embryos during in vitro cursiva culture.

A promising option for non-invasive PGT-A (niPGT-A) uses the spent blastocyst medium (SBM) in which the embryo is cultured (Figure 14.1). Initial isolation of embryo cell-free DNA (cfDNA) from SBM as a proof-of-concept demonstrated the method's potential to identify embryo chromosome copy number [11]. In contrast to TE biopsy, embryo cfDNA collection requires no specialized training and imposes negligible disruption to the embryo. Other non-invasive techniques to predict or assess IVF outcomes include time-lapse morphokinetics, proteomics, and metabolomics. However, unlike embryo cfDNA analysis, these techniques are indirect assessments or proxies for chromosome copy number.

Despite the promise of this non-invasive technology, several questions remain: 1) Can different embryo culture protocols be applied? 2) How long should the medium be in contact with the embryo before it is collected? 3) Are assisted hatching (AH) or previous vitrification needed to obtain sufficient cfDNA? 4) Is the amount of cfDNA collected influenced by the embryo quality? 5) How can contamination from cumulus cells be minimized? 6) Do the amplification methods, sequencing platforms, or bioinformatics analyses used impact the results? And the core question: can cfDNA from SBM be reliably used for aneuploidy testing in human blastocysts? In this chapter, we review the studies published in this field so far and discuss our own research.

STATE OF THE ART

In 2016, Shamonki et al. established that the embryonic culture medium contained cfDNA and that this DNA could be used to detect embryonic aneuploidy [11]. DNA was found in most of the media samples after amplification (55 out of 57). While array comparative genomic hybridization (aCGH) could provide a diagnosis in only two of these samples, results in both cases were concordant with TE biopsy results from the same embryo. Subsequent studies also evaluated their SBM results via comparison with an established reference method, usually the TE biopsy. Comparisons are enabled by several validation measures, including: the ploidy concordance rate, or the overall agreement between the SMB and reference samples as euploid vs. aneuploid; the false positive rate, i.e., when the SBM is aneuploid and the reference sample is euploid; and the false negative rate, i.e., when the SBM is euploid and the reference sample aneuploid. An additional measure is the informativity rate of SBM, typically defined as the percentage of diagnosable SBM samples.

Informativity rates and overall concordance rates for ploidy are variable, ranging from 55.6% to 97.6% and from 33.3% to 89.1%, respectively, in studies using TE biopsy as a reference [12–18] (Table 14.1). One study comparing SBM to polar body (PB) biopsy found an 81.8% informativity rate and a 72.2%

TABLE 14.1 Studies Comparing Ploidy Results from Spent Blastocyst Medium (SBM) vs. Trophectoderm (TE) or Polar Body (PB) Biopsy

STUDY	*NO. OF SBM*	*INFORMATIVE MEDIA*	*CONCORDANCE TE-SBM*	*FALSE POSITIVES*	*FALSE NEGATIVES*
Shamonki et al., 2016	57	96.5% (55/57)	33.3% (2/6)[1]	-	-
Feichtinger et al., 2017[2]	22	81.8% (18/22)	72.2% (13/18)	5.6% (1/18)	22.2% (4/18)
Vera-Rodríguez et al., 2018	56	91.1% (51/56)	33.3% (17/51)	-	66.7% (34/51)
Ho et al., 2018	41	97.6% (40/41)	65.0% (26/40)	-	-
Huang et al., 2019	52	92.3% (48/52)	89.1% (41/46)[3]	2.2% (1/46)	8.7% (4/46)
Yeung et al., 2019	168	D5: 55.6% (50/90) D6: 84.6% (66/78)	D5: 76% (38/50) D6: 71.2% (47/66)	D5: 12% (6/50) D6: 13.6% (9/66)	D5: 12% (6/50) D6: 15.2% (10/66)
Rubio et al., 2019[4]	115	D5: 81.8% (27/33) D6/7: 98.8% (81/82)	D5: 63% (17/27) D6/7: 84% (68/81)	D5: 29.6% (8/27) D6/7: 8.6% (7/81)	D5: 3.7% (1/27) D6/7: 2.5% (2/81)
Rubio et al., 2020[4]	1301	85.2% (1108/1301)	78.2% (866/1108)	12.4% (137/1108)	8.3% (92/1108)
Lledó et al., 2020[5]	92	92.4% (85/92)	74.7% (62/83) or 72.3% (60/83)[3]	12.0% or 15.7% (10/83 or 13/83)	13.3% or 12.0% (11/83 or 10/83)

Notes:

1 Although 55 SBM samples were informative, aCGH was only performed in the six samples with the highest DNA concentration.

2 SBM was compared to TE biopsy in all studies except this one, where SBM was compared to PB biopsy.

3 Concordance of all the informative SBM could not be analyzed because some of their respective TE biopsies did not provide a diagnosis.

4 Both studies found a third type of discordance called gender mismatch, i.e., SBM-TE pairs with concordant ploidies but different sexes (n = 5 [4.6%] in Rubio et al., 2019; n = 13 [1.2%] in Rubio et al., 2020).

5 Two different WGA methods were compared: MALBAC (listed first) and Sureplex (listed second).

TABLE 14.2 Studies Comparing Ploidy Results from Spent Blastocyst Medium (SBM) vs. Whole Blastocyst (WB)

STUDY	*NO. OF SBM*	*INFORMATIVE MEDIA*	*CONCORDANCE WB-SBM*	*FALSE POSITIVES*	*FALSE NEGATIVES*
Xu et al., 2016	42	100% (42/42)	85.7% (36/42)	9.5% (4/42)	4.8% (2/42)
Ho et al., 2018	41	97.6% (40/41)[1]	45.5% (15/33)	-	-
Huang et al., 2019	52	92.3% (48/52)[1]	93.5% (43/46)	6.5% (3/46)	-
Yin et al., 2020	75	78.7% (59/75)	89.8% (53/59)	10.2% (6/59)	-
Rubio et al., 2020[2]	81	90.1% (73/81)[1]	84.4% (54/64)	6.2 % (4/64)	9.4% (6/64)

Notes:

[1] Concordance of all the informative SBM could not be analyzed because some of their respective WB/ICM did not provide a diagnosis.

[2] SBM was compared to WB in all studies except this one, where SBM was compared to ICM.

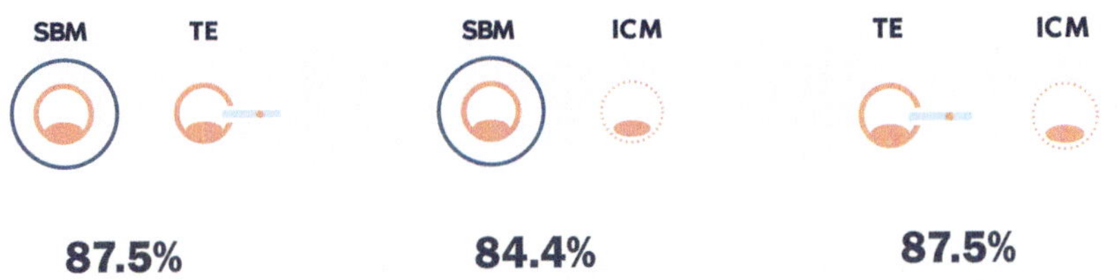

FIGURE 14.2 Concordance rates of ploidy analysis in cell-free DNA in spent blastocyst medium (SBM) with trophectoderm (TE) biopsies, and ICM (inner cell mass). (From Rubio et al., 2020.)

concordance rate [19] (Table 14.1). Four studies compared SBM to the whole blastocyst (WB). Xu et al. (2016) [20] used vitrified day 3 embryos that were thawed and cultured until day 5, resulting in an 85.7% concordance rate. Yin et al. (2020) [21] used day 5/6 thawed blastocysts that had previously undergone TE biopsy and were cultured for 24 hours after thawing, resulting in a 78.7% informativity rate and an 89.8% concordance rate. Ho et al. (2018) [13] and Huang et al. (2019) [14] found concordance rates of 45.5% and 93.5%, respectively (Table 14.2).

Our most recent study evaluated the contribution of the inner cell mass (ICM) and the TE to embryo cfDNA in the SBM [17]. TE, ICM, and SBM were analyzed in 81 aneuploid embryos, with informative results obtained in 97.5% (79/81) of TE biopsies, 90.1% (73/81) of ICM biopsies, and 90.1% (73/81) of SBM. All three sample types were informative in 79.0% of samples (64/81). SBM concordances with TE and with ICM were 87.5% and 84.4%, respectively (Figure 14.2). Of the ten discrepancies found between SBM and ICM, four corresponded to aneuploid SBM and euploid ICM, possibly due to mosaicism in the blastocyst (TE biopsies were also aneuploid). The six remaining discrepancies corresponded to euploid SBM and aneuploid ICM, most of them due to contamination of the SBM with maternal DNA that masked the aneuploidies.

Several studies have investigated the efficacy of adding blastocoel fluid (BF) to the SBM by collapsing the blastocyst and releasing its cavity's contents into the medium [22–27]. In addition to the artificial shrinkage of the blastocyst caused by this collapse, manipulations such as vitrification and/or AH were used. Informativity rates for SBM plus BF ranged from 87.5% to 100%, concordance rates between SBM

TABLE 14.3 Studies Comparing Ploidy Results from Spent Blastocyst Medium (SBM) Plus Blastocoel Fluid (BF) vs. Trophectoderm (TE) Biopsy or Whole Blastocyst (WB)

STUDY	*NO. OF SBM*	*INFORMATIVE MEDIA*	*CONCORDANCE TE-SBM+BF*	*FALSE POSITIVES*	*FALSE NEGATIVES*	*CONCORDANCE WB-SBM+BF*	*FALSE POSITIVES*	*FALSE NEGATIVES*
Kuznyetsov et al., 2018	47	100% (47/47)[1]	88.4% (38/43)	2.3% (1/43)	9.3% (4/43)	89.3% (25/28)	3.6% (1/28)	7.1% (2/28)
Li et al., 2018	40	97.5% (39/40)[1]	76.3% (29/38)	13.2% (5/38)	10.5% (4/38)	78.9% (30/38)	15.8% (6/38)	5.3% (2/38)
Jiao et al., 2019	62	98.4% (61/62)[1]	98.3% (58/59)	-	1.7% (1/59)	96.7% (59/61)	3.3% (2/61)	-
Zhang et al., 2019[2]	32	87.5% (28/32)[1]	-	-	-	70.4% (19/27)	14.8% (4/27)	7.4% (2/27)
Kuznyetsov et al., 2020	102	88.2% (90/102)	88.9% (80/90)	3.3% (3/90)	7.8% (7/90)	-	-	-
Chen et al., 2020	26	100% (26/26)	88.5% (23/26)	-	11.5% (3/26)	100% (26/26)[3]	-	-

Notes:

[1] Concordance of all the informative SBM+BF could not be analyzed because some of their respective TE/WB did not provide a diagnosis.

[2] This study found a third type of discordance called gender mismatch, i.e., SBM+BF-WB pairs with concordant ploidies but different sexes (n = 2 [7.4%]).

[3] This study compared SBM+BF to ICM, not to WB.

plus BF and TE ranged from 76.3% to 98.3%, and concordance rates between SBM plus BF and WB ranged from 70.4% to 100% (Table 14.3).

Where embryo cfDNA originates—whether the inner cell mass, the TE, or both compartments—remains unclear because examining the mechanisms underlying cfDNA secretion into the medium is challenging [22, 23]. One potential mechanism is apoptosis during preimplantation development. The embryo's rapid transformation, particularly during late preimplantation development, results not only in cell gain but also in cell loss, even in genetically healthy embryos [28]. Under this scenario, cfDNA may be more likely than the TE biopsy to show aneuploidies, since aneuploid cells may be more prone to apoptosis. DNA from the BF of expanded blastocysts that had already been diagnosed by TE chromosome analysis showed higher amplification failure in TE-euploid blastocysts than in TE-aneuploid blastocysts, suggesting a higher cfDNA concentration in the internal cavity of TE-aneuploid blastocysts that could be caused by increased apoptosis [29]. However, our finding that cfDNA levels were not significantly greater in media from aneuploid vs. from euploid embryos does not support this hypothesis [17].

Because DNA can be successfully amplified in low- and high-quality blastocysts, secreted cfDNA may in most cases reflect general embryo ploidy status. In one study, 35% of samples that were not contaminated with maternal DNA had complementary aneuploidy (i.e., they showed euploidy in the TE but aneuploidy in the SBM, or vice versa), a discrepancy that may be due to mosaicism or to aneuploidy in the PB only [12]. However, another study found that two-thirds of embryos with aneuploidies detected by PB analysis also showed at least one aneuploidy as detected in the culture medium, making it unlikely that embryonic cells in the culture medium consist solely of extruded aneuploid cells not fully representative of the mosaic embryo [19]. In addition, as noted above, our recent study found similar concordance rates of the cfDNA with TE biopsy and with the inner cell mass biopsy (87.5 and 84.5%, respectively) [17].

The high concordance rates between cfDNA and whole blastocyst analysis suggest that cfDNA may be more representative than TE biopsy of the full blastocyst's chromosome content [14]. On the other hand, TE cells should be the main source of secreted cfDNA since they are in direct contact with the medium and are more numerous than the cells of the ICM [30]. Further research is needed to determine the origin of the cfDNA in the SBM.

METHODOLOGY

Culture Conditions in the IVF Laboratory

Various strategies have been developed to culture the embryos *in vitro* and analyze the cfDNA released by them into the media. Most methods use a small volume of culture medium, since smaller culture drops have higher concentrations of embryo cfDNA [12–18]. We suggest using 10-μL drops for conventional incubators [16, 17] and 20–25-μL drops for in time-lapse systems. Culture systems may be single-step or two-step [31]. Time-lapse systems can be used if the embryos can be individually cultured in separate wells. Igenomix reports overall concordance rates of 85.1% for benchtop incubators and 86.7% for time-lapse incubators. In our multicenter study, we observed no significant differences in concordance rates between TE biopsies and SBM analysis in clinics using different brands of media and incubators [17] (Figure 14.3). Albumin percentage also does not influence the results, indicating that albumin supplements do not contain trace DNA that could interfere with embryonic cfDNA analysis [32]. Concordance rates can exceed 86% when culture conditions are controlled and external contamination is avoided [17].

Intracytoplasmic sperm injection (ICSI) is the most commonly used fertilization technique in studies analyzing cfDNA. We found that IVF and ICSI provide similar sensitivity (80.9% vs. 87.9%) and specificity (78.6% vs. 69.9%), as long as the oocytes/zygotes are carefully denudated before ICSI or on the day of the fertilization check in IVF [17]. Our multivariate analysis did not reveal any effects of patient clinical

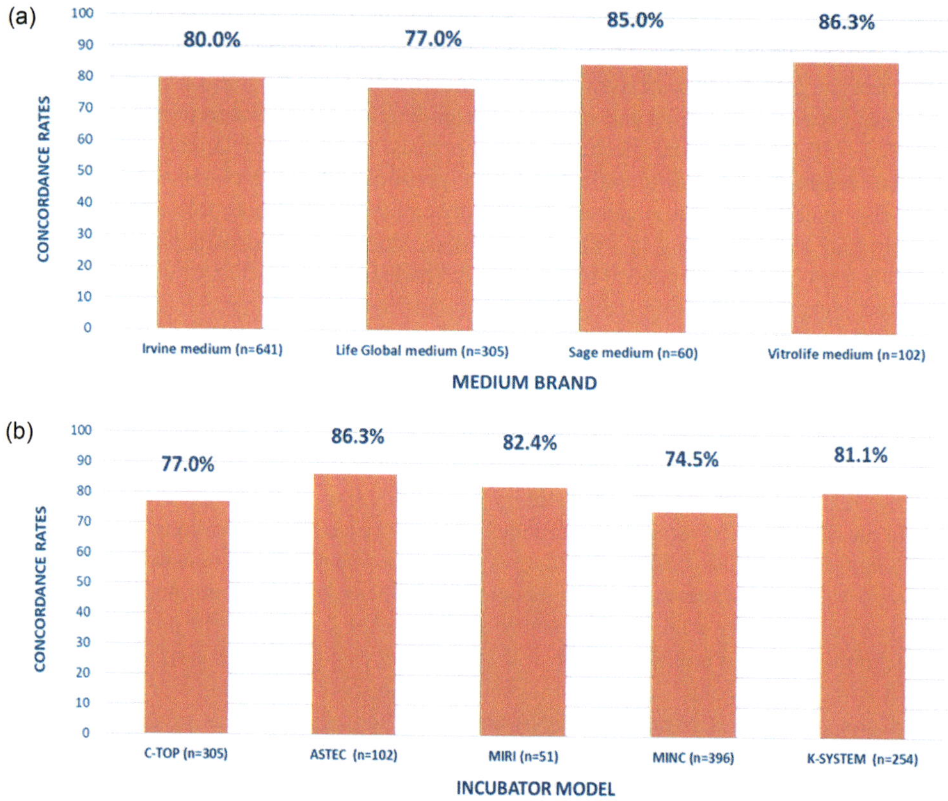

FIGURE 14.3 Concordance rates of ploidy analysis are similar across culture conditions. A) Results from four culture media. B) Results from five different incubators. (From Rubio et al., 2020.)

background, ovarian stimulation protocols, or embryo quality on testing accuracy or concordance. The day of media collection, however, had a strong effect, with embryonic cfDNA levels increasing during embryo development, consistent with previous studies. The time of media collection may thus be an important consideration when optimizing this technology. Proposed collection times range from the cleavage stage to the blastocyst stage (reviewed in [31]). Studies comparing results from day 5 and day 6 blastocysts agree that media collection on day 6 results in higher quantity of cfDNA and higher concordance rates [16, 17, 33].

The greatest methodological challenge of niPGT-A is avoiding contamination with DNA that does not come from the embryo. This exogenous DNA can come from the cumulus cells (maternal cells), or from external sources (e.g., technicians, contaminated media, working surfaces, or material) [16, 17, 34]. Strategies to minimize the carryover of cumulus cells into culture drops include gentle denudation after IVF or ICSI and extra serial washes. Individual handling of embryos, sterile working conditions, and the use of dedicated materials throughout the procedure also reduce the risk of contamination.

Variance in concordance rates across studies may be partly explained by methodological differences in embryo culture or embryo manipulation. Manipulations of the embryo before media collection may include AH on day 3 embryos, PB biopsy, TE biopsy, or embryo cryopreservation at different stages, all of which can impact the results (Table 14.4). In some studies, the blastocyst was collapsed to obtain BF before it was vitrified (Table 14.5). Thus, this approach, rather than being completely non-invasive, could be defined as "minimally invasive." To our knowledge, we are the only group to have published a study in which embryos were cultured without any previous manipulation (i.e., fresh embryos coming from fresh eggs, no AH, no embryo collapse), and which therefore can be considered as "truly" non-invasive [16, 17].

TABLE 14.4 Culture Conditions and Technologies Used for Chromosome Copy Number Analysis in cfDNA from Spent Blastocyst Medium (SBM)

STUDY	*VOLUME: μL IN DROP (μL ANALYZED)*	*PREVIOUS EMBRYO MANIPULATION*	*TIME IN CULTURE*	*WGA METHOD*	*CHROMOSOMAL ANALYSIS*
Shamonki et al., 2016	15 (15)	AH on D3	D3 -D5/6	Repli-G single cell kit (Qiagen)	aCGH (Agilent Tech)
Feichtinger et al., 2017	25 (5)	AH on D3	D0-D5/6	SurePlex (Illumina)	aCGH (Illumina)
Vera-Rodríguez et al., 2018	25 (20)	AH on D3	D3-D5	Sureplex (Illumina) + ReproSeq (Life Tech)	NGS (Life Tech)
Ho et al., 2018	25 (5)	AH on D3 vs. no AH	D1 to D5	Picoplex (Rubicon)	NGS (Life Tech)
Huang et al., 2019	15 (3.5)	AH on D3 Plus vitrification on D5/6	D5-D6 or D6-D7 Cultured 24 h after thawing	MALBAC (Yikon)	NGS (Illumina)
Yeung et al., 2019	30 (3)	AH on D3	D3-D5 D3-D6	SurePlex (Illumina)	NGS (Illumina)
Rubio et al., 2019	10 (10)	NO	D4-D5/D6/D7	Reproseq (Life Tech)	NGS (Life Tech)
Rubio et al., 2020	10 (10)	NO	D4-D6/7	Reproseq (Life Tech)	NGS (Life Tech)
Lledó et al., 2020	20 (7.5)	AH on D3	D3-D5/6	MALBAC (Yikon) or Sureplex (Illumina)	NGS (Illumina)
Xu et al., 2016	30 (5 to 20)	Vitrification on D3	D3- D5	MALBAC (Yikon)	NGS (Illumina)
Yin et al., 2020	25 (25)	Biopsy on D5/6 and vitrification	D5-D6 or D6-D7 Cultured 24 h after thawing	MALBAC (Yikon)	NGS (Illumina)

Abbreviations: AH: assisted hatching; D: day; WGA: whole genome amplification; NGS: next-generation sequencing.

TABLE 14.5 Culture Conditions and Technologies Used for Chromosome Copy Number Analysis in cfDNA When Spent Blastocyst Medium (SBM) Was Combined with Blastocoel Fluid (BF)

STUDY	*VOLUME: μL IN DROP (μL ANALYZED)*	*PREVIOUS EMBRYO MANIPULATION*	*TIME IN CULTURE*	*WGA METHOD*	*CHROMOSOMAL ANALYSIS*
Kuznyetsov et al., 2018	25 (25)	TE biopsy and vitrification with artificial shrinkage of the blastocyst	D4-D5/6	SurePlex (Illumina)	NGS (Illumina)
Li et al., 2018	25 (25)	Small breech in the ZP	D3-D5	MALBAC (Yikon)	NGS (Illumina)
Jiao et al., 2019	12 (10)	AH, vitrification D5/6; artificial shrinkage of the blastocyst	cultured for 14 h after thawing	MALBAC (Yikon)	NGS (Illumina)
Zhang et al., 2019	≈20-30 (≈20-30)	Vitrification D2/3; artificial shrinkage of the blastocyst	D4-D5 D5-D6	MALBAC (Yikon)	NGS (Illumina)
Kuznyetsov et al., 2020	25 (5)	AH on D4; artificial shrinkage of the blastocyst	D4-D6	SurePlex (Illumina)	NGS (Illumina)
Chen et al., 2020	15 (10)	TE biopsy and vitrification with artificial shrinkage of the blastocyst	cultured for 15 h after thawing	MALBAC (Yikon)	NGS (Illumina)

Abbreviations: AH: assisted hatching; D: day; WGA: whole genome amplification; NGS: next-generation sequencing.

DNA Whole Genome Amplification (WGA) and Next-Generation Sequencing (NGS)

After media collection, cfDNA undergoes WGA, sequencing, and analysis of the results using customized algorithms specific to cfDNA (Figure 14.4). Multiple technologies exist both for WGA and for the chromosomal analysis of the WGA products. While CGH arrays were used initially, NGS is now the gold standard (Tables 14.4 and 14.5).

NGS raw sequencing data can be analyzed using several algorithms. Our group developed a proprietary algorithm based on the results of our prospective multicenter study with 1,301 blastocysts [17]. We also developed a prioritization system for embryo transfer in which the media more likely corresponding to euploid blastocysts have higher priority for transfer. When using this system, the proper threshold must be set to define a chromosome as abnormal, as the criteria applied to TE biopsies and to SBM samples can differ [14].

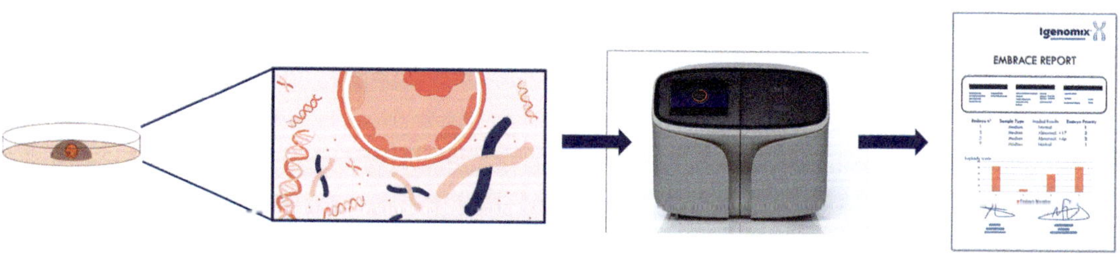

FIGURE 14.4 Workflow of the analysis of cell-free DNA (cfDNA) in the culture media: media collection, cfDNA whole genome amplification, sequencing and data analysis using customized algorithms.

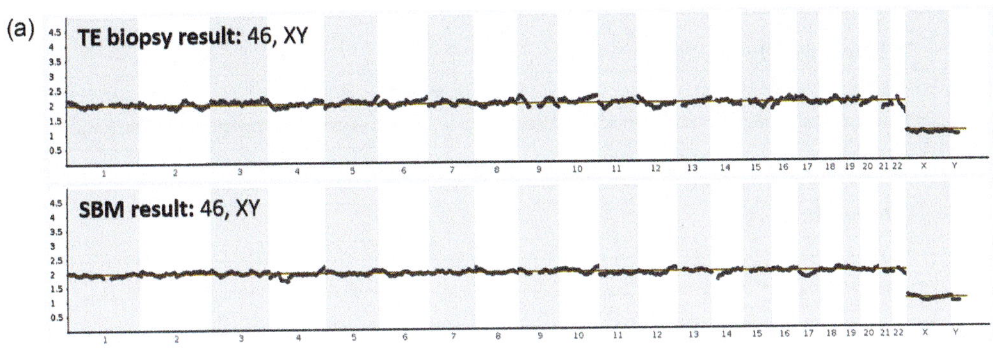

FIGURE 14.5A Concordant ploidy analysis results in a good quality blastocyst. The TE biopsy and the SBM were euploid, with XY sex chromosomes.

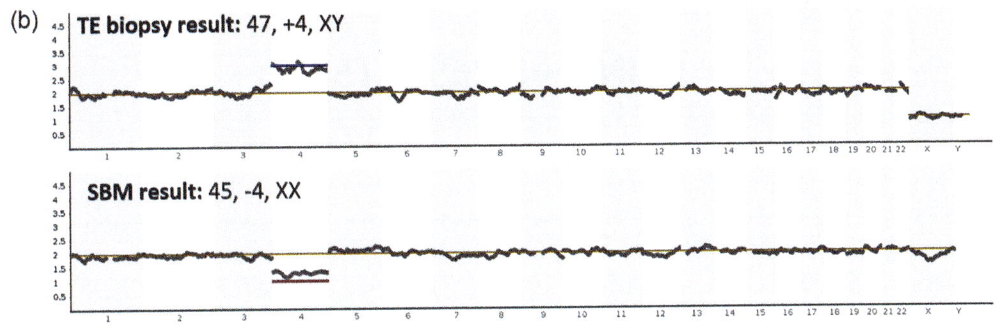

FIGURE 14.5B Discordant ploidy analysis results due to polar body contamination in an early blastocyst, evidenced by a complementary aneuploidy and female sex chromosomes in the SBM, as expected in polar bodies.

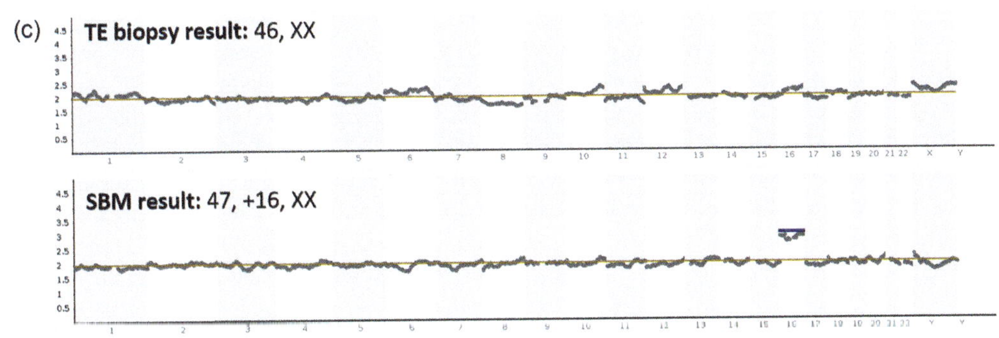

FIGURE 14.5C Discordant ploidy analysis results with euploid TE biopsy and aneuploid SBM. When a single aneuploidy is observed in the SBM, mosaicism in the blastocyst could explain the discrepancy.

Figure 14.5 shows representative SBM analyses and corresponding TE biopsies from our study [17], with examples of full concordance (Figure 14.5a); discrepancy with complementary aneuploidy, likely due to DNA from the PB remaining in the SBM (Figure 14.5b); discrepancy with aneuploid SBM, likely due to mosaicism in the blastocyst (Figure 14.5c); and discrepancy with a euploid SBM, possibly due to female contamination (Figure 14.5d).

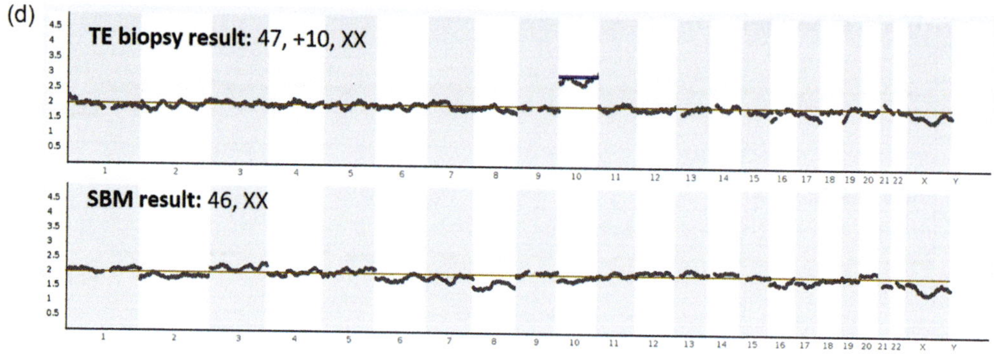

FIGURE 14.5D Discordant ploidy analysis results with aneuploid TE biopsy and euploid SBM. When the SBM is euploid and with XX sex chromosomes, the most probable cause of the discrepancy is contamination, usually from cumulus cells, although external contamination during the IVF procedure cannot be ruled out.

CLINICAL RESULTS

An efficient and reliable non-invasive approach could offer the benefits of invasive PGT-A without the accompanying technical and financial burdens. niPGT-A would also avoid the negative health effects associated with TE biopsy, including preeclampsia [35] and hypertensive disorders [36]. Furthermore, the non-invasive approach could be offered as a prioritization model, with cfDNA testing from culture media, in combination with morphology, providing a "ranking" of viable blastocysts most likely to result in a healthy baby. Such blastocyst prioritization would allow providers and couples to select the best blastocyst for the first embryo transfer and reserve the others for subsequent transfers rather than discarding them.

Our protocol to isolate and amplify embryo cfDNA without previous embryo manipulation involves collecting SBM from embryos with intact zona pellucida. In our pilot study at a single IVF center, this protocol resulted in a concordance rate with TE biopsies of 84% [16]. We performed 29 single-embryo transfers (SETs) with euploid TE biopsies and with media ploidy status unknown at the time of embryo transfer. After unmasking the media results and evaluating the clinical outcomes, we found higher ongoing pregnancy rates with euploid vs. aneuploid media (52.9% vs. 16.7%). We are currently performing a randomized study comparing SET with untested embryos selected by morphology vs. SET using cfDNA analysis of day 6 blastocysts. This study will help uncover the clinical impact that incorporating cfDNA analysis into embryo selection can have on patient outcomes (ClinicalTrials.gov#NCT4000152).

Two additional publications demonstrate the premise of niPGT-A. Xu et al. (2016) [20], conducted seven SETs in seven patients, five of whom became pregnant and delivered healthy babies. Fang et al. (2019) [37] transferred 52 euploid embryos in 50 transfers to 43 women. The biochemical pregnancy rate was 72% (36/50) and the clinical pregnancy rate was 58% (29/50); outcomes included 27 healthy babies and three miscarriages. More recently, it has been shown that the niPGT-A approach of our group can result in higher ongoing clinical pregnancy rates than conventional IVF/ICSI in women ≤38 years, 61.5% in niPGT-A versus 48.5% in IVF/ICSI. Interestingly, the results were similar to PGT-A for the same age range [38].

STRENGTHS AND LIMITATIONS, BARRIERS TO ACCESS

Researchers and clinicians agree that avoiding embryo biopsy for genetic evaluation would be economically and practically advantageous and would reduce the risk of embryo damage. Reported concordance rates between embryo cfDNA and blastocyst biopsy vary widely, and most sample sizes are relatively low. However, our analysis of 1,301 samples from eight different laboratories establishes the clinical potential of this non-invasive approach [17]. Furthermore, our comparison of embryo cfDNA with ICM biopsies (representing the embryo itself) in a subset of blastocysts showed concordance rates similar to those between cfDNA and TE biopsies. These findings pave the way for the clinical application of a non-invasive approach that would avoid embryo manipulation and decrease the methodological complexity and cost of analysis, thus allowing every IVF laboratory to offer embryo aneuploidy testing to all IVF patients. While contamination by maternal cumulus cells may contribute to false-negative results, this risk can be mitigated through careful removal of cumulus cells following oocyte pickup and through serial washing of the oocytes and embryos to eliminate all traces of maternal DNA.

The relatively low abundance of cfDNA in the culture media presents technical challenges for genetic analysis, but these can be successfully overcome by collecting the media at day 6 and by using WGA protocols modified to improve DNA amplification. In addition, cfDNA concentration in the culture drops can be increased by decreasing the drop volume to 10–15 μL. While such a low volume may sound concerning, it may in fact promote blastocyst development due to an increased concentration of the autocrine factors produced by the embryo [39]. As for culturing blastocysts to day 6, evidence suggests that this does not impact clinical outcomes. In one study, day 5 and day 6 vitrified blastocysts resulted in similar ongoing pregnancy rates if aneuploidy testing was performed [40]. Similarly, another study found no differences in the sustained implantation rates between day 5 and day 6 SET in PGT-A cases with deferred embryo transfer [41]. These results were corroborated by a metanalysis concluding that, when PGT-A is performed, rates of implantation, clinical pregnancy, live birth, and ongoing pregnancy are similar for euploid day 5 and euploid day 6 blastocyst transfers [42]. Finally, a recent study involving the transfer of frozen euploid embryos that had been assessed by PGT-A found that the day of cryopreservation did not significantly impact live birth rates [43].

CONCLUSIONS

A non-invasive PGT-A approach could be reliably applied under defined embryo handling and culture conditions with a customized NGS protocol and diagnostic algorithm. Differences in ovarian stimulation protocols, culture conditions, and embryo quality do not affect the accuracy of niPGT-A results. A prioritization system using niPGT-A to select embryos with higher chances of euploidy could be developed to increase live birth rates. Non-invasive PGT-A has several important advantages compared with the current standard approach of TE biopsy. In addition to minimizing invasiveness and potential harm to the embryo, this method would also expand aneuploidy testing to a larger number of clinics and increase its accessibility to a wider population of patients by reducing costs. While niPGT-A holds considerable clinical promise, further work is needed to clarify the origin of the embryo cfDNA and the mechanisms involved in its secretion.

FUTURE PERSPECTIVES

On the clinical side, unanswered questions include: Is this technology superior to current invasive techniques, in terms of live birth and miscarriage? Is this technology superior to no testing at all? Does it improve time to pregnancy? Does it reduce miscarriage? Is it cost-effective? These questions can only be answered by well-designed, large clinical trials. Class I evidence should come from randomized trials comparing the efficacy of niPGT-A across different patient age ranges. In addition, non-selection studies would help reveal how different cfDNA sequencing patterns correspond to different reproductive outcomes. Such data could inform the development of a scoring system for clinical use evaluating each sequencing pattern according to its likelihood of resulting in a healthy live birth.

REFERENCES

1. Franasiak JM, et al. *Fertil Steril.* 2014;101:656–63.
2. Rubio C, et al. *Biol Reprod.* 2019;101:1083–90.
3. Nagaoka SI, et al. *Nat Rev Genet.* 2012;13:493–504.
4. Soler A, et al. *Cytogenet Genome Res.* 2017;152:81–9.
5. Sugiura-Ogasawara M, et al. *Hum Reprod.* 2012;27:2297–303.
6. Margalioth EJ, et al. *Hum Reprod.* 2012;21:3036–43.
7. Silber S, et al. *Fertil Steril.* 2003;79:30–8.
8. Rodrigo L, et al. *Fertil Steril.* 2011;95:1005–12.
9. Neal SA, et al. *Fertil Steril.* 2018;110:896–904.
10. Somigliana E, et al. *Fertil Steril.* 2019;111:1169–76.
11. Shamonki MI, et al. *Fertil Steril.* 2016;106(6):1312–8.
12. Vera-Rodriguez M, et al. *Human Reprod.* 2018;33(4):745–56.
13. Ho JR, et al. *Fertil Steril.* 2018;110(3):467–75.
14. Huang L, et al. *PNAS.* 2019;116(28):14105–12.
15. Yeung QSY, et al. *J Assist Reprod Genet.* 2019;36(8):1609–21.
16. Rubio C, et al. *Fertil Steril.* 2019;112(3):510–9.
17. Rubio C, et al. *Am J Obstet Gynecol.* 2020;223(5):751.e1–13.
18. Lled6 B, et al. *Reprod Biomed Online.* 2020;42(3):555–563.
19. Feichtinger M, et al. *Reprod Biomed Online.* 2017;34(6):583–9.
20. Xu J, et al. *PNAS.* 2016;113(42):11907–11912.
21. Yin B, et al. *Reprod Biol Endocrinol.* 2021;19:41.
22. Kuznyetsov V, et al. *PLoS One.* 2018;13(5):e0197262.
23. Li P, et al. *Sci Rep.* 2018;8(1):9275.
24. Jiao J, et al. *Hum Reprod.* 2019;34(7):1369–1379.
25. Zhang J, et al. *J Assist Reprod Genet.* 2019;36(12):2505–2513.
26. Kuznyetsov V, et al. *Sci Rep.* 2020;10(1):7244.
27. Chen J, et al. *Fertil Steril.* 2020;114(3):e532–3.
28. Hardy K, et al. *PNAS.* 2001;98(4):1655–60.
29. Magli MC, et al. *Fert Steril.* 2019;111(1):77–85.
30. Gleicher N, et al. 2019;116:21976–21977.
31. Belandres D, et al. *J Assist Reprod Genet.* 2019;36:819–826.
32. Hammond ER, et al. *Hum Reprod.* 2016;31(8):1653–1661.
33. Meyers J, et al. *Reproductive Biomedicine Online.* 2020;114(3):e422.
34. Leaver M, et al. *Hum Reprod Update.* 2020;26(1):16–42.
35. Zhang WY, et al. *Fertil Steril.* 2019;112(2):283–290.
36. Makhijani R, et al. *Hum Reprod.* 2021;36(2):340–348.
37. Fang R, et al. *J Transl Med.* 2019;17:73.
38. Minasi MG, et al. *J Assist Reprod Genet* 2015;32:215–220.
39. Franco J, et al. *Hum Reprod.* 2021.
40. Kaye L, et al. *J Assist Reprod Genet.* 2017;34(7):913–919.
41. Tiegs AW, et al. *Hum Reprod.* 2019;34:1632–1639.
42. Bourdon M, et al. *Hum Reprod.* 2019;34:1948–1964.
43. Shear MA, et al. *RBMO.* 2020;41(6):981–989.

Mitochondria and Embryo Viability

15

Irene Corachan Garcia, Laura Iñiguez Quiles, and Antonio Diez-Juan

INTRODUCTION

Mitochondria play a critical role in the generation of metabolic energy in eukaryotic cells, using oxidative phosphorylation to derive energy (ATP) from carbohydrates and fatty acids. Mitochondria contain their own DNA, which encodes tRNAs, rRNAs, and some mitochondrial proteins (1). Ranging in size from 0.5 to 1.0 μm in diameter (2), these unique organelles have a double-membrane system consisting of inner and outer membranes separated by an intermembrane space (1). The outer mitochondrial membrane encloses the matrix (internal space) and contains a large number of proteins that form channels allowing small molecules to pass. The inner mitochondrial membrane, which is folded into structures (cristae) that increase the surface area, is less permeable, blocking the movement of ions and other small molecules. Both the inner and outer membranes contain specific transport proteins that can move molecules by a passive or active transport (2) (Figure 15.1).

ORIGIN OF MITOCHONDRIA: ENDOSYMBIOTIC THEORY

New data are continuously improving our understanding of mitochondrial evolution. Two hypotheses exist on the origin of mitochondria. The endosymbiotic hypothesis proposes that the mitochondrion originated after the eukaryotic cell arose; the autogenous hypothesis proposes that this organelle had its beginning at the same time as the cell (3). The endosymbiotic hypothesis is the more widely accepted of the two (3, 4). This hypothesis purports that mitochondria were originally prokaryotic cells capable of implementing oxidative mechanisms that later became endosymbionts; initially proposed by Lynn Margulis in 1970, this concept established that mitochondria evolved from free-living bacteria via symbiosis with a eukaryotic host cell (5, 6).

The oldest eukaryotic microfossils date back 1.45 billion years (Figure 15.2). At that time, the oceans were mostly anoxic due to marine H_2S-producing bacteria. Eukaryotes arose and diversified in this anoxic environment. The mitochondrial genome appears to have a monophyletic origin from a-Proteobacteria with Rickettsiales and several *Rickettsia*-like endosymbionts identified as the a-Proteobacterial order

DOI: 10.1201/9781003024941-15

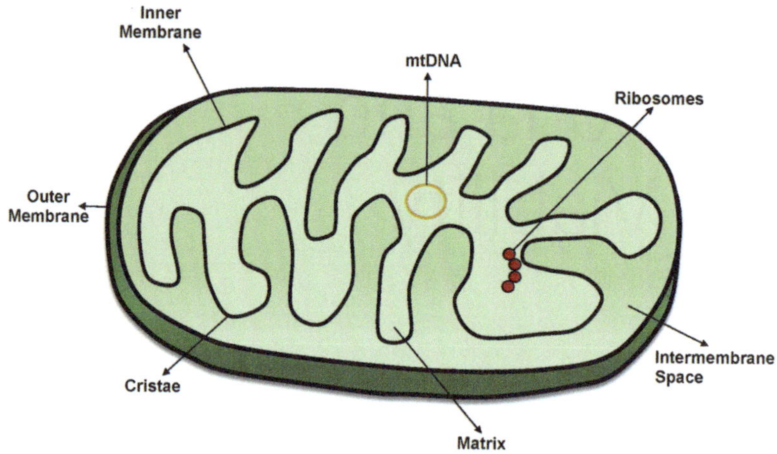

FIGURE 15.1 Diagram of a mitochondrion showing detail of the membranes and mitochondrial DNA.

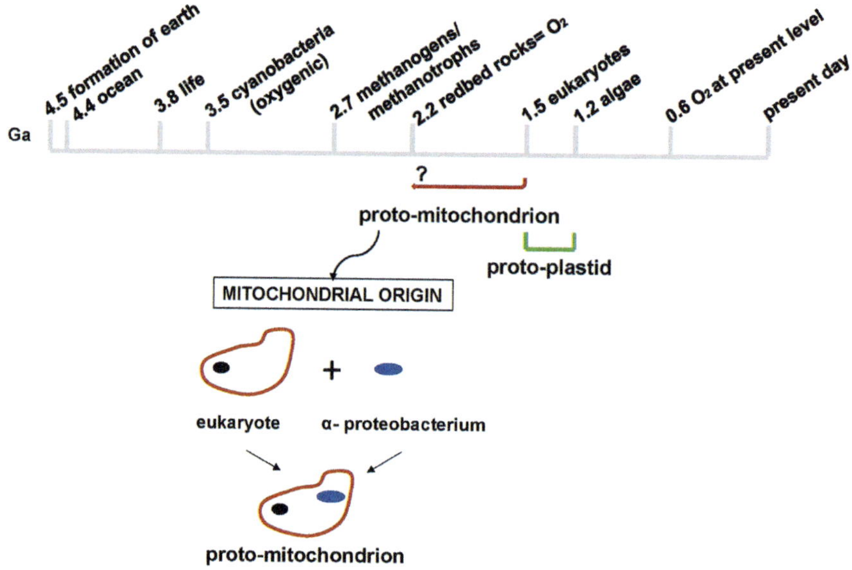

FIGURE 15.2 Timeline for the origin of life and major invasions giving rise to mitochondria and plastids. An α-proteobacterium is shown in blue and eukaryotic cell in yellow. Ga, billion years ago.

most closely related to mitochondria (3, 7, 8). New lines of research suggest that the host that acquired the mitochondrion was a prokaryote. Under this view, the ancestral mitochondrion was a facultative anaerobe, perhaps similar in physiology and lifestyle to modern Rhodobacteriales (4). However, phylogenetic analyses of both protein-coding genes and rRNA genes specified by mitochondrial DNA (mtDNA) demonstrate that the mitochondrial genome originated from the eubacterial, not the archaeal, domain of life (9, 10). The mitochondrial genome appears to have a monophyletic origin from α-proteobacteria, with the order Rickettsiales and several *Rickettsia*-like endosymbionts identified as the α-proteobacteria most closely related to mitochondria (3, 7, 8). In particular, *Rickettsia prowazekii*, the pathogen that causes epidemic typhus and is transmitted in the feces of lice, stands out as the most genetically similar (9). While these views of mitochondrial evolution continue to shape our understanding of its functions, the evidence is clear that these organelles have a diverse repertoire of roles in the cell.

MITOCHONDRIAL FUNCTIONS: MORE THAN ATP

The best understood mitochondrial functions include ATP production by oxidative phosphorylation, β-oxidation of fatty acids, and metabolism of amino acids and lipids. (11). Mitochondria also have other complex functions, including participating in a variety of cell-signaling cascades. Proteins such as GTPases, kinases, and phosphatases facilitate bidirectional communication between the mitochondria and the rest of the cell, helping to regulate metabolism, cell cycle control, development, and antiviral responses (12). A key example of the mitochondria's integration in signaling pathways is its role in the apoptotic cascade (13), with mitochondrial fragmentation and cristae remodeling being essential steps for cytochrome c release and cell death (12). Additionally, calcium signaling causes a dynamic change in the phosphorylation state of numerous proteins (14). Mitochondria act both as a calcium buffer and as the propagator of intracellular calcium waves during muscle contraction and synaptic vesicle release (12). Another example of the integration of mitochondria in signaling pathways is its role in the apoptotic cascade and cell death (13); indeed, mitochondrial fragmentation and cristae remodeling are essential steps for "cytochrome c" release and cell death (12).

Mitochondria are also crucial for DNA and RNA synthesis, containing multiple enzymes responsible for the metabolism of cytosolic ribonucleotides and deoxyribonucleotides, (11). For example, serine, a major source of one-carbon units, is required for the synthesis of glycine, thymidylate, methionine, and purines, as well as methylation reactions. Transfer of a methyl group from serine to tetrahydrofolate (THF) yields glycine and 5,10-methylene-THF (11). In the mitochondria, 5,10-methylene-THF is converted into formate in a succession of enzymatic reactions beginning with conversion into 5-methyl-THF and then into 10-formyl-THF. Cytosolic 10-formyl-THF is a crucial one-carbon-unit donor for the *de novo* synthesis of purine nucleotides (11). Mitochondrial dysfunction is linked to aberrant DNA and RNA synthesis. For example, mutations affecting electron transport chain (ETC) activity can have a detrimental effect on cytosolic dNTP levels, causing instability in the nuclear genome (11).

Oxidative Stress

Reactive oxygen species (ROS) are highly reactive molecules that include diverse chemical species such as the superoxide anion, the hydroxyl radical, and hydrogen peroxide (H_2O_2) (15). About 1–2% of the molecular oxygen consumed during physiological respiration is converted into superoxide radicals (13). Mitochondrial ROS are generated in the ETC through one-electron carriers (15, 16). Their direct interaction with mitochondrial proteins, lipids, and DNA results in lipid peroxidation, protein oxidation, and mitochondrial DNA (mtDNA) mutations (13, 16). Indeed, as the majority of ROS are products of mitochondrial respiration, mitochondria are the major targets for their damaging effects (13, 15).

Oxidative damage modifies the replication and transcription of mtDNA. This inhibits mitochondrial function, which in turn enhances ROS production. However, mitochondria have a defense system to detoxify ROS and repair ROS-induced damage (13). mtDNA is more susceptible than nuclear DNA to oxidative stress (15). Oxidative stress also plays an important role in aging (15), and the modulation of this system directly affects the susceptibility of cells to undergo apoptosis (13).

The study of how ROS influence the activation of the cell death program and other important pathways may continue to reveal mechanisms that inform therapeutic intervention in human diseases.

MITOCHONDRIAL DNA: CHARACTERISTICS

mtDNA is a covalently closed circular molecule (17) of 16.5 kb located in the mitochondrial matrix. Its genetic code differs slightly from the universal one, in that AUA codes for methionine and not isoleucine and UGA codes for tryptophan and not STOP. Further, the AGA/AGG pair used for arginine does not code for termination in bovine mitochondria. Moreover, mtDNA has no introns (18) (Figure 15.3a). In humans, 100–10,000 separate copies of mtDNA are usually present per cell. Energy-intensive tissues such as cardiac and skeletal muscle contain between 2,000 and 10,000 copies per cell, with an average mtDNA copy number per mitochondrion of between 1,000 and 10,000. Less energy-demanding tissues, like the lungs, have an average of between 200 and 3,000 copies per cell and 50 and 300 per mitochondrion. Oocytes contain an average of between 100 and 640 thousand copies per cell, with each mitochondrion containing one or two copies (reviewed by Van Blerkom [19]).

mtDNA encodes two rRNAs (12S and 16S), 22 tRNAs, and 13 mRNAs (18). It also has a regulatory sequence involved in replication and transcription, known as the D-loop (18). The proteins encoded by mtDNA are synthesized in the mitochondria. These proteins are essential subunits of the ETC complexes: seven subunits of complex I (20), one subunit of complex III (18), three subunits of complex IV (21), and two subunits of complex V (18). mtDNA gene mutations affect ETC functions, while mutations in the nuclear genes encoding mitochondrial proteins can affect multiple mitochondrial functions (11). Importantly, mtDNA mutates ten or more times faster than nuclear DNA due to its location near the ETC and its lack of protective histones (22). As a result, mtDNA is more prone to the deleterious effects of ROS (Figure 15.3b) (23).

MITOCHONDRIAL DNA: INHERITANCE AND BOTTLENECK SELECTION

Human mtDNA is maternally inherited (24). Although a zygote receives both maternal and paternal mtDNA at fertilization, the paternal mtDNA is specifically targeted for degradation and removed from the cytoplasm of the zygote during embryogenesis (25, 26). The number of mitochondria present in a metaphase II oocyte (19) represents only a small fraction of the maternal mtDNA pool due to a genetic bottleneck that occurs during oogenesis (27). In the primordial germ cells there is a large population of mtDNA that represents the maternal mtDNA pool. During germline development, this pool is subsampled to a relatively small number (24), resulting in only a small fraction of mtDNA being represented in the mature egg (27). This process limits mtDNA diversity in each oocyte and promotes homoplasmy. However, mtDNA differs across a cohort of oocytes because the maternal mtDNA is randomly segregated (27). Indeed, in mtDNA diseases, the level of mutant mtDNA differs between oocytes from the same patient (28). After the bottleneck, during oocyte maturation, both mitochondrial content and mtDNA copy number increase (27).

MITOCHONDRIAL REPLICATION AND MTDNA REPLICATION

mtDNA replication is mediated by several nuclear-encoded transcription and replication factors. These factors, along with mtDNA, form the mitochondrial nucleoid (29), which is responsible for the packaging, transcription, and replication of the mitochondrial genome (30). The nucleoid proteins comprise the mitochondria-specific polymerase gamma (which contains two subunits, POLGA and POLGB), mitochondrial RNA polymerase (mtRNA pol), mitochondrial transcription factors A (TFAM) and B (TFBM), mitochondrial single-stranded DNA-binding protein (mtSSB), helicase (Twinkle), key transcription factors (TFB1M, TFB2M), and the mitochondrial transcription termination factor (mTERF). All of these are

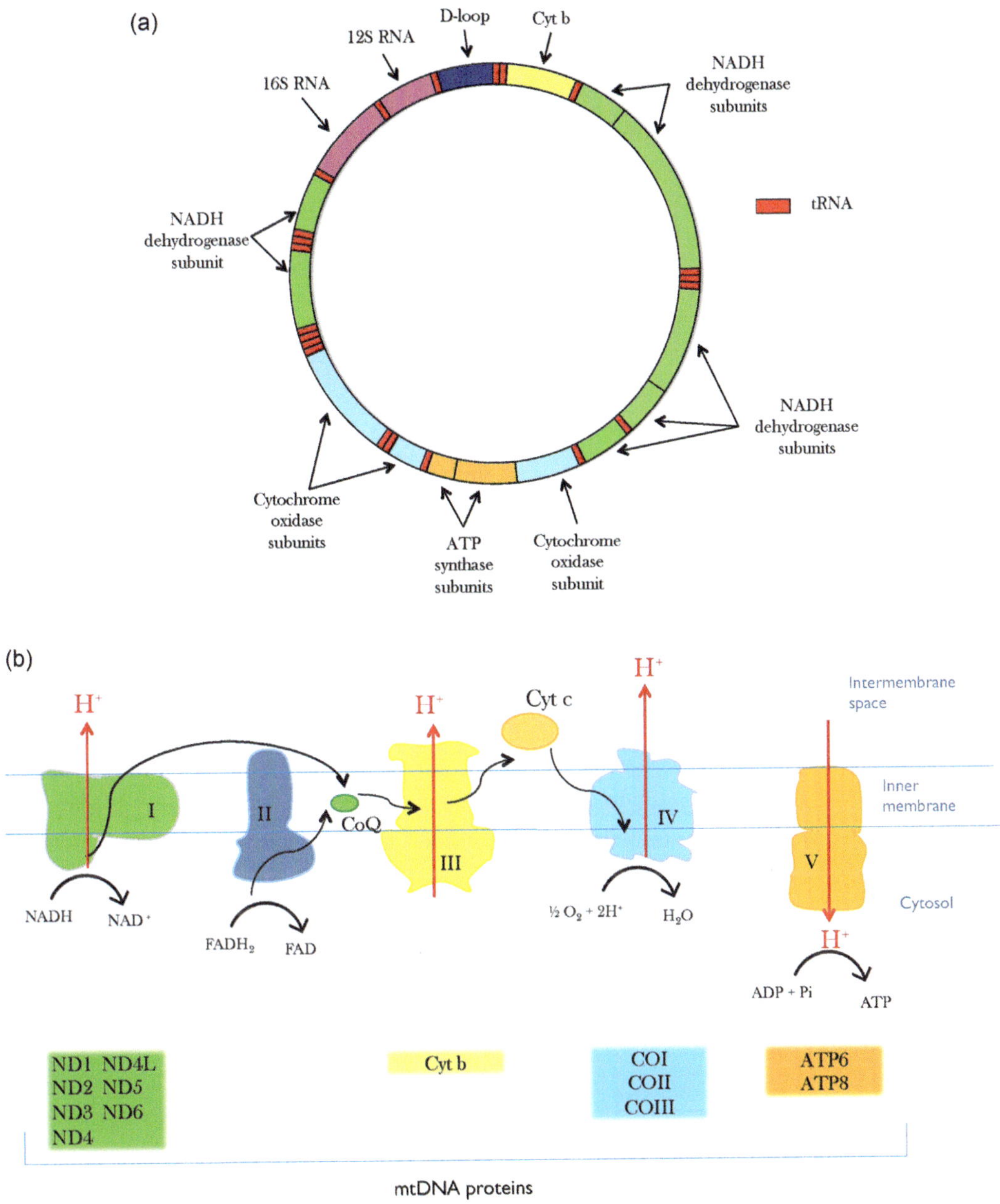

FIGURE 15.3 (a) mtDNA has a regulatory sequence, the D-loop (dark blue), and encodes 37 genes: 22 tRNAs (red), 2 rRNAs (pink), and 13 proteins. (b) The respiratory chain, located in the inner membrane, comprises 5 complexes. The 13 polypeptides encoded by mtDNA are located in the respiratory chain: 7 in complex I (green), 1 in complex III (yellow), 3 in complex IV (light blue), and 2 in complex V (orange). Complex II is entirely encoded by the nuclear genome (grey). Complexes I, III, and IV are H+ pumps that generate an electrochemical gradient across the membrane, where H+ are then pumped by complex V (ATP-synthase) to the matrix and ATP is produced. (Modified from 27.)

TABLE 15.1 Mitochondrial Nucleoid Proteins and Their Functions

NUCLEOID PROTEIN	*FUNCTION*
POLGA	mtDNA polymerase, catalytic subunit (29)
POLGB	mtDNA polymerase, accessory subunit; forms an heterotrimer 2:1 with POLGA (29)
mtRNA pol	Generates RNA primer for mtDNA replication (29)
TFAM	Initiates replication (29)
TFBM	Forms heterodimer with RNA polymerase, allowing specific transcription initiation (31)
mTERF	Terminates transcription (32)
mtSSB	Stabilizes mtDNA and stimulates Twinkle activity (29)
Twinkle	Facilitates mtDNA unwinding (29)
ATAD3	Backbone to the nucleoid (29)

located in the central region of the nucleoid; in contrast, ATAD3 is arranged peripherally and acts as the backbone of the nucleoid (reviewed by St. John [29]). Table 15.1 summarizes the functions of mitochondrial nucleoid components.

TFAM initiates replication, binding to mtDNA to induce structural changes that expose the promoter region. Next, mtRNA pol synthesizes an RNA, which POLGA uses as a primer to initiate mtDNA replication. This process is supported by POLGB, which stabilizes POLGA and increases efficiency, and by mtSSB and Twinkle, which mediate mtDNA unwinding. Mitochondrial genome replication differs in its timing from the replication of the organelle itself and is controlled by signals modulated by cellular energetic demand.

Based on studies of mouse models, the total amount of mtDNA remains stable from the one-cell ovum through the early blastocyst stage (31), with no further mtDNA replication occurring. Therefore, during cleavage, the number of mtDNA molecules per cell is reduced with each cell division. At the blastocyst stage, mtDNA replication resumes (32, 33), likely beginning first in the trophectoderm (TE) and then in the inner cell mass (ICM). In pigs, mitochondrial-specific polymerase levels increase at the morula stage in the outer edges of embryos, the region of cells that likely become TE, but not in the inner embryo, representing future ICM cells (33) (Figure 15.4). Accordingly, approximately 80% of the ATP in mouse embryos is generated in TE (34).

EARLY MITOCHONDRIAL NUMBER AND MORPHOLOGY CHANGES

Mitochondria undergo morphological changes during embryo development. Unlike differentiated cells, oocytes and fertilized eggs have structurally undeveloped mitochondria. Mitochondria of fetal oogonia are elongated, with a dense matrix and tubular cristae. In oocytes of primordial follicles, mitochondria are rounded, with a less dense matrix and membranes arranged in fewer cristae. During oocyte growth and maturation, mitochondria are predominantly spherical to oval, with a diameter of $\leq$ 1 μm and an increasingly dense matrix (35, 36). While their appearance makes them seem inactive, some actively generate ATP by oxidative phosphorylation (19). Their structure remains unchanged until about the eight-cell stage, when a decrease in matrix density begins. During blastocyst differentiation, expansion,

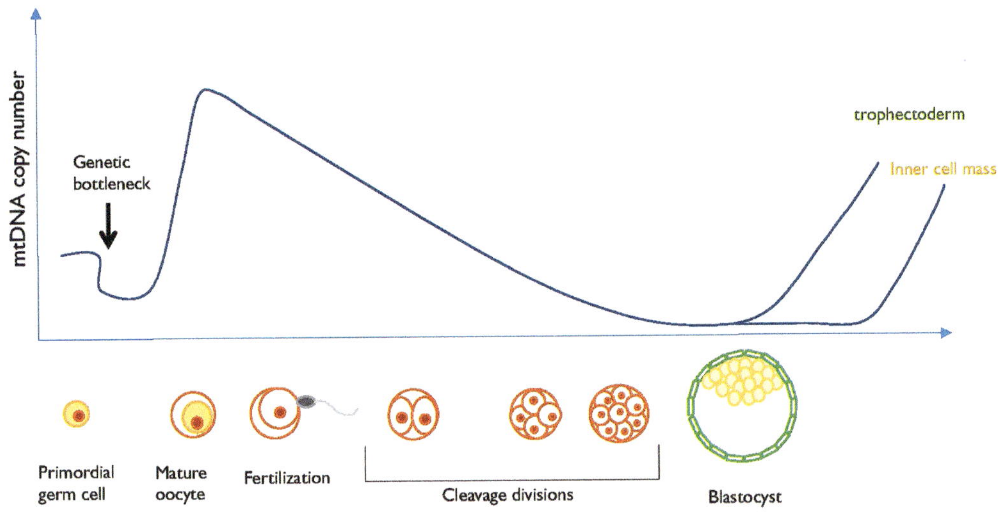

FIGURE 15.4 Changes in mtDNA copy number during development. During oogenesis, a genetic bottleneck occurs and the number of mtDNA molecules is reduced. The remaining mtDNA that is then amplified in the mature oocyte represents only a small fraction of the initial maternal pool. From this time, no further mtDNA replication takes place until the blastocyst stage. The number of mtDNA molecules is thus reduced with each cell division. At the blastocyst stage, mtDNA replication is resumed, likely in the trophectoderm first, then in the inner cell mass.

and hatching, mitochondria become progressively more elongated, with a lighter matrix, and have more numerous cristae, a clear sign of increased metabolic activity (35, 36).

MITOCHONDRIAL METABOLISM IN EARLY EMBRYO DEVELOPMENT

Early embryos exhibit changes in their metabolism throughout development. Initial cleavage divisions are controlled by maternal mRNA. Embryos at this period have a metabolic preference for pyruvate, demonstrating the importance of mitochondrial metabolism during this time (37) (Figure 15.5). By the time of embryonic genome activation, at about the eight-cell stage in humans, anabolic conditions switch to catabolic metabolism and glucose is the main energy source.

Changes in pyruvate and glucose uptake in human preimplantation embryos reflect metabolic changes. Pyruvate uptake increases from day 2.5 until day 4.5, then starts to lessen (38). In contrast, glucose uptake occurs throughout, increasing on successive days of development (39). This increase is minor from day 2.5 to 4.5 but becomes greater on day 5.5, at the blastocyst stage, as is expected (38).

Pyruvate and glucose uptake levels also reveal a relationship between metabolism and embryo quality. Pyruvate uptake is higher in embryos that develop successfully than in those that arrest at cleavage stages. Differences in glucose uptake occur during the final upsurge in consumption. At that point, embryos that are going to arrest fail to show the final upsurge in glucose consumption (38). Further, embryos that form high-quality blastocysts have higher glucose consumption on days 5 and 6 than those of bad quality, but they exhibit no difference in pyruvate uptake. This pattern is in line with the metabolic changes outlined above, and indicates that glucose, which allows for both aerobic and anaerobic glycolysis, is the more important energy source for human blastocysts (39).

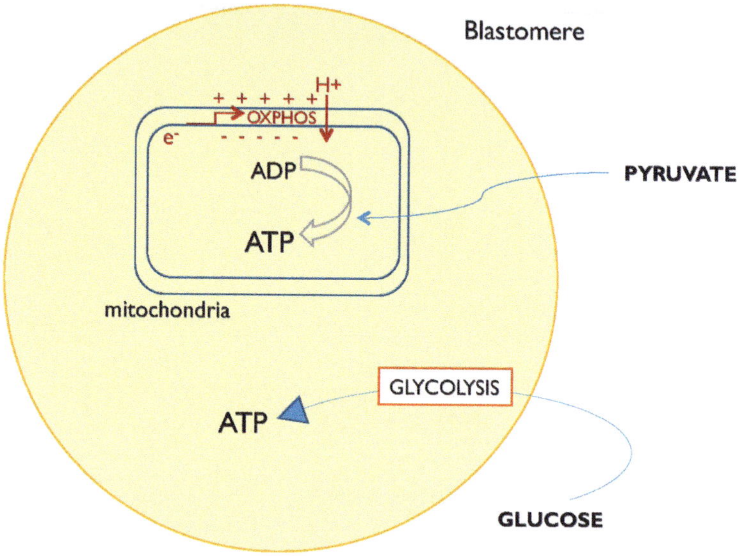

FIGURE 15.5 Blastomere energy production. Early embryo development requires ATP, which can be produced by two mechanisms: glycolysis, using glucose as the substrate, and mitochondrial oxidative phosphorylation (OXPHOS), using aerobic substrates such as pyruvate. During the first stages of development, mitochondria are the main energy producers; at the blastocyst stage, there is a shift to ATP generation by glycolysis.

MITOCHONDRIAL FUNCTION IN THE EARLY EMBRYO

Oxidative Stress and Early Embryo Development

As noted above, mitochondria produce ROS during oxidative phosphorylation. Excess ROS can induce oxidative stress in the embryo, which can lead to apoptosis and embryo fragmentation during early development. One function of the low mitochondrial activity in embryos may thus be to protect embryonic DNA from mitochondrial ROS.

Mitochondria play a crucial role in apoptosis during early development. Pro-apoptotic factors such as cytochrome c and the apoptosis-inducing factor are released from the intermembrane space into the cytosol, triggering caspase activation and in mouse zygotes undergoing *in vitro* culture, treatment with hydrogen peroxide (H_2O_2) induces cell death (40). Apoptosis is associated with oxidative stress: H_2O_2 production is correlated with apoptosis and DNA fragmentation in human embryos, and is also significantly higher in fragmented embryos than in non-fragmented ones (41). Similarly, H_2O_2 treatment induces cell death in cultured mouse zygotes. However, the *in vivo* environment differs from the *in vitro* environment used in the studies mentioned above. A key difference is the presence of defense mechanisms against ROS (41, 42). Antioxidants are found in the cumulus cells (43), follicular fluid, ovarian tissue, fallopian tubes, and endometrium. They protect the oocyte and embryo from oxidative stress by preventing ROS formation, neutralizing ROS, and promoting DNA repair (44). *In vitro* culture environments can also differ. A comparison of ROS levels on the first day of culture after conventional *in vitro* fertilization (IVF) vs. intracytoplasmic sperm injection (ICSI) found that, in both types of cycles, high ROS levels were associated with lower pregnancy rates (44). However, increased ROS levels were associated with low blastocyst rates, low fertilization rates, low cleavage rates, and high embryonic fragmentation only in ICSI cycles. This finding is consistent with the presence of cumulus cells, which may have antioxidant activity, in conventional IVF but not in ICSI.

Mitochondrial Ca^{2+} Signaling and Fertilization

Fusion of egg and sperm leads to metaphase II oocyte activation, resumption and completion of meiosis, extrusion of the second polar body, and pronuclear formation. This activation produces cortical granule exocytosis, zona pellucida and plasma membrane remodeling, and initiation of the maternal mRNA translation. Intracellular Ca^{2+} signaling has a well-defined role during fertilization. Cytosolic Ca^{2+} levels increase after sperm-egg fusion, followed by a series of oscillations (45). In mice, sperm-triggered Ca^{2+} waves in the fertilized egg stimulate mitochondrial oxidative phosphorylation (46). In turn, mitochondrial oxidative phosphorylation is necessary to maintain those Ca^{2+} oscillations: when cultured in a medium without the substrates required for oxidative phosphorylation, an egg can be fertilized but Ca^{2+} waves are inhibited (46).

Fertilization requires ATP, yet ATP levels do not decrease during this process, indicating that the energy supply matches the energy demand. Ca^{2+} signaling triggered by the entrance of the spermatozoa into the oocyte activates mitochondrial ATP production (45), providing the energy necessary to undergo subsequent developmental stages; low levels of oxidative phosphorylation are maintained when energy is not needed (47). The Ca^{2+} increase upon fertilization thus acts as a bridge between a cell's energy requirements and its energy producers, i.e., mitochondria. In turn, mitochondria are essential to maintaining the Ca^{2+} oscillations that are needed to complete meiosis and start embryo development (48).

MITOCHONDRIA AS A STRESS SENSOR

Mitochondria are able to sense internal or environmental changes, responding for example to toxins and diet-derived nutrients. They signal stress through membrane depolarization, alterations in adenine nucleotide levels, ROS production, Ca^{2+} fluxes, permeability transition pore opening, and perhaps secretion of proteins/peptides (49). They also regulate bioenergetic, thermogenic, oxidative, and/or apoptotic responses to reestablish homeostasis.

Acute exposure to stress mediators is associated with increased mitochondrial biogenesis and changes in the activity of the respiratory chain complexes, controlled ROS production, thermogenesis, and apoptosis. Under chronic stress, the damage can exceed mitochondrial responses, leading to abnormal mitochondrial biogenesis, respiratory chain dysfunction, decreased ATP production, increased ROS generation, lipid peroxidation, mitochondrial and nuclear DNA damage, and increased cell apoptosis (50) (Figure 15.6).

The "quiet embryo hypothesis" posits that, by minimizing oxygen consumption (quiet metabolism) from the zygote to the morula stage, the embryo limits ROS formation and therefore damage to the genome, transcriptome, and proteome (51, 52). The more "noisy" the embryo, the greater the demand for nutrients and energy and the greater the level of damage. The embryo is likely dependent on energy accumulation during oocyte maturation, and only in a stress situation (reduced metabolic fuel) does the cellular machinery increase mtDNA copy number in an effort to produce more mitochondria. Thus, mitochondrial dysfunction is associated with mitochondrial hyperproliferation (51, 53).

Mitochondrial biogenesis is mainly activated by increased energetic demands. This step requires the regulation of numerous processes such as mtDNA synthesis, the import and synthesis of nuclear-encoded proteins, and the assembly of proteins from nuclear and mitochondrial DNA. This regulation involves a set of nuclear transcription factors and coactivators, including PPAR coactivator 1a (PGC-1a) as the principal member. Increased energetic demands increase the cellular adenosine monophosphate (AMP)/ATP ratio, which is sensed by AMP-activated protein kinase (AMPK). AMPK phosphorylates and activates PGC-1a, inducing mitochondrial maturation and increasing mtDNA copy number and cristae density (53).

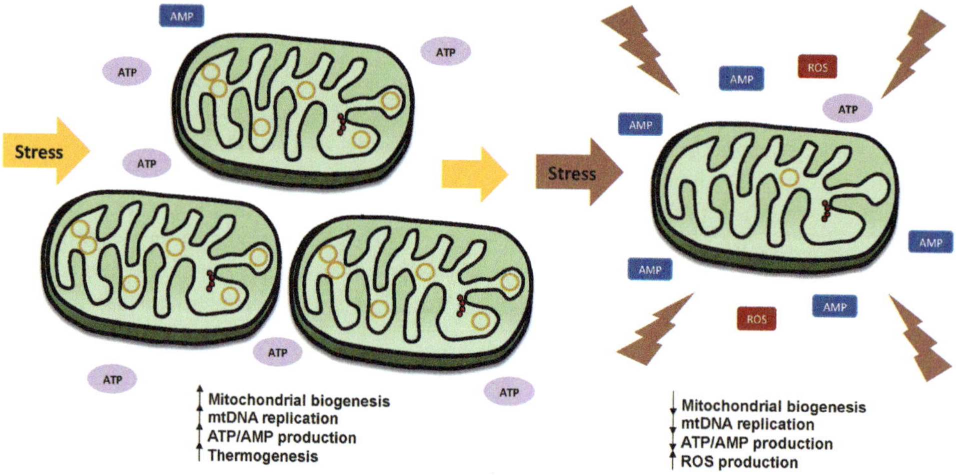

FIGURE 15.6 The mitochondrial stress response under (left) acute or (right) chronic stress. Acute stress is associated with increases in mitochondrial biogenesis and in the enzymatic activity of selected subunits of the respiratory chain complexes, to meet the increased energy demands of the cell. Prolonged stress to mitochondrial homeostasis can exceed mitochondrial reserves and lead to abnormally decreased mitochondrial biogenesis, respiratory chain dysfunction, decreased ATP production, increased ROS generation, and mtDNA damage.

An increase in mtDNA copy number in embryos is symptomatic of metabolic stress. This stress could be related to intrinsic factors during oocyte maturation or could be a response to impaired respiratory capacities due to mtDNA mutations (51, 53, 54). High mtDNA copy number in euploid embryos is indicative of lower embryo viability in terms of implantation potential (54), and increases in mtDNA content are associated with embryo loss (51). Aneuploid blastocysts contain significantly greater amounts of mtDNA; oxidative stress is also implicated in the shortening of telomeres that contributes to aneuploidy (53). Given the relationships between mtDNA content and embryo implantation ability and embryo chromosomal status, mtDNA quantification represents a potential new biomarker: an embryo's mtDNA copy number is an index of energetic stress and could thus be used to predict its implantation capacity (23, 51, 54).

MITOCHONDRIA IN EMBRYONIC IMPLANTATION AND DEVELOPMENT

An adequate energy supply is essential to embryo viability. This energy is provided by the accumulated mitochondria present in the oocyte, and only in cases of reduced metabolic fuel is there an increased mtDNA copy number (53, 54).

Because mtDNA content remains stable during the first days of embryonic development, the total amount of mtDNA must be split among cells during embryo division. By day 6 of development, mtDNA copy number is low in each cell (22, 54), increasing only when mtDNA replication resumes after implantation (55). Disrupting the mouse gene for mitochondrial transcription factor A (low mitochondrial number in oocytes) does not affect fertilization or early embryonic development: the embryo continues with implantation and gastrulation, However, these embryos die by embryonic day (E) 10.5, indicating that the replication of mitochondria is not initiated until well after implantation (19, 53, 56), suggesting that oxidative phosphorylation is not required at this stage, or that the maternal contribution of functional mitochondria is sufficient (56). These embryos do not die until later, by day 10.5, indicating that mitochondrial replication is not initiated until well after implantation (19, 53, 56).

During the cleavage and early blastocyst phases, mitochondria exhibit morphological changes. During cleavage (8–16 cells) in mouse, rabbit, and human embryos, mitochondrial geometry varies from spherical to elliptical and the cristae become more numerous. Finally, in the expanded blastocyst stage, mitochondrial organization and morphology resemble that in differentiated cells, with many well-formed cristae, particularly in trophectodermal cells (19). Mitochondria may complete this maturation process after the embryo has undergone the first cellular differentiation into TE and ICM (23, 57).

mtDNA replication in blastocysts begins in the TE, likely because of its earlier differentiation and loss of pluripotency to form the placenta. Thus, the TE accumulates enough mtDNA to provide sufficient ATP to fuel implantation. In contrast, the ICM maintains its pluripotency until later in development, so mtDNA replication begins later in these cells, reducing their mtDNA copy number (23, 57).

In the cleavage stage, the oocyte mitochondria disperse into the blastomeres, with little or no replication. No significant mtDNA replication occurs between fertilization and the blastocyst stage; the majority of mtDNA present during the cleavage stage is derived from the oocyte (23). The low number of mtDNA copies in the ICM is maintained until organogenesis, suggesting that mtDNA replication is limited to pluripotent cells (57). Later in development, each cell type has specific energy requirements, necessitating different levels of ATP production via oxidative phosphorylation. Neurons, cardiomyocytes, and muscle cells, for example, need many mtDNA copies; in contrast, endothelial cells, which use glycolysis rather than oxidative phosphorylation, need fewer copies (57).

Many aspects of the maternal-embryonic transition remain to be completely understood as the full functionality of the embryonic genome emerges (58). Critical among them, however, is the proper functioning of mitochondria and the mitochondrial genome during the early stages of embryonic development. These organelles are crucial for providing the energy needed to fuel essential processes related to metabolism, synthesis, cell division, and differentiation (23).

CONCLUSIONS

- Mitochondria originated as ancient endosymbionts, and still retain part of their own genome. Mitochondrial genome inheritance is non-Mendelian.
- In the absence of cell reserves, mitochondria are indispensable to support life.
- mtDNA copy number is regulated during early embryo development.
- mtDNA copy number is tightly regulated in response to stress signals.
- mtDNA copy number is an indicator of embryonic implantation potential.

BIBLIOGRAPHY

1. Cooper GM. *The Cell: A Molecular Approach.* 2nd ed. ASM Press; 2000.
2. Adam SA. *Mitochondrion,* 2002. http://wwwencyclopediacom/topic/mitochondriaaspx#2.
3. Gray MW. *Cold Spring Harb Perspect Biol.* 2012;4(9):a011403.
4. Martin W, & Mentel M. *Nature Education.* 2010;3(9):58.
5. Margulis L. *Origin of Eukaryotic Cells.* Yale University Press; 1970.
6. Dyall SD, et al. *Science.* 2004;304(5668):253–7.
7. Andersson SG, et al. *Philos Trans R Soc Lond B Biol Sci.* 2003;358(1429):165–77; discussion 77–9.
8. Emelyanov VV. *Biosci Rep.* 2001;21(1):1–17.
9. Gray MW, et al. *Genome Biol.* 2001;2(6):REVIEWS1018.

10. Cavalier-Smith T. *Proc Biol Sci.* 2006;273(1596):1943–52.
11. Desler C, et al. *J Nucleic Acids.* 2010;2010:701518.
12. McBride HM, et al. *Curr Biol.* 2006;16(14):R551–60.
13. Ott M, et al. *Apoptosis.* 2007;12(5):913–22.
14. Hopper RK, et al. *Biochemistry.* 2006;45(8):2524–36.
15. Cui H, et al. *J Signal Transduct.* 2012;2012:646354.
16. Lenaz G. *Biochim Biophys Acta.* 1998;1366(1–2):53–67.
17. Giles RE, et al. *Proc Natl Acad Sci U S A.* 1980;77(11):6715–9.
18. Anderson S, et al. *Nature.* 1981;290(5806):457–65.
19. Van Blerkom J. *Semin Cell Dev Biol.* 2009;20(3):354–64.
20. Hirst J. *Annu Rev Biochem.* 2013;82:551–75.
21. Li Y, et al. *J Bioenerg Biomemb.* 2006;38(5–6):283–91.
22. Cummins J. *Rev Reprod.* 1998;3(3):172–82.
23. Fragouli E, et al. *Semin Reprod Med.* 2015;33(6):401–9.
24. Stewart JB, et al. *PLoS Genet.* 2014;10(10):e1004670.
25. Sato M, et al. *Autophagy.* 2012;8(3):424–5.
26. Sutovsky P, et al. *Reprod Biomed Online.* 2004;8(1):24–33.
27. Mishra P, et al. *Nat Rev Mol Cell Biol.* 2014;15(10):634–46.
28. Marchington DR, et al. *Am J Hum Genet.* 1998;63(3):769–75.
29. St John J, et al. *Hum Reprod Update.* 2010;16(5):488–509.
30. St John J. *Biochimica et Biophysica Acta.* 2014;1840(4):1345–54.
31. Piko L, et al. *Dev Biol.* 1987;123(2):364–74.
32. Thundathil J, et al. *Mol Reprod Dev.* 2005;71(4):405–13.
33. Spikings EC, et al. *Biol Reprod.* 2007;76(2):327–35.
34. Houghton FD. *Differentiation.* 2006;74(1):11–8.
35. Motta PM, et al. *Hum Reprod.* 2000;15 Suppl 2:129–47.
36. Sathananthan AH, et al. *Hum Reprod.* 2000;15 Suppl 2:148–59.
37. Jansen RP, et al. *Mitochondrion.* 2004;4(5–6):577–600.
38. Hardy K, et al. *Hum Reprod.* 1989;4(2):188–91.
39. Gardner DK, et al. *Fertil Steril.* 2001;76(6):1175–80.
40. Liu L, et al. *Biol Reprod.* 2000;62(6):1745–53.
41. Yang HW, et al. *Hum Reprod.* 1998;13(4):998–1002.
42. Guerin P, et al. *Hum Reprod Update.* 2001;7(2):175–89.
43. Bedaiwy MA, et al. *Fertil Steril.* 2004;82(3):593–600.
44. Agarwal A, et al. *Reprod Biol Endocrinol.* 2005;3:28.
45. Armant DR. *Adv Exp Med Biol.* 2015;843:151–71.
46. Dumollard R, et al. *Development.* 2004;131(13):3057–67.
47. Ramalho-Santos J, et al. *Hum Reprod Update.* 2009;15(5):553–72.
48. Nixon VL, et al. *Curr Biol.* 2002;12(9):746–50.
49. Barbour JA, et al. *Int J Cell Biol.* 2014;2014:156020.
50. Manoli I, et al. *Trends Endocrinol Metab.* 2007;18(5):190–8.
51. Fragouli E, et al. *PLoS Genet.* 2015;11(6):e1005241.
52. Leese HJ. *Reproduction.* 2012;143(4):417–27.
53. Meldrum DR, et al. *Fertil Steril.* 2016;105(3):548–59.
54. Diez-Juan A, et al. *Fertil Steril.* 2015;104(3):534–41.e1.
55. Steffann J, et al. *Clin Genet.* 2015;88(5):416–24.
56. Larsson NG, et al. *Nat Genet.* 1998;18(3):231–6.
57. St John JC. *Cell and Tissue Research.* 2012;349(3):795–808.
58. Cummins JM. *Reprod Biomed Online.* 2002;4(2):176–82.

16 Preimplantation Genetic Diagnosis for Single Gene Disorders

Ana Cervero, Jose Antonio Martínez-Conejero, Lucía Sanz-Salvador, Claudia Gil-Sanchís, Maribel Sánchez-Piris, and Laura Iñiguez Quiles

INTRODUCTION

Preimplantation genetic testing for monogenic/single gene defects (PGT-M) is an alternative to prenatal diagnosis for those couples with a known risk of transmitting a genetic condition to their offspring. Oocytes and/or preimplantation embryos obtained by *in vitro* fertilization (IVF) are analyzed and only those embryos free of the disorder under study are transferred to the uterus to achieve pregnancy. Therefore, the main advantages of PGT-M are (i) to increase the likelihood of having an unaffected pregnancy by selecting from multiple embryos and (ii) to circumvent invasive prenatal diagnosis which is frequently followed by the difficult decision of pregnancy termination if results are unfavorable.

Early research in PGT-M was pioneered by Edwards and Gardner (1), who, in 1968, biopsied rabbit blastocysts to determine gender. The first successful application of PGT-M in humans was performed by Handyside and colleagues in 1990 (2), who carried out sexing of embryos by polymerase chain reaction (PCR) to avoid males affected with an X-linked disorder. Gender was determined in single blastomeres by PCR using primers for amplifying Y-chromosome specific DNA sequences and those embryos identified as female were selectively transferred to the uterus (2). Later, successful PGT-M was reported for cystic fibrosis (3), based on the amplification of a DNA fragment containing the causative mutation and its detection by fragment analysis (3).

Today, PGT indications have been expanded and they are performed in laboratories worldwide. The precise number of PGT cycles that have been performed to date can only be estimated. The European Society for Human Reproduction and Embryology (ESHRE) PGT consortium has collected data about PGT cycles since 1999, showing thousands of PGT cycles performed allowing the birth of hundreds of healthy children. The last such analysis between January and December 2010, reported a total of 5,732 cycles of IVF cycles with PGT, including PGT for sex selection. Of these, 1,574 cycles were performed for single gene disorders, including human leukocyte antigen (HLA) typing (4). Although the ESHRE

DOI: 10.1201/9781003024941-16

data represent only a partial record of the PGT-M cases conducted worldwide, they are indicative of the general trend in the field.

PGT-M involves a multidisciplinary team and requires deep collaboration between the assisted reproduction unit and the genetic laboratory which can either be co-located within the same institute or geographically separate. In this latter situation, an effective transportation and logistics operation must be in place; IVF treatment (controlled ovarian stimulation, oocyte retrieval and *in vitro* fertilization, embryo culture, and embryo transfer) is carried out in an assisted-reproduction unit and only the biopsied embryonic samples are transported to the genetic unit where the PGT-M analysis is performed.

PGT-M is regulated or even prohibited in many countries based on national or local laws. In many countries, PGT-M is limited to serious conditions with a high likelihood of transmission and sometimes, especially for adult-onset disorders, HLA matching, or a previously untested disorder, it is necessary to apply to a specific national committee to get the approval for performing the PGT-M (5).

INDICATIONS FOR PGT-M

PGT-M is recommended when couples are at risk of transmitting a known genetic abnormality to their children and therefore, its indications are similar to conventional prenatal diagnosis. PGT-M has been mainly used to diagnose and prevent well-defined autosomal recessive, autosomal dominant, or X-linked single gene disorders, although most single gene disorders that can be diagnosed can be also identified in the embryo. The conditions for which PGT-M has been applied increase annually. Other uses for PGT-M include gender selection, compatible HLA typing (aiming for a "savior sibling"), the identification of hereditary cancers with variable penetrance (e.g., BRCA 1, 2 status), and late-onset genetic diseases all of which have been considered controversial.

Single Gene Disorders

According to the latest ESHRE PGT consortium data, the most common indications for autosomal recessive disorders are cystic fibrosis, spinal muscular atrophy, and hemoglobinopathies (4) which involve the presence of two mutated copies from each healthy carrier parent.

For the autosomal dominant conditions, one mutated copy of the gene is enough for a person to be affected. Myotonic dystrophy type 1, neurofibromatosis, and Huntington's disease are the most frequently requested indications (4). In some cases, an affected person inherits the condition from an affected parent. In others, the condition may result from a new mutation in the gene and occur in people with no history of the disorder in their family. When the patient has a *de novo* mutation it is necessary to identify the molecular variant causing the disease. Once the mutation has been characterized, this variant can be analyzed in the embryo.

X-linked recessive disorders are mainly transmitted by healthy carrier mothers to their sons. In this case, an affected male will not have risk of affected offspring; all his daughters will be obligate carriers but none of his sons will be affected. PGT-M for the X-linked disorders is mainly carried out for Duchenne's muscular dystrophy, hemophilia, and fragile X syndrome (4). Although initially sexing with FISH was widely applied for X-linked disorders in order to select female embryos for transfer (6), specific diagnosis of the molecular defect has important advantages and has replaced FISH. First, it facilitates the identification and subsequent transfer of those healthy male embryos that by FISH diagnosis would be discarded. Second, female carriers may be identified that can be excluded from transfer or not, according to patient wishes or center policy. Indeed, this is significant for those X-linked dominant disorders (e.g., fragile X syndrome) where it is possible that carrier females may manifest symptoms of the disease.

Cancer Predisposition and Late-Onset Diseases

PGT-M is also offered to those couples where one partner carries a mutation predisposing to cancer or another late-onset disease (7). For cancer predisposition syndromes that are not fully penetrant and for which some form of therapeutic measures may be available, prenatal diagnosis and termination of pregnancy remain controversial and PGT-M appears as an attractive option, preventing the difficult decision of termination of an established pregnancy. In contrast, the use of the PGT-M for diseases that will not develop until adulthood or for mutations that only confer a heightened risk raises issues of how to weigh the possible benefits of the PGT-M for the future child against the risks of PGT-M and IVF for the patient. Despite these ethical and legal issues (8), the number of PGT-M cycles reported for these types of condition is increasing (4), and the procedure has already been carried out for several diseases, including the common syndromes of genetic predisposition to colon and breast cancer (7, 9).

For some adult-onset conditions (e.g., Huntington's disease – HD), patients do not want to perform the genetic study because they do not want to know their genetic status in advance, but they want to make sure that their children do not have the mutation. Unlike prenatal diagnosis, PGT-M for Huntington's disease allows to keep the status of the carrier of the mutation blinded (10) either by non-disclosure or the less ethically troublesome exclusion tests. In the former, the mutation is analyzed but the results are not revealed to the patient. This approach has several practical and ethical issues including the need to not give any details of the IVF cycle in order to avoid a potential clue about the patient´s carrier status. Moreover, a mock transfer could be necessary if there are not any embryos for transfer so that the patients don´t guess they are carriers. In the exclusion test, those embryos inheriting the haplotype coming from the affected grandparent are ruled out for the transfer. The drawback in this case is that unaffected embryos could be rejected if the patient is not a carrier.

HLA Matching

Another controversial but relatively well-established indication is HLA matching in which PGT is employed to conceive a child (with or without a specific genetic risk themselves) who may donate compatible cord blood or hematopoietic stem cells for transplantation to save an affected sibling (11). Hematopoietic stem cell transplantation (HSCT) from an HLA-identical donor is the best therapeutic option for genetic diseases affecting the hematopoietic and/or immune system in children (e.g., β-thalassemia, Fanconi anemia), and can also be an effective therapeutic option for acquired diseases (e.g., leukemia, acquired medullary aplasia) (12, 13). The frequent unavailability of HLA-identical donors for affected children within the corresponding families or in HSC banks has made the combination of IVF with HLA-typing for the selection of HLA-identical embryos a realistic therapeutic approach for such families.

PGT for HLA typing alone is performed for acquired diseases, such as severe aplastic anemia or leukemias, or can be performed in conjunction with a single gene disorder, in order to select an embryo free of the inherited condition and HLA-matched to an existing affected child (11–13). This approach was applied for Fanconi anemia for the first time in 2001 (11) and since then has been performed for a number of different diseases affecting the hematopoietic system. Worldwide, current HLA testing on preimplantation embryos is usually performed using short tandem repeat markers (STRs), since multiple STRs throughout the HLA region allow 100% accuracy HLA-typing and detect possible recombination events (14, 15).

It should be noted that the clinical results for this indication are worse than those obtained for the others, mainly due to the low number of embryos available for transfer. In the standard case, where PGT/HLA testing is used to select embryos that are both free of a specific disease and an HLA match, the chance of an embryo being both healthy and a suitable match is only 18.75% in the case of autosomal recessive conditions, such as beta-thalassemia. The other important limitation is that the majority of patients requesting preimplantation HLA typing are of advanced reproductive age, so the outcome of the procedure has limited success, with many patients requiring two or more attempts before they become pregnant and deliver an HLA identical offspring. Therefore, patients should have realistic expectations for the overall success of this approach and should be informed about the possible risks and complications. Since the development of the HLA testing, establishing pregnancy, and pregnancy itself are time-consuming, the diseased child will sometimes die before the HSCT become available (16). Moreover, as indicated by Kahraman et al. (16), the stem-cell dose obtained from umbilical cord blood is frequently insufficient, and extra time is needed for the child to gain sufficient weight to be able to donate his or her bone marrow cells. All of these limitations might increase the time it will take for the sick sibling to undergo the transplant, aside from the fact that nine months are required for delivery of a successfully implanted embryo.

Despite ethical objections, including the instrumentalization of the future child (the new child is considered by certain people as an instrument to cure another child), the results show that this clinical procedure is an option with documented positive outcomes for couples with affected children requiring HLA compatible stem cell transplantation (14–16).

It is important to remark that as with some other PGT indications, the law regulating the PGT for HLA matching depends on the country and in some of them, HLA-PGT cases have to be approved case by case by a national committee after evaluating the clinical and therapeutic characteristics and weighing carefully the potential risks and benefits to all those involved (5).

Isoimmunization

Hemolytic disease of the newborn, also known as erythroblastosis fetalis, isoimmunization, or blood group incompatibility, occurs when fetal red blood cells, which possess an antigen that the mother lacks, cross the placenta into the maternal circulation, where they stimulate antibody production. ABO incompatibility is the most common cause of hemolytic disease of the newborn followed by the Rhesus and Kell systems (17).

Although the incidence of severe RhD alloimmunization has decreased with prophylactic anti-D immunoglobulin administration during and after pregnancy, sensitization still occurs in a small group of women. In such women, Rh disease will continue to be a significant problem and for their babies who may be affected. PGT-M may be utilized to avoid materno-fetal blood group incompatibility in women who are Rh negative and are highly sensitized with antibodies against the Rh factor. If Rh genotyping in the male shows that he is heterozygous, it is feasible to perform a PGT-M to avoid possible erythroblastosis fetalis and intrauterine blood exchange transfusion (18). In the same way, PGT-M can also be used in women sensitized by other blood factors, such as the Kell antigen or other antigens present on the platelet surface (19).

DIAGNOSTIC METHODS

The different approaches that are employed in PGT-M are based on haplotyping. Genetic markers, short tandem repeats (STRs), or single nucleotide polymorphisms (SNPs) located close to the gene of interest are analyzed in DNA samples from the reproductive couple and relevant relatives during the preclinical work-up or pre-PGT-M test. The haplotype which is present in the family members with the familial pathogenic variant is named the at-risk haplotype, whereas the haplotype without the familial pathogenic variant is referred to as the wild-type or low-risk haplotype. During pre-PGT-M work-up, the analysis of polymorphic markers in DNA samples from patients and other relatives identifies which alleles are expected in the embryos, and the specific marker alleles which co-segregate with the mutation. This combined approach improves accuracy, minimizing potential errors caused by undetected allele drop out (ADO) or contamination (20). ADO refers to the amplification failure (or extreme preferential non-amplification) of one of the two alleles, making a heterozygous locus appear homozygous, and potentially leading to misdiagnosis. The clinical test can be either direct, when the pathogenic variant plus linked genetic markers are assessed, or indirect, when testing is based on haplotyping only. Recently, the ESHRE

PGT Consortium published a paper providing recommendations on the technical aspects of PGT-M, covering recommendations on basic methods for PGT-M and testing strategies (21).

PCR and Fragment Analysis

Multiplex PCR using targeted primers designed specifically for the mutation of interest combined with primers for closely linked STR markers has been traditionally the gold standard to perform the PGT-M (22).

Genotyping of the amplified products can be performed by means of different strategies, with minisequencing the most frequently used method for the detection of point mutations (23). In the minisequencing technique, a primer extension reaction is performed, allowing rapid and accurate detection of point mutations. The minisequencing primer is designed to anneal one base before the target site, and it can only be elongated with one specific dideoxynucleotide. The four different dideoxynucleotides are labeled with different fluorochromes, and the products can be analyzed on an automated DNA sequencing system. Other strategies such as amplification refractory mutation system (24), restriction enzyme digestion (25), and real-time PCR (26) have been also applied in PGT-M. Small deletions and duplications can also be detected by sizing of PCR products from specific regions containing the mutation under study.

The use of multiplex PCR for linkage markers alone (so-called preimplantation genetic haplotyping or PGH) has become widespread in PGT for monogenic disorders (21, 27) and HLA typing (28). The main advantage is that such protocols can be used for several couples, independent of the mutation they carry, thus saving time and resources in pre-PGT-M work-ups. However, the ability to use such indirect testing depends on the availability of appropriate family samples to determine the "at risk haplotype." In cases where no such samples are available or in *de novo* mutation cases, it is necessary to identify the disease-causing mutation and analyze it directly in the embryos.

WGA Amplification

The creation of a robust and accurate multiplex protocol requires careful design and optimization and validation before its clinical use. As a result, investment of time and resources is needed. The implementation of whole genome amplification (WGA) was a technical improvement that became widespread and proved to be a practical and efficient alternative to perform PGT-M (29). WGA amplifies the entire genome producing enough amplified DNA for multiple downstream applications. Many different standard PCR assays may be performed for haplotyping and the direct analysis of mutations in case of monogenic diseases avoiding the necessity of optimizing multiplex PCR protocols (30). Moreover, WGA allows combining the PGT for a single gene disorder or HLA typing with array comparative genomic hybridization (aCGH) or next-generation sequencing (NGS) for the detection of chromosomal imbalances using the same sample with the aim of improving clinical PGT results (31). Finally, WGA facilitates repeat testing of samples for use in proficiency testing, validation, or in the case of run failure.

Despite the large quantities of amplified DNA produced, it is well-established that WGA methods yield relatively high ADO rates overall when a single blastomere is biopsied (32). This problem can be circumvented with the application of enough linked markers to avoid misdiagnosis and/or reduced with the use of trophectoderm biopsies instead of single cells because the former provides multiple cells in the biopsy sample which is known to result in lower ADO rates (31).

Karyomapping

Karyomapping uses a high-density single nucleotide polymorphism (SNP) array to accurately identify DNA haplotypes in samples even when the starting DNA is minimal. By genotyping the parents at several hundred thousand SNP sites throughout the genome, a dense set of informative SNP markers are identified for each of the four parental chromosomes (33). The phase of the alleles for each informative SNP locus along each chromosome and linkage of the risk alleles with the parental chromosomes can then be established by reference to the genotype of a relative with a known disease status. The parental origin of each chromosome in the embryo is then ascertained by comparison with the genotype of the reference (33). The principal advantage of this platform is that it is applicable to almost any familial single gene disorder, or any combination of loci, within the chromosome regions covered by informative SNP loci, eliminating the need for developing patient- or disease-specific tests (33). The main disadvantages of the karyomapping approach are that diagnosis is challenging when insufficient informative SNP markers are available (e.g., in some telomeric genes), or when pseudogenes are involved. Moreover, it cannot be used on its own for *de novo* mutation cases or when other tested family members are not available to provide samples or are not informative owing to recombination (33). In such cases, direct mutation testing from at least one embryo is necessary to establish phase (34).

Currently, it is possible to determine both monogenic diagnosis and aneuploidy detection (plus HLA haplotyping) by PGT-M using the same WGA product (35). For this reason, a single assay using the same platform to detect simultaneously both chromosomal and monogenic disorders is desirable. Since karyomapping defines unique sets of SNP markers for each of the four parental chromosomes, it is such a method (34), by allowing accurate identification of the region of interest containing the mutation and simultaneous high-resolution molecular cytogenetic analysis. Meiotic trisomies can be identified by the presence of both haplotypes from one parent in segments of the chromosome, resulting from the inheritance of two chromosomes with different patterns of recombination, in combination with a single haplotype from the other parent. Moreover, monosomies or deletions can be identified by the absence of one of the parental haplotypes (33, 34).

Several studies have reported the clinical use of karyomapping, showing the simultaneous detection of monogenic and chromosomal disorders (36, 37). However, this approach has several limitations (as described above) which need to be overcome as more carrier screening tests are performed which identify co-carriage of mutations in the same gene among couples with no family history or affected children—the historic source of referrals for PGT-M. At present karyomapping is not commercially validated for aneuploidy screening and does not readily detect mitotic trisomies nor simple copy number variation. Nonetheless, use of karyomapping can be a powerful tool to investigate parental origin and phase of origin of meiotic chromosome errors.

Next-Generation Sequencing

Next-generation sequencing (NGS) provides high throughput and base pair resolution data, providing the analysis of multiple genetic loci and samples from different couples simultaneously.

Several studies have been published showing the possibility of using NGS to test single cells (38, 39). In 2013, Treff et al. published a specific protocol to test DNA from a trophectoderm biopsy with NGS that was consistent with two conventional methodologies of PGT-M (40). However, the major concern relating to NGS technology is that an insufficient sequencing depth may result in false positive or failure to identify a mutation (false negative) due to the presence of sequencing artifacts and ADO, respectively. Consequently, sequencing at high coverage is required, which is expensive for routine clinical applications. Moreover, NGS has technical limitations in testing for dynamic mutations. Therefore, several strategies, aiming at affordable and rapid protocols are being developed. One of them is to increase the read depth specifically around the site of the variant of interest only (41).

CLINICAL RESULTS

Regarding clinical results of PGT, the last data collection carried out for the ESHRE consortium showed that 81% of PGT-M cycles resulted in an embryo transfer. The clinical pregnancy rate obtained for the cycles was 28% per oocyte retrieval and 36% per embryo transfer, giving an overall implantation rate of 27%. Finally, the delivery rate was 24% per oocyte retrieval and 31% per embryo transfer and the miscarriages rate was 10% (4). However, it is important to note that there may be significant differences in the clinical results between the different IVF centers.

Extensive evidence has revealed a high incidence of chromosomal abnormalities in human embryos obtained by assisted reproduction techniques leading to miscarriages and implantation failures (42, 43). Couples presenting for PGT to avoid transmission of single gene disorders are not without these same problems despite the fact that they are often categorized as "fertile." For this reason, they can also benefit from the simultaneous analysis of the disease and the presence of aneuploidies, selecting for transfer those PGT-M "unaffected" and simultaneously euploid embryos.

Several studies have shown significant improvement in rates of pregnancy and live births following testing for aneuploidy in patients undergoing IVF for infertility (43–45). The use of WGA product provides a straightforward solution for performing any required test in the same biopsy material, allowing the simultaneous analysis of PGT-M and aneuploidy screening. Using this approach, several studies have reported the clinical use simultaneous detection of monogenic and chromosomal disorders using different technologies (34, 38, 43, 44). One of the first articles was published by Rechitsky et al. where the simultaneous detection of cytogenetic disorders and cystic fibrosis was described (46). In 2015, the first systematic study of PGT-M combined with the aneuploidy screening was published demonstrating an increase in the pregnancy rate from 45.4% in the conventional PGT-M to 68.5% with combined aneuploidy screening and three-fold miscarried reduction (5.5% vs. 15%) (47).

Our data indicate that 43.6% of the embryos diagnosed at blastocyst stage as normal for the genetic disease have some chromosome abnormalities that can be rise to miscarriages or implantation failure (48). Specifically, 16.3% of those PGT-M normal embryos exhibited trisomies or monosomy X that could lead to miscarriage if transferred. A further 20.2% of PGT-M normal embryos were carrying some monosomies that could result in implantation failure. Therefore, these results, in line with other studies, indicate that the combination of accurate PGT-M and detection of chromosome aneuploidy may improve implantation and rates of healthy live births.

LIMITATIONS AND BARRIERS TO ACCESS

Not all people who wish to have PGT-M can do so. It is essential that the disorder which is intended to be diagnosed has a comprehensive and accurate genetic characterization to at least identify the gene responsible for the condition. Some conditions, including autism and some immunological disorders where the causes remain unidentified, are not suitable for PGT-M.

As PGT-M involves both an IVF procedure and a selection procedure based on genetic testing, it is extremely important to predict the number of unaffected embryos expected for transfer prior to starting the procedure. This number will depend on the embryo quality and the theoretical risk according to the genetic disorder (e.g., recessive, dominant, sex-linked). The embryo's potential to implant depends mainly on the woman's age and the absence of additional factors that facilitate the production of incompetent gametes in men or women. It is well-known that chromosome aneuploidy is the major cause of IVF failure and miscarriage. To benefit patients, a sufficient number of embryos should be analyzed to obtain non-affected embryos for transfer. Recent advances in vitrification procedures have made it feasible to batch oocytes or embryos to reach a minimum number of embryos for analysis (49). Table 16.1 shows the rates of transferable embryos considering the inheritance patterns and the potential presence of aneuploidies.

One final significant point regarding PGT-M concerns patient and clinician expectation: PGT-M provides no guarantee of a completely healthy baby. Instead, PGT-M minimizes the risk for the disease for which the couple has a high risk of transmitting to their offspring. However, there remains a

TABLE 16.1 Rates of Transferable Embryos Considering the Inheritance Patterns and the Potential Presence of Aneuploidies

INDICATION	*UNAFFECTED EMBRYOS*	*PGS NORMAL EMBRYOS**	*TRANSFERABLE EMBRYOS*
Autosomal dominant	1/2	1/2	1/4
Autosomal recessive	3/4	1/2	3/8
X-linked recessive	3/4	1/2	3/8
X-linked dominant	1/2	1/2	1/4
HLA matching	1/4	1/2	1/8
HLA + AD	1/8	1/2	1/16
HLA + AR	3/16	1/2	3/32
HLA + X-linked	3/16	1/2	3/32

* The 50% of blastocyst are expected to be PGS abnormal on average according to Igenomix internal data.

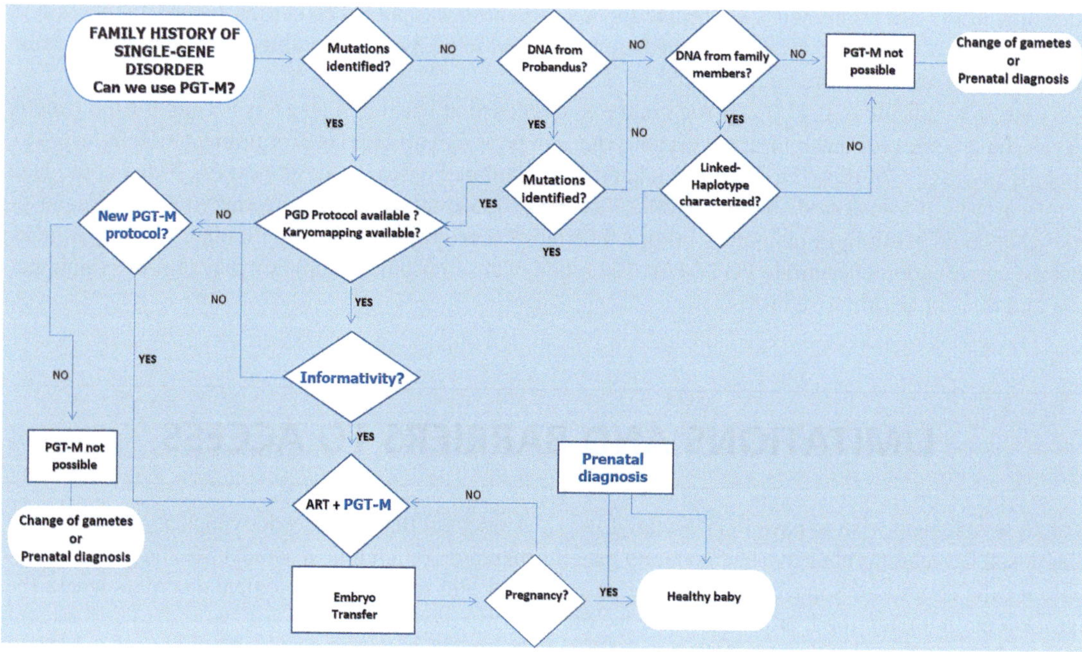

FIGURE 16.1 Flowchart for the different stages or procedures that should be followed in a regular PGT-M case.

small risk of misdiagnosis due to the existence of mosaicism or technical limitations, and therefore couples should always be offered the possibility of prenatal diagnosis to confirm results. Moreover, sometimes couples assume that they will have a normal healthy child following successful PGT-M for a specific disorder forgetting that a more cryptic disorder may be present which was not what was causing the disease identified previously in the family. Furthermore, no other genes or mutations even within the same gene are routinely analyzed and, if chromosomes are not analyzed, an aneuploid but genetically unaffected embryo may be transferred. The entire procedure should always be explained in detail before the treatment to ensure that couples are informed about the potential risks and limitations (Figure 16.1).

CONCLUSIONS

PGT-M can be used to screen embryos for almost any kind of genetic disorder in which the genetic cause is characterized, increasing the number of indications and the total number of PGT-M cases year by year. With improving technologies, best practice guidelines and the adoption of external quality assessment programs and laboratory accreditation (20, 50–52) PGT-M analysis has reached a high level of accuracy and has enabled the possibility of performing multiple diagnoses from the same sample. Diagnosis of a monogenic disease can be combined with HLA typing and/or with the detection of chromosomal abnormalities allowing the improvement of reproductive outcomes. The explosion of preconception carrier screening for couples with no family history of specific genetic disease as a result of best practice guidelines, reduced cost, and improved access to pan-ethnic expanded carrier screening panels, as well as government funding for PGT-M in many countries, is certain to increase the number of PGT-M cases further.

FUTURE PERSPECTIVES

The development of genome-wide technology, bioinformatics, and databases and biobanks has allowed the application of PGT to include predicting the risk of polygenic disorders (PGT-P) (53). In PGT-P, embryos are categorized according to the likelihood of common disease (e.g., diabetes, cardiovascular disease, hypothyroidism, cancer) (54). The clinical utility of PGT-P is a subject of intense debate (55) and the application for this emerging technology raises ethical considerations that need to be addressed (56).

The presence of DNA in the blastocoel fluid and spent embryo culture media is a potential alternative for a non-invasive genetic assessment of preimplantation embryos (57). Several studies have been published using both types of samples and the spent culture media seems to be a better source of embryonic DNA (58). However, the amplification rates are lower and ADO rates are still higher than those observed using conventional cellular biopsy and consequently, the risk of misdiagnosis is elevated (59). Linkage analysis strategies may help to overcome these technical problems, but the potential presence of maternal DNA contamination could be a significant issue. Therefore, although there have been published promising results, more work needs to be done to define optimal methods for the collection and analysis of this extra-embryonic DNA before the clinical application can be a reality.

REFERENCES

1. Gardner RL, et al. *Nature*. 1968;218(5139):346–9.
2. Handyside AH, et al. *Nature*. 1990;344(6268):768–70.
3. Handyside AH, et al. *N Engl J Med*. 1992;327(13):905–9.
4. De Rycke M, et al. *Hum Reprod*. 2015;30(8):1763–89.
5. Soini S. *Med Law*. 2007;26(2):309–23.
6. Harper JC, et al. *Hum Reprod*. 1994;9(4):721–724.
7. Rechitsky S, et al. *Reprod Biomed Online*. 2002;5(2):148–55.
8. Clancy T. *Fam Cancer*. 2010;9(1):9–14.
9. Spits C, et al. *Prenat Diagn*. 2007;27(5):447–56.
10. Sermon K, et al. *Eur J Hum Genet*. 2002;10(10):591–8.
11. Verlinsky Y, et al. *JAMA*. 2001, 27;285(24):3130–3.
12. Kuliev A, et al. *Reprod Biomed Online*. 2005;11(3):362–70.

13. Verlinsky Y, et al. *Reprod Biomed Online.* 2007;14(2):214–23.
14. Van de Velde H, et al. *Hum Reprod.* 2009;24(3):732–40.
15. Fiorentino F, et al. *Hum Reprod.* 2006;21(3):670–684.
16. Kahraman S, et al. *Reprod Biomed Online.* 2014;29(3):340–51.
17. Avent ND, et al. *Blood.* 2000;95(2):375–87.
18. Avner R, et al. *Mol Hum Reprod.* 1996;2(1):60–2.
19. Verlinsky Y, et al. *Fertil Steril.* 2003;80(4):1047–51.
20. Harton GL, et al. *Hum Reprod.* 2011;26(1):33–40.
21. ESHRE PGT-M Working Group, et al. *Hum Reprod Open.* 2020;2020(3).
22. Fiorentino F, et al. *Hum Reprod* 2006;21(3):670–684.
23. Fiorentino F, et al. *Mol Hum Reprod.* 2003;9(7):399–410.
24. Moutou C, et al. *Mol Hum Reprod.* 2007;13(2):95–101.
25. Spits C, et al. *Fertil Steril.* 2006;86(2):310–320.
26. Vrettou C, et al. *Hum Mutat.* 2004;23(5):513–521.
27. Dreesen JC, et al. *Mol Hum Reprod.* 2000;6(5):391–6.
28. Fiorentino F, et al. *Eur J Hum Genet.* 2005;13(8):953–958.
29. Handyside AH, et al. *Mol Hum Reprod.* 2004;10(10):767–72.
30. Renwick P, et al. *Reprod Biomed Online.* 2010;20(4):470–6.
31. Rechitsky S, et al. *Fertil Steril.* 2015;103(2):503–12.
32. Spits C, et al. *Hum Mutat.* 2006;27(5):496–503.
33. Handyside AH, et al. *J Med Genet.* 2010;47(10):651–8.
34. Konstantinidis M, et al. *Reprod Biomed Online.* 2015;31(3):394–403.
35. Handyside AH, et al. *Semin Reprod Med.* 2012;30(4):255–8.
36. Natesan SA, et al. *Reprod Biomed Online.* 2014;29(5):600–5.
37. Thornhill AR, et al. *J Assist Reprod Genet.* 2015;32(3):347–56.
38. Navin N, et al. *Genome Med.* 2011;3(5):31.
39. Wang J1, et al. *Cell.* 2012;20;150(2):402–12.
40. Treff NR, et al. *Fertil Steril.* 2013;99(5):1377–1384.
41. Yan L, et al. *Proc Natl Acad Sci U S A.* 2015;112(52):15964–9.
42. Rodrigo L, et al. *Biomed Res Int.* 2014;2014:517125.
43. Rubio C, et al. *Fertil Steril.* 2013;99(5):1400–7.
44. Scott RT Jr, et al. *Fertil Steril.* 2013;100(3):697–703.
45. Yang Z, et al. *Mol Cytogenet.* 2012;5(1):24.
46. Rechitsky S, et al. *Fertil Steril.* 2015;103(2):503–12.
47. Rechitsky S, et al. *Reprod Biomed Online.* 2013;26(5):420–30.
48. Cervero A, et al. Clinical outcome of preimplantation genetic diagnosis for single-gene disorders could be improved with simultaneous comprehensive chromosome screening. Paper presented at the 14th International Conference on PGDIS, Chicago, May 2015.
49. Mukaida T, et al. *Best Pract Res Clin Obstet Gynaecol.* 2012;26(6):789–803.
50. Harper JC, et al. *Hum Reprod.* 2010;25(4):1051–65.
51. ESHRE PGT Consortium Steering Committee, et al. *Hum Reprod Open.* 2020;2020(3).
52. SenGupta SB, et al. *Reprod Biomed Online.* 2016;32(3):263–70.
53. Treff NR, et al. *Reprod.* 2020;160(5):A13–A17.
54. Treff NR, et al. 2020;11(6):648.
55. Karavani E, et al. 2019;179(6):1424–1435.e8.
56. Lázaro-Muñoz G, et al. *Genet Med.* 2021;23(3):432–434.
57. Leaver M, et al. *Hum Reprod Update.* 2020;26(1):16–42.
58. Capalbo A, et al. *Fertil Steril.* 2018;110(5):870–879.
59. Liu WQ, et al. *Ann Med.* 2017;49(4):319–328.

Preimplantation Genetic Testing for Polygenic Disorders

17

Nathan R. Treff, Diego Marin, and Laurent C. A. M. Tellier

INTRODUCTION

Many common human diseases are polygenic in nature, including several among the top ten leading causes of death (1). Recent development of population level genome-wide DNA sequence and medical record repositories have made it possible to develop genomic predictors of polygenic disease (2–10). Typically, DNA sequence data (genotypes) from a set of cases and controls are used as a "training set." The training set often involves a vast number of datapoints (i.e., 500,000 genomes x 800,000 genotypes = 400 trillion datapoints). The analysis of the training set requires the use of machine learning (computational analysis). The computer is supplied an algorithm to produce a DNA sequence-based predictor. The machine learning algorithm may give varying weight to different positions in the genome in order to maximize the predictor's ability to separate cases and controls. The algorithm may also incorporate methods such as LASSO to account for sparsity of data (i.e., 800,000 genotype positions/3 billion human genome positions = 0.03%). The predictor is then applied to a new "test" set of cases and controls in order to determine its validity and establish its performance characteristics (5, 11, 12) (Figure 17.1).

When a DNA sequence-based predictor of polygenic disease is applied to individuals, it is often referred to as polygenic risk scoring (13). Stratification of individuals based on polygenic risk scores can be performed before diseases typically manifest in adulthood (5, 14). Interest in polygenic risk scoring has therefore rapidly grown from a solid foundation in basic research to contemplation of systematic implementation by national healthcare policymakers (15). With the low cost of genome-wide DNA analysis, many countries are considering universal population testing, with the goal of early identification of high-risk individuals leading to overall reduction in healthcare costs and improvement in human health (16). Indeed, early detection and prevention of disease is a concept that has driven national research endeavors including the original human genome project (17).

One example is the application to breast cancer risk assessment. Breast cancer typically manifests in adulthood. Polygenic scoring can identify individuals in the top 3% of genetic risk, well before manifestation of disease (i.e., a lump on mammography) since DNA is available as early as the preimplantation stage of human development (5, 18). Individuals in the top 3% of genetic risk have been shown to

DOI: 10.1201/9781003024941-17

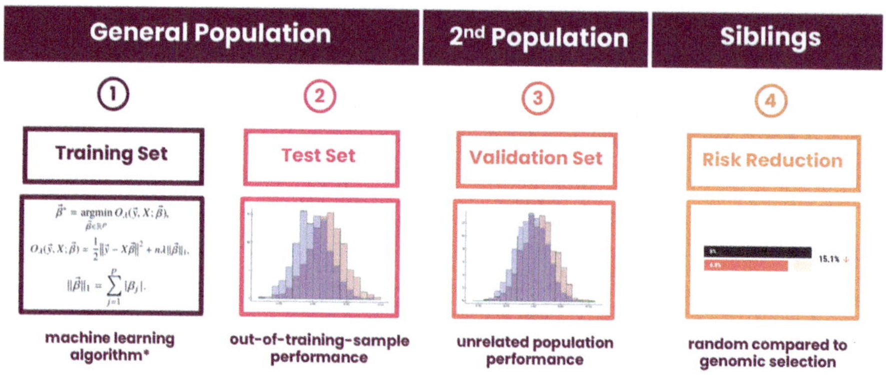

FIGURE 17.1 The process of developing and validating polygenic risk scoring by using machine learning algorithms on a training set of known cases and controls, testing performance on samples left out of the training set, samples from an unrelated cohort, and validation of relative risk reduction on sibling pairs.

have ~5x greater chance of developing breast cancer compared to individuals with average risk (19), as has been validated by multiple research groups (3, 5, 20). Meanwhile, some BRCA variants which today are routinely screened in PGT are associated with a breast cancer prevalence which is ~4x that of the average population prevalence. In other words, persons identified to be at high polygenic risk for breast cancer have risk equivalent to, or sometimes exceeding, persons carrying monogenic mutations which today are routinely tested for in IVF. Both types of high-risk individuals may benefit more than the average person, from earlier and more frequent mammograms. The current strategy involves annual mammograms for the entire population from the age of 40 and over (21–24). Early knowledge of high risk through DNA analysis could allow for better prevention at a reduced cost to society by directing more frequent and early mammograms to high-risk individuals, and fewer, less frequent mammograms to low-risk individuals.

In parallel to advancements made in the ability to accurately predict polygenic disease risk from adult DNA, comprehensive and accurate analysis of DNA from the preimplantation embryo has also been made possible. The amount of DNA obtained from an embryo biopsy is significantly less (~5 cells) than the amount that can be obtained from an adult (i.e., blood or saliva, millions of cells). This limitation has been one of the main focuses of research in the field of preimplantation genetic testing (PGT). Several methods for DNA analysis have been used for clinical PGT since its inception over 30 years ago. These include polymerase chain reaction (PCR), fluorescence *in situ* hybridization (FISH), whole genome amplification (WGA), array comparative genomic hybridization (CGH), quantitative real-time PCR (qPCR), single nucleotide polymorphism (SNP) array, and next-generation sequencing (NGS) (25) (Figure 17.2). Embryo biopsy whole-genome amplification combined with SNP array and parental DNA analyses has demonstrated higher accuracy of embryonic genotyping compared to adult genotyping, despite the limited quantity of DNA (26). Mendelian inheritance rules allow improvement in the interpretation of genotyping data on embryos by considering what is possible given the parental genotypes.

Given the high degree of accuracy in the embryo compared to adult DNA genotyping, and that polygenic risk scoring is based on DNA alone, the utility of polygenic risk scoring in sibling embryo selection and disease reduction (PGT for polygenic disease, PGT-P) can be evaluated by comparing random selection to genetic selection in adult siblings with known disease status. Indeed, evaluation of over 11,000 adult sibling pairs with known disease status demonstrated significant relative risk reduction with genetic selection compared to random selection for several common medical conditions including breast cancer, diabetes, and heart disease (27). For example, choosing one sibling with DNA testing resulted in a 17% reduction in breast cancer disease prevalence, relative to choosing one sibling randomly. This study demonstrated the utility of PGT-P with only two euploid sibling embryos to choose from. As expected, when more than 2 siblings are available for testing, relative risk reduction increases (18).

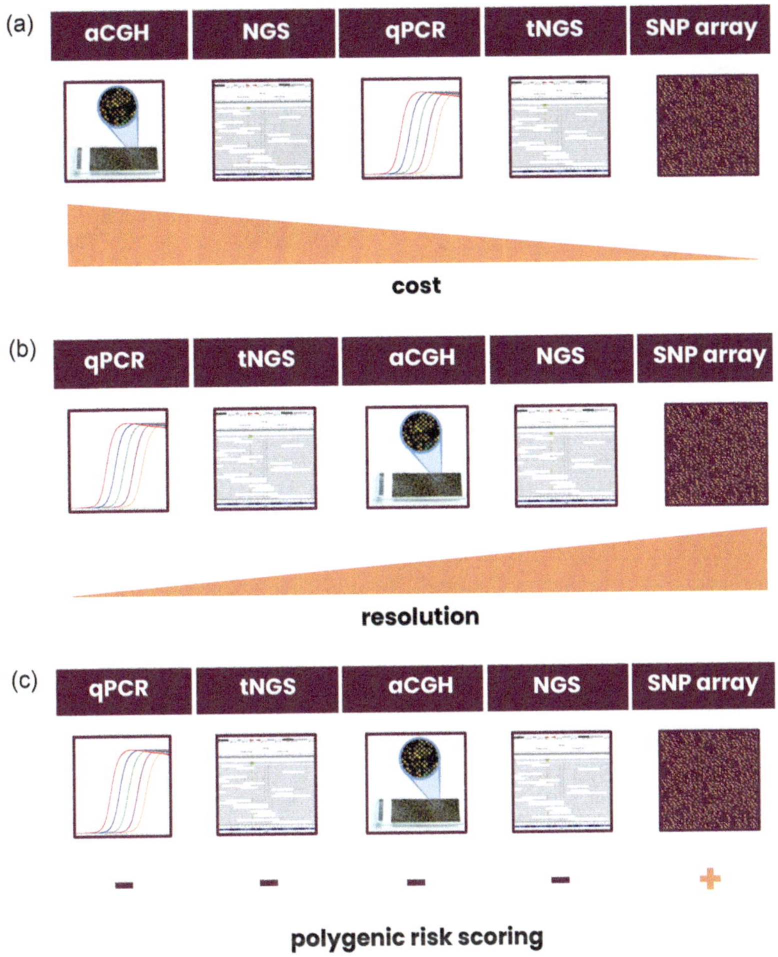

FIGURE 17.2 Comparison of PGT platforms in terms of cost to perform each test (a), the resolution of genetic analysis (b), and the ability to obtain polygenic risk scores (c).

THE METHOD AND IMPLEMENTATION

Family history of a medical condition available for testing is a common indication for PGT-P. In some cases, such as breast cancer, history of a polygenic disease is routinely discussed as part of standard clinical care during preparation for *in vitro* fertilization (IVF). However, family history for other polygenic diseases is not routinely discussed, such that patients may never be made aware of available options to reduce the risk of diseases that run in their families. Education around polygenic risk scoring has lagged behind scientific achievements but is likely to improve as more clinical applications and demonstration of utility continue. Other indications for PGT-P may include the availability of "surplus" embryos. For example, patients using an oocyte donor might expect to have several euploid embryos available for selection. Another example might be patients that produce two or more euploid embryos.

Patients electing to utilize PGT-P undergo genetic counseling prior to their IVF cycle, in order to review the test limitations and capabilities. Patients also provide saliva samples in order to improve the accuracy of genotyping their embryos for polygenic risk scoring. Trophectoderm biopsies are obtained

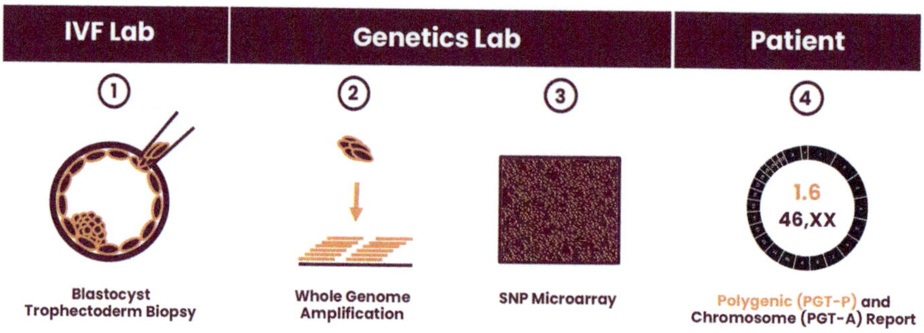

FIGURE 17.3 A schematic of the clinical workflow for evaluating polygenic disease risk in the preimplantation embryo.

and placed into PCR tubes at the embryology laboratory and sent to the genetic testing laboratory. The biopsied embryos are cryopreserved for eventual use following genetic testing.

Trophectoderm biopsies are processed by whole genome amplification and single nucleotide polymorphism (SNP) microarray analysis. Results of analysis of over 800,000 SNPs from each embryo include genotyping and copy number evaluation (26). PGT for aneuploidy (PGT-A) involves machine learning based analysis of both genotypes and copy numbers, resulting in a molecular karyotype with > 99% accuracy. In parallel, genotypes are used to create polygenic risk scores for several conditions. A composite polygenic risk score (a cPRS, known as a genomic index) provides a ranking for each euploid embryo, allowing patients and providers to easily utilize results for embryo selection. Patients have the option of obtaining PGT-A results first and then, after determining how many euploid embryos are available, can add polygenic risk scoring (Figure 17.3).

LIMITATIONS AND BARRIERS TO ACCESS

Disease predictors are highly dependent on availability of population biorepositories with DNA sequence data (genotypes) and electronic medical records (phenotypes). While several large repositories are available today, not all ethnicities are adequately represented (14, 28). This poses an important limitation on current applications, as only patients with certain ancestries can utilize PGT-P. As national policymakers continue to realize the great potential of genetic testing to prevent and better manage disease, further development of world-wide biobanks can be expected.

Infertility is estimated to affect 8–12% of the reproductive aged population (29). In the United States in 2018, 40% of treatment cycles involved PGT (30). The most common application of PGT involves testing for aneuploidy (PGT-A) to reduce the risk of miscarriage and improve the chance of success in the first attempt. In contrast, approximately 1–2% of IVF involves testing embryos for monogenic disorders (PGT-M) (31). Patients at risk of transmitting a monogenic disorder to their children are often not infertile but have the option to utilize IVF and PGT to reduce the risk of disease. Many monogenic disorders involve early-onset and severe disease. However, growing use of testing for adult-onset disease of lesser severity has received ethical review by several medical professional groups including the American Society for Reproductive Medicine (32).

Several studies have demonstrated that current polygenic disease risk scoring, from DNA alone, can identify individuals at risk equivalent to monogenic disorder risk (3, 5, 20). Given the determination that reproductive liberty ethically allows for PGT of adult-onset conditions (32), and the significant relative disease risk reduction achieved with PGT-P (27), patients without infertility may begin to consider utilization of IVF and PGT-P in a manner similar to the application of PGT-M. Evidence of utility in the general population has also been established by evaluating risk reduction from sibling selection within a large and relatively healthy cohort (27). Current application of PGT-P has primarily focused on providing additional embryo selection options for couples already undergoing IVF to treat infertility. Infertility itself represents an indication for PGT-P, given that patients with infertility have a higher risk of cancer, diabetes, and heart disease when compared to the general population (33).

Pleiotropy, or the association of a genetic variant with multiple phenotypes, is an important consideration in PGT, which hitherto has not been addressed in the application of PGT. With the advent of population biobanks, this concern is now being addressed for the first time, by evaluating the impact of selection between siblings upon multiple phenotypes concurrently. It has been demonstrated, in over 11,000 adult sibling pairs with known disease status, that ranking based on a composite polygenic risk score, known as a genomic index, reduces incidence on a panel of diseases concurrently, and in most disease instances reduces disease prevalence better than selection against that same disease individually (27). Continued research can be expected to further elucidate this positive impact of pleiotropy upon the efficacy of PGT for multiple concurrent risks.

FUTURE PERSPECTIVES

IVF with PGT-P has the potential to significantly reduce the prevalence of disease in humans (34). While PGT is already performed in 40% of IVF cycles in the United States, only 1.8 percent of births are from IVF (35), thus limiting the potential for reducing the global burden of disease. As awareness increases for the availability of PGT-P, more patients may incorporate polygenic risk scoring in their IVF treatment plan. In most cases, this may derive from first knowing of the option, but also having knowledge of family history of a polygenic disease. Much like the utility of expanded carrier screening, where patients may often not be aware of a family history but are found to be carriers of a mutation that increases risk, polygenic risk scoring may be useful to many would-be parents. Patients may have an unknown high risk of disease that would be uncovered by DNA analysis ahead of attempting pregnancy. This additional information may also illustrate the potential benefits of PGT-P on an individualized basis.

CONCLUSIONS

Capturing patient perspectives on the use of PGT-P may help improve and expand applications. An ongoing clinical trial, "The Embryo Health Study" (www.clinicaltrials.gov ID NCT04528498), will begin to evaluate the percentage of patients electing to obtain additional information on polygenic risk in embryos during IVF with PGT-A. Data on eligibility, and mode of use, such as whether patients prefer to obtain PGT-P results only after PGT-A has been completed or at the same time as PGT-A is reported, will be obtained. The role of known family history and the degree to which embryos have previously known risk (i.e., affected sibling or parents) will be instrumental in determining how to communicate available options to patients in the future (Figure 17.4, [36–43]). Long term follow-up will also allow for the development of appropriate counseling and understanding of perspectives in families with children born following IVF with PGT-P.

The science of polygenic risk scoring is growing exponentially in parallel with the development of population level DNA sequence repositories. These efforts are instrumental to improving early detection and prevention of disease. Stratification of individuals with high genetic risk through low-cost DNA testing has the potential to improve healthcare and reduce the global burden of disease in humans. This concept also applies to embryo selection to reduce the risk of disease in children born through IVF.

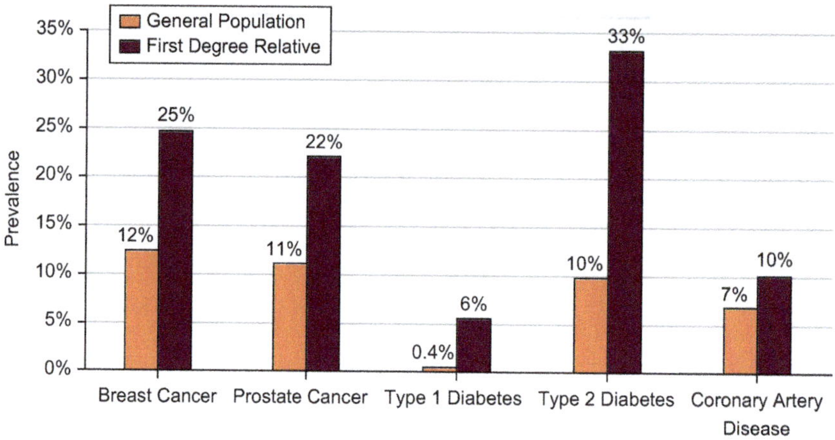

FIGURE 17.4 A comparison of disease risk in the general population and in individuals (embryos) with a first degree affected relative (sibling or parent).

Risk reduction by PGT-P has demonstrated utility not only in situations where there is a known family history of disease, but also in an otherwise healthy population. Infertility itself is a risk factor for polygenic diseases such as cancer, diabetes, and heart disease, making the IVF population ideal for PGT-P. Applications will expand as new DNA repositories are made available and more diseases and more diverse populations are represented.

REFERENCES

1. World Health Organization. *Global Status Report on Noncommunicable Diseases 2014*. Geneva: World Health Organization; 2014.
2. Torkamani A, et al. *Nat Rev Genet*. 2018;19(9):581–90.
3. Khera AV, et al. *Nature Genet*. 2018;50(9):1219–24.
4. Bycroft C, et al. *Nature*. 2018;562(7726):203–9.
5. Lello L, et al. *Sci Rep*. 2019;9(1):15286.
6. Inouye M, et al. *J Am Coll Cardiol*. 2018;72(16):1883–93.
7. Abraham G, et al. *Nat Commun*. 2019;10(1):5819.
8. Kachuri L, et al. *Nat Commun*. 2020;11(1):6084.
9. Mars N, et al. *Nat Med*. 2020;26(4):549–57.
10. Mavaddat N, et al. *Am J Hum Genet*. 2019;104(1):21–34.
11. de Los Campos G, et al. *Trends Genet*. 2018;34(10):746–54.
12. Lello L, et al. *Genetics*. 2018;210(2):477–97.
13. Chatterjee N, et al. *Nat Rev Genet*. 2016;17(7):392–406.
14. Duncan L, et al. *Nat Commun*. 2019;10(1):3328.
15. Commission E. *JRC F7 - Knowledge Health and Consumer Safety*, 2019;EUR 29815 EN.
16. Barwell J, et al. *Clin Med (Lond)*. 2019;19(4):273–7.
17. Collins F, et al. *Science*. 1993;262(5130):43–6.
18. Treff NR, et al. *Front Endocrinol (Lausanne)*. 2019;10:845.
19. Bhattacharya A. Gee whizz. *The Economist*. November 9, 2019.
20. Lello L, et al. *Scientific Reports*. 2020;10(1):13190.
21. Tung NM, et al. *J Clin Oncol*. 2020;38(18):2080–106.
22. Armstrong N, et al. *Clin Epidemiol*. 2019;11:543–61.
23. Maxwell KN, et al. *J Clin Oncol*. 2016;34(34):4183–5.
24. Bray F, et al. *CA Cancer J Clin*. 2018;68(6):394–424.
25. Treff NR, et al. *Annu Rev Genomics Hum Genet*. 2017;18:189–200.

26. Treff NR, et al. *Eur J Med Genet.* 2019;62(8):103647.
27. Treff NR, et al. *Genes.* 2020;11(6):648.
28. Martin AR, et al. *Nat Genet.* 2019;51(4):584–91.
29. Martinez GM, et al. *Natl Health Stat Report.* 2018(113):1–17.
30. SART. *Preliminary National Summary Report for 2018*, 2018. Available from: https://www.sartcorsonline.com/rptCSR_PublicMultYear.aspx?reportingYear=2018.
31. Theobald R, et al. *Hum Reprod.* 2020;35(4):986–98.
32. ASRM. *Fertil Steril.* 2018;109(6):989–92.
33. Cedars MI, et al. *Hum Reprod Open.* 2017;2017(2):hox008.
34. Treff NR, et al. *Reproduction.* 2020;160(5):A13–a7.
35. Sunderam S, et al. *MMWR Surveill Summ.* 2019;68(4):1–23.
36. Rojas K, et al. *Clin Obstet Gynecol.* 2016;59(4):651–72.
37. Shiyanbola OO, et al. *Cancer Epidemiol Biomarkers Prev.* 2017;26(12):1753–60.
38. Lloyd T, et al. *BMC Med.* 2015;13:171.
39. Steinberg GD, et al. *Prostate.* 1990;17(4):337–47.
40. Redondo MJ, et al. *Pediatr Diabetes.* 2018;19(3):346–53.
41. Hariri S, et al. *Genet Med.* 2006;8(2):102–8.
42. Bachmann JM, et al. *Circulation.* 2012;125(25):3092–8.
43. Benjamin EJ, et al. *Circulation.* 2019;139(10):e56–e528.

Should Genome Editing Replace Embryo Selection Following PGT?

18

Nada Kubikova and Dagan Wells

CURRENT APPROACHES TOWARDS EMBRYO SELECTION

Assisted reproductive treatment coupled with embryo selection using preimplantation genetic testing for monogenic disease (PGT-M) is considered a valuable reproductive option for couples who are at high risk of transmitting heritable genetic disorders to their children. The demand for PGT-M has been steadily increasing over the last three decades as patient awareness of the strategy has grown and as technical advances have improved accuracy and reduced the time required to develop tests for individual families. Traditional PGT-M methods typically involve creation of individualized protocols, each customized for the amplification and detection of the disease-causing mutation(s) and informative linked polymorphisms relevant to a particular family. In many cases, combinations of mutations and polymorphisms are unique to an individual family, meaning that the PGT-M protocol developed, following extensive work in the laboratory, may only be of use for a single couple.

For the first 15 years of PGT-M, amplification of diagnostically relevant DNA fragments was accomplished using the polymerase chain reaction (PCR). However, more recently, whole genome amplification (WGA) methods have enabled reliable amplification of the majority of the embryonic genome from small numbers of cells biopsied from preimplantation embryos (1, 2). This innovation has permitted a move away from patient-specific tests, providing more generic PGT-M protocols (3, 4). For example, a method known as karyomapping permits the simultaneous detection of embryos affected by single gene defects and cytogenetic abnormalities of meiotic origin, accomplished using the principle of linkage analysis (3). Parental and embryonic DNA, the latter having undergone WGA, is genotyped for ~300,000 single nucleotide polymorphisms (SNPs) scattered throughout the genome using a microarray. An additional member of the family, such as an affected child of the couple undergoing PGT-M, is also required to be tested and allows setting of phase—i.e., it permits determination of the unique combination of parental SNP alleles (haplotype) associated with each chromosome copy. Particular attention is paid to the alleles in the vicinity of the gene responsible for the familial disorder, and the inheritance of haplotypes associated

DOI: 10.1201/9781003024941-18

with mutant and normal copies of the gene can be examined in the embryos (4). Embryos found to carry high-risk haplotypes (inherited along with mutant genes) are excluded from transfer. In terms of the cytogenetic status of embryo samples, the haplotype information from individual chromosomes reveals which parental chromosomes have been inherited (and how many), while quantitative data concerning the amount of DNA hybridized to the SNP probes on a given chromosome provides a complementary indication of its copy number.

Genetic methodologies are evolving rapidly and offer the possibility of delivering PGT with even higher accuracy and lower cost than was possible using earlier techniques. Chief among the new wave of methods has been next-generation sequencing (NGS), which has been widely applied in the context of preimplantation genetic testing for aneuploidy (PGT-A), helping to distinguish aneuploid embryos from those with a normal chromosome copy number (5–8). Most NGS methods for aneuploidy detection have focused on a "shotgun" strategy, randomly sequencing a small fraction of the genome (typically < 0.1%) and then evaluating the proportion of reads mapping to each chromosome, providing a measure of relative chromosome copy number. Since so little of the genome is sequenced, the probability of obtaining data concerning the status of a specific gene is negligible. However, more targeted analyses, enriching or amplifying DNA sequences from specific genes prior to NGS can be employed, allowing clinical diagnosis of gene mutations in preimplantation embryos (9–13). As costs continue to fall, PGT strategies using NGS are becoming increasingly popular and are slowly replacing earlier techniques. This is particularly the case for diseases displaying a high degree of mutation diversity because NGS provides a straightforward means of sequencing the entirety of the investigated gene (or even multiple genes), providing a universal solution applicable to any couple undergoing PGT-M for the associated condition(s).

MAIN LIMITATIONS OF EMBRYO SELECTION

Despite its success and wide implementation, PGT-M has a number of limitations. Firstly, the probability of finding an unaffected embryo and the likelihood of achieving a viable pregnancy are strongly influenced by the number of embryos available for testing and the number of embryos generated in an IVF cycle is limited. Many of the embryos are discarded due to either inadequate development or because they are found to be affected by the inherited mutation (14). Inevitably, there are occasions when the only viable embryo(s) produced in an IVF cycle is(/are) found to be affected by the familial disorder. When both parents are carriers of a recessive mutation, the proportion of embryos suitable for transfer to the uterus is expected to be reduced by 25% with respect to a routine IVF cycle. In the case of a dominant disease, where inheritance of a single copy of a mutant gene generates a disease phenotype, only 50% of the embryos are expected to be free of the disease (15). Published data indicates that 16–20% of PGT-M cycles do not reach the point of uterine transfer and that fewer than 25% of cycles result in a healthy ongoing pregnancy (16, 17).

The relative infrequency of embryos that are both developmentally competent and unaffected by the familial condition is the principal reason why the majority of PGT M cycles fail to produce a pregnancy after what is often a physically, psychologically, and financially demanding treatment. Thus, the development of novel strategies that succeed in preventing germline transmission of mutations, while avoiding the discard of affected but otherwise potentially viable embryos, is desirable. A technology that permitted the "rescue" of affected embryos, for example by removing/correcting a mutation, would lead to a greater number of embryos considered suitable for transfer, improving the chances of a pregnancy and reducing the risk of inherited disease. Such a technology would be advantageous for patients having PGT-M and would be particularly important for couples whose opposition to pregnancy termination includes an unwillingness to discard affected preimplantation embryos. Without a means of repairing defects in affected embryos, the reproductive options for patients opposed to discarding embryos are extremely limited and currently include the use of donor oocytes, sperm or embryos, and adoption, all of which sacrifice the possibility of having a genetically related child (18).

NEW PROSPECTS FOR ELIMINATION OF INHERITED DISEASES BY GERMLINE GENOME EDITING

For the last 30 years, carriers of disease-causing mutations wishing to increase their chances of having unaffected, genetically related children, have needed to seek PGT or prenatal diagnosis. More recently, advances in genome engineering technologies have begun to suggest that other approaches, involving germline genome editing (GE), might represent feasible alternatives. While heritable genome editing does not currently exist as a widely available, validated reproductive strategy, it could do so in the future. In theory, the application of GE for the avoidance of inherited disease could allow the modification of a mutant gene in order to restore a wild-type DNA sequence. Such an approach would shift the paradigm of current reproductive strategies away from the diagnosis and exclusion and towards "cure" (18).

From a technical perspective, conducting genome editing interventions at the time of fertilization or during early preimplantation embryonic stages can be viewed as desirable, since it is easier to ensure that the active components used for GE will be delivered to all cells. This is important because many inherited conditions affect multiple tissues or organs and a return to normal function, eliminating the disease phenotype, may require the mutant gene to be corrected in a high proportion of cells. It can be extremely difficult to access the genomes of cells later in life, when the cells are potentially numbered in millions and populate internal organs that cannot easily be accessed. A fertilized oocyte or early embryo that underwent GE could ultimately give rise to an individual carrying the edited gene in all cells of their body. However, this would not only include all of the somatic cells, but the germ cells also, meaning that the edited gene could be transmitted to future generations, potentially becoming a permanent feature of the human genome. This fact has caused some to raise ethical concerns about the application of GE in instances where the germline may be altered. Genome editing interventions, performed on human embryos, are without a doubt ethically controversial; however, some might argue that the current practice of discarding affected but otherwise viable embryos is wasteful and perhaps no better from ethical, moral, as well as certain religious perspectives (14). With GE, embryos previously diagnosed as affected could become eligible for transfer, thus rescuing them from being destroyed and increasing the number of IVF embryos available to the couple, which would likely lead to improved chances of a pregnancy during that treatment cycle. In an ideal scenario where GE has been shown to be safe, accurate, and efficient, it could be combined with natural cycle IVF, in which only a single oocyte is collected and fertilized with the partner's sperm, with the result that affected pregnancies are entirely avoided without discarding a single embryo (18).

CRISPR-CAS9 AS THE LEADING APPROACH TOWARDS GENOME EDITING OF PREIMPLANTATION EMBRYOS

Genome editing technologies utilize programmable nucleases in order to cut the DNA in a highly specific manner. In most cases, endogenous cellular repair pathways are then co-opted in order to introduce the desired alteration to the DNA sequence in the vicinity of the cut site (19). The ability to genetically modify living cells is fundamental to the idea of using GE to eliminate genetic disease. However, achieving specific and targeted changes has been, and still remains, a considerable challenge. The first historical attempts to introduce specific genetic modifications into eukaryotic cells relied on providing a DNA template with the aim of altering a target site in the genome through homologous recombination (HR) (20). In all eukaryotic organisms, HR is one of several cellular processes involved in the repair of double-stranded DNA breaks (DSBs). Homologous recombination resolves DSBs by using the undamaged (homologous) copy of the gene as a template for repair. This mechanism is exploited by GE, which introduces an artificial DNA molecule of homologous (i.e., very similar) sequence to the target, but containing the desired alteration. The intention is for cells to utilize the artificial DNA molecule as a template for HR-mediated repair of the site where a DSB has been occurred, thereby incorporating any sequence alterations that it contains (19). Unfortunately, the chances of a DSB occurring spontaneously at the targeted site are low. Furthermore, HR is not the preferred method of DSB repair in the majority of cells, only occurring relatively rarely. Thus, incorporation of the artificial (donor) DNA template only occurs at very low efficiency, making this strategy unsuitable for instances when the number of cells being treated is limited. (20)

An improvement came in the early 1990s with the introduction of zinc finger nucleases (ZFNs) and transcription-activator-like effector nucleases (TALENs), which could be engineered in order to induce DSBs at specified sites. ZFNs utilize zinc finger protein domains that bind to a 3-bp motif in a modular manner, making them ideal as building units for the creation of sequence-specific DNA binding nucleases (19). TALENs, on the other hand, recognize a single base in each repeat domain, allowing up to four different domains to be mixed and matched to generate a novel DNA binding protein. However, both of these programmable nucleases are associated with an appreciable incidence of "off-target" effects, defined as non-specific cleavage of the DNA at locations other than the intended target site, potentially resulting in cytotoxicity (19, 21). Furthermore, since the target specificity is determined by modification of the DNA binding domain, the application of these nucleases is limited to cases where successful engineering of binding domains is possible, at significant cost of time and resources.

The field of GE expanded dramatically in 2012 with the development of CRISPR-Cas9 (clustered regularly interspaced short palindromic repeat-Cas9) technology that utilizes RNA-guided programmable endonuclease (RGEN) (22, 23). Here, the target specificity is determined by a single stranded guide RNA (sgRNA) that forms a ribonucleoprotein (RNP) complex with the bacterial endonuclease Cas9. In bacteria, the CRISPR-Cas9 system has evolved to facilitate adaptive immune responses to protect the cells from bacteriophage infection and horizontally acquired gene elements (24). The majority of efforts to edit DNA using this system both *in vitro* and *in vivo* utilized a type II Cas9 derived from *Streptococcus pyogenes* (Spy) (24). In this system, the guide RNA contains a 23bp sequence complimentary to the target DNA sequence that is located adjacent to a protospacer motive known as PAM (defined as any nucleotide followed by GG—e.g., NGG) (Figure 18.1b). The RNP can be readily modified by customizing the guide RNA, a process analogous to PCR primer design, allowing targeting of most sites in the genome quickly and cheaply, and eliminating the need for elaborate enzyme engineering and assembly. Upon successful recognition of target sequence and PAM, Cas9 flanks the targeted region and cleaves the DNA strand, generating a DSB. In the context of removing an unwanted mutation at a preimplantation stage, the most straightforward approach is to microinject the RNP components into the MII oocyte at the same time as intracytoplasmic sperm injection (ICSI) (Figure 18.1a). Alternatively, microinjection at the zygote stage is also relatively simple. It is intended that the desired editing occurs prior to the first cell division, thus maximizing the likelihood of delivering the modification to all cells of the future individual.

Following the induction of a DSB, most cells attempt to resolve the damage predominantly by two conserved mechanisms, non-homologous end joining (NHEJ) and, less frequently, homology directed repair (HDR) (Figure 18.1b). During the process of reconnecting the two ends of the cleaved DNA strand, NHEJ frequently introduces insertions or deletions (indels), which typically result in disruption of the targeted gene and a consequent loss of function (Figure 18.1a). This type of repair mechanism is not ideal for use in genome editing, unless the intention is to disrupt the gene function and not the correction of the mutant copy of the gene. For correction of a mutation, HDR is a preferred mechanism since it rebuilds the site of breakage using a homologous DNA molecule as a template, potentially restoring the wildtype sequence. In cases where an embryo is heterozygous for a mutation, the second copy of the gene would remain uncut (as it does not contain the targeted mutation) and could serve as a template for repair. This is

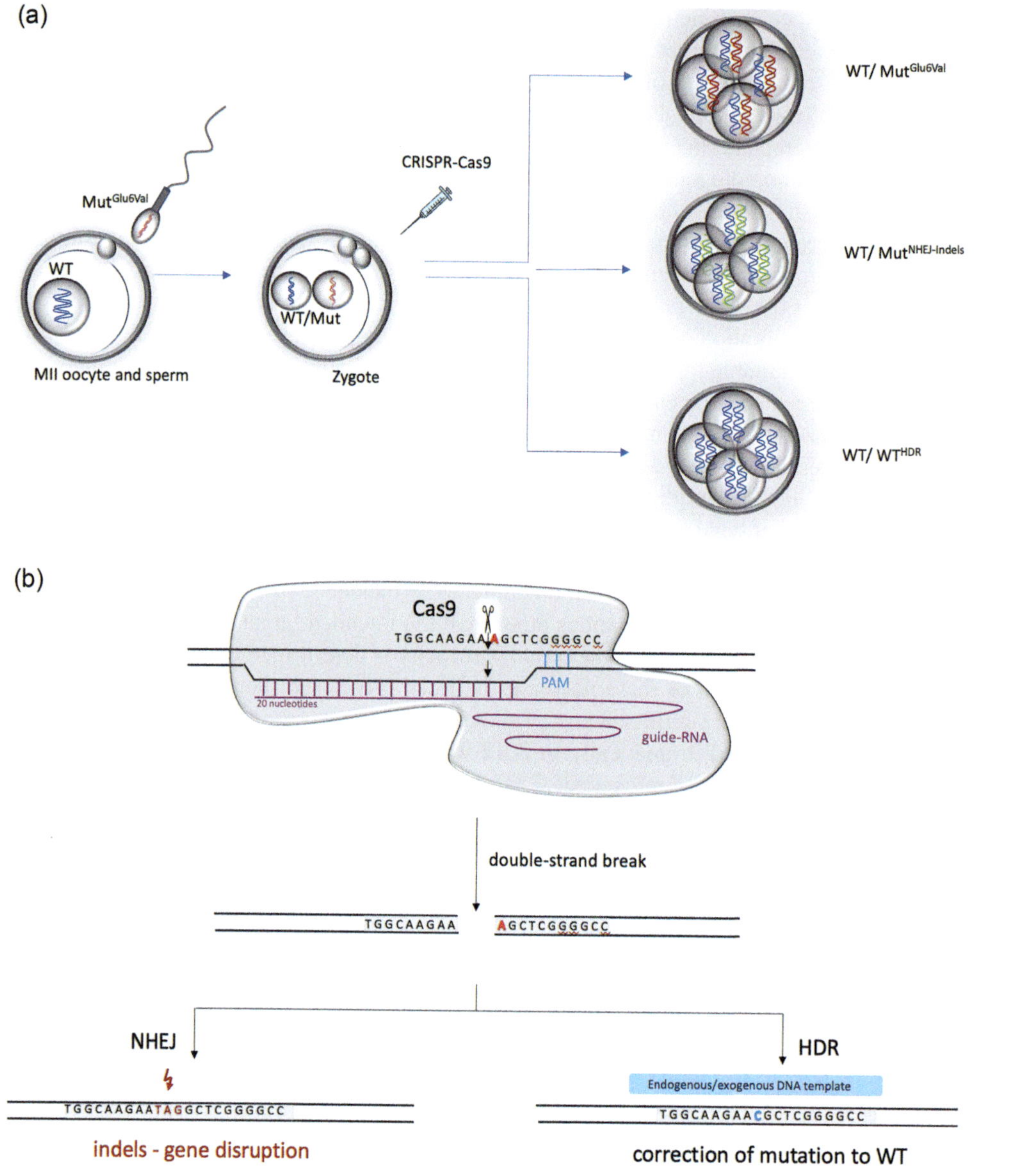

FIGURE 18.1 Schematic representation of CRISPR-Cas9 germline genome editing. (a) Microinjection of CRISPR-Cas9 components after fertilization at zygote stage using an example of a male carrier of Glu6Val mutation found in the *HBB* gene (the sickle cell anemia mutation). Targeting of only one of the parental alleles (the paternal Glu6Val mutation) can result in three possible outcomes observed during the cleavage stage: 1) The editing is unsuccessful and does not alter the genotypes of the blastomeres that continue to carry one maternal wild-type copy of the *HBB* gene and one paternal copy containing the Glu6Val mutation (heterozygous recessive). 2) Induced DSB is repaired via NHEJ, resulting in cleavage-stage blastomeres that carry a copy of maternal wild-type allele and the second paternal allele containing insertions and deletions (indels) induced by NHEJ. 3) Induced DSB is repaired via HDR where maternal copy of the gene (wild-type) serves as a template for strand replacement and correction. The homozygous wild-type genotype is restored and the disease allele eliminated. (b) Detailed representation of the RNP complex binding to target DNA upon the recognition of the complimentary DNA sequence and PAM motif and the resulting genotypes upon DNA cleavage and repair via NHEJ and HDR. NHEJ: non-homologous end joining, WT: wild type, HDR: homology directed repair.

essentially how HDR works when a DSB occurs naturally—the undamaged copy of the gene being used for the restoration of the damaged homologous copy—but it can also occur in the experimental context (i.e., *in vitro*) (Figure 18.1a). However, it is also possible, and potentially more efficient, to supply an exogenous (synthetic) homologous DNA fragment for the cell to use as a template. In cases where the intention is to correct a mutation, the exogenous DNA would be similar or identical to the wild-type sequence. Needless to say, the addition of a synthetic DNA template is indispensable in cases where neither cellular copy of the gene has a wild-type sequence (e.g., homozygous recessive mutation).

WHAT HAS BEEN DONE IN RESEARCH AND CLINICALLY SO FAR?

In 2017, a proof of concept study demonstrated the possibility of targeted correction of germline mutations in human preimplantation embryos for the first time, using a CRISPR-Cas9-based technology to remove a dominant heterozygous *MYBPC3* mutation, responsible for an inherited form of cardiomyopathy (25). This was a research study and the embryos produced were not transferred to the uterus.

In the same year, a group of UK-based scientists used a similar CRISPR-Cas9-based approach to edit a group of human zygotes in order to investigate the regulation of early human development. This study was carried out on embryos donated for research purposes and was conducted under license from the UK regulator (HFEA). In this case, the study aimed to knockout gene function rather than correct a mutation (26). The authors selected the *POU5F1* gene, encoding the OCT4 protein, an interesting target given its essential role in the maintenance of pluripotency within the inner cell mass. The CRISPR-Cas9 system was employed to introduce DSBs within *POU5F1* and the researchers relied upon error-prone NHEJ to create indels during the repair process, disrupting the coding region of the gene and producing premature stop codons. Using this strategy, it was shown that OCT4 is essential for the formation and maintenance of human blastocysts (26).

In 2019, only seven years after CRISPR-Cas9-based genome editing was first described as a molecular biology tool (27, 28), the birth of the first "CRISPR babies" Lulu and Nana was announced at a summit in Hong Kong. Gene editing had been applied at an embryonic stage in order to disrupt normal copies of the *CCR5* gene, with the intention of conferring resistance to the HIV virus (the father of the children was HIV-positive). The announcement was met with condemnation from many quarters, not least from the scientific community. It was considered that the safety of the method had not been adequately demonstrated. In particular, there were concerns over the potential for inadvertent editing of unintended areas of the genome (i.e., off-target effects) and that there could be other unforeseen consequences of using CRISPR-Cas9 during early embryonic development. Additionally, many considered GE unnecessary, since there are well-validated "sperm washing" procedures, which avoid HIV transmission when using IVF. The two girls born as a result of the procedure are understood to have non-functional copies of *CCR5*. While this may provide the intended resistance to HIV infection, some evidence suggests there may also be negative health consequences of loss of *CCR5* function, a problem that will be passed on to subsequent generations.

This case stimulated wide scientific, societal, and ethical discussions, as well as triggering calls for a moratorium on human germline genome editing across the world. It has also led to initiation of efforts to introduce a new regulatory framework, since rapid technological advancements had outpaced the ability to legislate and modify existing controls. Considering how little is understood about the underlying biology of early human development and the impact of introducing DNA damage in the form of DSBs on the developmental capacity of human embryos, there is a broad consensus that it is premature to consider techniques such as CRISPR-Cas9 for clinical application to human embryos. There are many technical questions that remain to be answered with respect to the safety of CRISPR-Cas9, as well as ethical discussions and public consultations that need to take place, prior to any further clinical application of GE at the preimplantation stage. Additionally, biological uncertainties, especially concerning the functionality of key pathways involved in the resolution of DNA damage at early embryonic stages, also need to be resolved.

CURRENT BARRIERS TO GENOME EDITING

Relative Infrequency of HDR Following Genome Editing

CRISPR-Cas9 based genome editing is an indispensable molecular biology tool with the potential to reduce the burden of genetic disease, although its clinical application remains a subject of considerable international debate, centered around efficacy, safety, and complicated ethics, especially when applied at early developmental stages (29, 30). Several groups have conducted initial studies aiming to assess the feasibility of this approach in human preimplantation embryos (31–33). Although it seems possible to introduce a DSB at the targeted site with relatively high efficiency, it appears that only a minority of embryos resolve the DNA damage by HDR and therefore succeed in correcting the mutant gene. A significant proportion of the edited embryos, likely the majority, harbor additional indels. Far from correcting a defective gene, this results in further disruption to the DNA sequence. One study used CRISPR-based editing to cleave the *HBB* gene in human tripronuclear zygotes (3PN) and found that although the achieved editing efficiency was high (52%), the majority of the edited zygotes contained indels and only 14% of the cells utilized the exogenous HDR repair template supplied by the researchers (31). Similar low frequencies of HDR were observed in another study that used the same system to introduce the naturally occurring CCR5Δ32 allele of the *CCR5* gene into 3PN zygotes by providing an exogenous DNA template to facilitate HDR (32).

It has become evident that the CRISPR-Cas9 technique faces significant challenges in terms of its technical feasibility and clinical applicability at the preimplantation stage. As well as the problems of the persistent low efficiency of HDR, the possibility of additional on-target indel mutations produced by NHEJ, and continuing concern over the potential for off-target editing of homologous and paralogous sequences, there is also the potential complication of mosaicism when GE is applied at early embryonic stages (i.e., where some cells are successfully edited and others are not). Such mosaicism can occur even when editing is attempted in zygotes, as there can be a delay before editing occurs, during which the first mitotic division may occur. It would be highly advisable to carry out PGT on any embryo that had undergone GE, in order to confirm that editing had been successful. However, PGT results might be more difficult to interpret, and potentially less accurate, in the presence of mosaicism. It remains to be determined whether mosaicism in gene editing can be reduced by modulating the cell cycle stage and/or embryonic stage at which DSBs are induced and whether this approach can yield 100% uniformity in the desired genotype. An additional challenge that will need to be overcome lies in the lack of standardized tools and methodology to assess these effects comprehensively.

Does CRISPR-Cas9 Editing Induce Genomic Instability in Preimplantation Embryos?

Another major challenge of genome editing applied at early embryonic stages is the possibility of induction of large deletions and/or complex chromosomal rearrangements, extending over many kilobases of DNA from the cut site. There are growing concerns that CRISPR-Cas9 applied to embryonic cells could induce a state of genomic instability and mitotic arrest, both of which would be deleterious to the developing embryo. In the study of Ma et al. (2017), wild-type oocytes were fertilized with sperm derived from a donor heterozygous for a dominant mutation in the *MYBPC3* gene. Fertilization was achieved using ICSI, and a mixture of Cas9 protein, sgRNA specifically targeting the male mutant locus, and an exogenous oligonucleotide HDR template for repair, were injected along with the sperm (25). The HDR oligo harbored two SNP variants not present in the WT or mutant sequence, in order to allow confirmation of when (if) HDR took place. Since half of the sperm used would be assumed to carry the mutant allele, a 50:50 ratio of normal to affected embryos would be expected in the absence of editing. Rather unexpectedly, the group observed a homozygous WT phenotype in 66.7% of the zygotes injected at S-phase of the cell cycle. Furthermore, 72.4% of M-phase injected oocytes (n = 58) also exhibited this pattern. Only a minority of the embryos showed signs of NHEJ in the paternal allele (as indicated by the detection of characteristic indels) or HDR repair using the provided exogenous template. The authors concluded that the excess of the uniformly homozygous embryos was the result of HDR using the maternal WT allele as a template to resolve the lesion caused by the Cas9 enzyme in the paternal strand. This conclusion was received with skepticism by several other groups, as alternative phenomena could explain the observed result: 1) Failure to amplify the paternal allele (i.e., allele drop-out) or extreme preferential amplification of the maternal strand. This could potentially be caused by a polymorphism affecting one of the PCR primer annealing sites in the paternal allele. Another cause of ADO could be successful induction of a DSB using CRISPR-Cas9 followed by a failure to repair the strands; 2) CRISPR-Cas9 associated GE resulting in a large deletion at the site of editing, leading to a loss of one or both primer annealing sites or other PCR-refractory changes (e.g., a large insertion, inversion or translocation), ultimately resulting in a failure to amplify the edited allele (34–38); and 3) Whole chromosome loss.

The fact that several groups have observed formation of large deletions and/or complex structural rearrangements at sites of editing could be of concern for several reasons. First, the lack of available methodology to accurately characterize the complexity of the on-target mutations, and second, the appreciable frequency with which these potentially detrimental events have been reported. Adikusuma et al. (2018) detected deletions ranging in size between several hundred bases up to 2.3 kilobases in approximately half of the mouse preimplantation embryos subjected to CRISPR-based germline editing when using the long-range PCR approach (34). Similarly, Owens et al. (2019) reported a two kilobase deletion after targeting the Runx1 locus in mouse embryonic stem cells in 23% of the clones. These deletions escaped detection by short fragment PCR, of the type used for characterization of edited sites in many other studies, and were only detected after a long-range PCR had been carried out to amplify the target (38). Kosicki and colleagues recently reported up to six kilobases deletions and complex rearrangements in 5–20% of the mouse embryonic stem cell clones after targeting the line with CRISPR-Cas9 (37). Furthermore, the study conducted by Cullot et al. (2019) who used the CRISPR-targeted system to target the *UROS* locus in HEK293T and K562 cells suggested that the editing introduced a megabase scale chromosomal truncation in approximately 10% more cells compared to the untreated controls (35). Of most relevance to the application of GE to human embryos, the recent study of Alanis-Lobato et al. (2020) evaluated the on-target mutations that would be missed by conventional genotyping techniques in *POU5F1*-targeted human preimplantation embryos. They observed loss-of-heterozygosity in edited cells that spanned regions beyond the *POU5F1* on-target locus, as well as segmental gains and losses of chromosome 6 (the site of the *POU5F1* gene) present in approximately 22% of the samples (39). This body of work highlights the importance of continuous basic research to evaluate the safety of genome editing techniques in human embryos, and will hopefully inform debates about the potential clinical use of this technology.

Off-Target Consequences of Genome Editing

If genome editing is ever to replace embryo selection in the future, it is essential to examine the potential of GE technologies to induce mutations in unintended regions of the genome. Such off-target edits are anticipated at DNA sequences that share close homology with the targeted locus. The approaches that investigate the off-target effects of GE have thus far relied on the use of *in silico* modeling as they are rapid, easy to use, and inexpensive. However, they have only a predictive value and their estimates are far from definitive. With the prospect of clinical implementation, one might foresee that whole genome sequencing of the edited embryo might become an inevitable and essential accessory to exclude the possibility of any off-target mutagenesis and to ensure the delivery of safe and efficacious treatment (18). Needless to say, a requirement for whole genome sequencing would entail further technical challenges and substantially increased costs.

ETHICS

An exploration of the ethical debate around the application of GE to human embryos is beyond the scope of the current chapter. Indeed, an adequate discussion of the ethics of germline GE would require an entire chapter dedicated to that subject alone. However, it is worth noting that there are multiple issues related to personal freedom, as well as religious, moral, and societal considerations of great relevance to the use of GE in instances where modification of the germline is involved. Leaving technical issues and safety to one side, there are questions of what should be considered a "disorder" and where the line is drawn between a cure undertaken for legitimate medical reasons and an enhancement that could be considered eugenic. In most jurisdictions, it is yet to be determined who would be responsible for making and enforcing such decisions. Concerns have also been raised over informed consent. Clearly, it is impossible to obtain such consent from embryos and their descendants who would be affected by germline GE. There are also questions of justice and equity, especially the concern that GE might only be accessible to the wealthy, representing a disparity in access to healthcare and potentially increasing other aspects of social inequality. Some have also worried that the use of GE could lead individuals affected by genetic disorders to be viewed differently, potentially increasing discrimination in the workplace and medical/life insurers, as well as within the wider society.

FUTURE PERSPECTIVES IN THE AREA OF GERMLINE GENOME EDITING

Thus far, the use of CRISPR-Cas9-based editing in the human germline has had a pre-clinical focus, aimed at establishing whether disease-causing mutations could be disrupted/repaired. The studies provide a proof-of-principle as well as methodological framework for the technique's utility in preimplantation embryos. Collectively, the body of published work suggests that it is possible to successfully modify the genome of early human embryos with appreciable efficiency. However, diverse mutational profiles occur as a result of predominant NHEJ. This likely precludes the use of the technique in its current format for correction of the vast majority of mutations associated with inherited diseases in human embryos. In the future, novel GE methods might permit the genome to be modified with higher precision, bypassing the need to introduce a DSB and thereby avoiding the creation of further mutations via error-prone NHEJ. A new generation of GE tools using catalytically-impaired CRISPR-Cas9, known as "base editors," have recently been developed and offer hope in this regard (40–43). Base editors have potential to convert one base into another without inducing a DSB, an important advance considering how little is understood about the capacity of preimplantation embryos to carry out DNA repair, particularly prior to activation of the embryonic genome. If base editors prove to be reliable, they could provide a powerful addition to the growing number of genome engineering and informatics tools, possibly bringing us one step closer to clinical application in the context of the human germline.

CONCLUSIONS

There has been much excitement over the development of new genome editing methods. Such techniques offer the hope that mutations causing serious inherited conditions could one day be corrected during the preimplantation stage, although considerable public consultation, thorough ethical debate, advances in regulation and oversight, as well as assessment of safety, still need to take place before such methods enter clinical practice. The main safety concerns relate to the possibility of off-target editing and uncertainty over whether GE could induce a state of genomic instability in embryonic cells. There have been indications that the cells of human preimplantation embryos are less able to resolve DSBs than other cell types and are consequently prone to developing segmental aneuploidy and other gross chromosomal abnormalities (44). It is clear that there are limitations in our current understanding of the basic biology of human development, as well as uncertainty about how DSBs induced by CRISPR-Cas9 are processed within embryonic cells and their ultimate impact on the genome and embryo viability and health.

Clinical utilization of genome editing should not be considered until more research has been undertaken to solidify knowledge in areas where important deficiencies currently exist. It is imperative that the eventual application of genome editing succeeds in conferring a significant reduction in the risk of inherited disease for the children born following the procedure. While no medical procedure is entirely risk-free, the likelihood of complications resulting from the editing procedure should be low, particularly considering that the impact may extend to future generations and therefore affect multiple individuals. For use in human reproduction, safety, efficacy, and ethical application are paramount. Applying similar scientific rigor and public engagement as was done in the United Kingdom when considering the clinical introduction of mitochondrial replacement therapy (a process that in some respects can also be considered to introduce a heritable alteration to embryos) will be crucial for determining therapeutic potential and introducing the methodology safely. Nonetheless, even in its current form GE represents a powerful research tool, which can greatly enhance the study of human development, expanding our knowledge in this key area.

REFERENCES

1. Wells D, et al. *Nucleic Acids Res.* 1999;27(4):1214–8.
2. Zheng Y, et al. *J Zhejiang Univ Sci B.* 2011;12(1):1–11.
3. Handyside AH, et al. *J Med Genet.* 2010;47(10):651–8.
4. Natesan SA, et al. *Genet Med.* 2014;16(1):1–8.
5. Fiorentino F, et al. *Hum Reprod.* 2014;29(12):2802–13.
6. Fiorentino F, et al. *Fertil Steril.* 2014;101(5):1375–1382.e2.
7. Wells D, et al. *J Med Genet.* 2014;51(8):553–62.
8. Zheng H, et al. *Mol Cytogenet.* 2015;8(1):38.
9. Chen S-C, et al. *Sci Rep.* 2016;6(1):25488.
10. Ren Y, et al. *J Genet Genomics.* 2016;43(9):541–7.
11. Treff NR, et al. *Fertil Steril.* 2013;99(5):1377–1384.e6.
12. Yan L, et al. *Proc Natl Acad Sci.* 2015;112(52):15964–9.
13. Kubikova N, et al. *Reprod Biomed Online.* 2018;37(2):136–44.
14. Wells D, et al. *Prenat Diagn.* 2019;39(5):344–50.
15. Barrangou R, et al. *Nat Biotechnol.* 2016;34(9):933–41.
16. Gutiérrez-Mateo C, et al. *Fertil Steril.* 2009;92(5):1544–56.
17. De Rycke M, et al. *Hum Reprod.* 2017;32(10):1974–94.
18. Kubikova N, et al. In: García-Velasco JA, Seli EBT-HRG, editors. Academic Press; 2020:255–69.
19. Kim JS. *Nat Protoc.* 2016;11:1573–78.

20. Capecchi MR. *Nat Rev Genet.* 2005;6:507–12.
21. Hye JK, et al. *Genome Res.* 2009;19(7): 1279–88.
22. Cho SW, et al. *Nat Biotechnol.* 2013;31:230–232.
23. Mali P, et al. *Science (80-).* 2013;339(6121):823–826.
24. Kennedy EM, et al. *Virology.* 2015;479(480):213–20.
25. Ma H, et al. *Nature.* 2017;548(7668):413–9.
26. Fogarty NME, et al. *Nature.* 2017;550:67.
27. Jinek M, et al. *Science (80-).* 2012;337(6096):816–21.
28. Gasiunas G, et al. *Proc Natl Acad Sci USA.* 2012;109(39):E2579–86.
29. The National Academies of Sciences, Engineering and Medicine. *Human Genome Editing: Science, Ethics, and Governance.* 2017, Washington, DC: National Academies Press.
30. Lovell-Badge R. *Development.* 2019;6;146(3):dev175778.
31. Liang P, et al. *Protein Cell.* 2015;6(5):363–372.
32. Kang X, et al. *J Assist Reprod Genet.* 2016;33(5):581–588.
33. Tang L, et al. *Mol Genet Genomics.* 2017;292(3):525–533.
34. Adikusuma F, et al. *Nature.* 2018;560(7717):E8–E9.
35. Cullot G, et al. *Nat Commun.* 2019;10(1):1136.
36. Egli D, et al. *Nature.* 2017;560(7717):E5–E7.
37. Kosicki M, et al. *Nat Biotechnol.* 2018;36(8):765–771.
38. Owens DDG, et al. *Nucleic Acids Res.* 2019;47(14):7402–7417.
39. Alanis-Lobato G, et al. *Proc Natl Acad Sci USA.* 2021;118(22):e2004832117.
40. Li G, et al. *Protein and Cell.* 2017;8(10):776–9.
41. Zhou C, et al. *Protein and Cell.* 2017;8(10):772–775.
42. Gaudelli NM, et al. *Nature.* 2017;551(7681):464–71.
43. Komor AC, et al. *Nature.* 2016;533(7603):420–4.
44. Babariya D, et al. *Hum Reprod.* 2017;32(12):2549–60.

Molecular Diagnosis of Endometrial Receptivity

19

Maria Ruiz-Alonso, Diana Valbuena, and Carlos Simón

INTRODUCTION

Synchrony is the simultaneous occurrence of two or more events. This phenomenon is necessary for many biologic processes, especially one crucial to reproduction: embryonic implantation. There are two main actors in this process, but for too long, research has focused on just the embryo. Thus, embryology has evolved significantly over the past 30 years, while analysis of the other player, the endometrium, has been largely ignored.

This lack is unfortunate because, despite the improvement in assisted reproductive techniques, the efficacy of IVF remains low, with live-birth rates of ~25–30% per started cycle [1]. These low rates could be caused, in part, by an imbalance between the embryo and endometrium. Recently, some research effort has shifted to analyze the relevance and the active involvement of the endometrium during implantation, as driven by transcriptomic analysis and the development of other new techniques. Currently, the status of the endometrium can be objectively assessed by transcriptomic profiles that define each menstrual stage [2, 3], including the receptive period or window of implantation (WOI) [4]. After more than ten years of basic research following this line on endometrial transcriptomic, the Endometrial Receptivity Analysis (ERA) was published [5].

Personalized strategies for embryo transfer are now possible due to clinical application of the ERA, which can be used to synchronize an implantable blastocyst and an endometrium ready to receive it. This chapter will comprehensively review this concept, including data supporting the clinical utility of predicting endometrial receptivity by molecular diagnostics.

ENDOMETRIAL RECEPTIVITY ANALYSIS (ERA)

After several years of research, our group found that the period of endometrial receptivity is characterized by the expression of 238 endometrial genes related to successful embryonic implantation. To achieve this, we performed transcriptomic analysis of pre-receptive and receptive endometria under three different conditions (natural menstrual cycles, controlled ovarian stimulation, or a refractory period induced by intrauterine device). Genes with an absolute fold-change > 3 and a false discovery rate < 0.05 were

DOI: 10.1201/9781003024941-19

selected and used to create a customized array. The gene expression data obtained with this array, from endometrial biopsies collected during the menstrual cycle, were employed to train a computational predictor based on mathematical algorithms. The computational predictor was able to classify an endometrial sample as proliferative, pre-receptive, receptive, or post-receptive, based on the combined expression level of these 238 genes. This ERA was the first transcriptomic test developed to evaluate endometrial receptivity [5].

Single-cell RNA sequencing (scRNA-seq) has confirmed these transcriptomic findings [6], providing a molecular and cellular map of changes through the menstrual cycle. We analyzed 73,180 single cells from endometrial tissue samples of 29 healthy ovum donors using two methods, microfluidics with Fluidigm technology, and nanodroplets with 10× genomics [6]. This study identified six major endometrial cell types (epithelial cells, including a previously uncharacterized epithelial ciliated cell type, stromal fibroblasts, endothelial cells, macrophages, and lymphocytes) based on canonical markers and highly differentially expressed genes. We also found that the human WOI opens with abrupt and discontinuous transcriptomic activation in the epithelia, confirming at the single-cell level our transcriptomic signature of endometrial receptivity previously identified in bulk endometrial tissue [5].

Endometrial mRNA is the basis of the ERA process. It should be extracted from an endometrial biopsy collected during the standard WOI (after 120 hours of endogenous progesterone exposure or exogenous progesterone administration). Then gene expression is analyzed and evaluated by the computational predictor. The ERA algorithm will classify the endometrium as receptive or non-receptive (proliferative, pre-receptive, or post-receptive). A non-receptive result reflects that the WOI is displaced, meaning that the endometrium requires more or less time than the standard five days of progesterone (P) to reach receptivity, providing the information about the personalized WOI (pWOI) of the patient independent of endometrial histology (Figure 19.1).

Based on six years of clinical application and data accumulated from > 20,000 endometrial samples, a new computational predictor was created using next-generation sequencing (NGS) [7]. The new ERA predictor defines a shorter and more optimal WOI timeframe. To define this receptivity signature, the new predictor was trained by selecting well-defined and curated endometrial profiles, i.e., only receptive samples from patients who became pregnant were used. For the non-receptive stages, training was performed using only those cases in which receptivity was reached after following the specific recommendation associated with the non-receptive profile. The specificity and sensitivity reached with this new predictor were 0.97 and 0.90, respectively.

Accuracy and Reproducibility of the ERA Test

To test the accuracy of this test, the ERA result was compared with a histological criteria method performed by two independent pathologists. Endometrial biopsies were collected throughout the menstrual cycle and results measured by the quadratic weighted Kappa index, calculating the concordance with the real endometrial stage (based on the cycle day collected). The concordance reached by the ERA was 0.922 (0.815–1.000), while those of the pathologists were 0.618 (0.446–0.791) and 0.685 (0.545–0.824). Interobserver variability between pathologists yielded a Kappa index of 0.622 (0.435–0.839) [8].

The reproducibility of the ERA was demonstrated using second endometrial biopsies obtained from the same women under the same conditions as the first study cycle. These second biopsies were obtained between 29 and 40 months after the first and showed no variations between cycles [8]. The consistency of the ERA result is also revealed by its prediction rate. That is, when the ERA shows a pre-receptive endometrium that needs one more day with progesterone, more than 89% of patients reach receptivity after adding that extra day. These results were obtained by performing a paired biopsy verification with 1,616 endometrial samples (n = 808 paired biopsies). When the first sample resulted as being non-receptive, the predictor established the timing in which it was expected to reach receptivity. This prediction was validated with a second endometrial biopsy at the predicted WOI timing. The success proportion for pre-receptive samples needing one more day with progesterone was 89% (95% confidence interval [86%,

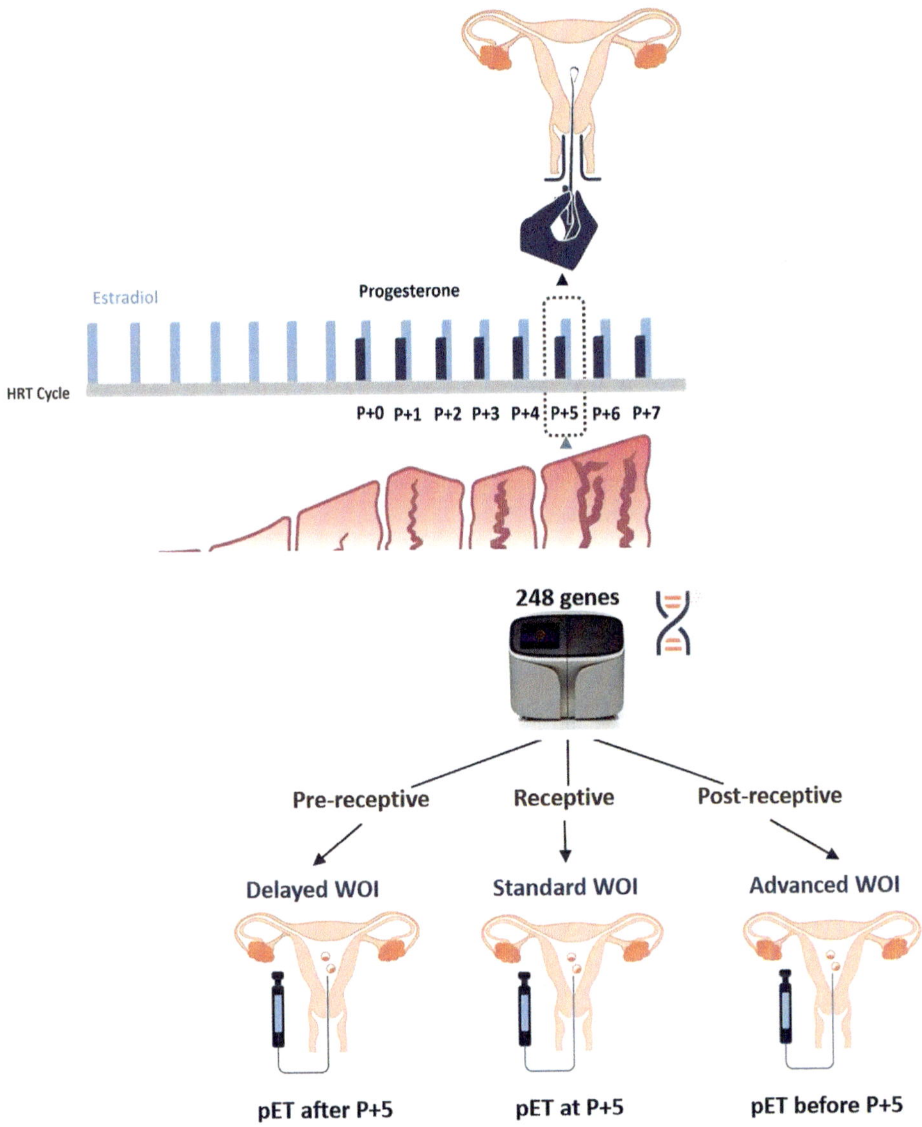

FIGURE 19.1 Schema of the ERA procedure and the major possible results: pre-receptive, receptive, and post-receptive.

91%]). This rate would be impossible to achieve if the WOI varied randomly between different cycles [7]. More recently, the consistency of the pWOI was tested in one patient through four different biopsies taken during different cycles and at various times differing from that recommended by the ERA of the first biopsy. All three additional biopsies corroborated our initial finding [9].

Challenging the ERA Test: Use of a Progesterone Receptor Antagonist to Create a Refractory Endometrium

Conceptually, the acquisition of endometrial receptivity is based on the activation of the progesterone receptor (PR) by its ligand, P. Our group collaborated in a study to evaluate the receptivity status of patients receiving mifepristone, an antagonist that blocks P action by binding PR [10]. We analyzed endometrial biopsies collected at LH+7 from 11 control women and 7 patients treated with a single dose of mifepristone (200 mg) at LH+2 together with proliferative-stage samples from 7 patients. As expected, the control samples were predicted to be pre-receptive, receptive, or late-receptive, while the mifepristone-treated samples exhibited proliferative (or in one case pre-receptive) transcriptomic profiles, indicating the ERA can precisely detect transcriptomic changes affecting the receptive status of patients treated with PR antagonists. Gene expression analysis demonstrated 60 differentially expressed genes, reflecting inactivation of the progesterone and glucocorticoid receptors.

HOW TO USE ERA IN ROUTINE CLINICAL PRACTICE

To perform an ERA test, a small endometrial biopsy must be taken from the uterine fundus using a pipelle catheter (Cornier Devices, CCD Laboratories, Paris, France) or similar device. This requires a piece of tissue around 70 mg. In difficult cases, where the cervix impedes entrance of the biopsy catheter, an endometrial sample can be obtained by vigorous aspiration with the transfer catheter.

The endometrial biopsy must be transferred immediately to a cryotube containing 1.5 mL of RNA-later (Sigma-Aldrich, St. Louis, MO, USA), a solution that prevents RNA degradation during shipment to the laboratory. The tube must be shaken vigorously for a few seconds so that the RNA-later fully penetrates the sample. The sample tube must be kept refrigerated (not frozen) at 4°C for at least four hours until shipment, which can be at room temperature (< 35°C).

Clinical Protocols for Endometrial Receptivity Testing

For ERA, the endometrial biopsy can be obtained in a natural cycle after monitoring the LH surge, after hCG administration, or in a hormonal replacement therapy (HRT) cycle. Consistency between the type of cycle and the time at which the endometrial biopsy is obtained and the personalized embryo transfer (pET) performed is key.

In natural cycles, biopsies should be obtained seven days after the luteinizing hormone surge (LH+7) or after the hCG trigger (hCG+7) and, in HRT cycles, after five days of P administration (P+5, which is equivalent to 120 hours of P exposure). This maximizes the probability of obtaining a receptive result in the first biopsy, which will occur in ~70% of patients [11].

Sampling during an HRT cycle is preferred due to its simplicity and consistency. The classic protocol for endometrial preparation begins with the confirmation of ovarian quiescence by ultrasound. Then, estrogen is administered, beginning on the first or second day of menstruation. In Europe, popular preparations include estradiol valerate at a dose of 6 mg/day or estradiol hemihydrate patches delivering 150 µg/48 hours, while in the USA oral estrace at 200 mg/8 hours is favored. Between days 7 and 10 of HRT priming, when endogenous P is < 1 ng/mL and ultrasound assessment reveals a trilaminar endometrium > 6 mm, exogenous P administration is initiated. Vaginal micronized progesterone (or similar) at a dose of 400 mg/12h or 50 mg intramuscular progesterone daily is typically used in Europe or the USA, respectively. The day on which progesterone treatment begins is referred to as P+0, and the biopsy must be taken on day P+5, approximately 120 +/– 3 hours after the first P treatment (Figure 19.2).

It is important to have proper control of endogenous progesterone before ovulation (in natural cycles) or before exogenous P administration (in HRT cycles). The endogenous P level must be < 1 ng/mL 24 hours before the first P intake in HRT cycles or at hCG+0 / LH+0 in natural cycles. This measure is necessary to avoid premature activation of the PR receptor, which could lead to an artifact.

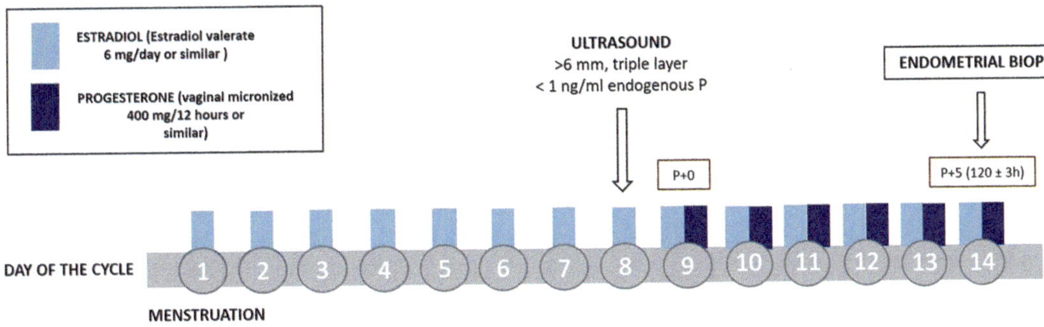

FIGURE 19.2 Standard HRT protocol scheme for ERA biopsy collection.

Interpreting ERA Results

A) *Receptive*: A receptive endometrial profile is divided into three sub-signatures— optimal receptive, early receptive, and late receptive.
- An optimal receptive profile indicates the endometrium is ready to receive the embryo. In this case, the recommendation is to proceed with pET in a subsequent cycle that replicates the endometrial preparation protocol and timing of the cycle when the ERA test was performed.
- An early receptive profile indicates the endometrium is entering the receptive phase but requires 12 additional hours of P exposure to acquire an optimally receptive profile.
- A late receptive profile indicates the endometrium is at the end of the WOI, so P exposure should be reduced by 12 hours.

B) *Non-receptive*: Our algorithm revealed that the gene expression profile of a non-receptive endometrium is usually due to a physiological displacement of the WOI. Non-receptive profiles include a proliferative profile, which generally indicates that the endometrium has not been exposed to endogenous or exogenous progesterone, or pre-receptive or post-receptive transcriptomic profiles.
- A pre-receptive diagnosis indicates that the transcriptional activation necessary to achieve receptivity has not yet occurred. The patient is predicted to reach receptivity by extending progesterone treatment one or two days beyond the time at which the biopsy was taken.
- A post-receptive diagnosis indicates that the endometrium has already passed the ideal WOI, so one or two days' less P administration is required to achieve receptivity.

Due to the consistency of the algorithm, over 90% of cases will not require a second biopsy to validate the predicted WOI. Exceptions are specific pre-receptive cases requiring two more days of P administration and post-receptive cases.

THE ASSESSMENT OF ENDOMETRIAL RECEPTIVITY IN RECURRENT IMPLANTATION FAILURE

Recurrent implantation failure (RIF) remains a poorly characterized yet intriguing clinical challenge of unknown etiology [12, 13]. RIF is most commonly defined as the failure of three embryo transfers in which the patient received one or two high-grade quality embryos per cycle [13], or after two embryo transfer failures in patients receiving donated oocytes.

The causes of RIF can be grouped into several categories. These include pathological alterations of the endometrium/endometrial cavity, e.g., hyperplasia, submucous myomas, or endometrial polyps, endometritis, and synechiae (which can be found in 18%–27% of cases) [14]. Others include hydrosalpinx [15], which may act via direct toxicity to the embryo or adversely affect endometrial receptivity [16], embryonic chromosomal abnormalities [17, 18], obesity [19], lifestyle, or other factors, e.g., hereditary or acquired thrombophilias [20]. An immunological factor has been suggested but not proven to diagnose or guide treatment for this condition [21].

RIF highlights the need for synchrony in biological processes. It is critical to ensure the adequacy of the embryo and endometrium individually but also paramount to determine the ideal timing to synchronize their pairing. Initial proof that RIF can be originated, at least in part, by the lack of synchronization between embryo and endometrium and not from an endometrial dysfunction came from a prospective interventional clinical trial [11]. In this study, ERA was performed for 85 patients with RIF of unknown etiology who had experienced three or more previous failed cycles (4.8 ± 2.0) during which a total of at least four morphologically high-grade embryos were unsuccessfully transferred. There was also a control group of 25 patients, each with ≤ 1 previous failed attempts. Of the RIF patients, 25.9% had a displaced WOI (with 84% being pre-receptive and 16% post-receptive), while only 12% of patients in the control group had a non-receptive profile. These results were translated into clinical practice by personalizing the embryo transfers according to the moment at which a patient's endometrium was predicted to be receptive. Following this, the pregnancy and implantation rates of these RIF patients were 50.0% and 38.5%, respectively, rates similar to those obtained in patients without RIF. These results suggest that personalizing ET in RIF patients can normalize pregnancy and implantation rates [11].

This concept was supported in a follow-up case report on a patient who previously experienced seven failed embryo transfers (four with her own oocytes and three with oocyte donation) [22]. This patient was found to have a delayed WOI detected by ERA and was predicted to be receptive on P+7. Their previous transfers were all based on the standard WOI (P+5); however, after performing pET at P+7 a successful pregnancy was achieved. This case report was presented together with the analysis of 17 women with one to six previous implantation failures. These IVF patients had received transfer of donated oocytes at moments in their cycles later determined as non-receptive by ERA. Upon receiving ET based on their personalized WOIs, 60% of these patients achieved a clinical pregnancy, while only 19% achieved such during transfer at non-receptive stages [22]. These results call into question whether RIF of endometrial origin is actually a "disease"; instead, it may result from a failure to tailor the clinical approach to each patient.

Data from independent groups support our clinical findings (Table 19.1). Mahajan et al. [23] published a retrospective study in an Indian population by analyzing patients with RIF and patients with just one previous failure. They found that RIF patients had a higher incidence of displaced WOIs (27.5% versus 15%), as expected based on our previous research. Following pET, the ongoing PR and IR rates of the RIF group were 42.4% and 33%, respectively, similar to rates achieved by those patients with just one previous failure. In this study, they also found that within a subgroup of RIF patients exhibiting morphologically atrophic endometria (< 6 mm), 25% had non-receptive profiles. This group achieved a PR of 66.7% when the timing of embryo transfer was based on the ERA. In a study published in 2017, Hashimoto et al. demonstrated the clinical applicability of ERA in a Japanese population [24]. They retrospectively analyzed 50 RIF patients, of which 24% had displaced WOIs. Once pET was performed guided by ERA, these patients reached a 50% PR. Similar results were obtained in the Czech Republic in a retrospective study of 85 patients (74 RIF and 11 controls). In that study, 36.5% of RIF patients had a displaced WOI and 69.2% of those became pregnant following pET [25].

Several case reports on very difficult patients also support the utility of ERA. Ota al. described a Japanese patient that achieved pregnancy with pET guided by ERA after 11 previous failed attempts [26]. Simrandeep et al. published a report on three Indian patients who had failed several transferences of good quality blastocysts. The etiology of these failures was unknown (i.e., uterine, embryonic, and immunological factors were ruled out) [27]. In two of these cases, the patients had previously received ERA at a different medical center and exhibited WOI displacements. However, pET was not applied, leading

TABLE 19.1 Scientific Publications Reporting Clinical Outcome Based on Applying pET Guided by ERA

STUDY TYPE	YEAR	TITLE	AUTHORS	JOURNAL
Letter to editor	2018	Intrapatient variability in the endometrial receptivity assay (ERA) test—letter to editor.	Cho K et al.	*J Assist Reprod Genet.* 2018;35(5):929–930.
	2018	Variations in the endometrial receptivity assay (ERA) may actually represent test error—letter to editor.	Dahan MH et al.	*J Assist Reprod Genet.* 2018;35(10):1923–1924.
	2018	Inter-cycle consistency versus test compliance in endometrial receptivity analysis test.	Stankewicz T et al.	*J Assist Reprod Genet.* 2018.
Case report	2014	Live birth after embryo transfer in an unresponsive thin endometrium.	Cruz F & Bellver J	*Gynecol Endocrinol.* 2014;30(7):481–484.
	2018	Different endometrial receptivity in each hemiuterus of a woman with uterus didelphys and previous failed embryo transfers.	Carranza F et al.	*J Hum Reprod Sci.* 2018;11(3):297–299.
	2019	Why results of endometrial receptivity assay testing should not be discounted in recurrent implantation failure?	Simrandeep K et al.	*The Onco Fertility Journal.* 2019;2(1):46–49.
	2019	The reproductive outcomes for the infertile patients with recurrent implantation failures may be improved by endometrial receptivity array test.	Ota T et al.	*Journal of Medical Cases.* 2019;10(5):138–140.
Retrospective	2014	What a difference two days make: "personalized" embryo transfer (pET) paradigm: a case report and pilot study.	Ruiz-Alonso M et al.	*Hum Reprod.* 2014;29(6):1244–1247.
	2015	Endometrial receptivity array: Clinical application.	Mahajan N	*J Hum Reprod Sci.* 2015;8(3):121–129.
	2017	Efficacy of the endometrial receptivity array for repeated implantation failure in Japan: A retrospective, two-centers study.	Hashimoto T et al.	*Reprod Med Biol.* 2017;16(3):290–296.
	2017	Window of implantation transcriptomic stratification reveals different endometrial subsignatures associated with live birth and biochemical pregnancy.	Diaz-Gimeno P et al.	*Fertil Steril.* 2017;108(4):703–710.e3.
	2018	The role of the endometrial receptivity array (ERA) in patients who have failed euploid embryo transfers.	Tan J et al.	*J Assist Reprod Genet.* 2018;35(4):683–692.
	2018	Does the endometrial receptivity array really provide personalized embryo transfer?	Bassil R et al.	*J Assist Reprod Genet.* 2018;35(7):1301–1305.
	2018	Window of implantation is significantly displaced in patients with adenomyosis with previous implantation failure as determined by endometrial receptivity assay.	Mahajan N et al.	*J Hum. Reprod. Sci.* 2018;11(4):353.
	2019	Personalized embryo transfer helps in improving in vitro fertilization/ICSI outcomes in patients with recurrent implantation failure.	Patel JA et al.	*J Hum Reprod Sci.* 2019;12(1):59–66.
	2019	Endometrial receptivity analysis—a tool to increase an implantation rate in assisted reproduction.	Hromadova L et al.	*Ceska Gynekol.* 2019;84(3):177–183.
	2019	What is the clinical impact of the endometrial receptivity array in PGT-A and oocyte donation cycles?	Neves AR et al.	*J Assist Reprod Genet.* 2019;36:1901.
	2020	Evaluation of the endometrial receptivity assay and the preimplantation genetic test for aneuploidy in overcoming recurrent implantation failure.	Cozzolino M et al.	*J Assist Reprod Genet.* 2020;37(12):2989–2997.
	2020	Comparing endometrial receptivity array to histologic dating of the endometrium in women with a history of implantation failure.	Cohen AM et al.	*Syst Biol Reprod Med.* 2020;66(6):347–354.
	2021	Clinical utility of the endometrial receptivity analysis in women with prior failed transfers.	Eisman LE et al.	*J Assist Reprod Genet.* 2021;38(3):645–650.
	2021	Evaluation of pregnancy outcomes of vitrified-warmed blastocyst transfer before and after endometrial receptivity analysis in identical patients with recurrent implantation failure.	Kasahara Y et al.	*Fertility & Reproduction.* 2020;3(2):35–41.
	2021	The use of propensity score matching to assess the benefit of the endometrial receptivity analysis in frozen embryo transfers.	Bergin K et al.	*Fertil Steril.* 2021;116(2):396–403.
Prospective	2013	The endometrial receptivity array for diagnosis and personalized embryo transfer as a treatment for patients with repeated implantation failure.	Ruiz-Alonso M et al.	*Fertil Steril.* 2013;100(3):818–824.
	2021	Routine endometrial receptivity array in first embryo transfer cycles does not improve live birth rate.	Riestenberg C et al.	*Fertil Steril.* 2021;115(4):1001–1006.
RCT	2020	A 5-year multicentre randomized controlled trial comparing personalized, frozen, and fresh blastocyst transfer in IVF.	Simon C et al.	*Reprod. BioMed. Online.* 2020;41(3):402–415.

to a subsequent failure. Once pET was properly carried out, both patients achieved successful clinical pregnancies [27]. ERA was also reported to assist in cases exhibiting endometrial anomalies, including endometrial atrophy [28] or uterus didelphys [29].

Other studies have compared the clinical outcomes of patients found to be receptive at the standard WOI to those of patients who reached receptivity with different lengths of P administration. Patel et al. retrospectively analyzed 210 RIF patients who underwent pET. They found no statistical differences in clinical outcome between the standard WOI group (receiving transfer on day P+5) and patients receiving personalized transfers on receptive days differing from P+5 [30]. These results highlight the importance of performing embryo transfers not based on a standard day of the cycle but at the moment when the endometrium has reached a receptive transcriptional phenotype.

Tan et al. observed that when the embryo factor is corrected by transferring just euploid ones, there were a higher implantation and ongoing pregnancy rates in patients with a corrected displaced WOI compared to those without displacement (although observed differences were not statistically significant). The same happened when analyzing just RIF patients who received euploid embryos, showing higher implantation and ongoing pregnancy rates in those cases having a displaced WOI versus those receptive at P+5 (66.7 vs. 44.4% and 58.3 vs. 33.3%, respectively; although not statistically significant) [31].

Other authors have evaluated pET in a retrospective cohort study comparing patients with an ERA indication treated by pET to those without an ERA indication and found no statistical differences [32–34]. We cannot forget that patients with an ERA indication have typically been the most difficult cases with no known explanation for their RIF condition. Thus, it is notable that pET normalized their clinical outcomes to resemble those of patients without ERA indication.

ASSESSING ENDOMETRIAL RECEPTIVITY IN PATHOLOGICAL CONDITIONS AND AT THE FIRST APPOINTMENT

Obesity

Though many women with obesity (BMI > 30 kg/m^2) conceive naturally, this status does impact reproduction. Rich-Edwards [35] reported that women with obesity have a three-fold greater risk of anovulation than women with normal BMI. The time to conception is two-fold longer in women who are overweight (BMI > 25 kg/m^2), and even when ovulatory, those with central distribution of fat or obesity associated with polycystic ovary syndrome (PCOS) have lower rates of conception [36]. It is particularly problematic that women with obesity who are treated by ART experience reduction of implantation, clinical pregnancy, and live-birth rates together with increased miscarriage rates [37–40]. While some early publications argue that BMI does not exert a negative effect on endometrial receptivity [41–43], mounting evidence contradicts this [44–47], and a uterine factor may be responsible for the reduced fertility associated with obesity. However, the molecular mechanism for this remains poorly understood.

A prospective cohort study using ERA in patients with different BMIs was carried out by our group to evaluate how the WOI might be affected in an overweight/obese population [48]. Ninety-one infertile patients, including all BMI categories defined by the WHO obesity classification system [49], underwent an endometrial biopsy at P+5 in HRT cycles. We found that the incidence of a non-receptive endometrium tended to be higher in the obese and morbidly obese patients (22.5% and 37%, respectively) than in normal weight and overweight patients (9.1% and 7.7%, respectively), but this did not reach statistical significance. In addition, obese subjects exhibited a significant alteration in endometrial gene expression during the optimal WOI, highlighting that energetic metabolism is also important in determining the WOI and that obesity increases the risk of WOI displacement [48]. Recently, another prospective study [50] of 170 infertile women demonstrates similar features. These patients were classified according to BMI into non-obese (18.5–29.9 kg/m^2; n = 73) and obese (> 30 kg/m^2; n = 97) categories. It was found that the non-receptivity rate increases with BMI, and the non-obese and obese patients had statistically different rates (9.7% vs. 25.3 %; p = 0.02).

Endometriosis

Endometriosis is an estrogen-dependent disorder affecting an estimated 10% of women of reproductive age [51], and its prevalence in women experiencing pain or infertility can be as high as 50% [52]. Endometriosis is defined by the presence of endometrial tissue outside the uterus; symptoms may range from practically none to chronic pelvic pain, dysmenorrhea, and cyclic urinary or bowel complaints. It has an impact on women's physical, mental, and social well-being and is commonly believed to relate to infertility.

Infertility is a side effect of endometriosis, although the etiology of this defect is multifactorial, affecting some or all of the following: ovarian reserve and/or oocyte quality, embryo quality, and endometrial receptivity. Some studies have reported a number of different endometrial markers in women with endometriosis [51, 53–56] without elucidating if they are the cause of the effect of the disease. Furthermore, altered endometrial gene expression in patients with endometriosis has been related to impaired embryo implantation [56, 57].

A prospective functional study of the transcriptomic signature of the endometrium during the WOI in patients with different stages of endometriosis (minimum, mild, moderate, and severe) compared to women without endometriosis was conducted using ERA [58]. No differences in expressed genes were found among the different endometriosis stages, but clustering analysis revealed that gene expression was linked more closely to the day when the biopsy was performed than to the stage of endometriosis. This indicates that, according to the ERA diagnosis, the transcriptomic signature during the WOI is similar in infertile patients regardless of whether endometriosis is present.

Clinical Outcomes after Personalized Embryo Transfer at the First IVF Cycle

We investigated the effectiveness of pET versus regular ET at the first appointment [59] by performing a prospective open label randomized clinical trial (RCT). The trial had three arms comparing fresh embryo transfer under stimulation protocol (ET), frozen embryo transfer at P + 5 in HRT cycles (FET), and pET guided by ERA with frozen embryos in HRT cycles (pET) (Figure 19.3). This RCT involved 16 reproductive centers across Europe, America, and Asia and included 458 patients ≤ 37 years undergoing IVF with blastocyst transfer at their first appointment. Clinical outcomes by intention-to-treat analysis were comparable across transfer types, except that the cumulative pregnancy rate (CPR) was significantly higher in the pET (93.6%) compared to the FET (79.7%) (p = 0.0005) and ET (80.7%) (p = 0.0013) groups. The analysis per protocol showed that the PR at the first attempt was statistically higher in the pET (72.5%) group than in the FET (54.3%; p = 0.01) and ET (58.5%; p = 0.05) groups. The pET group also had a higher IR (57.3%) than the FET (43.2%; p = 0.03) and ET (38.6%; p = 0.004) groups. On the other hand, there were no significant differences in the live-birth (LB) rate at the first attempt (pET 56.2% vs. FET 42.4% p = 0.09; and versus ET 45.7% p = 0.17), but the cumulative LB rate (after 12 months) was statistically higher in the pET group (71.2%) than in the FET (55.4%; p = 0.04) and ET (48.9%; p = 0.003) groups. Obstetrical outcomes, type of delivery, and neonatal outcomes were similar in all groups. Thus, this study showed how performing pET in patients at the beginning of their assisted reproductive treatment could improve their outcomes.

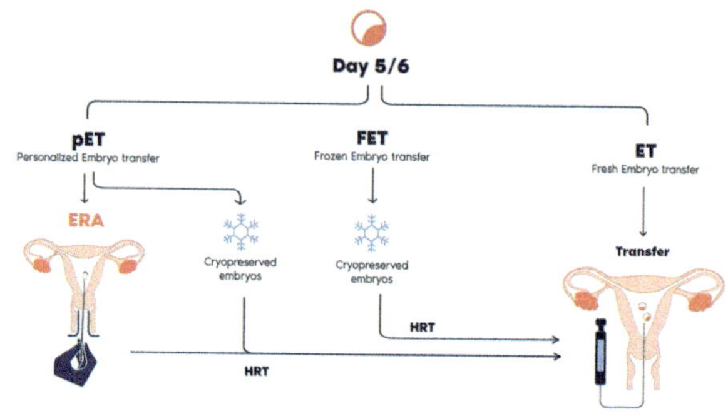

	pET (n = 80)	FET (n = 92)	ET (n = 94)	pET vs FET		pET vs ET	
				Relative risk (95% CI)	P-value	Relative risk (95% CI)	P-value
Pregnancy rate (PR)	58 (72.5%)	50 (54.3%)	55 (58.5%)	1.56 (1.07-2.29)	0.01	1.42 (0.98-2.08)	0.057
Implantation rate (IR)	63/110 (57.3%)	60/139 (43.2%)	58/150 (38.6%)	1.37 (1.03-1.82)	0.03	1.54 (1.15-2.05)	0.004
Live birth rate (LBR)	45 (56.2%)	39 (42.4%)	43 (45.7%)	1.35 (0.97-1.86)	0.09	1.26 (0.91-1.74)	0.17
Cumulative PR	76/80 (95%)	65/92 (70.6%)	59/94 (62.8%)	4.18 (1.65-10.56)	0.0001	5.49 (2.14-14.06)	0.0001
Cumulative LBR	57 (71.2%)	51 (55.4%)	46 (48.9%)	1.47 (1.01-2.13)	0.04	1.71 (1.17-2.49)	0.003

FIGURE 19.3 Clinical outcome of patients at their first clinical appointment obtained in fresh embryo transfer under stimulation protocol (ET), frozen embryo transfer at P + 5 in HRT cycles (FET), and pET guided by ERA with frozen embryos in HRT cycles (pET). Source: Based on results from the RCT study from Simon et al. 2020.

LIMITATIONS

The endometrial biopsy can present technical difficulties during collection. Thus, up to 5% of the samples received are unsuitable to process due to degradation or an inability to obtain enough RNA. mRNA is a highly sensitive genetic material that can be degraded for different reasons (high temperature for shipment and/or storage, failure to use RNAse inhibitors, or a lack of sterile conditions). Also, the ERA only assesses the endometrium at the transcriptional level, whereas other possible problems may be present, such as an altered uterine microbiome, which may impair the clinical results of an otherwise receptive endometrium.

In regards to using the ERA to guide pET, the conditions used to perform the ERA biopsy must be replicated in the subsequent transfer cycle. Thus, it is only possible to perform pET in a frozen embryo transfer in cases where the patient will receive their own oocytes.

Since the publication of our seminal paper identifying the transcriptomic signature of endometrial receptivity [5], several companies have launched commercial endometrial transcriptomic tests under different acronyms with different evidence reported. These tests are based on different technologies analyzing other genes than those present on ERA, so there could be no concordance in the results [60]. Thus, the reported clinical outcome when applying pET guided by ERA is particular to it and cannot be extrapolated to other tests.

CONCLUSIONS

It is important to assess the integrity and normality of the two main players involved in the implantation process—endometrium and embryo—as well as the synchrony between these two partners. This

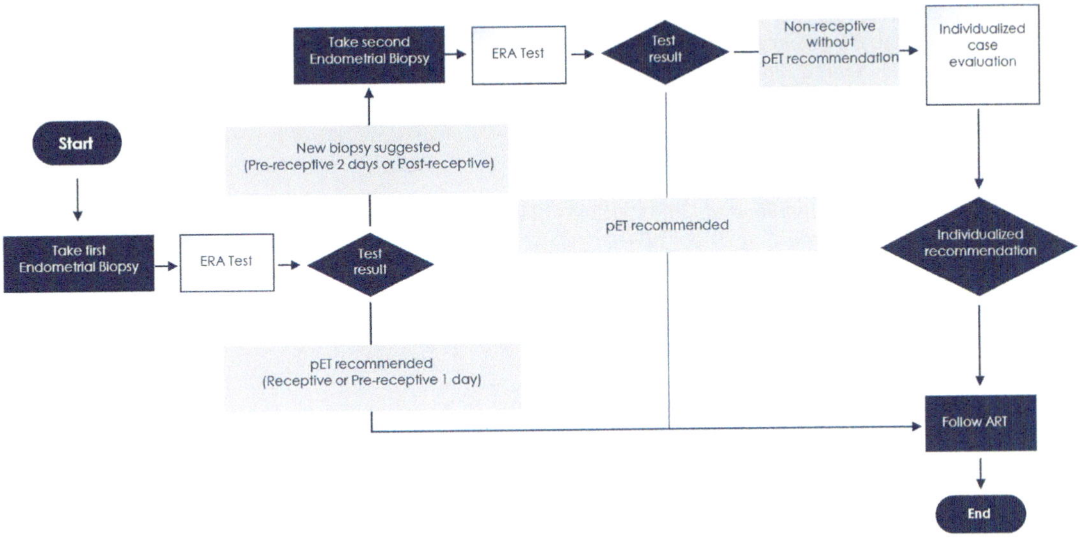

FIGURE 19.4 Decision tree to identify endometrial receptivity by the ERA test.

synchrony is one of the critical factors that contributes to a successful reproductive outcome. Even with morphologically and chromosomally normal embryos, human implantation efficacy is low, reaching only 50%–65% in cases involving a variety of endometrial thicknesses and patterns [61]. Thus, the contribution of the endometrium is clearly an important limiting factor and should be incorporated into the ART treatment process.

Several treatments or analyses have been proposed to evaluate the endometrial factor, most lacking proven scientific evidence. However, we should consider all evidence-based procedures that demonstrate a significant increase in clinical outcomes to provide better options for our patients. The ERA test is currently clinically available and able to personalize an embryo transfer by synchronizing the blastocyst with a receptive endometrium (Figure 19.4). In recent years, different scientific publications from around the world have emerged, assessing the clinical results of the ERA as applied during ART. This evidence will be useful for future patients and their clinicians to evaluate all their options along their assisted reproductive journey.

REFERENCES

1. Adamson G, et al. *Fertil Steril.* 2018;110:1067–80.
2. Díaz-Gimeno P, et al. *Int J Dev Biol.* 2014;58(2–4):127–37.
3. Garrido-Gómez T, et al. *Fertil Steril.* 2013;99(4):1078–85.
4. Kao LC, et al. *Endocrinology.* 2002;143(6):2119–38.
5. Díaz-Gimeno P, et al. *Fertil Steril.* 2011;95(1):50–60.
6. Wang W, et al. *Nat Med.* 2020;26:1644–53.
7. Clemente-Ciscar, M, et al. *Human Reprod.* 2018;33:8–8.
8. Díaz-Gimeno P, et al. *Fertil Steril.* 2013;99(2):508–17.
9. Stankewicz T, et al. *J Assist Reprod Genet.* 2018;35:1307–8.
10. von Grothusen C, et al. *Hum Reprod.* 2018;33:1889–97.
11. Ruiz-Alonso M, et al. *Fertil Steril.* 2013;100(3):818–24.
12. Margalioth EJ, et al. *Hum Reprod.* 2006;21(12):3036–43.
13. Simon A, et al. *Fertil Steril.* 2012;97(5):1039–43.
14. Demirol A, et al. *Reprod Biomed Online.* 2004;8(5):590–4.

15. Zeyneloglu HB, et al. *Fertil Steril.* 1998;70(3):492–9.
16. Meyer WR, et al. *Hum Reprod.* 1997;12(7):1393–8.
17. Pehlivan T, et al. *Reprod Biomed Online.* 2003;6(2):232–7.
18. Rubio C, et al. *Fertil Steril.* 2013;99(5):1400–7.
19. Broughton DE, et al. *Fertil Steril.* 2017;107:840–7.
20. Penzias AS. *Fertil Steril.* 2012;97(5):1033–8.
21. Stephenson MD, et al. *Fertil Steril.* 2000;74(6):1108–13.
22. Ruiz-Alonso M, et al. *Hum Reprod.* 2014;29:1244–1247.
23. Mahajan N. *J Hum Reprod Sci.* 2015;8:121–9.
24. Hashimoto T, et al. *Reprod Med Biol.* 2017;16:290–6.
25. Hromadova L, et al. *Ceska Gynekologie-Czech Gynaecol.* 2019;84:177–83.
26. Ota T, et al. *J M Cases.* 2019;10:138–40.
27. Simrandeep K., et al. *Onco Fert J.* 2019;2:46–9.
28. Cruz F, et al. *Gynecol Endocrinol.* 2014;30:481–4.
29. Carranza F, et al. *J Hum Reprod Sci.* 2018;11:297–9.
30. Patel JA, et al. *J Hum Reprod Sci.* 2019;12:59–66.
31. Tan J, et al. *J Assist Reprod Genet.* 2018;35:683–92.
32. Bassil R, et al. *J Assist Reprod Genet.* 2018;35:1301–5.
33. Neves AR, et al. *J Assist Reprod Genet.* 2019;36:1901.
34. Cozzolino M, et al. *J Assist Reprod Genet.* 2020;37:2989–97.
35. Rich-Edwards JW, et al. *Am J Obstet Gynecol.* 1994;171(1):171–7.
36. Hassan MA, et al. *Fertil Steril.* 2004;81(2):384–92.
37. Wang JX, et al. *BMJ.* 2000;321:1320–1.
38. Fedorcsák P, et al. *Hum Reprod.* 2004;19(11):2523–8.
39. Lintsen AME, et al. *Hum Reprod.* 2005;20(7):1867–75.
40. Maheshwari A, et al. *Hum Reprod Update.* 2007;13(5):433–44.
41. Styne-Gross A, et al. *Fertil Steril.* 2005;83(6):1629–34.
42. Dechaud H, et al. *Eur J Obstet Gynaecol Reprod Biol.* 2006;127(1):88–93.
43. Dokras A, et al. *Obstet Gynecol.* 2006;108(1):61–9.
44. Bellver J, et al. *Fertil Steril.* 2007;88(2):446–51.
45. Dessolle L, et al. *Hum Reprod.* 2009;24:3082–9.
46. Provost MP, et al. *Fertil Steril.* 2016;105(2):364–8.
47. Bellver J, et al. *Fertil Steril.* 2013;100(4):1050–8.
48. Comstock IA, et al. *Fertil Steril.* 2017;107(3):740–748.e2.
49. World Health Organization. *Obesity: Preventing and Managing the Global Epidemic.* No. 894. Geneva: World Health Organization; 2000.
50. Bellver J, et al. *Reprod. Sci.* 2021.
51. May KE, et al. *Hum Reprod Update.* 2011;17(5):637–53.
52. Meuleman C, et al. *Fertil Steril.* 2009;92(1):68–74.
53. Daftary GS, et al. *Mol Endocrinol.* 2002;16(3):571–9.
54. Lessey BA, et al. *J Clin Endocrinol Metab.* 1994;79(2):643–9.
55. Taylor HS, et al. *Hum Reprod.* 1999;14(5):1328–31.
56. Wei Q, et al. *Fertil Steril.* 2009;91(5):1686–91.
57. Revel A. *Fertil Steril.* 2012;97(5):1028–32.
58. Garcia-Velasco JA, et al. *Reprod Biomed Online.* 2015;31(5):647–54.
59. Simón C, et al. *Reprod Biomed Online.* 2020;41:402–15.
60. Dahan MH, et al. *Obstet Gynecol Sci.* 2021; Jul 15.
61. Rubio C, et al. *Fertil Steril.* 2017;107(5):1122–9.

The Vaginal Microbiome

20

Shahriar Mowla, Phillip R. Bennett, and David A. MacIntyre

INTRODUCTION

The complex communities of microorganisms colonizing the human body play important symbiotic roles in many aspects of human physiology including metabolism, pathogen resistance, and immune homeostasis (1–4). The assemblage of these microorganisms, which includes bacteria, viruses, fungi, and archaea, in a specific niche is referred to as the "microbiota" (5). The term "microbiome" is often used interchangeably with microbiota but more correctly refers to the entire habitat, including the microbiota, their genomes, and surrounding biochemical milieu (6). Similar to other body sites that have been studied in greater detail (e.g., oral cavity and gut) the lower female reproductive tract hosts a thriving microbiota, the composition of which is subject to change throughout a woman's lifespan. There is growing recognition that specific vaginal microbiota are associated with states of good reproductive health and can therefore be considered "optimal." In contrast, other microbiota compositions are more frequently observed prior to or during a shift toward a pathogenic state and are thus considered "suboptimal." The endogenous and exogenous factors that influence colonization by different vaginal microorganisms, and how the host responds to them are critical in determining their role in the niche. In this chapter we will explore the current understanding of the factors that shape the composition of the vaginal microbiome, how it can be studied and characterized, and the evidence linking vaginal microbiota-host interactions to reproductive success or failure.

CHARACTERIZATION OF VAGINAL MICROBIOTA—A HISTORICAL CONTEXT

Specific vaginal bacteria began to be linked to states of reproductive health and pathology during the Bacterial Revolution of the late 19th century. Investigating the concept that vaginal bacteria may be involved in post-partum sepsis, Albert Döderlein discovered and described for the first time the *Lactobacillus* species, present in high abundance in the vaginal secretions of many women (7). Soon after it was noted that in the absence of the *Lactobacillus* species, the vagina would instead be often colonized by "diphtheroids, enterococci and staphylococci" (8). Culture and microscopy would continue to be the main tools used

DOI: 10.1201/9781003024941-20

for investigating vaginal microorganisms for at least the next century. These studies enabled important associations to be made between "abnormal" vaginal microbial communities, typically described as those depleted of *Lactobacillus* species with increased diversity of facultative or strict anaerobes, and a range of poor reproductive health outcomes including increased risk of sexually transmitted infections (9, 10) and pregnancy complications such as miscarriage and preterm birth (11–13).

Culture and microscopy continue to be mainstay techniques for clinical diagnosis of abnormal vaginal microbiota, particularly in the case of bacterial vaginosis (BV). The Amsel Criteria, first described in 1983, is based upon a combined assessment of both clinical symptoms (e.g., fishy odor, pH, white discharge) and microscopic features of vaginal secretions that include presence of vaginal epithelial cells coated by gram-variable bacteria referred to as Clue cells (14). The Nugent Criteria is often used for diagnosis of BV in research settings and relies upon microscopic evaluation and grading for the presence or absence of three key bacterial morphotypes; i) large gram-positive rods, ii) small gram-variable rods and iii) curved gram-variable rods (15). These morphotypes were thought to correspond to the genera *Lactobacillus, Gardnerella*, and *Mobiluncus* respectively, although recent molecular analyses of the *Mobiluncus* morphotypes indicates that they are more likely to be other BV-associated bacteria such as BV-associated bacterium-1 (BVAB1) (16), now referred to as *Candidatus* Lachnocurva vaginae (17).

MOLECULAR-BASED PROFILING OF VAGINAL MICROBIOTA

The last decade has seen an explosion in the application of culture independent, molecular-based methods for microbial identification and characterization. Two main sequencing strategies that both rely upon next-generation sequencing (NGS) platforms have emerged as the most widely used approaches. "Metagenomics" refers to the use of shotgun sequencing of DNA within a sample followed by assemblage and mapping of the sequencing to a reference database followed by annotation (6). By providing information about all genomes within a sample, metagenomics can be used for both analysis of the compositional structure of a microbial community and its potential function. The drawbacks of this approach are high economic and computational costs and the difficultly of applying it to low biomass samples where the majority of amplified and sequenced DNA is derived from the host (e.g., the placenta or endometrium). "Metataxonomics" addresses some of these issues through the amplification of microbial taxonomic marker genes (e.g., bacterial 16S ribosomal RNA genes or fungal 18S rRNA genes). These genes are composed of highly conserved regions that can be targeted by PCR primers for amplification and are interspaced by hyper-variable regions that provide genetic diversity useful for classifying bacterial taxonomy to species and in some cases strain level (18). As the majority of metataxonomic profiling approaches rely upon the sequencing of short amplicons, care needs to be taken when considering the hypervariable regions selected, the choice of primers used, and the informatics pipelines used for analyses, all of which can introduce bias. An alternative to sequencing approaches gaining in popularity is the IS-profiling technique, which exploits species-specific differences in the 16S-23S ribosomal interspace (IS) fragment length that can be used for identification and profiling (19). An advantage of this technique is that it can be easily implemented in most molecular laboratories and is amenable to high-throughput screening of samples; however, it has been reported that resolution of vaginal microbial communities provided by this technique is not as detailed as that provided by sequencing approaches (20).

One of the first large scale applications of NGS-profiling of the vaginal microbiota formed part of the National Institutes of Health-funded Human Microbiome Project (HMP) that was completed in 2012 (21). In this study, samples were collected from the introitus, midpoint, and posterior fornix of the vagina of 113 women. This study showed that compared to other body sites such as the gut, oral cavity, and the skin, the vagina typically has low microbial diversity and richness. The study also confirmed earlier findings obtained using culture and microscopy indicating that *Lactobacillus* species are the dominant species found in the vagina of reproductive aged women.

In 2011, a study by Ravel and colleagues provided a deeper characterization of the vaginal microbiota in a large (n = 396) and ethnically diverse cohort of women of reproductive age (22). Investigating samples collected from equal numbers of asymptomatic Asian, Black, Hispanic, and White women, the study showed that bacterial profiles of the vagina could be classified into 5 groups termed "community state types" (CSTs) that were characterized by the presence or absence of four *Lactobacillus* species; CST I- *L. crispatus* dominated, CST II- *L. gasseri* dominated, CST III- *L. iners* dominated, and CST V- *L. jensenii* dominated. A fifth group termed CST IV, was characterized by a lack of *Lactobacillus* species and increased relative abundance of typically strict anaerobes that had previously been linked to BV including *G. vaginalis, Prevotella, Dialister*, and *Atopobium* species. Importantly, the study also reported that the bacterial composition of the vaginal microbiome is influenced by ethnicity of the host. Vaginal community states from White (89.7%) and Asian (80.2%) women were most commonly those dominated by *Lactobacillus* species. In comparison, only 59.6% of Hispanic and 61.9% of Black women had vaginal microbiota dominated by *Lactobacillus* species. Numerous studies have since confirmed the relationship between ethnicity and vaginal microbiome composition (23–26) and the broad compositional characteristics of the original CSTs, sometimes now referred to as "vagitypes" or "vaginal microbial communities," have been consistently observed in other populations (23, 27–30). As the volume of vaginal microbiome studies increases, it becomes clearer that additional, rarer vaginal CSTs exist, and efforts are now being made to standardize how CSTs are calculated and described, to improve comparability of data across studies (31).

FACTORS INFLUENCING THE COMPOSITION OF THE VAGINAL MICROBIOME

Despite its comparative simplicity, the compositional structure of the vaginal microbiota is dynamic (27) and changes throughout a women's lifespan. These changes often occur in concert with circulating hormonal levels, particularly estrogen (E2). Soon after birth, colonization of the vagina begins under the influence of high circulating maternally derived E2 concentrations (32). This leads to increased storage of glycogen in the vaginal epithelia, which when broken down into complex sugars by the host α-amylase, can be used preferentially as energy sources by *Lactobacillus* species, which were seeded there during vaginal delivery. As maternal E2 is metabolized, glycogen levels are depleted and a shift towards a high-diversity community structure depleted in *Lactobacillus* species is observed, which is accompanied by increased pH (33, 34). As puberty is reached, the vaginal microbiota composition becomes more similar to reproductive-age women, characterized by increased *Lactobacillus* species' dominance, and reduced diversity (35). The highly estrogenic state of pregnancy, caused by feto-placental production of E2, further supports *Lactobacillus* species' dominance of the vagina (36, 37). Post-delivery, disturbance of the biochemical microenvironment of the vaginal mucosa caused by passing of amniotic fluid and bleeding during delivery combined with a rapid decline in maternal E2 (38) sees vaginal microbiota increase in richness and diversity (36). This shift towards CST-IV profiles has been shown to persist in some women for up to one year following delivery (39). Consistent with a role in shaping *Lactobacillus* species' dominance in the vaginal niche, reduced E2 following the menopause correlates with decreased glycogen levels and vaginal epithelium thinning that in turn associates with depletion of the *Lactobacillus* species and increased diversity (40). In contrast, in women receiving hormone replacement therapy, *Lactobacillus* species continue to be a dominant feature of the microbiota (41, 42). Other host factors that influence the composition of the vaginal microbiota include menses (27, 43–45), hormonal contraceptives (46–48), antibiotics (49–52), and sexual activity (46, 53, 54) (Figure 20.1). In a recent study of 26 college-aged women who self-sampled daily for ten weeks, changes in vaginal microbiota composition were also associated with diet and exercise (45). This is an intriguing finding as it highlights the potential of lifestyle change as a means for non-invasively manipulating vaginal microbiota composition.

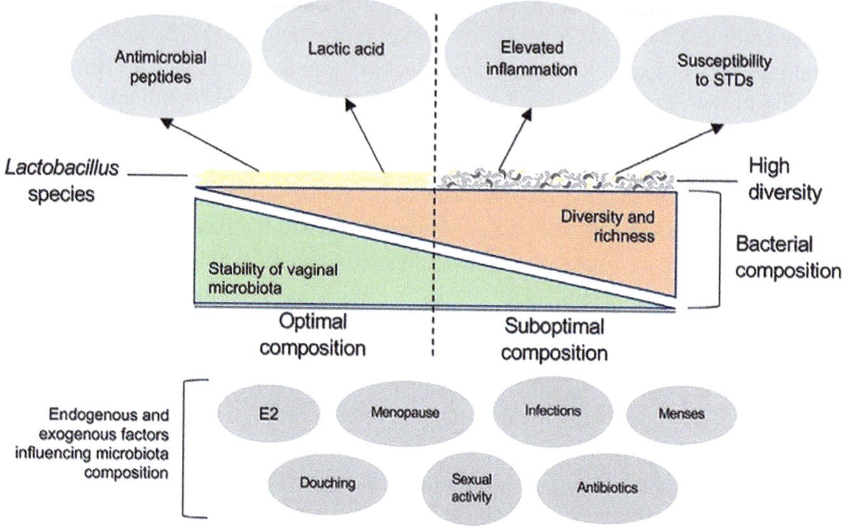

FIGURE 20.1 Overview of factors influencing vaginal microbiota composition and stability and its relationship with health and disease states.

VAGINAL MICROBIOTA, PATHOGEN COLONIZATION, AND HOST RESPONSE

There are distinct anatomical, immunological, and morphological differences between the lower (vagina and ectocervix) and upper (uterus and endocervix) reproductive tract that reflect their unique functions (Figure 20.2). For example, the vaginal epithelium consists of multiple protective layers of non-keratinized stratified squamous epithelial cells that can vary in thickness from a few to 45 cells (55). In comparison, the epithelium of the upper reproductive tract consists of a single layer of columnar epithelial cells. In the vagina, cells in the apical layer undergo cornification, a process by which they lose their nuclei and intracellular organelles (56). Regular exfoliation of these cells into the vaginal lumen (55) assists in the removal of invading pathogens (57, 58). Mitosis in the basal layer promotes renewal of the epithelium (55) with an estimated turnover of approximately 96h (59).

In comparison to the upper reproductive tract, the lower reproductive tract has evolved alongside a high bacteria load. This is largely due to the cervix which acts as both a physical and immunological barrier between the upper and lower reproductive tract (60, 61). Mucus produced by the cervix is predominantly composed of water and highly glycosylated proteins termed "mucins" that can act as sites of attachment and adhesion as well as energy substrates for microbiota (62). Mucus production by the cervix is regulated by E2. As E2 levels rise, viscosity typically decreases which is important for allowing the passage of spermatozoa through the mucus (63). During pregnancy the cervical mucus forms a plug that seals the cervix helping to prevent ascending infection from the vagina (64, 65). The mucosal layer is also enriched by proteins and metabolites produced by both host and microbial cells that regulate its function. For example, specific isoforms of lactic acid produced by *Lactobacillus* species can alter viscosity and thus the viral trapping capacity of cervical mucus (66–68). Lactic acid also decreases mucosal pH to a level (4.0 ± 0.5) which is microbicidal for many pathogens. Secretion of biosurfactants and bacteriocins and competitive exclusion through displacement and inhibitory adhesion are additional strategies by which *Lactobacillus* species help to protect against pathogen colonization in the lower reproductive tract (22, 69–71). While *Lactobacillus* species are widely considered to be important components of a healthy vaginal microbiome, increasing evidence suggests that *L. iners*, a common species frequently isolated

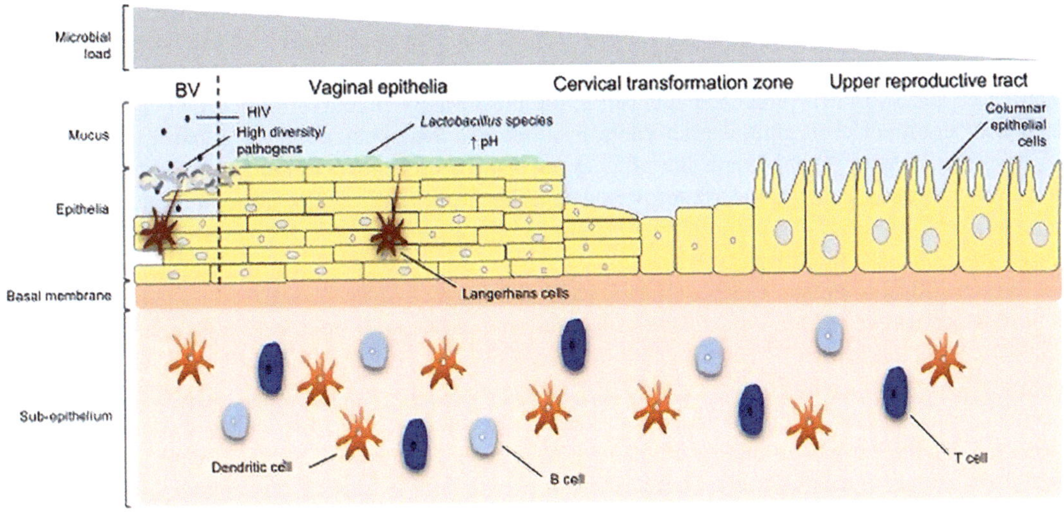

FIGURE 20.2 Microbiota-host interactions at the epithelial mucosal interface throughout the reproductive tract.

from the vagina, may act as an intermediatory between optimal and sub-optimal vaginal microbiomes (72). The genome of *L. iners* is unusually small (1.3 Mbp) compared with *L. crispatus* (3-4 Mbp) and contains genes encoding for iron-sulfur proteins, unique σ-factors as well as a pore-forming toxin called inerolysin (73). Complex nutritional requirements have also likely contributed to its apparent tolerance by *L. iners* of potentially pathogenic species that provide it with metabolic breakdown products. This means that *L. iners* is frequently detected in both healthy and pathogenic states (72).

Pattern recognition receptors, such as Toll-like receptors (TLRs) and NOD-like receptors (NLRs) expressed by epithelial and immune cells, sense and respond to pathogen invasion in the vagina by triggering a complex cascade of signals leading to the activation of the innate immune response. This involves the production of cytokines and chemokines, which may have anti- or pro-inflammatory effects, as well as anti-microbial peptides including secretory leukocyte protease inhibitor (SPLI), β-defensins, and other immune factors (55, 74–77). Host tolerance of commensal microbiota is thought to be achieved through the suppression of activation of pattern-recognition receptors and subsequent inhibition of pro-inflammatory cytokines (74). In contrast, pathogens and species often associated with BV including *G. vaginalis* (78) and *A. vaginae* (79) are associated with increased production of pro-inflammatory cytokines and chemokines such as IL-6, IL-8, RANTES, and MIP1β, including during pregnancy (80). These immune mediators regulate diverse developmental, physiological, and inflammatory processes throughout the reproductive tract including the epithelial permeability through restructuring of tight junctions (81, 82). Damage to the lower reproductive tract epithelial layer caused by microbial-associated inflammation (83, 84) can increase susceptibility to infections including HIV (85).

THE VAGINAL MICROBIOME AND REPRODUCTIVE HEALTH OUTCOMES

Given its important role in mediating immune response and inflammation within the reproductive tract, there is growing interest in understanding the relationship between the vaginal microbiome and reproductive health and disease states. In women of reproductive age, vaginal microbiota communities dominated by *Lactobacillus* species provide protection against a wide range of pathologies and diseases including bacterial vaginosis (86, 87), pelvic inflammatory disease (88), candidiasis (89), and sexually transmitted infections such as HIV, HSV-2 (90–92), and human papillomavirus, the latter being causally linked to the development of cervical cancer (93, 94). In a recent study we showed that regression of untreated cervical intraepithelial neoplasia 2, the pre-invasive precursor of cervical cancer, is associated with dominance of vaginal microbiota by *Lactobacillus* species (95). Vaginal microbiota may also have the capacity to modulate the efficacy of some therapeutic interventions. For example, the topical pre-exposure prophylactic, Tenofovir, is superior in preventing HIV acquisition in women with *Lactobacillus*-dominated vaginal microbiota compared to women colonized predominately by *G. vaginalis* and BV-associated bacteria (96).

There is also substantial evidence that the vaginal microbiota influences the reproductive success of the host. Much recent attention has focused on investigating potential links between vaginal microbiota composition and infertility (97–100). Some studies have even proposed vaginal microbiota as a predictive biomarker for *in vitro* fertilization success (101, 102). Despite differences in design and techniques across studies in this area, there is general agreement that *Lactobacillus* species' dominance of the vaginal niche is associated with successful embryo implantation whereas *Lactobacillus* species depletion and overgrowth of facultative anaerobes are associated with poorer rates of success. A summary of studies that have used molecular based approaches to characterize the relationship between vaginal microbiota composition and *in vitro* fertilization success are presented in Table 20.1. The importance of the *Lactobacillus* species' dominance continues into pregnancy. Vaginal microbiota depleted of the *Lactobacillus* species have recently been linked with first trimester miscarriage (103) and early data suggests that women who suffer unexplained recurrent miscarriage have higher prevalence of potential pathogens including *Atopobium*, *Prevotella*, and the *Streptococcus* species (104).

The vast majority of research on the pregnancy vaginal microbiome to date has been focused on understanding its relationship with preterm birth (110), which continues to be the leading cause of childhood mortality worldwide (111). Vaginal pathogen infection (112) or pathologies such as BV (113, 114) and aerobic vaginitis (112) have long been recognized to increase preterm birth risk. As reviewed in detail elsewhere (110, 115), recent application of metataxonomic profiling approaches in pregnancy have given new insights into the relationship between vaginal microbiota composition and preterm birth. Numerous studies across different global populations have now shown that in women of White ethnic backgrounds, there is a positive association between increased vaginal microbial diversity and risk of PTB (24, 116–121). In contrast, studies of Black pregnant women have not shown any relationship between vaginal microbiota composition and preterm risk (122) or have reported decreased vaginal diversity in women who subsequently experience preterm birth (123, 124). As mentioned earlier in this chapter, these findings suggest that the influence of ethnicity on the vaginal microbiota extends to its association with pathologies, including preterm birth.

Recent work from our laboratories has further explored the relationship between the vaginal microbiota and a specific sub-group of preterm birth, namely preterm pre-labor rupture of the fetal membranes (PPROM) (125, 126). PPROM proceeds around 30–40% of preterm birth cases (127) and is associated with increased risk of poor maternal and neonatal outcomes as the uterine cavity and fetus become exposed to potentially pathogenic vaginal bacteria (128–134). Our findings showed that vaginal dysbiosis characterized by *Lactobacillus* species' depletion increased bacterial diversity and presence of potential pathogens is significantly more prevalent in women who subsequently experience PPROM (29). Moreover, the event of membrane rupture itself additionally promotes sub-optimal vaginal microbiota colonization, which may be further exacerbated by specific antibiotic treatment (erythromycin) widely used clinically in an attempt to prolong pregnancy and reduce neonatal morbidity following PPROM (50). This seems to have consequences for the mother and neonate following delivery. In this study cohort, the vaginal microbiota of women with chorioamnionitis with funisitis were enriched for *Prevotella*, *Sneathia*, *Peptostreptococcus*, and *Catonella* species, and had raised C-reactive protein (a systemic marker of inflammation) and white blood cell counts. Moreover, in cases developing early onset neonatal sepsis, maternal vaginal swabs collected closest to the time of delivery were enriched for *Streptococcus agalactiae*, *Fusobacterium nucleatum*, *Escherichia coli*, *Catonella*, and *Sneathia* species. Interestingly,

TABLE 20.1 Summary of Studies Using Molecular-Based Approaches to Characterize the Relationship between Vaginal Microbiota Composition and *In Vitro* Fertilization Success

YEAR	*AUTHORS, REFERENCE*	*NO. OF SUBJECTS*	*TARGET FOR MOLECULAR CHARACTERISATION OF VAGINAL MICROBIOTA*	*SAMPLING POINTS THROUGHOUT TREATMENT CYCLE*	*RELATIONSHIP WITH OUTCOME?*	*SUMMARY OF RELATIONSHIP REPORTED BETWEEN VAGINAL MICROBIOTA AND IVF SUCCESS REPORTED*
2012	Hyman et al. (105)	30	Full length 16S rRNA gene	Prior to treatment cycle, oocyte retrieval, embryo transfer, and 6–8 weeks' gestation	Y	Low diversity, *Lactobacillus* species dominated vaginal microbiota positively correlated with successful IVF treatment.
2012	Mangot-Bertrand et al. (106)	307	Nugent score and targeted qPCR	Oocyte retrieval	N	No significant difference in embryo implantation rate between BV negative and positive patients.
2016	Haahr et al. (107)	130	Targeted qPCR	2–4 weeks prior to treatment cycle	Y	High concentrations of *G. vaginalis* and *A. vaginae* associated with low clinical pregnancy rates.
2018	Haahr et al. (101)	120	V4 region of the 16S rRNA gene and qPCR	Prior to treatment cycle	Y	High diversity (Shannon index >0.93) associated with low pregnancy and live birth rates. Abnormal vaginal microbiota (diagnosed by qPCR) associated with poor reproductive outcome.
2018	Wee et al. (97)	31	V1-V3 region of the 16S rRNA gene	During hysteroscopy examination outside of treatment cycle	Y	Vaginal microbiome composition is associated with successful fertility outcomes. Trend association of *Gardnerella* in the cervix and *Ureaplasma* in the vagina with women with history of infertility.
2019	Koedooder et al. (102)	192	IS-pro technique (16S-23S rRNA gene)	Prior to treatment cycle	Y	Women with low relative abundance of vaginal *Lactobacillus* less likely to have successful embryo implantation.
2019	Vergaro et al. (108)	150	qPCR (gDNA)	Day of embryo transfer	N	No association detected.
2020	Fu et al. (109)	67	V3-V4 region of the 16S rRNA gene	1 month and 7 days before embryo transfer	Y	Vaginal *Lactobacillus* positively correlated with successful pregnancy after IVF. Patients with recurrent implantation failure contained significantly less *Lactobacillus* and more pathogens compared to controls.
2020	Kong et al. (99)	475	V4 region of the 16S rRNA gene	During treatment cycle	Y	Lower abundance of *Lactobacillus* and increased *Gardnerella* and *Prevotella* in patients not achieving pregnancy following IVF.

maternal colonization by *Lactobacillus crispatus* was associated with protection against early onset neonatal sepsis following PPROM.

CONCLUSIONS

There is now extensive data supporting the notion that vaginal microbiota plays an integral role in the maintenance of reproductive health throughout a woman's lifespan. Broadly, vaginal microbial community compositions characterized by *Lactobacillus* species' depletion and increased diversity are often associated with disease risk and pathology. In contrast, *Lactobacillus* species' dominance of the vaginal niche, particularly by *L. crispatus*, is associated with states of good health. However, as our understanding of vaginal microbiota-host interactions improves, it becomes more obvious that a "one size fits all" definition of optimal vaginal microbiota composition is unlikely to exist. Instead, host factors such as ethnicity and immune response are likely to be important determinants of the microbiota-host interactions that shape disease risk. This highlights the need for the development of new methodologies and tools that permit point-of-care assessment of vaginal microbiota composition and host responses that could facilitate more targeted and personalized strategies for the manipulation of the microbiota toward an optimized state of health for that individual (135).

REFERENCES

1. Wu H-J, et al. *Gut Microbes*. 2012;3(1):4–14.
2. Cummings JH, et al. *Clin Nutr*. 1997;16(1):3–11.
3. Dethlefsen L, et al. *Nature*. 2007;449:811.
4. Hooper LV, et al. *Science*. 2001;292(5519):1115–8.
5. Lederberg J, et al. *'Ome Sweet' Omics--A Genealogical Treasury of Words*. The Scientist; 2001.
6. Marchesi JR, et al. *Microbiome*. 2015;3:31.
7. Lash AF, et al. *J Infect Dis*. 1926;38(4):333–40.
8. Cruickshank R. *J Hyg (Lond)*. 1931;31(3):375–81.
9. Martin HL, et al. *J Infect Dis*. 1999;180(6):1863–8.
10. Sewankambo N, et al. *Lancet*. 1997;350(9077):546–50.
11. Hay PE, et al. *BMJ*. 1994;308(6924):295–8.
12. Leitich H, et al. *Am J Obstet Gynecol*. 2003;189(1):139–47.
13. Martius J, et al. *Obstet Gynecol*. 1988;71(1):89–95.
14. Amsel R, et al. *Am J Med*. 1983;74(1):14–22.
15. Nugent RP, et al. *J Clin Microbiol*. 1991;29(2):297–301.
16. Srinivasan S, et al. *PLoS One*. 2013;8(10):e78633.
17. Holm JB, et al. *Front Cell Infect Microbiol*. 2020;10:117.
18. Chakravorty S, et al. *J Microbiol Methods*. 2007;69(2):330–9.
19. Budding AE, et al. *FASEB J*. 2010;24(11):4556–64.
20. Dols JA, et al. *BMC Infect Dis*. 2016;16:180.
21. The NIHHMPWG, et al. *Genome Res*. 2009;19(12):2317–23.
22. Ravel J, et al. *Proc Natl Acad Sci*. 2011;108(Supplement 1):4680–7.
23. Borgdorff H, et al. *PLoS One*. 2017;12(7):e0181135.
24. Callahan BJ, et al. *Proc Natl Acad Sci U S A*. 2017;114(37):9966–71.
25. Fettweis JM, et al. *Microbiology (Reading)*. 2014;160(Pt 10):2272–82.
26. Serrano MG, et al. *Nat Med*. 2019;25(6):1001–11.
27. Gajer P, et al. *Sci Transl Med*. 2012;4(132):132ra52.
28. Martin DH. *J Infect Dis*. 2016;214 Suppl 1:S36–41.

29. Brown RG, et al. *Transl Res.* 2019;207:30–43.
30. Brown SE, et al. *Sex Transm Dis.* 2019;46(11):753–8.
31. France MT, et al. *Microbiome.* 2020;8(1):166.
32. Dominguez-Bello MG, et al. *Proc Natl Acad Sci U S A.* 2010;107(26):11971–5.
33. Hammerschlag MR, et al. *Am J Obstet Gynecol.* 1978;131(8):853–6.
34. Hammerschlag MR, et al. *Pediatrics.* 1978;62(1):57–62.
35. Hickey RJ, et al. *MBio.* 2015;6(2):1–14.
36. MacIntyre DA, et al. *Sci Rep.* 2015;5:8988.
37. Romero R, et al. *Microbiome.* 2014;2(1):4.
38. Nott PN, et al. *Br J Psychiatry.* 1976;128:379–83.
39. DiGiulio DB, et al. *Proc Natl Acad Sci U S A.* 2015;112(35):11060–5.
40. Brotman RM, et al. *Menopause.* 2014;21(5):450–8.
41. Hummelen R, et al. *PLoS One.* 2011;6(11):e26602.
42. Muhleisen AL, et al. *Maturitas.* 2016;91:42–50.
43. Eschenbach DA, et al. *Clin Infect Dis.* 2000;30(6):901–7.
44. Srinivasan S, et al. *PLoS One.* 2010;5(4):e10197.
45. Song SD, et al. *mSphere.* 2020;5(4).
46. Bradshaw CS, et al. *Clin Infect Dis.* 2013;56(6):777–86.
47. Crucitti T, et al. *PLoS One.* 2018;13(7):e0201003.
48. van de Wijgert JH, et al. *AIDS.* 2013;27(13):2141–53.
49. Ahrens P, et al. *PLoS One.* 2020;15(7):e0236036.
50. Brown RG, et al. *BMC Med.* 2018;16(1):9.
51. Mayer BT, et al. *J Infect Dis.* 2015;212(5):793–802.
52. Stokholm J, et al. *Clin Microbiol Infect.* 2014;20(7):629–35.
53. Eschenbach DA, et al. *J Infect Dis.* 2001;183(6):913–8.
54. Plummer EL, et al. *Sci Rep.* 2019;9(1):19749.
55. Anderson DJ, et al. *Am J Reprod Immunol.* 2014;71(6):618–23.
56. Eckhart L, et al. *Biochim Biophys Acta.* 2013;1833(12):3471–80.
57. Muenzner P, et al. *Science.* 2010;329(5996):1197–201.
58. Vornhagen J, et al. *J Clin Invest.* 2018;128(5):1985–99.
59. Averette HE, et al. *Am J Obstet Gynecol.* 1970;108(1):8–17.
60. Cole A. *J Low Genit Tract Dis.* 2007;11(2):129–30.
61. Han L, et al. *Contraception.* 2017;96(5):310–21.
62. van Putten JPM, et al. *J Innate Immun.* 2017;9(3):281–99.
63. Kumar N, et al. *J Pharm Bioallied Sci.* 2011;3(4):496–503.
64. Becher N, et al. *Acta Obstet Gynecol Scand.* 2009;88(5):502–13.
65. Cortés ME, et al. *Linacre Q.* 2014;81(3):209–18.
66. Lai SK, et al. *Adv Drug Deliv Rev.* 2009;61(2):86–100.
67. Olmsted SS, et al. *Sex Transm Dis.* 2003;30(3):257–61.
68. Witkin SS, et al. *mBio.* 2013;4(4):1–7.
69. O'Hanlon DE, et al. *PLoS One.* 2013;8(11):e80074.
70. Velraeds MM, et al. *Appl Environ Microbiol.* 1996;62(6):1958–63.
71. He Y, et al. *Front Med (Lausanne).* 2020;7(284):1–8.
72. Macklaim JM, et al. *Proc Natl Acad Sci USA.* 2011;108(Suppl 1):4688–95.
73. Petrova MI, et al. *Trend Microbiol.* 2017;25(3):182–91.
74. Doerflinger SY, et al. *J Infect Dis.* 2014;209(12):1989–99.
75. Fichorova R, et al. *Biol Reprod.* 1999;60(2):508–14.
76. Herbst-Kralovetz M, et al. *Am J Reprod Immunol.* 2008;59(3):212–24.
77. Kaushic C, et al. *Methods.* 2011;55(2):114–21.
78. Eade CR, et al. *PLoS One.* 2012;7(11):e50106.
79. Libby EK, et al. *Microbes Infect.* 2008;10(4):439–46.
80. Kindinger LM, et al. *Sci Transl Med.* 2016;8(350):350ra102.
81. De Tomasi JB, et al. *J Immunol Res.* 2019;2019:7693183.
82. Boomsma CM, et al. *Reprod Biomed Online.* 2009;18(1):85–94.
83. Myer L, et al. *J Infect Dis.* 2005;192(8):1372–80.
84. Galvin SR, et al. *Nat Rev Microbiol.* 2004;2(1):33–42.
85. Lederman MM, et al. *Nat Rev Immunol.* 2006;6(5):371–82.
86. Antonio MA, et al. *J Infect Dis.* 2005;192(3):394–8.

87. Eschenbach DA, et al. *J Clin Microbiol.* 1989;27(2):251–6.
88. Ness RB, et al. *Obstet Gynecol.* 2004;104(4):761–9.
89. Wang S, et al. *Front Microbiol.* 2017;8:564.
90. Lewis FMT, et al. *Obstet Gynecol.* 2017;129(4):643–54.
91. van de Wijgert J. *PLoS Med.* 2017;14(12):e1002478.
92. Brotman RM. *J Clin Invest.* 2011;121(12):4610–7.
93. Mitra A, et al. *Microbiome.* 2016;4(1):58.
94. Norenhag J, et al. *BJOG.* 2020;127(2):171–80.
95. Mitra A, et al. *Nat Commun.* 2020;11(1):1999.
96. Klatt NR, et al. *Science.* 2017;356(6341):938–45.
97. Wee BA, et al. *Aust N Z J Obstet Gynaecol.* 2018;58(3):341–8.
98. Zhao C, et al. *mSystems.* 2020;5(4):e00450–20.
99. Kong Y, et al. *Front Med (Lausanne).* 2020;7(217).
100. Haahr T, et al. *BJOG.* 2019;126(2):200–7.
101. Haahr T, et al. *J Infect Dis.* 2018;219(11):1809–17.
102. Koedooder R, et al. *Hum Reprod.* 2019;34(6):1042–54.
103. Al-Memar M, et al. *BJOG.* 2020;127(2):264–74.
104. Zhang F, et al. *Exp Ther Med.* 2019;17(5):3307–16.
105. Hyman RW, et al. *J Assist Reprod Genet.* 2012;29(2):105–15.
106. Mangot-Bertrand J, et al. *Eur J Clin Microbiol Infect Dis.* 2013;32(4):535–41.
107. Haahr T, et al. *Hum Reprod.* 2016;31(4):795–803.
108. Vergaro P, et al. *Reprod Biomed Online.* 2018;38(6):883–91.
109. Fu M, et al. *mBio.* 2020;11(3):e03242–19.
110. Bayar E, et al. *Semin Immunopathol.* 2020;42(4):487–99.
111. Causes of Child Mortality. http://www.who.int/gho/child_health/mortality/causes/en/. (accessed 7 July 2016).
112. Donders GG, et al. *BJOG.* 2009;116(10):1315–24.
113. Flynn CA, et al. *J Fam Pract.* 1999;48(11):885–92.
114. McGregor JA, et al. *Am J Obstet Gynecol.* 1993;169(2 Pt 2):463–6.
115. Bennett PR, et al. *Obstet Gynecol Clin North Am.* 2020;47(4):503–21.
116. Kindinger LM, et al. *Microbiome.* 2017;5(1):6.
117. Stafford GP, et al. *Front Physiol.* 2017;8:615.
118. Hocevar K, et al. *Front Med (Lausanne).* 2019;6:201.
119. Fettweis JM, et al. *Nat Med.* 2019;25(6):1012–21.
120. Freitas AC, et al. *Microbiome.* 2018;6(1):117.
121. Tabatabaei N, et al. *BJOG.* 2019;126(3):349–58.
122. Romero R, et al. *Microbiome.* 2014;2:18.
123. Stout MJ, et al. *Am J Obstet Gynecol.* 2017;217(3):1–18.
124. Nelson DB, et al. *Am J Perinatol.* 2016;33(9):887–93.
125. Brown RG, et al. *BJOG.* 2019;126(7):916–25.
126. Brown RG, et al. *BMC Med.* 2018;16(1):9.
127. Parry S, et al. *New Eng J Med.* 1998;338(10):663–70.
128. Rocha G, et al. *J Matern Fetal Neonatal Med.* 2007;20(10):745–9.
129. Lu H, et al. *J Stroke Cerebrovasc Dis.* 2016;25(4):807–12.
130. Vigneswaran R. *J Paediatr Child Health.* 2000;36(4):293–6.
131. Yoon BH, et al. *Am J Obstet Gynecol.* 2000;182(3):675–81.
132. Drassinower D, et al. *J Matern Fetal Neonatal Med.* 2015;29(17):48–52.
133. van Dillen J, et al. *Curr Opin Infect Dis.* 2010;23(3):249–54.
134. Puri K, et al. *PLoS One.* 2016;11(9):e0162734.
135. Pruski P, et al. *Methods.* 2018;149:74–84.

The Uterine Microbiota

21

Jonah Bardos, Carlos Simón, and Inmaculada Moreno

INTRODUCTION

The human body is colonized with over ten times more bacteria than the number of human cells.[1] Most of the earlier medical research on micro-organisms was focused on disease-causing bacteria and only recently has there been a coordinated focus on studying the resident bacteria, viruses, and fungi, collectively called the microbiome. The outcome of the host–microbe interaction can be one of three options: mutually beneficial, commensal, or parasitic. In 2008, the NIH studied the microbiome in nasal passages, the oral cavity, skin, gastrointestinal (GI) tract, and the urogenital tract in 300 healthy volunteers.[2] Prior to the last decade, few studies had focused on uterine microbiome as it was originally thought to be a sterile environment. An astounding 9% of the total human microbiota is found in the urogenital tract.[3]

A perfect relationship between the uterus and the embryo, mediated by a tightly controlled interaction between the embryo and the endometrium, is required for successful implantation. With almost 140 million births annually worldwide, it is important to continue researching and discovering key factors of healthy pregnancies. Any factors affecting this communication, such as an altered microbiome, may lead to poor reproductive outcomes.

Over the past five years, there has been a new focus on analyzing the uterine microbiome after a pilot study in 2016 showed that women with a pathogenic (defined as non-Lactobacillus dominated, NLD) uterine environment had an almost 40% decrease in pregnancy rates.[4] Another study in 2017 utilized targeted sequencing and microarray data focused on the hypervariable regions of 16S rRNA and showed that the uterine microbiome at the time of embryo transfer during IVF can be characterized by sequencing the tip of the transfer catheter from the embryo transfer.[5] Since then, many authors have been attempting to characterize the natural microbiome of the uterus to determine what effect the uterine microbiome has on pregnancy outcomes.[6] Studies have shown conflicting results regarding the "normal" uterine microbiome; however, it does appear that *Lactobacillus* is consistently present among the physiological endometrial microbiome.[7]

Early reproductive microbiome work, prior to the next-generation sequencing boom, involved culture-based approaches.[8] These approaches failed to identify many organisms. With the advent of the ease and reduced cost of next-generation sequencing, it was discovered that only about 1–2% of the bacteria present had been identified in culture or by microscopy.[4,9,10] More recent reproductive studies utilized 16S ribosomal RNA (rRNA) hypervariable region gene sequencing, which can determine the genus and species level of bacteria, but not sub-speciation.[11] Advantages of 16S rRNA studies include decreased cost and the requirement of smaller amounts of DNA as amplification of bacterial DNA is performed during the process. Amplification requirement can also be a disadvantage as the process can also introduce an

DOI: 10.1201/9781003024941-21

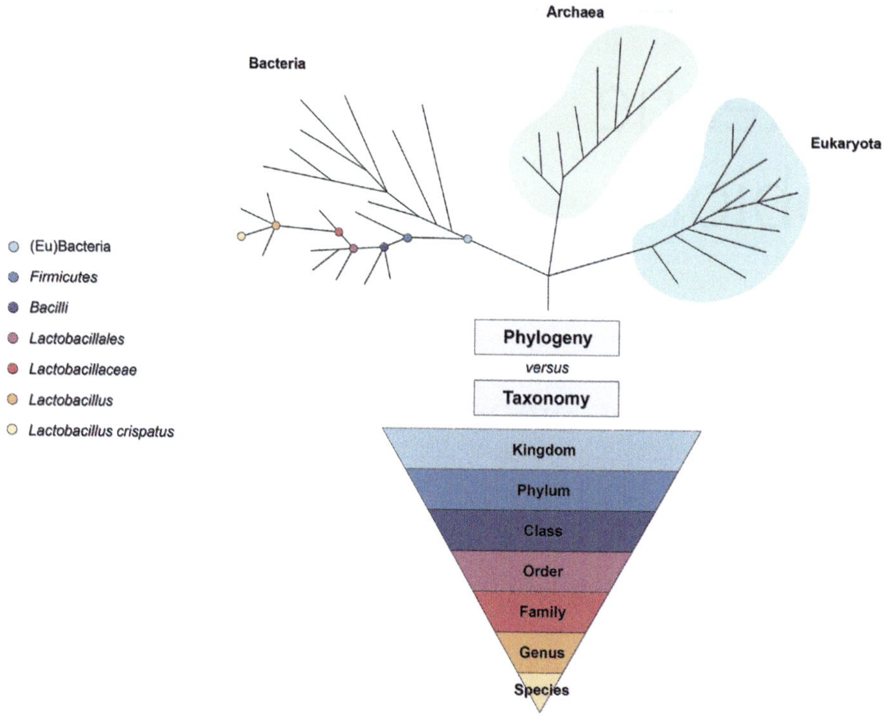

FIGURE 21.1 Phylogeny vs. taxonomy and examples of taxonomic naming systems for reference use throughout this chapter.

amplification bias when dealing with only one/two hypervariable regions, making it harder to accurately determine the relative abundance of the bacterial species. Additionally, 16S technology utilizes a mapping database, meaning that the species must have previously been characterized by others. When compared with 16S-based sequencing, whole-genome shotgun metagenomics can help with the identification of lower taxonomic resolution, meaning detecting low-abundance microbial communities, and can better differentiate between closely related species.[12] Essentially, previous work with cultures and 16S was potentially missing key components of the reproductive tract microbiome. Additionally, microbiome presence alone does not definitively determine dysbiosis, and functional studies explaining how the particular bacterial profile affects the local signaling are needed. However, there are few studies published to date on shotgun sequencing of the endometrium and showing the functional impact of the endometrial microbiota at the time of embryo transfer or during early pregnancy in IVF patients.[13,14]

In this chapter, we utilize different taxonomical terms, referring to phyla, genera, and species. Taxonomy is the science of naming and classifying organisms, including plants, animals, and microorganisms. It is useful for the reader to refer to Figure 21.1 as a reference of taxonomical terminology.

NORMAL REPRODUCTIVE TRACT MICROBIOME

Endometrial Microbiome

For many decades, the uterus was thought to be a sterile environment, with bacterial colonization occurring only during an infection or a pathological process. Despite being adjacent to the bacterially colonized vagina, it was thought that the cervical mucus maintained uterine sterility. In 1996, Egbase et al. challenged this dogma and demonstrated that the reproductive tract microbiota can have an effect on IVF outcomes.[15] Current knowledge suggests that the uterus indeed harbors its own microbiome. In 2016, Franasiak et al. were the first to measure uterine microbiome from catheter tips at the time of an embryo transfer.[12] As with other parts of the body, it is important to determine if there is a healthy, "normal," uterine microbiome. Over the past 15 years, many studies have examined uterine microbiota, and over 60 have looked at the reproductive tract microbiome, although most studies to date have involved women with pathology.[7,16] Prior to 2014, most studies had attempted to use culture techniques, which have subsequently been shown to ignore the majority of the pathogens present, as only bacteria whose metabolic needs are met in culture will grow.[17] Additionally, highly abundant and fast-growing bacteria will dominate and suppress others.[18] Thus, it was not until the expansion of next-generation sequencing that a better picture of the reproductive microbiome was possible.

In studies of healthy women, the most consistent phyla in endometrial samples have been Firmicutes, Bacteroidetes, Proteobacteria, and Actinobacteria.[4–6,12,19–23] The most common genera found in multiple studies have been *Lactobacillus* and *Streptococcus*, both of which can be found in the vagina and the cervix.[7]

Uterine microbial transmission is not clearly established, leading to the question of where the uterine microbiome comes from (Figure 21.2). There are currently a number of theories. The principal theory is ascension from the vagina. While there is a known cervical plug that does protect the uterine environment, during intercourse, semen is able to enter the uterus through small channels in the cervical mucus. Studies have shown evidence of a uterine pump moving radiotagged isotopes from the vagina into the uterus within 15 minutes of intercourse.[24] Other possible methods include hematogenous spread from the gut and transmembrane gut leakage into the peritoneal cavity with retrograde ascension *via* the fallopian tubes. Dendritic cells and leukocytes traffic bacteria found in the gut and can hematogenously spread bacteria to other locations, such as the uterus.[25] One study showed that, when genetically labeled *Enterococcus faecium* was placed in the oral cavity of a mouse, it could be detected in the placenta.[26] Given these possible origins of the uterine microbiome, it is important to understand the microbiome of various anatomical locations that may seed the endometrium (Figure 21.3).

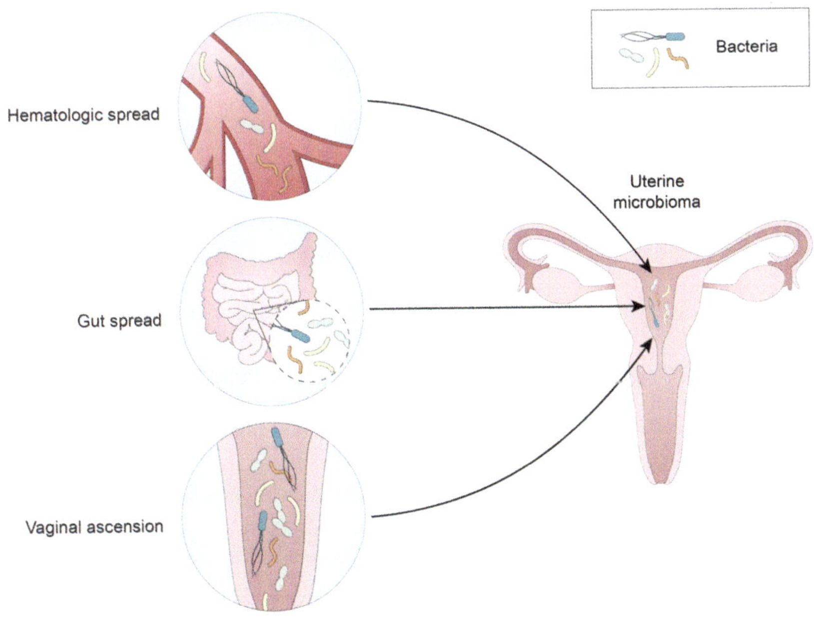

FIGURE 21.2 Three possible theories of where the uterine microbiome originates: hematologic spread, gut spread, and vaginal ascension.

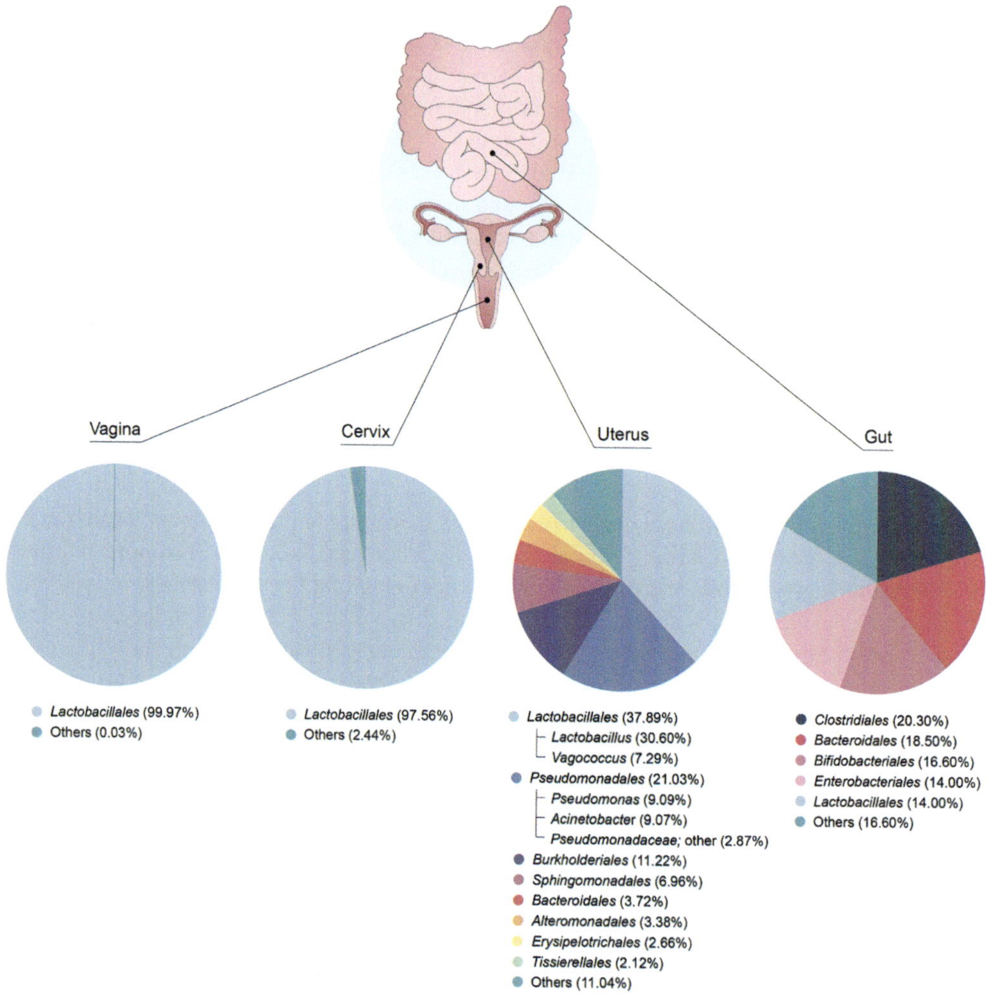

FIGURE 21.3 Order level breakdown of various bacteria found in the possible locations where the uterine microbiome originates; genus level is broken down where appropriate.

Vaginal Microbiome

A leading theory for the colonization of the uterine cavity is bacterial ascension from the vagina to the uterine cavity, so it is important to review the vaginal microbiome. Over the past decade, studies involving both 16S-sequencing and metagenomics have examined the vaginal microbiome which harbors predominantly *Lactobacillus* spp. in high concentrations.[6,27,28] The exact reason for the presence of native *Lactobacillus* has not been elucidated, although there are a few potential benefits. *Lactobacillus* is known to secrete lactic acid, creating a low pH. This low pH is protective against cervico-vaginal infections.[29,30] There are other beneficial aspects of the native microbiome, including competitive exclusion. Competitive exclusion is where the native microbiome adapts to be the best nutrient scavenger in that environment, competing with potential invaders for nutrients and, in turn, starving other bacteria, including pathogens.

The literature describes five major types of vaginal microbiota or community state types (CST) in reproductive-age women. CST I, II, III, and V are all predominantly *Lactobacillus*. CST I appears to be more common in Caucasian women and is protective against bacterial vaginosis (BV). CST IV, more common in African-American and Hispanic women, is predominantly a mixture of strict and facultative anaerobes including *Gardnerella*, *Atopobium*, *Mobiluncus*, and *Prevotella*.[31,32] CST IV-B, a subtype of CST IV, is associated with BV.[33,34]

RELEVANCE OF THE PERI-CONCEPTIONAL NICHE FOR PREGNANCY OUTCOMES

The immune environment during the peri-conceptional period is controlled by many factors, including cytokines, chemokines, and leukocyte lineages. These factors are intertwined with ovarian steroid hormones, seminal fluid, diet, nutrition, metabolism, obesity, infections, and the microbiome to influence the balance between embryotrophic and embryotoxic milieus and ultimately between a successful or unsuccessful pregnancy.

Sex steroids, essential for endometrial lining preparation, play an important role in modifying the immune response. In the proliferative phase, estrogen stimulates proliferation and differentiation of endometrial cells as well as pro-inflammatory cytokines such as GM-CSF and TNF. During the luteal phase, progesterone suppresses the pro-inflammatory cytokines by downregulation of expression of the genes *GM-CSF 1* and *IL1*. Progesterone also enhances the expression of *IL8* and attracts uterine natural killer (NK) cells.[35] One theory in patients with endometriosis suggests that the endometrial tissue may have progesterone resistance resulting in increased inflammation and poor pregnancy outcomes.[36] Another theory proposes that a dysbiotic gut microbiota may be a potential cause of infertility by negatively impacting the estrobolome which, in turn, alters the levels of deconjugated estrogen and increases local inflammation[37] Excess inflammatory cytokines such as TNF, during the implantation period, can act to skew the adaptive immune response toward cytotoxicity.[38] Excess inflammatory factors are found during a local infection or times of nutritional or metabolic stress. This inflammatory response is normally beneficial, acting to suppress pregnancy during unfavorable conditions. However, this inflammation can lead to failed implantation or pregnancy loss. Excessive inflammation may also be found when the host microbiome is altered, leading to negative effects on implantation and pregnancy outcomes.

Embryo Signaling and Implantation

While researchers still do not fully understand how implantation works, many aspects have been elucidated, including hormonal control and the local role of additional cytokine factors. Changes in the natural microbiome may affect the local cytokine/chemokine profile affecting embryo implantation. Approximately 6–7 days after fertilization, the blastocyst interacts with the uterine wall in a process called apposition. The interaction between the uterine cell wall and the embryo activates cytokine signaling and initiates remodeling of the epithelial cell layer. Matrix metalloproteinases, important for trophoblast invasion, are regulated by cytokines and downregulated by tissue inhibitors of metalloproteinases (TIMPs).[39] TIMPs are stimulated by TGFβ, a signaling molecule of colonizing microbes. This is another example of how the local microbiome can affect reproductive outcomes. Implantation is the result of an intricate bi-directional interaction between the embryo and the endometrium. This interaction is mediated by a host of factors regulating the invading cells, from cytokines and growth factors to steroid hormones and proteinases.[40] Abnormal placentation can have clinical implications due to either excessive or inadequate invasion. Excessive invasion can lead to abnormally strong attachment such as an accreta, increta, or percreta, depending on the depth of invasion. Inadequate invasion has been implicated in pre-eclampsia, and intrauterine growth restriction (IUGR).[41,42] Embryos display cytokine receptors from conception until implantation, which play a key role in embryo development, gene expression, embryo competence, and viability.[43,44] Embryotoxic cytokines, such as TNF, IFNG, and TRAIL, are stimulated during inflammation or local infections *via* Toll-like receptors and inhibit embryo development.[45] The delicate cytokine balance, controlled by many factors, including the local microbiome, may play an important role in reproductive outcomes.

How Microbiome Affects the Body: Implications for the Uterus

Researchers have examined germ-free mice and discovered that, when mice lack bacteria, there are profound effects on mouse lymphoid tissue. These studies suggest that innate bacteria are important for the development of immunity.[46] The majority of immune response–bacterium interactions occurs in the gut and the GI tract has become the most studied area of the human microbiome.[47] The gut contains the most abundant population of micro-organisms with more than 5,000 bacteria taxa.[48] The gut microbiome is colonized at birth and once a child is about 2.5 years old their microbiome remains relatively stable until age 65.[49] Although the gut microbiome is stable, it can be altered by various host factors such as antibiotics, travel, or high-fat diets.[50] In contrast, the uterine microbiome undergoes changes with menses.

Dysbiosis in the gut has been linked to many diseases, including liver disease, GI cancer, metabolic disease, respiratory disease, mental health disorders, and autoimmune disorders.[51] Studies suggest that alterations in the local gut microbiome increase immune system sensitivity to pathogens, and cause increased local inflammation.[52] The most potent causes of dysbiosis are gut pathogens. Dysbiosis of gut microbiota has profound effects on the T cell-mediated autoimmune and inflammatory responses, as the gut microbiota is known to provide signaling for proper development, differentiation, and epigenetic influences of immune cells. Additionally, the gut microbiota can alter the local chemokine profile, causing inflammatory responses.[53]

In addition, a dysbiotic gut microbiota may be a potential cause of infertility by negatively impacting the estrobolome, which in turn alters the levels of deconjugated estrogen and increases local inflammation.[43]

Another theory of how local changes in the microbiome affect disease state is through a concept called competitive exclusion. Host bacteria ("normal microbiome") have the chance to adapt to their environment. The adaptation process allows them to specialize in utilization of locally available nutrients. This adaptation depletes nutrients that would otherwise be available for invading pathogens, through a process called colonization resistance. The theory of competitive exclusion could support the potential importance of a healthy uterine microbiome.[54] Additionally, symbiotic bacteria may also compete for the cell's receptors, thereby blocking access to those receptors for invading pathogens. One study utilizing an *in vitro* model found that the presence of lactobacilli in the reproductive tract prevented *Neisseria gonorrhoeae*, the bacterium causing gonorrhea, from attaching to endometrial cells.[55]

The host microbiome may also help to support a healthy and intact epithelial barrier, preventing pathogen access to the cell. Gut microbiome studies suggest that the microbiome can affect epithelial cell differentiation and maintenance, and can modulate epithelial cell permeability.[56,57] Given that the uterus frequently undergoes shedding and regrowth, microbial support could play an important role in maintenance of uterine epithelial tissue. Conversely, dysbiosis could be a cause of thin endometrial lining and abnormal placentation.

ABNORMAL REPRODUCTIVE TRACT MICROBIOME

Studies have found that different gynecological conditions are associated with different microbiota. Infertility has been associated with increased abundance of *Atopobium*, *Pelomonas*, and *Sneathia*, among many other genera.[58] Chronic endometritis (CE) is typically defined as a chronic inflammation of the uterine lining and is associated with the presence of plasma cells following endometrial biopsy.[59,60] Multiple studies have suggested that chronic endometritis is associated with recurrent pregnancy loss.[61,62]

Numerous microbes have been found in patients with CE, including *Neisseria gonorrhea*, *Gardnerella vaginalis*, *Chlamydia trachomatis*, *Escherichia coli*, *Streptococcus* spp., *Staphylococcus* spp., and *Enterococcus faecalis* and non-microbial causes such as retained tissue. Additionally, endometriosis has been hypothesized to alter the endometrium through increased inflammation and progesterone resistance which can affect implantation, increase the risk of miscarriage, and result in poor pregnancy outcomes, including pregnancy-induced hypertension and preterm birth.[63] It would appear that, regardless of the cause, increased inflammation at the endometrial level may affect implantation and pregnancy outcomes.

Alterations of the vaginal microbiota may also be associated with various pregnancy outcomes (Table 21.1). During pregnancy, there is a consistently elevated level of estrogen associated with an increase in vaginal *Lactobacillus* dominance.[64] Several studies have suggested that altered microbiota are associated with preterm birth,[65–67] preterm premature rupture of membranes (PPROM), chorioamnionitis, and early or late miscarriages.[68–73] One study of the vaginal microbiome on the day of embryo transfer suggested that lower bacterial diversity was associated with better IVF outcomes. A recent study suggested that higher frequencies of vaginal *Lactobacillus crispatus* are associated with higher chance of pregnancy when utilizing intracytoplasmic sperm injection (ICSI).[74] Neither of these studies included sampling of the upper reproductive tract, so it is hard to draw conclusive results.[75] Another small study in 2020, studying both the vaginal and uterine microbiomes, suggested that a Lactobacillus dominant vaginal environment was associated with improved IVF outcomes. That study also found that *Kocuria dechangensis*, a Gram-positive aerobic bacterium in the vagina, is associated with poor reproductive outcomes.[76] The literature appears to support the concept that alterations in the vaginal microbiome are associated with various poor outcomes and that examining the vaginal microbiome, along with the upper reproductive tract microbiome, at the time of an IVF transfer may shed additional information into the causes of these poor outcomes (Table 21.1).

Effects of Microbiome on Uterine Pathophysiology

Based on mouse and human gut studies, there are likely four ways in which local microbiome alterations can affect the clinical sequelae of pregnancy (Figure 21.4).[18] First, an abnormal microbiome can cause alterations in regional signaling pathways by initiating an abnormal local inflammatory response through the uterine Toll-like receptors. When Toll-like receptors are activated, they can alter the cytokine milieu, swaying the local response to a pro-inflammatory and anti-tolerance immune cell response. Second, as mentioned previously, a changed microbiome can result in an alteration of the endometrial epithelial barrier integrity.[77] Certain pathogens cause a localized decrease in matrix degradation proteins, which may affect placentation. Third, the effect could occur *via* competitive advantage. The host microbiome has adapted to be the best nutrient scavenger in that area and can starve out invading species *via* competitive exclusion[78] The above three theories suggest that the native microbiome acts in a protective manner, blocking invading pathogens. Fourth, host microbes can secrete metabolites, such as short-chain fatty acids, that suppress the growth of certain invading species.[30] Taken together, the natural microbiome can have an impressive effect on the local interactions between the embryo and the endometrium with implications for implantation, placentation, and embryonic growth. Host microbiome therefore can ultimately play an important role in clinical sequelae of a resulting pregnancy and pregnancy outcomes (Table 21.2).

Landmark studies by Moreno and collaborators demonstrated that the presence of pathogens in the endometrium during the window of implantation was associated with poor reproductive outcomes, whereas Lactobacillus dominance conferred a protective benefit, resulting in increasing implantation rates.[4,79] However, given that 16S rRNA methodology only allows for taxonomic identification at the genus level, it is unclear whether certain species or subspecies of *Lactobacillus* may be capable of conferring this benefit, or whether the presence of any type of *Lactobacillus* prevents the growth of pathogenic bacteria, leading to a normal or healthy microbiota. A recent study on the endometrial microbiota and chronic endometritis reported that *Lactobacillus crispatus* was less abundant in patients with CE, suggesting that there may be certain *Lactobacillus* spp. that are protective.[80] Another study in 2019 by Garcia-Grau et al.

TABLE 21.1 Studies Showing Association of the Cervicovaginal Microbiota with Infertility and Pregnancy Outcomes

POPULATION	*SAMPLE*	*RESULTS*	*REFERENCE*
Women undergoing IVF	Vaginal swabs	Abnormal vaginal microbiota with *Gardnerella vaginalis* and *Atopobium vaginae* negatively associated with clinical pregnancy outcomes in IVF patients.	84
Women with idiopathic infertility, diagnosed infertility, BV and controls	Cervicovaginal fluid	Infertile women presented increased BV-associated microbiota profile (decreased *Lactobacillus* abundance and increased *G. vaginalis*, *A. vaginae*, and *Ureaplasma parvum*)	85
Women undergoing infertility treatment	Vaginal swabs	Infertile patients presented increased prevalence of asymptomatic vaginosis with higher abundance of BV-associated bacteria (decreased *Lactobacillus* and increased *Enterococcus* spp., *Escherichia coli*, *Candida* spp.) than healthy controls.	86
ART patients	Vaginal, cervical, and endometrial biopsies	Bacterial communities were generally consistent across the reproductive tract. Vaginal *Ureaplasma* and cervical *Gardnerella* are positively associated with infertility	87
Infertile women with indication of ART	Vaginal swabs	The vaginal microbiome can influence the results of ART. The stability of the microbiota and *Lactobacillus* dominance are associated with pregnancy.	88
Primary idiopathic infertility	Vaginal swabs	Microbiota was different between infertile patients and healthy controls. *Lactobacillus crispatus* associated with successful implantation.	89
Women undergoing IVF	Vaginal swabs	Abnormal vaginal microbiota determined by q-RT-PCR and NGS predict the reproductive outcome.	90
Women with RIF	Vaginal swabs	*Lactobacillus* dominance was positive correlated with pregnancy rate. RIF patients presented higher abundance of *Streptococcus*, *Atopobium*, *Prevotella*, and *Bifidobacterium* spp. than the controls.	91
Women undergoing IVF	Vaginal aspirate	Decreased *Lactobacillus* abundance and increased *Gardnerella* and *Prevotella* spp. associated with no pregnancy outcome after IVF.	92
Infertile patients with implantation failure	Vaginal aspirate	*Lactobacillus* dominance associated with pregnancy. *G. vaginalis*, *Escherichia fergusonii*, and *Streptococcus dentisani* associated with failed cycles.	88
Women with secondary infertility	Vaginal swabs	Infertile women presented increased abundance of *Atopobium*, *Aerococcus*, and *Bifidobacterium* spp., and decreased abundance of *Lactobacillus* and *Leuconostoc*.	93

Abbreviations: ART, assisted reproductive technologies; BV, bacterial vaginosis; IVF, *in vitro* fertilization; NGS, next-generation sequencing; RIF, repeated implantation failure; q-RT-PCR, quantitative real-time polymerase chain reaction.

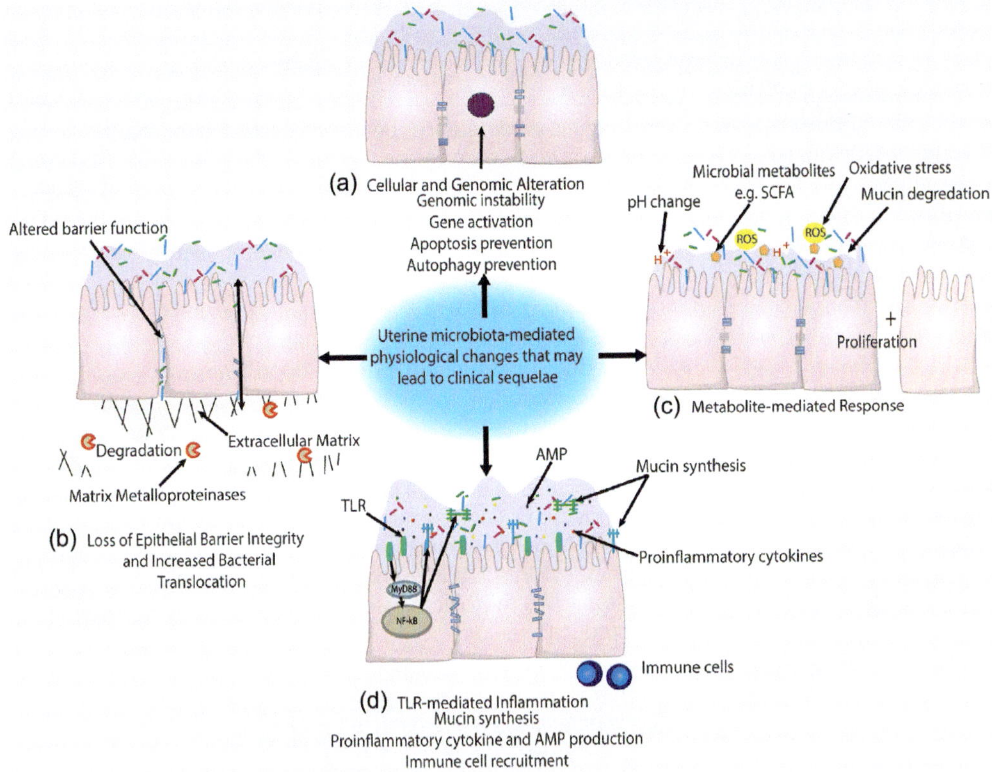

FIGURE 21.4 Potential impact of a dysbiotic microbiota on the endometrial physiological niche and uterine function. (Reprinted from Baker JM, Chase DM and Herbst-Kralovetz MM, Uterine Microbiota: Residents, Tourists, or Invaders? *Frontiers in Immunology* 2018;9:208. Copyright (2018), with permission from Creative Commons.)

found *Gardnerella vaginalis* causing uterine dysbiosis after repeated sampling over an 18-month period in a patient with recurrent reproductive failure.

Recently, a study by Moreno et al. looked at the endometrial microbiota in early pregnancy. The study looked at one patient who had had a prior miscarriage of a euploid embryo at eight weeks. In a subsequent pregnancy, the researchers compared the microbiome composition between the ongoing pregnancy and the miscarriage. They found that the microbial profile prior to the miscarriage had greater bacterial diversity and lower *Lactobacillus* abundance than the endometrial fluid from the healthy pregnancy. They then did functional metagenomics which detected different *Lactobacillus* species between the two samples. In this case, an unstable functional pattern, characterized by transposases and insertion elements, were seen prior to the miscarriage, with only a small proportion of the functional genes observed in the sample belonging to *L. crispatus*. *Lactobacillus iners* was the most prevalent microbe found in the endometrium during early pregnancy after a dilation and curettage, suggesting, in this small study, that the presence of this species could protect against growth of other dysbiotic pathogens.[14]

Potential Therapies for Endometrial Dysbiosis

Over 20 years ago, researchers suggested treating possible uterine microbiome alterations with antibiotics.[81] Additionally, studies have suggested that identifying and treating chronic endometritis with antibiotics can improve reproductive outcomes in patients with recurrent implantation failure.[59,82] Several studies have looked at using an oral probiotic, specifically various *Lactobacillus* isolates, to alter the composition of the reproductive tract microbiome. The outcome of a case report mentioned earlier showed that this

TABLE 21.2 Studies Showing Association of the Endometrial Microbiota with Infertility and Pregnancy Outcomes

POPULATION	*SAMPLE*	*RESULTS*	*REFERENCE*
Women undergoing IVF	Embryo transfer catheter	No significant differences between microbiota in ongoing and non-ongoing pregnancies.	12
Women with RIF	Endometrial fluid	Endometrial microbiota not dominated by *Lactobacillus* is associated with failed reproductive outcomes.	4, 79
Women undergoing IVF	Endometrial fluid	Endometrial *Lactobacillus* is significantly lower in IVF patients than in controls. Abundance of pathogens such as *Atopobium, Bifidobacterium, Gardnerella, Megasphaera, Sneathia, Prevotella, Staphylococcus*, and *Streptococcus* associated with IVF patients.	94
Women undergoing IVF	Endometrial fluid	The endometrial bacterial profiles of pregnant cases with dysbiotic endometrium were comparable with those of non-pregnant cases. Species-level resolution may help identifying non-*Lactobacillus* microbiota associated with reproductive failure.	95
Women with RIF	Endometrial fluid and vaginal swabs	EF microbiota showed significant variation in community composition in RIF vs. control women. *Burkholderia* spp. were detected only in EF of RIF women, were not detected in the EF microbiota of any samples in the control group but were detectable in one-quarter of the RIF group.	96
Women undergoing IVF	Endometrial fluid	*Lactobacillus* dominance is associated with successful pregnancy. Patients with non-*Lactobacillus*-dominant microbiota achieve pregnancy after antimicrobial therapy.	91

Abbreviations: EF, endometrial fluid; IVF, *in vitro* fertilization; RIF, repeated implantation failure.

concept may be another tool for patients facing recurrent implantation failure. Another possible supplementation includes prebiotics. Prebiotics, such as lactoferrin, can help an existing healthy microbiome sustain growth. One study looked at lactoferrin after antibiotic therapy and showed that it can help restore a *Lactobacillus*-dominant uterine environment.[83] It is important to note that these are emerging areas of preliminary research and that further studies are warranted to pave the way for personalized treatment of endometrial dysbiosis within assisted reproductive technologies (ART).

CONCLUSIONS AND FUTURE AREAS OF RESEARCH

Current data have yet to support the characterization of a normal uterine microbiome. However, it is hard to believe that microbes cannot reach the the endometrial mucosa, as this is in close proximity to the vagina that presents a well-characterized microbiota and the bacteria could penetrate the uterine cavity through the cervix, as it is the case of sperm. It is still unclear whether those microbes are transient and whether a host microbiome controls that transience. Studies to date have been limited by the use only of 16S rRNA gene sequencing, potentially missing any species not currently in the bioinformatics database.

It is highly likely that a normal endometrial microbiome does exist and, given what we know about gut microbiomes, it probably plays a role in immunoregulation, endometrial remodeling, pregnancy implantation, and placentation. Future research should include utilizing shotgun metagenomics for a better understanding of all the pathogens at play. As the cost of NGS decreases, a focus on multi-omics techniques, including meta-transcriptomics, metabolomics, and proteomics, will help to elucidate the true effect of the microbiomes on the uterine microenvironment.

REFERENCES

1. NIH Hmp Working Group, et al. *Genome Res.* 2009;19(12):2317–23.
2. NIH Human Microbiome Project. *About the Human Microbiome.* Accessed November 08, 2018. https://hmp-dacc.org/hmp/overview/.
3. Sirota I, et al. *Semin Reprod Med.* 2014;32(1):35–42.
4. Moreno I, et al. *Am J Obstet Gynecol.* 2016;215(6):684–703.
5. Tao X, et al. *Hum Microb J.* 2017;3:15–21.
6. Chen C, et al. *Nat Commun.* 2017;8:875.
7. Koedooder R, et al. *Hum Reprod Update.* 2019;25:298–325.
8. Selman H, et al. *J Assist Reprod Genet.* 2007;24(9):395–9.
9. Verhelst R, et al. *BMC Microbiol.* 2004;4:16.
10. Wade W. *J Royal Soc Med.* 2002;95: 81–3.
11. Franasiak JM, et al. *J Assist Reprod Genet.* 2016;33(1):129–36.
12. Li F, et al. *Gigascience.* 2018;7(10):Giy107.
13. Moreno I, et al. *Am J Obstet Gynecol.* 2020;222(4):296–305.
14. Garcia-Grau I, et al. *Pathogens.* 2019;8(4):205.
15. Egbase PE, et al. *Hum Reprod.* 1996;11(8):1687–9.
16. Baker JM, et al. *Front Immunol.* 2018;9:1–16.
17. Nadkarni, et al. *FEMS Microbiol Lett.* 2009;296:45–51.
18. Hiergeist, et al. *ILAR J.* 2015;56:228–40.
19. Fang RL, et al. *Am J Transl Res.* 2016;8(3):1581–92.
20. Verstraelen H, et al. *Peerj.* 2016;4:E1602.
21. Walther-António MR, et al. *Genome Med.* 2016;8(1):122.
22. Miles SM, et al. *Fertil Steril.* 2017;107(3):813.E–20.E.
23. Khan KN, et al. *Eur J Obstet Gynecol Reprod Biol.* 2016;199:69–75.
24. Kunz G, et al. *Adv Exp Med Biol.* 1997;424:267–77.
25. Donnet-Hughes A, et al. *Proc Nutr Soc.* 2010;69:407–15.
26. Jiménez E, et al. *Res Microbiol.* 2008;159(3):187–9.
27. Srinivasan S, et al. *PLoS One.* 2010;5(4):e10197.
28. Kroon S, et al. *Fertil Steril.* 2018;110(3):327–36.
29. Ronnqvist PD, et al. *Acta Obstet Gynecol Scand.* 2006;85:726–35.
30. Aldunate M, et al. *Front Physiol.* 2015;6:164.
31. Ravel J, et al. *Proc Natl Acad Sci USA.* 2011;108:4680–7.
32. Gajer P, et al. *Sci Transl Med.* 2012;4:132ra52.
33. Borgdorff H, et al. *Sex Transm Dis.* 2015;42:143–52.
34. Marrazzo JM, et al. *Plos One.* 2010;5:E11139.
35. Robertson SA, et al. *Int J Dev Biol.* 2014;58:205–17.
36. Vannuccini S, et al. *Hum Reprod Update.* 2016;22:104–15.
37. Baker JM, et al. *Maturitas.* 2017;103:45–53.
38. Moldenhauer LM, et al. *Biol Reprod.* 2009;81(1):86.
39. Raga F, et al. *J Clin Endocrinol Metab.* 1999;84:636.
40. Pagidas K, et al. In: Farquharson and Stephenson, eds, *Early Pregnancy.* Cambridge University Press; 2017:27–42.
41. Norwitz ER. *Reprod Biomed Online.* 2006;13(4):591–9.
42. Krebs C, et al. *AJOG.* 1996;175(6):1534–42.

43. Sharkey AM, et al. *Biol Reprod.* 1995;53:974–81.
44. Guzeloglu-Kayisli O, et al. *Semin Reprod Med.* 2009;27(1):62–79.
45. Chin PY, et al. *Reproduction Fertility and Development* 2010;22(9):90.
46. Hooper LV, et al. *Science.* 2012;336:1268–73.
47. Brown EM, et al. *Ann Rev Immunol.* 2019;37.
48. Thursby E, et al. *Biochem J.* 2017;474:1823–36.
49. Grice EA, et al. *Annu Rev Genomics Hum Genet.* 2012;13:151–70.
50. David LA, et al. *Genome Biol.* 2014;15:R89.
51. Wang B, et al. *Engineering.* 2017;3:71–82.
52. Sochocka M, et al. *Mol Neurobiol.* 2019;56(3):1841–51.
53. Hooper LV, et al. *Science.* 2012;336:1268–73.
54. Kamada N, et al. *Nat Immunol.* 2013;14:685–90.
55. Spurbeck RR, et al. *Infect Immun.* 2008;76:3124–30.
56. Cario E, et al. *Gastroenterology.* 2007;132:1359–74.
57. Yu Y, et al. *Am J Physiol Gastrointest Liver Physiol.* 2016;311:G521–32.
58. Molina NM, et al. *Biomolecules.* 2020;10(4):593.
59. McQueen DB, et al. *Fertil Steril.* 2014;101(4):1026–30.
60. Maria M, et al. *Am J Reprod Immunol.* 2009;61:322–9.
61. D'Ippolito S, et al. *Am J Reprod Immunol.* 2018;80:e13065.
62. McQueen DB, et al. *Fertil Steril.* 2015;104(4):927–31.
63. Farland LV, et al. *Obstet Gynecol.* 2019;134(3):527–36.
64. Digiulio DB, et al. *Proc Natl Acad Sci U S A.* 2015;112:11060–5.
65. Elovitz MA, et al. *Nat Commun.* 2019;10:1305.
66. Hyman RW, et al. *Reprod Sci.* 2014;21:32–40.
67. Stout MJ, et al. *Am J Obstet Gynecol.* 2017;217:356.E1–18.
68. Donders GG, et al. *Am J Obstet Gynecol.* 2000;183:4317.
69. Hay PE, et al. *BMJ.* 1994;308:295–8.
70. Llahi-Camp JM, et al. *Hum Reprod.* 1996;11:1575–8.
71. Ralph SG, et al. *Am J Obstet Gynecol.* 1993;169:460–2.
72. Hillier Sl, et al. *NEJM.* 1995;333(26):1737–42.
73. Paramel Jayaprakash T, et al. *Plos One.* 2016;11:E0166794.
74. Koedooder R, et al. *Hum Reprod.* 2019;34(6):1042–54.
75. Hyman RW, et al. *JARG.* 2012;29:105–15.
76. Riganelli L, et al. *Front Cell Infect Microbiol.* 2020;10:350.
77. Hooper LV, et al. *Science.* 2012;336:1268–73.
78. Kamada N, et al. *Nat Immunol.* 2013;14:685–90.
79. Moreno I, et al. *Microbiome.* 2022;10(1):1.
80. Liu Y, et al. *Fertil Steril.* 2019;112(4):707–717.e1.
81. Egbase PE, et al. *Lancet.* 1999;354(9179):651–2.
82. Cicinelli E, et al. *Hum Reprod.* 2015;30(2):323–30.
83. Kyono K, et al. *Reprod Med Biol.* 2018;18(1):72–82.
84. Haahr T, et al. *Hum Reprod.* 2016;31(4):795–803.
85. Campisciano G, et al. *J Cell Physiol.* 2017;232(7):1681–8.
86. Babu G, et al. *J Clin Diagn Res.* 2017;11(8):DC18–22.
87. Wee BA, et al. *Aust N Z J Obstet Gynaecol.* 2018;58(3):341–8.
88. Bernabeu A, et al. *J Assist Reprod Genet.* 2019;36(10):2111–9.
89. Amato V, et al. *Open Forum Infect Dis.* 2019;7(1):ofz525.
90. Haahr T, et al. *J Infect Dis.* 2019;219(11):1809–17.
91. Fu M, et al. *Mbio.* 2020;11(3):e03242–19.
92. Kong Y, et al. *Front Med (Lausanne).* 2020;7:217.
93. Zhao C, et al. *mSystems.* 2020;5(4):e00450-20.
94. Kyono K, et al. *Reprod Med Biol.* 2018;17(3):297–306.
95. Hashimoto T, et al. *J Assist Reprod Genet.* 2019;36(12):2471–9.
96. Kitaya K, et al. *Mediators Inflamm.* 2019;2019:4893437.

22 Endometritis
New Times, New Opportunities

Ettore Cicinelli, Rossana Cicinelli,
Carla Mariaflavia Santarsiero, and Amerigo Vitagliano

BACKGROUND

In the past decade, there has been growing interest toward the evaluation of endometrial "health" in reproductive issues. Receptivity characteristics of the endometrium have been investigated at a molecular level through whole-tissue transcriptomic profiling, with the purpose of identifying the optimal implantation window in frozen-thawed embryo transfer cycles (1). Moreover, recent studies have shed light on the potential relevance of chronic endometritis (CE), a chronic inflammatory condition of the endometrium, as a causative factor of infertility (2–9).

Today, we know that the uterine cavity is not sterile, but harbors a specific microbiome involved in the uterine physiology. It was recently postulated that an alteration of the uterine microbiome may cause inflammation and increase the risk of pregnancy complications (e.g., miscarriage, preterm labor), as well as the development of some gynecological pathologies like endometrial polyps, intra-uterine adhesions, endometriosis, and cancer (10).

CE can actually be considered to be a "new pathology," as it was completely ignored until recent times by physicians. Clinically, CE has a subtle behavior with mild or no symptoms. However, different studies showed that unrecognized CE may reduce the pregnancy rates and increase the miscarriage rates in both natural conception (4, 6, 11, 12) and assisted reproduction (2, 5, 9). Likewise, the treatment of CE may improve the ongoing pregnancy rates in women with reproductive impairment (2, 5, 9, 11–14).

However, many aspects of CE need to be elucidated. CE still represents an "uncertain matter" in terms of etiopathogenesis, diagnosis, reproductive impairment, and treatment.

ETIOPATHOGENESIS

As stated above, today, we know that the uterine cavity is not sterile but harbors a specific low-biomass microbiome (15–17). Placental tissues derived from elective, term cesarean deliveries contained bacteria

DOI: 10.1201/9781003024941-22

in 70% of placental membranes (18), with a general dominance of *Lactobacillus* species. Studies based on traditional endometrial cultures and on molecular biology-based methods showed that CE is often associated with an abnormal endometrial microbiome, with the local proliferation of common Gram-positive (e.g., *Streptococcus, Staphylococcus*), Gram-negative (e.g., *Escherichia coli, Klebsiella pneumoniae, Neisseria gonorrhoeae*) or, less commonly, intracellular (Mycoplasmi, Ureaplasmi, Chlamidiae) or anaerobic bacteria (*Bifidobacterium, Prevotella*) (19–21).

The origin, characteristics, and mechanisms modulating the composition of the normal uterine microbiome, its alterations, and relationships with nearby body niches represent a field of great scientific interest. In contrast to the vaginal microbiota (22), there is a lack of understanding of the microbial community in the uterine cavity. Recent studies into the microbiome of the female genital apparatus suggest, in analogy with different niches of the human body, the existence of a strict relationship between the microbiome of the vagina and that of the uterus. It has been reported that living microbes may be ubiquitous in the female upper reproductive tract, including in the uterus (22–24). With respect to the microbial taxa largely shared among different parts of the reproductive tract of the same individual, which demonstrate gradual changes from the vagina to the peritoneum, the microbes harbored in the female upper reproductive tract have been speculated to migrate from the lower genital tract or to be transferred from the peritoneal fluid (25). Synchronous variation and increasing convergence of the uterine and vaginal microbiomes with advancing age have been shown (10). Therefore, the ecological stability of the vaginal and uterine microbiota plays an important role in the health of the female reproductive system. The uterine cavity and the vagina are physiologically adjacent channels. Hence, theoretically, the bacteria that colonize the vagina have the opportunity to migrate upward to the uterus *via* the cervix. In general, we can understand that a healthy endometrial cavity must correspond to a healthy vaginal cavity.

These concepts open up new preventive and therapeutic strategies for CE and for many obstetrical and gynecological pathologies. The integration of the vaginal microbiome with "healthy bacteria," as well as the transplantation of bacteria from individuals with healthy vaginal and endometrial microbiome, may represent a fascinating perspective. In a recent paper from Wang and coworkers, transplanting certain strains of vaginal bacteria into the vagina of rats induced or reduced endometritis-like symptoms, thus providing new evidence for the protective effects of specific families of vaginal bacteria toward the endometrium. This study clarified the interdependent relationship of vaginal bacterial translocation with uterine microecology and endometrial health, which will undoubtedly increase our understanding of female reproductive health (10). The effects of increases or decreases in frequencies of certain bacterial taxa on endometritis-like changes are characteristics of bacterial vaginosis (26) and postpartum vaginal microbiota (27).

The microbiome effects might be partially responsible for the clinical linkage of these two conditions with parturition. According to Wang and colleagues, the age-related changes in composition of the vaginal and endometrial microbiome may explain the high incidence of endometrial diseases in elderly women and underline the significance of paying attention to the potential threats to the upper reproductive system caused by microbial translocation.

Very recently, the only infectious nature of CE has been questioned (28). A new theory about an impaired inflammatory state of the endometrium (IISE) relying on a multi-faceted pathogenesis (including hormonal, infectious, and autoimmune stimuli) has been formulated. Basing on IISE theory, it has also been proposed to abandon the entity termed CE (28). However, in a recent retrospective, a non-concurrent case–control study on consecutive women with CE who received or did not receive antibiotic therapy and were diagnosed with CE through hysteroscopy (HYS), histology (HIS), and immunohistochemistry (IHC) for CD138, we found that the cumulative cure rate was considerably higher in treated vs. untreated women (81.3% vs. 6%, respectively; $p < 0.0001$) so that the infectious nature of CE was supported (29). Accordingly, two recent large studies confirmed the potential curative effect of antibiotics on CE. In particular, the retrospective cohort study by Xiong et al. (including 640 women undergoing frozen-thawed embryo transfer cycles) added new evidence to the negative effects of CE on IVF outcomes and showed the potential benefits of a standard oral antibiotic regimen for CE on embryo implantation (11). The other study, by Song and coworkers, included a total of 132 women with CE confirmed by CD138 immunohistochemistry (30). Women were randomly selected to receive a standard antibiotic therapy (oral levofloxacin 500 mg and tinidazole 1,000 mg daily for 14 days; treatment group) or no treatment (controls). The authors found that the rate of CE cure in the treatment group (89.3%) after one course of antibiotic treatment was significantly higher ($P < 0.001$) than that in the control group (12.7%). These studies reinforce the concepts that CE in most cases is actually due to a bacterial infection and its interaction with an altered endometrial microbiota.

However, despite published studies and advances in uterine microbiome knowledge, the definition of the normal endometrial microbiome, the assignment of taxa, the stable or temporary nature, the host–pathogen relationships, the relationships between vaginal and endometrial microbiomes, the effects of abnormal colonization of the uterine cavity, and the effects in terms of gynecological and obstetrical pathologies are all matters for future investigations.

DIAGNOSIS

The diagnosis of CE is a very challenging matter for physicians. Clinical examination and transvaginal ultrasound are non-specific for CE. The most common symptoms include abnormal uterine bleeding, leukorrhea, pelvic pain, and dyspareunia (7). The diagnostic workup for CE relies on microbial culture and histology combined with Immunohistochemistry (IHC) and hysteroscopy. In recent years, molecular biology has been proposed to evaluate changes in the endometrial microbiome.

Endometrial Culture

The data in the literature show that the results from traditional cultures are laboratory dependent and have limited value. Notably, in roughly 40% of CE cases diagnosed by hysteroscopy, endometrial cultures are negative for bacteria (probably due to the involvement of non-cultivable bacteria). Moreover, even when cervical disinfection is performed before endometrial sampling, any contamination of endometrial specimens with vaginal and/or endocervical content cannot be excluded. Consequently, microbial culture might not be a reliable approach to detect CE (7). However, the results of cultures and subsequent therapy guided on the basis of antibiogram have been shown to be effective in improving reproductive outcomes (2, 12, 38).

Histology and Immunohistochemistry

Today, the diagnostic gold standard for CE is the demonstration of the presence of plasma cells (PC) within endometrial tissue. Histologically, the diagnosis of CE is performed with hematoxylin and eosin (H&E) staining of PCs. However, the value of H&E detection is limited by high inter-observer and intra-observer variability (31, 32). The introduction of immunochemistry (IHC) stains for CD138 (syndecan-1), a proteoglycan present on the surface of PCs, has improved reliability of PC counting, and at the moment CD138 is the most widely employed technique for diagnosis of CE. The main problem with this technique is that currently there is still no consensus regarding the cut-off number of PCs (per section) needed for a positive CE diagnosis, so that the reported prevalence of CE among similar studies (as established based on PC counts) shows huge variability (33). Very recently, Xiong and coworkers found no difference in terms of reproductive outcomes in women affected by mild CE as diagnosed by less than 5 PC/HPF (high power field) with CD138 staining when treated with antibiotics, suggesting that 5 PC/HPF could be the diagnostic limit for CE (11). Obviously, this finding needs to be confirmed by other studies. Another limitation of CD138-based PC counting could be related to the blind nature of sampling when performed without hysteroscopic control. Blind sampling with a biopsy cannula may result in a sampling from a normal area of the endometrial mucosa giving false negative results (34). On the other hand, it is known that syndecan-1 can stain endometrial cells, giving false positive results. In our considerable experience, we found, in comparison with hysteroscopy, a very high agreement between CE and positive CD138 (high specificity) but not an optimal sensitivity. In a recent study conducted by our group (29), we found that, after three courses of antibiotic therapy, 19% of women still showed signs of CE with hysteroscopy but only 6.1% with CD138 staining, indicating that some cases of CE may be missed if using only CD138 counts. Recently, a new marker for PCs, MUM1 (Multiple Myeloma antigen), has been proposed for CE diagnosis, based on a possibly higher specificity for MUM1, compared with syndecan-1 (35).

Hysteroscopy

Hysteroscopy is a technique which must only be carried out by gynecologists and it allows a thorough exploration of the uterine cavity. Fluid hysteroscopy has been shown to be an effective technique to detect CE, based on the detection of the so-called subtle or small lesions (7). The most characteristic lesions are micropolyps that can be easily recognized by fluid hysteroscopy as small or subtle lesions floating on the endometrial surface. Recently, using a Delphi procedure, we have defined the main diagnostic criteria for CE using hysteroscopy, namely micropolyps, hyperemia, mucosal edema, hemorrhagic spots, and "strawberry" pattern, with high inter-observer agreement (36).

For the diagnosis of CE, hysteroscopy shows some distinct advantages:

1) It allows a complete endometrial evaluation
2) Information is immediate and not lab dependent
3) Using hysteroscopy, it is possible to make eye-directed biopsies on most altered endometrial areas, theoretically improving histological detection.

Nevertheless, the major problem for hysteroscopy is that is strongly operator dependent and the diagnosis is limited by high inter-observer variability.

Molecular Microbiome

In recent years, biomolecular methods have been developed in order to provide a complete definition of the endometrial microbiome. The technique is based on a quantitative real-time-PCR (qPCR), using specific primers for the variable regions of the 16S rRNA gene of the most common bacteria causing CE. Recent studies, based on molecular biology-based methods and endometrial culture, showed that CE is often associated with an abnormal endometrial microbiome, with the local proliferation of common Gram-positive (e.g., *Streptococcus*, *Staphylococcus*), Gram-negative (e.g., *E. coli*, *K. pneumoniae*, *N. gonorrhoeae*) or less commonly intracellular (Mycoplasmi, Ureaplasmi, Chlamidiae) or anaerobic bacteria (*Bifidobacterium*, *Prevotella*) (1, 7, 10, 16, 19–21, 25). When comparing biomolecular approaches with traditional diagnostic techniques employed for CE, the agreement was 76.9%, suggesting that the biomolecular approach may be reliably employed as the only technique necessary for diagnosing CE (37). However, further clinical trials are needed to confirm the actual effectiveness of the sole metagenomic approach for diagnosing and treating CE. On the other hand, it is possible to speculate that, by using a biomolecular study of the endometrial microbiome, it will be possible to investigate the correlations between specific infectious profiles and endometrial pathologies in depth, thus offering the possibility of more effective and targeted personalized therapies and preventive strategies.

TREATMENT

The impact of antibiotic treatment on the rate of negative test results for CE is shown in Table 22.1. The impact of CE and its response to antibiotic therapy on clinical outcome has received much interest in recent years and, at present, represents a hotly debated issue. The impact on reproduction has been examined in several cohort studies. Cicinelli et al. conducted a retrospective study which showed that, after antibiotic therapy, the pregnancy rate (PR) in women who became CE negative was higher than those who remained CE positive after antibiotic treatment (76.3% versus 20%, $P < 0.0001$, respectively), with a similar finding for live birth rate (LBR) in the same women (65.8% versus 6.6%, $P < 0.0001$, respectively) (38).

TABLE 22.1 The Impact of Antibiotic Treatment on the Rate of Negative Test Results for CE

AUTHOR AND YEAR	*TREATMENT*	*CURE RATE AFTER ANTIBIOTIC TREATMENT*
Cicinelli et al. 2017 [51]	Treatment on the basis of antibiogram: Gram-negative bacteria: ciprofloxacin 500 mg twice a day for ten days Gram-positive bacteria: amoxicillin plus clavulanate 1000 mg twice a day for eight days *Mycoplasma* and *U. urealyticum*: josamycin 1000 mg twice a day for 12 days In case of persistence: minocycline 100 mg twice a day for 12 days Negative cultures: ceftriaxone 250 mg in a single dose plus doxycycline 100 mg orally twice a day for 14 days with metronidazole 500 mg orally twice a day for 14 days	82.3%
McQueen 2014 et al. [6]	Ofloxacin 400 mg and metronidazole 500 mg twice a day for 14 days Alternative: doxycycline alone, doxycycline and metronidazole, or ciprofloxacin and metronidazole	100%
Xiong et al. 2021 [11]	First course: doxycycline 100 mg twice a day for 14 days Second course: levofloxacin lactate 200 mg twice a day plus metronidazole 500 mg three times a day for 14 days Third course: levofloxacin lactate 200 mg twice a day plus metronidazole 500 mg three times a day for 14 days	89% after the first or the second course of antibiotic treatment
Kitaya et al. 2017 [39]	First course: doxycycline 100 mg twice a day for 14 days Second course: oral metronidazole 250 mg twice a day for 14 days plus ciprofloxacin hydrochloride 200 mg twice a day for 14 days	First course: 92.3% Second course: 99.1%
Cicinelli et al. 2014 [12]	Treatment on the basis of antibiogram: (see Cicinelli et al. 2017 [51])	75.4%
Cicinelli et al. 2014 [12]	Treatment on the basis of antibiogram: (see Cicinelli et al. 2017 [51])	71%
Song et al. 2021 [31]	Oral levofloxacin 500 mg and tinidazole 1000 mg daily for 14 days	89.3%
Cicinelli et al. 2021 [34]	Treatment on the basis of antibiogram: (see Cicinelli et al. 2017 [51])	31.25% after one cycle, Additional 31.25% after two cycles Additional 18.75% after three cycles

In another study involving women with recurrent implantation failure (RIF), Kitaya et al. reported that the LBR in the first embryo transfer (ET) cycle and cumulative three ET cycles in the cured CE group following antibiotic treatment (32.8% and 38.8%, respectively) was significantly higher than those from the non-CE group (22.1% and 27.9%, respectively) (39). McQueen et al. observed that the cumulative LBR was 88% (21/24) for the treated CE group versus 74% (180/244) for the group without CE. In the CE group, the LBR for those before treatment was 7% (7/98) versus 56% (28/50) after treatment (6). A recent, randomized, controlled trial by Song et al. (9) brings to light the potential effectiveness of empirical double-regimen oral antibiotic therapy (i.e., levofloxacin 500 mg and tinidazole 1000 mg daily for 14 days) for CE treatment. To the credit of the authors, this was the first study on antibiotic therapy for CE with rigorous methodology, including blinding on the pathologists. Interestingly, the authors found an 89.3% CE negative rate after a single course of antibiotics in the treatment group, compared to 12.7% in the control group. Such a difference between comparators was statistically significant (RR = 7.06; 95% CI, 3.51–14.72) and suggested a promising effect of broad-spectrum antibiotic therapy based on oral quinolones and nitroimidazoles against the pathogens involved in CE. In particular, levofloxacin would be active against most Gram-positive and Gram-negative bacteria including streptococci, staphylococci, and Enterobacteriaceae (40). Tinidazole would provide coverage against most intracellular and anaerobic bacteria including *Ureaplasma*, *Mycoplasma*, and bifidobacteria (11). On this basis, the combination of levofloxacin and tinidazole may offer appropriate antibiotic coverage against most of the pathogens involved in CE.

Whereas this therapeutic strategy seems effective against CE, some may perceive the indiscriminate administration of broad-spectrum antibiotics as potentially conducive to the development of antibiotic resistance. This phenomenon, which results from selection for mutations in the genes encoding proteins that are targeted by antibiotics, is increasingly common with respect to quinolones and represents a major public health concern (41).

Modulation of Endometrial Microbiome and Transplantation of Vaginal Bacteria

Although the use of antibiotics to treat CE is, at present, the most logical approach, new strategies, such as pro- and prebiotic administration, and microbiota transplants are gaining popularity for improving and maintaining the optimal composition of the microbiota. Bacterial manipulation could be a choice of clinical intervention for the maintenance of homeostasis of the uterine microenvironment and prevention of disease, similar to the use of vaginal microbiota transplantation for the treatment of intractable BV (42, 43).

The rationale is to modulate the microbial communities in a way that is beneficial to human health. However, several important questions related to these new clinical strategies remain unresolved, such as indicators for its use, the comparative efficacy of monostrain and multistrain probiotics, the choice of excipients, and methods of administration and delivery (44–46). Furthermore, at the moment, the pharmacological mechanisms of probiotic and prebiotic action are poorly understood, data on their clinical efficacy are not robust as they are based on studies on a small sample size, and the heterogeneity in the strains of bacteria used, the duration of treatment, and the lifestyle of the patients did not provide enough evidence to support the use of probiotics for medicinal purposes (10, 45, 47).

Intrauterine Infusion of Antibiotics

Due to the chronic nature of CE, along with the challenging management and resulting infertility, alternative therapeutic approaches like the intrauterine infusion of antibiotics have been proposed alone or in combination with oral antibiotic therapy. The therapeutic perspective consists of the possibility of concentrating the antibiotic effect in the uterine cavity with less general application and greater local efficacy.

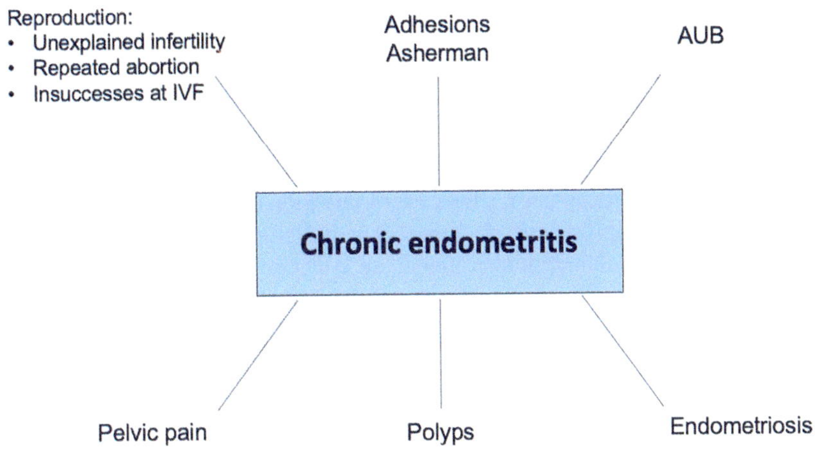

FIGURE 22.1 Clinical aspects related to chronic endometriosis.

However, the results at the moment do not demonstrate a superiority of this new approach compared to the traditional one (48) but this remains a promising option for treating CE instead of systemic administration of antibiotics.

Connection with Other Pathologies

As reported above, the occurrence of dysbiosis (altered microbiome) in CE and pelvic inflammatory disease has been connected with pregnancy loss as well as adverse pregnancy outcomes, Notably, in recent years, different effects of dysbiosis have been hypothesized. In particular, many studies support the hypothesis that dysbiosis may contribute to the development of endometrial polyps, endometriosis, and also cancer (10, 49, 50). Alterations to the bacterial composition in a microbiota, or dysbiosis, have been shown to favor a pro-carcinogenic state through altered immune responses, dysregulated hormone metabolism, and modulation of the cell cycle (50). Chronic endometritis has been positively correlated with endometriosis by clinical and population studies. The risk of having CE in women diagnosed with endometriosis is about three times higher than that in the general population (51). Chronic endometritis could cause altered uterine contractility with the occurrence of non-propagating waves and increased intrauterine pressure that causes predisposition to blood retrograde intraperitoneal spillage. According to the genetic-epigenetic etiopathogenetic theory for endometriosis, the endometrial cells could easily implant in an inflamed peritoneal environment (51, 52). This implies that, in the treatment of many gynecological pathologies, the presence of CE should be ruled out and treated before surgery, as it is possible that in some cases indication for surgery may disappear (like pain or AUB) and that in other cases symptoms may persist after surgery (Figure 22.1).

CONCLUSIONS

In conclusion, evidence reported in the literature supports the detrimental role of CE in reproduction and probably an etiopathogenetic role for many gynecological pathologies. However, at present, recognition of CE remains a challenging issue and all diagnostic techniques show some limitations. Hysteroscopy offers a reliable diagnosis but is strongly operator dependent and probably overestimates the diagnoses of CE. Today, plasma cell counting by CD138 immunostaining is the most employed diagnostic technique for

CE. In recent years, biomolecular methods have been developed in order to provide a complete definition of the endometrial microbiome. The technique is based on a quantitative real-time-PCR (qPCR), using specific primers for the variable regions of the 16S rRNA gene of the most common bacteria causing CE. By using a biomolecular study of the endometrial microbiome, it will be possible to investigate in-depth the correlations between specific infectious profiles and endometrial pathologies, thus offering the possibility of a more effective and targeted personalized therapy and preventive strategy.

Data in the literature demonstrate that, in the case of infection caused by a pathogen, antibiotic treatment is required and improves reproductive outcomes. The literature shows that both standard antibiotic treatment and antibiogram-guided antibiotic treatment are effective in improving reproductive outcomes. The active research in this field opens up new therapeutic perspectives of CE in terms of modulation of microbiome, use of probiotics and, more recently, transplantation of the normal microbiome. Modulation of the uterine microbiota to restore and maintain its microbial composition is a promising field of research, with high clinical relevance.

Knowledge of the uterine microbiome may allow us to increase our understanding of the female reproductive tract and to develop more personalized reproductive therapies, highlighting the potential importance of using microbiological assessment in infertile patients. The emerging relationship between altered microbiome and many clinically relevant gynecological pathologies, like endometriosis and endometrial polyps, suggests that, in the cure of these pathologies, the presence of CE should be excluded preliminarily, especially when surgical treatment is planned.

REFERENCES

1. Garrido-Gómez T, et al. *Fertil Steril.* 2013;99(4):1078–85.
2. Cicinelli E, et al. *Hum Reprod.* 2015;30:323–30.
3. Johnston-MacAnanny EB, et al. *Fertil Steril.* 2010;93:437–41.
4. Kitaya K. *Fertil Steril.* 2011;95(3):1156–8.
5. Vitagliano A, et al. *Fertil Steril.* 2018;110(1):103–112.e1.
6. McQueen DB, et al. *Fertil Steril.* 2014;101:1026–30.
7. Cicinelli E, et al. *Fertil Steril.* 2008;89(3):677–84.
8. Kitaya K, et al. *Am J Reprod Immunol.* 2016;75:13–22.
9. Song D, et al. *Reprod Biomed Online.* 2018;36:78–83.
10. Wang J, et al. *Nat Commun.* 2021;12:4191.
11. Xiong Y, et al. *Fertil Steril.* 2021;116(2):413–421.
12. Cicinelli E, et al. *Reprod Sci.* 2014;21(5):640–7.
13. Kaku S, et al. *BMC Womens Health.* 2020;20(1):114.
14. Kuroda K, et al. *Immun Inflamm Dis.* 2020;8(4):650–658.
15. Moller BR, et al. *Acta Obstet Gynecol Scand.* 1995;74:216–9.
16. Mitchell CM, et al. *Am J Obstet Gynecol.* 2015;212:611.e1–9.
17. Aagaard K, et al. *Sci Transl Med.* 2014;6:237ra65.
18. Steel JH, et al. *Pediatr Res.* 2005;57(3):404–11.
19. Baker JM, et al. *Front Immunol.* 2018;9:208.
20. Chen C, et al. *Nat Commun.* 2017;8:875.
21. Moreno I, et al. *Fertil. Steril.* 2017;108:32–39.
22. Serrano MG, et al. *Nat Med.* 2019;25:1001–1011.
23. Chen C, et al. *Nat Commun.* 2017;8:875.
24. Winters AD, et al. *Sci Rep.* 2019;9:9905.
25. Li F, et al. *Gigascience.* 2018;7(10):giy107.
26. Ravel J, et al. *Microbiome.* 2013;1(1):29.
27. Romero R, et al. *Microbiome.* 2014;2(1):4.
28. Drizi A, et al. *Menopause Rev.* 2020;19(2):90–100.
29. Cicinelli E, et al. *Fertil Steril.* 2021;115(6):1541–1548.

30. Song D, et al. *Fertil Steril.* 2021;115(6):1549–1556.
31. Adegboyega PA, et al. *Hum Pathol.* 2010;41(1):33–7.
32. Kitaya K, et al. *Exp Ther Med.* 2013;5(2):485–488.
33. Liu Y, et al. *Fertil Steril.* 2018;109(5):832–839.
34. Cicinelli E, et al. *Fertil Steril.* 2021;116(2):345–346.
35. Parks RN, et al. *Int J Surg Pathol.* 2019;27(4):372–379.
36. Cicinelli E, et al. *Fertil Steril.* 2019;112(1):162–173.e2.
37. Moreno I, et al. *Am J Obstet Gynecol.* 2018;218(6):602.e1–602.e16.
38. Cicinelli E, et al. *Am J Reprod Immunol.* 2018;79(1).
39. Kitaya K, et al. *Am J Reprod Immunol.* 2017;78(5).
40. Cicinelli E, et al. *Fertil Steril.* 2021;115(6):1445–1446.
41. Bush NG, et al. *Molecules.* 2020;25(23):5662.
42. Lev-Sagie A, et al. *Nat Med.* 2019;25(10):1500–1504.
43. Cohen CR, et al. *N Engl J Med.* 2020;383(8):791–792.
44. Molina NM, et al. *Biomolecules.* 2020;10(4):593.
45. Chenoll E, et al. *Front Cell Infect Microbiol.* 2019;9:114.
46. Molina NM, et al. *Hum Reprod.* 2021;36(4):859–879.
47. Wang J, et al. *Nat Commun.* 2021;12(1):4191.
48. Pantos K, et al. *Sci Rep.* 2021;11(1):15581.
49. Cicinelli E, et al. *J Minim Invasive Gynecol.* 2019;26(7):1346–1350.
50. Chambers LM, et al. *Curr Oncol Rep.* 2021;23(8):92.
51. Cicinelli E, et al. *Fertil Steril.* 2017;108(2):289–295.e1.
52. Koninckx PR, et al. 2019;11(3):209–216.

Decidualization Resistance

23

A New Condition Identified in Severe Preeclampsia

Irene Muñoz-Blat, Nerea Castillo-Marco, Teresa Cordero, Carlos Simón, and Tamara Garrido-Gómez

INTRODUCTION

The endometrial cycle has two dominant phases: the proliferative phase that follows menstruation and precedes ovulation, and the secretory phase characterized by the transformation into being receptive to embryo adhesion and differentiation to regulate embryo invasion (1) (Figure 23.1a). Following ovulation, estradiol induces increased progesterone receptor expression in stromal cells. Consequently, cells become responsive to progesterone and a coordinated cascade of modifications occurs in the tissue, leading to decidualization (2, 3). Decidualization is a differentiation process of the endometrial cells in response to a hormonal stimulus released before pregnancy and is essential to a successful reproductive outcome. Curiously, in humans, decidualization occurs each menstrual cycle, regardless of whether implantation occurs (1).

A successful pregnancy may depend on a healthy preconception period with an optimal decidualization process (1, 4). The main modifications occurring in initial decidualization include (i) phenotypic transformation of human endometrial stromal cells (hESCs) and (ii) secretory activity of the uterine glands, which ultimately results in: (iii) extracellular matrix remodeling; (iv) immune cell recruitment, distribution, and function of uterine natural killer (uNK) and regulatory T cells; (v) angiogenesis, and (vi) oxidative stress resistance (5–8) (Figure 23.1b). The main function of the decidualized endometrium is to become more permissive for the conceptus. The epithelium releases paracrine effectors to transduce the decidualization signal to the stromal compartment. Epithelial–stromal crosstalk is essential to orchestrate decidualization and, consequently, to maintain an optimal uterine microenvironment during pregnancy (2, 9, 10). Secretion of prolactin (PRL) and insulin growth factor binding protein 1 (IGFBP-1) increases as a key indicator of this process. These molecules drive decidualization in connection with a complex network of intracellular and secreted proteins that regulate the differentiation process (11). In addition to the

DOI: 10.1201/9781003024941-23

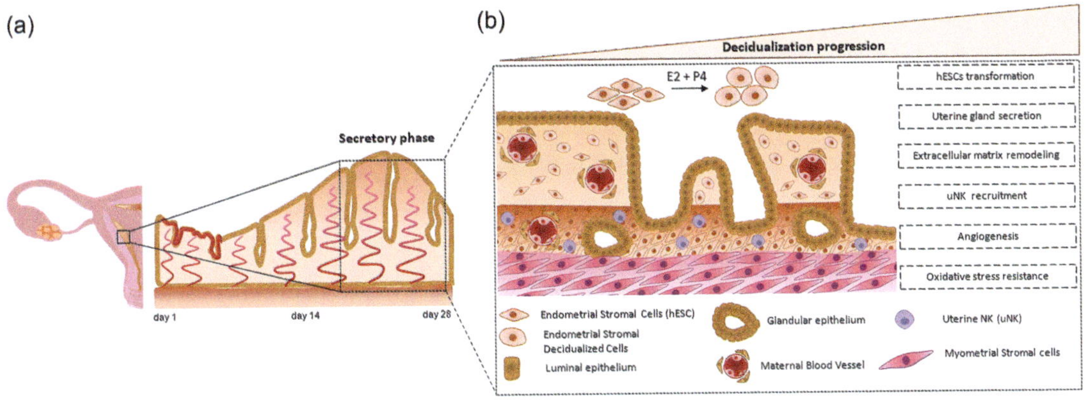

FIGURE 23.1 Decidualization progression in the late secretory menstrual phase. (a) Endometrium during the secretory phase (days 14 to 28). (b) Cell types involved in the decidualization process and the main tissue transformations regulated by the action of hormones estradiol (E2) and progesterone (P4).

classical decidualization markers, proteomic analysis of the secretome of decidualized hESCs vs. controls revealed the existence of 60 differentially expressed proteins that modulate this process (12).

In pregnancy, endometrial adaptations drive decidua formation, entailing drastic tissue remodeling, involving physical and humoral changes along with maternal immune cell recruitment (13). Thus, decidualization is key to supporting blastocyst implantation and subsequent placenta formation through proper decidua specialization. The decidua is divided into three parts in relation to the developing conceptus during pregnancy: the decidua basalis is beneath the site of implantation, the decidua capsularis covers and coats the blastocyst, and the rest of the decidua is referred to as the decidua parietalis (Figure 23.2a). The decidua basalis supports the formation of the discoid placenta in middle and late pregnancy, whereas the decidua capsularis and decidua parietalis fuse after 16 weeks of gestation (Figure 23.2b). The decidua is crucial for embryo protection from the maternal immune response at the maternal–fetal interface and from oxidative stress, resulting in changes in oxygen tension (13, 14), regulation of the local immune response to control trophoblast invasion (15), and to allow angiogenesis to prepare the maternal vasculature to support embryo development during pregnancy (16, 17). Different cell types perform specific tasks in this complex process. The main cell types are the decidualized hESCs, which provide nutritional support during blastocyst implantation. Hematopoietic cells, macrophages, uNK cells, and monocytes are important components of the immune cells that constitute the decidualized tissue (18). Other structures appear to be associated with these cell types, such as uterine glands and small blood vessels, including spiral arteries, which support the maternal blood flow essential to conceptus growth (17). A recent study demonstrated (by single-cell resolution) the presence of different cell populations that provide the proper microenvironment essential for establishing the maternal–fetal interface (19). These scientific data support the pivotal role of a healthy decidua in a successful pregnancy.

STATE OF THE ART—DECIDUALIZATION DEFECTS IN INFERTILITY AND PREGNANCY COMPLICATIONS

Throughout the 20th century, reproductive failure was attributed to endocrinopathies, such as polycystic ovary syndrome, as well as embryonic defects or placental disorders. The trend was to consider the endometrium as an inert reservoir dependent on the hypothalamic–pituitary–gonadal axis and without an active function in implantation and the pregnancy progress (5). With the emergence of studies linking new

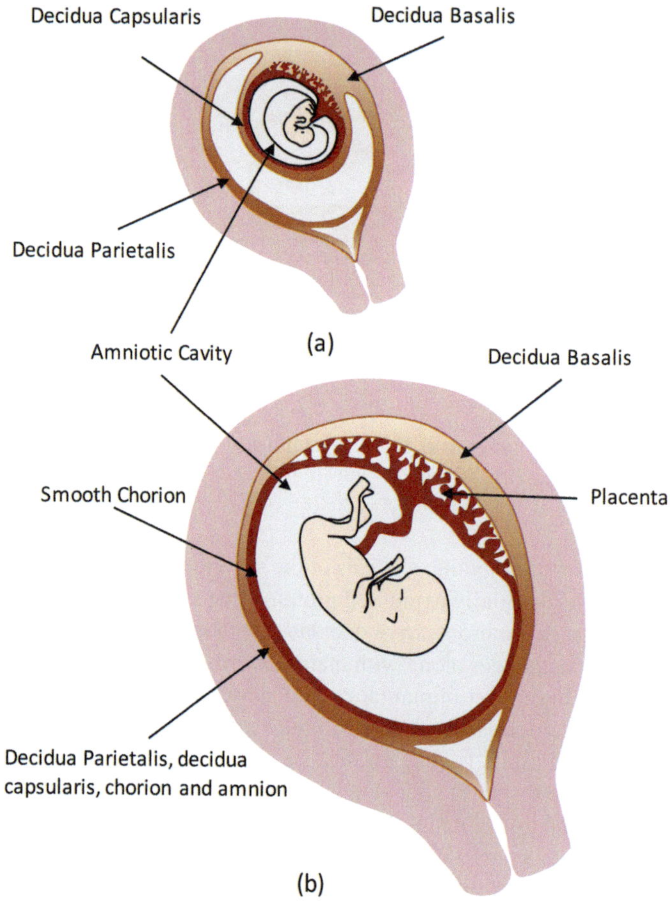

FIGURE 23.2 Anatomical location of different parts of the decidua during pregnancy. (a) At 8 weeks' gestation, three anatomically and functionally discrete regions of the decidua (basalis, capsularis, and parietalis) are defined based on their relationship to the fetoplacental unit. (b) A virtual space exists between the decidua capsularis and decidua parietalis until about 15–16 weeks of gestation, at which point these tissues come together and fuse.

regulatory pathways with hormonal coordination, a fresh point of view arose, giving the endometrium the importance, it deserves. In fact, many gestational diseases are interrelated based on their risk factors, suggesting that they could have a common origin (1). In women who experience a preterm birth (PTB) at 28 weeks, the risk of recurrence in subsequent pregnancies increases along with the risk of developing other pathologies, such as preeclampsia (20). Additionally, decidualization disruptions are described in infertility cases associated with endometriosis (21), recurrent pregnancy loss (22, 23), age–reproductive decline (24), and pregnancy complications, such as intrauterine growth restriction (IUGR) (25) and preeclampsia (PE) (26, 27) (Figure 23.3). In this sense, a healthy pregnancy outcome depends on pre-pregnancy events occurring as early as decidualization.

Overwhelming evidence indicates that aberrant decidualization could drive endometriosis, and this defect might explain the risk of infertility associated with the condition. Endometriosis affects 6% to 10% of women of reproductive age and causes formation of ectopic endometrial implants outside of the uterus. Although Sampson's theory supports the observation that endometriosis is caused by retrograde menstruation (28), some women who have this type of menstruation do not develop endometriosis. Similarly, a study of hESCs treated with progesterone over time showed a differential gene expression pattern between endometriosis patients and controls (29). This observation gave rise to the term "progesterone resistance."

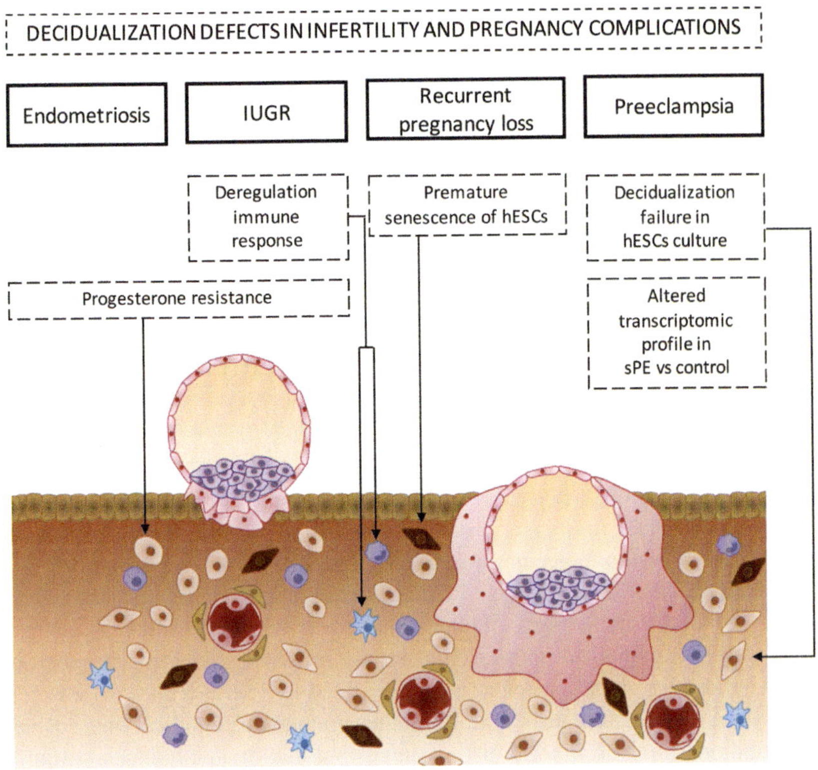

FIGURE 23.3 Decidualization and blastocyst invasion including different cell types that participate in this process and decidualization defects related to infertility conditions and pregnancy complications.

In endometriosis, progesterone resistance precedes deregulation in stopping epithelial proliferation and impaired stromal decidualization. Consequently, these patients have an increased risk of infertility and miscarriage (5, 21).

In this context, in recurrent pregnancy loss, there is evidence of aberrant decidualization, but, in these cases, it is related to premature hESC senescence. Recently, it was hypothesized that the decidual response changes cycle by cycle during women's reproductive years, leaving evidence of cumulative effects of menstruations, miscarriages, and births. This could explain the epidemiology of sporadic and recurrent miscarriage. An increasing number of pregnancy losses may correlate with a lower capacity of the uterus to adapt (1). Recently, Lucas et al. correlated recurrent pregnancy loss with a pro-senescence decidual response during the peri-implantation window. Their single-cell transcriptomic analysis shows that decidualization is a multistep process that eventually leads to chronic senescence, which is a cellular state incompatible with decidua formation. An *in vitro* approximation suggests that decidual cells require recruitment of immune uNK cells to eliminate senescent decidual cells, avoiding the default pathway. The authors identified two biomarkers: high levels of iodothyronine deiodinase (DiO_2) are positively correlated with the presence of these senescent cells, and high expression of the *SCARA5* gene is associated with properties of hESCs (30).

In IUGR, decidualization failures occur due to a defective progesterone response in combination with a deregulated immune response against extravillous trophoblasts (EVTs). This results in the uterovascular remodeling disruptions observed in this pregnancy complication (25). In addition, increased phosphorylated IGFBP-1 is observed in decidual tissue at the time of delivery and maternal plasma in the late first trimester in pregnant women diagnosed with IUGR compared with normal pregnancies (31).

In the obstetric field, the most feared pregnancy complication is preeclampsia, affecting 8% of first-time pregnancies (32). This is an important obstetric complication with short- and long-term life consequences for both mother and child (33). Our research focuses on the maternal contribution to the etiology of preeclampsia and this is where we will go in-depth to explain the current research trends. Importantly, preeclampsia symptoms cease when the placenta is removed, removing the decidua accelerates recovery (34, 35). Thus, placentation plays a key role in preeclampsia pathogenesis (20). However, the etiology of preeclampsia remains unclear. The unsolved question is: why is placentation defective in preeclampsia (36)? Many studies have attempted to answer this question, paying increased attention to the decidual role in the maternal–fetal interface equation during pregnancy.

THE PIVOTAL ROLE OF MATERNAL "SOIL" IN PREECLAMPSIA

At 8–10 weeks after conception, the placental EVTs transform into invasive cells. These cells undergo a partial epithelial-mesenchymal transition, caused by replacement of the epithelial-like adhesion molecules with vascular-like adhesion molecules, mimicking a vascular adhesion phenotype (37). Following this transition, EVTs invade the full thickness of the decidualized endometrium and reach the inner third of the myometrium (38). In the second wave of trophoblast invasion, trophoblast cells colonize the lumens of the spiral arteries (39, 40). Shallow endovascular invasion induces defective remodeling of maternal spiral arteries (41) and dysfunctional placental perfusion, leading to the clinical disease (20). This dysfunctional vasculature is a characteristic common to other pregnancy complications, such as intrauterine growth restriction and one-third of preterm births, suggesting that this pathological event is neither sufficient nor specific enough to cause preeclampsia (PE) (20).

Opening the door to new explanations of PE pathogenesis, Zhou and colleagues published a study where an abnormal gene expression pattern in cytotrophoblast (CTBs) cells immediately isolated from PE placentas was restored to control levels after 48 hours of culture. Their cell invasion and differentiation abilities were autocorrected (42). These surprising findings suggest that the maternal "soil"—decidua—is a prime candidate for the shallow invasiveness of the CTBs (42). In support of this theory, decidua removal immediately postpartum in severe preeclampsia (sPE) patients has remarkable clinical benefits (34). Specifically, a pilot study based on 32 cases of sPE showed decreased arterial pressure when the curettage was made, in comparison with women who did not receive this treatment. In addition, platelet counts increased in the intervention group (34). Additional studies on the positive effect of postpartum curettage on the resolution of PE signs have observed the same phenotype, with an improvement in blood pressure and platelet counts after delivery (35, 43).

These results opened the door to a new line of thought that the maternal soil could be a pivotal contributor to the pathogenesis of PE. In this regard, decidualization failure theory began as a logical deduction based on the crosstalk between EVTs and the decidua during the placentation stage (44, 45). An important limitation to gaining a deeper understanding of the implication of an aberrant decidua in this pathology is the difficulty in obtaining representative samples from the early stages of pregnancy, when inadequate invasion of EVTs occurs. For this reason, many studies have used third-trimester placentas and basal plate decidua. Transcriptional analysis of the decidua basalis isolated from preeclamptic patients at delivery revealed 455 differentially expressed genes (DEGs). This wide genetic deregulation seems to play an important role in PE pathogenesis (46). This maternal contribution theory was supported by a transcriptomic analysis of decidua obtained from chorionic villus samples (CVS) at 11.5 gestational weeks. This analysis identified 396 DEGs in the decidua from women who developed PE symptoms six months later, compared with normal pregnancies. Using a bioinformatic approach, their transcriptomic data were compared with those from other studies, and 154 of the 396 DEGs identified overlapped with normal endometrial maturation before and after implantation. Additionally, 112 of these 154 DEGs had the opposite expression pattern in the endometrium of PE patients than during normal maturation of the endometrium, with 16 of them being downregulated specifically in decidual NK (dNK) cells. This research reveals impairment of the endometrial maturation during the secretory menstrual phase and early pregnancy in women who develop PE. Additionally, the authors proposed the dNK to be critical for spiral artery remodeling (47). These findings reinforce the idea that the maternal soil contributes to PE physiopathology.

METHODOLOGY—DECIDUALIZATION STUDIES IN SEVERE PREECLAMPSIA

Following the idea that proper decidualization is key for a successful pregnancy, our studies focused on the progression of this process in sPE. In our recent publication, we hypothesized that aberrant decidualization could be a pivotal contributor to the phenotypic alterations in placentation associated with sPE (26). To test this theory, we used two approaches. The first approach was an *in vitro* decidualization analysis of isolated hESCs obtained from women who had a previous pregnancy affected by sPE, compared with healthy women who had uneventful pregnancies one to five years earlier (Figure 23.4). In detail, endometrial samples were collected in the early luteal phase (cycle day 15–17). *In vitro* decidualization was induced by adding cyclic adenosine monophosphate (cAMP) and medroxyprogesterone (MPA) to the culture medium for five days. Decidualization was evaluated morphologically by staining filamentous (F)-actin with rhodamine phalloidin and comparing decidualized hESCs from healthy (n = 13) vs. sPE (n = 13) cases. In addition, levels of PRL and IGFBP1 secreted by hESCs were measured and compared between groups. Finally, transcriptional analysis was performed on a subset of *in vitro* decidualized hESCs from healthy women (n = 7) and sPE patients (n = 5) using a microarray strategy.

The second approach was based on laser-microdissection isolations of portions of the decidua basalis or decidua parietalis from the maternal–fetal interface (Figure 23.5). Cells were captured from tissue sections of biopsy specimens from women with sPE (n = 5) vs. control women (n = 4) (i.e., gestational age-matched samples from women who had a spontaneous preterm birth with no signs of infection [nPTB]) at delivery time. After isolation, transcriptomic analysis of these samples was performed. In addition, the decidualization status of decidual cells was determined by measuring the levels of stage-specific antigens PRL and IGFBP1. Finally, re-decidualization and CTB invasion in the presence of conditioned medium from these decidualized cells were analyzed.

RESEARCH RESULTS—DECIDUALIZATION DEFECTS DURING AND AFTER SEVERE PREECLAMPSIA

Our results demonstrate that defective decidualization is associated with sPE, pointing to a maternal contribution to the etiology of the disease (26). Endometrial biopsies from non-pregnant donors were used for hESC isolation and their *in vitro* decidualization system was used to analyze this differentiation process in terms of stage-specific antigens, morphology, and transcriptional profiles between control and PE cases. These important experiments revealed that hESCs from patients with previous sPE were unable to decidualize. Transcriptomic analysis identified 129 DEGs in sPE cases compared with controls. They include down-regulation of genes that play important roles in decidualization such as *IGFBP-1*, *CNR1*, and *IL-1B*. Functional annotation of these genes demonstrated alterations in numerous pathways associated with failure of the differentiation process, such as cytokine–cytokine interactions, estrogen and TFG-ß signaling, and inflammation.

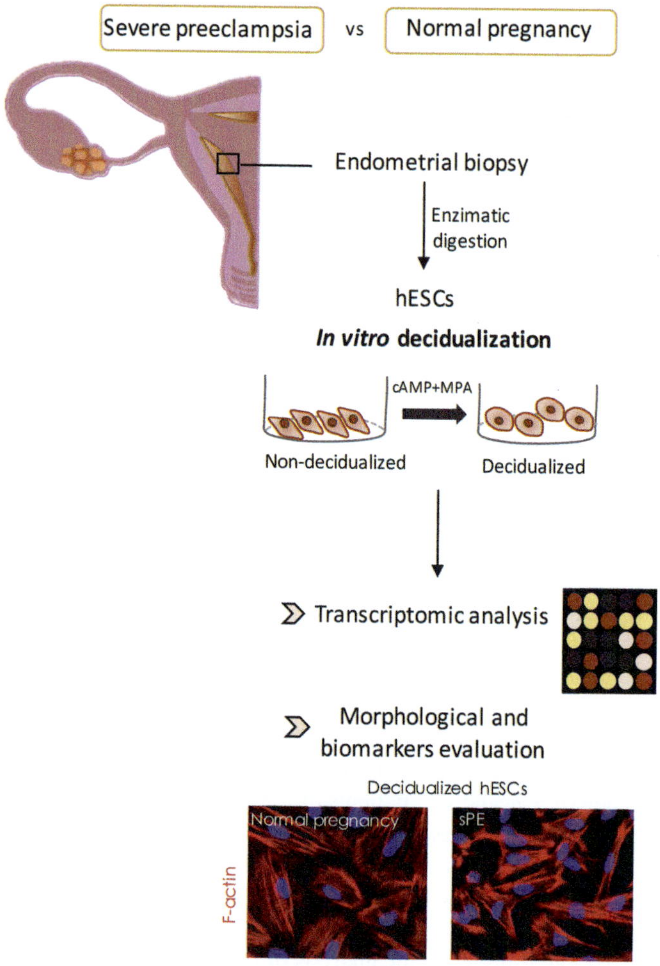

FIGURE 23.4 First approach of the experimental design used by Garrido-Gómez et al. (2017). Endometrial biopsies of non-pregnant women who developed sPE and women with previous healthy pregnancies (n = 13/group). Samples were used for *in vitro* decidualization and transcriptomic analysis.

We were interested to determine whether this deficient decidualization was present at the end of pregnancy. By laser microdissection, portions of the decidua basalis and parietalis were isolated from the sPE and control (nPTB) donors. Transcriptomic profiling revealed more than two hundred DEGs in sPE related to dysfunctionality of the decidua. Furthermore, trying to re-decidualize hESCs *in vitro* failed. Ultimately, CTB invasion was tested using conditioned medium from *in vitro* cultures of stromal cells from sPE decidua samples. CTBs were isolated from second-trimester human placentas and cultured on Matrigel. Conditioned medium from stromal cells of sPE patients did not allow CTB invasion, suggesting that decidualization resistance is associated with shallow invasion of CTBs in sPE (26).

Our transcriptomic results revealed altered gene expression during *in vitro* decidualization of hESCs from former patients with sPE. Among the 129 genes with altered expression, the gene encoding Annexin A2 (*ANXA2*) was further analyzed (26). This gene is associated with the first steps of embryo adhesion during the implantation process (48). *ANXA2* is widely expressed in the placenta and its defective activity could cause fibrinolytic deficiency linked with increasing thrombosis and PE predisposition (49). We studied the role of *ANXA2* in defective decidualization to evaluate its potential for being a maternal biomarker for sPE prediction (50). Deficient decidualization was detected in an *in vitro* model using small interfering RNA (siRNA) for *ANXA2* in hESCs from control individuals. In addition, blocking *ANXA2* expression

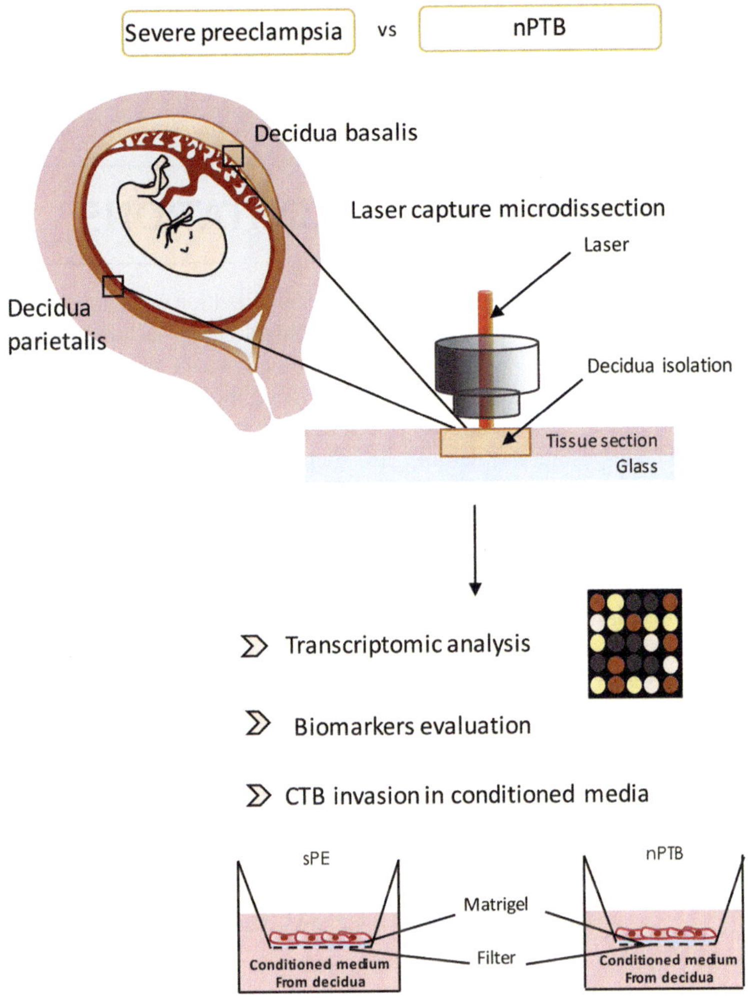

FIGURE 23.5 Second approach of the experimental design used by Garrido-Gómez et al. (2017). Decidua isolation with laser microdissection at the time of delivery in women with sPE (n = 5) and spontaneous preterm birth with no signs of infection (nPTB) as control pregnancies (n = 4). Decidua was used for transcriptomic analysis, biomarker evaluation, and invasion assays of cytotrophoblast cells (CTBs) with conditioned medium from decidual cells.

during pregnancy in an *in vivo* knockout mouse model resulted in deficient implantation and placentation. Thus, these results support the occurrence of decidualization resistance in women with sPE at the time of delivery and propose *ANXA2* as a possible biomarker to determine the risk of PE (50). Together, these results support the maternal contribution to PE.

Presumably, a successful pregnancy is determined before implantation and placentation, and is influenced by decidualization. Decidualization resistance involves the inability of the maternal soil to undergo the proper tissue modifications, leading to aberrations in placental invasion and adverse pregnancy outcomes. Cellular changes in the decidualization process of hESCs at the morphological, biochemical, and secretory levels, and immune cell recruitment (uNK and regulatory T cells) are the best-established modifications (1–6). When the decidualization process fails at some point, it results in a deficient decidua, leading to shallow EVT invasion, arterial remodeling, and, consequently, sPE. We recently hypothesized that development of sPE begins before pregnancy, with an altered baseline profile of decidualization resistance that can be identified during the menstrual cycle (33). This would allow us to detect women at risk even before pregnancy, making it possible to apply preventive and treatment strategies at the earliest stage. Advances in this direction could open promising avenues that could entirely change the current and discouraging clinical reality of sPE, shedding light on prevention, diagnosis, and treatment.

STRENGTHS AND LIMITATIONS

Our working hypothesis takes a step forward, suggesting decidualization resistance as a possible maternal contributor to severe preeclampsia. Understanding this decidualization failure could allow us to know more about sPE pathophysiology, as well as to discover biomarkers for predicting risk and therapeutic targets to reverse this condition. Clinically, this would be especially valuable since sPE prevention is key to reducing short- and long-term consequences in both mother and child. Furthermore, this knowledge could shed light on other late-gestational pathologies, such as fetal growth restriction or preterm birth, since defective decidualization could be a common defect between them.

Nevertheless, there are some limitations that could appear between research and clinical translation. Experimentally, bulk tissue analysis could fail to identify targets that are deregulated in only a few cell populations, due to the depth of the analysis. In this case, new strategies will need to be utilized, such as sequencing at single-cell resolution. This approach allows us to characterize the endometrium by identifying and describing specific cell populations (51). On the other hand, sampling could be a sticking point, since endometrial biopsy is an invasive and inconvenient intervention for women. Thus, new non-invasive approaches need to be explored. A promising approach is to use blood samples to obtain genetic information non-invasively, obtaining information that could be used, not only to predict sPE risk, but also to monitor patient health during pregnancy.

BARRIERS TO ACCESS

The complexity of developing a predictive tool lies in the scope of the prospective study where the predictive capacity of decidualization-defective markers will be tested. Additionally, the main affected tissue, the endometrium, cannot be easily interrogated due to the invasiveness of the procedure and the large sample size needed for prospective clinical trials. Once we have demonstrated the role of decidualization defects in the origin of sPE, the next step is to expand our results to a clinical trial to assess the potential of using this failure to determine a woman´s risk of developing sPE. The main barrier is identification of a set of biomarkers that distinguish the inherent variability of healthy pregnancies from molecular deregulation associated with pregnancy complications. Moreover, these biomarkers must segregate cases and controls with high sensitivity and specificity. Biomarker identification is challenging due to the complexity of endometrial dynamism, its highly coordinated regulation of decidualization, and interindividual variability. However, one solution would be to expand the number of subjects involved in the clinical study and to explore the use of biological samples obtained by non-invasive methods.

FUTURE PERSPECTIVES

The main challenge in preeclampsia is identifying a woman´s risk before clinical symptoms develop so that strategies for reducing the associated morbidity and mortality can be applied. For this purpose, the first step is to identify robust biomarkers that could be detected early in the first trimester. Since decidualization begins even before pregnancy, disruptions in this process could be identified preconceptionally. However, it is necessary to overcome the biopsy sample barriers and reduce patient inconvenience in diagnostic tests. In this line, a recent publication shows a proof of concept where a bioinformatic tool can distinguish between circulating RNA transcripts from plasma of preeclamptic and healthy pregnant women (52). The latest advances in technology and an understanding of maternal contribution open promising avenues to the development of future predictive, early diagnostic, and monitoring tools for pregnancy complications.

CONCLUSIONS

Proper decidualization contributes to a successful pregnancy. We have analyzed the changes in this endometrial process linked with pregnancy complications, such as infertility associated with endometriosis, recurrent pregnancy loss, IUGR, and preeclampsia. Our research is focused on severe preeclampsia, which seems to have a maternal component through decidualization resistance, revealing a further complexity to the etiology of this disease. Advances in this direction could be key to preventing, diagnosing, and treating sPE. Our findings encourage us to accept the challenge of developing a predictive tool to identify the risk of developing preeclampsia.

ACKNOWLEDGMENT

Work summarized in this review was supported by the project grant PI19/01659 from the Spanish Carlos III Institute awarded to Tamara Garrido-Gómez. Irene Muñoz-Blat was supported by the Ph.D. program PRE2019-090770 and funding was provided by the grant RTI2018-094946-B-100 from the Spanish Ministry of Science and Innovation with Carlos Simón as principal investigator. Nerea Castillo-Marco was supported by the Ph.D. program FDGENT/2019/008 from the Spanish Generalitat Valenciana.

REFERENCES

1. Ng SW, et al. *Int J Mol Sci.* 2020;21(11).
2. Wetendorf M, et al. *Mol Cell Endocrinol.* 2012;357(1–2):108–18.
3. Hamilton KJ, et al. *Reprod Biol.* 2014;14(1):3–8.
4. Cha J, et al. *Nat Med.* 2012;18(12):1754–67.
5. Gellersen B, et al. *Endocr Rev.* 2014;35(6):851–905.
6. Okada H, et al. *Reprod Med Biol.* 2018;17(3):220–7.
7. Evans J, et al. *Nat Rev Endocrinol.* 2016;12(11):654–67.
8. Schatz F, et al. *Hum Reprod Update.* 2016;22(4):497–515.
9. Wang X, et al. *Placenta.* 2017;60 Suppl 1:S20–S6.
10. Wang X, et al. *Nat Commun.* 2018;9(1):4421.
11. Gellersen B, et al. *J Endocrinol.* 2003;178(3):357–72.
12. Garrido-Gomez T, et al. *J Clin Endocrinol Metab.* 2011;96(3):706–16.
13. Mori M, et al. *Semin Immunopathol.* 2016;38(6):635–49.
14. Vinketova K, et al. *J Pregnancy.* 2016;2016:8689436.

15. Sharma S, et al. *Am J Reprod Immunol.* 2016;75(3):341–50.
16. Albrecht ED, et al. *Hum Reprod.* 2003;18(10):2039–47.
17. Su RW, et al. *Adv Anat Embryol Cell Biol.* 2015;216:189–213.
18. Dunn CL, et al. *Reprod Biomed Online.* 2003;7(2):151–61.
19. Vento-Tormo R, et al. *Nature.* 2018;563(7731):347–53.
20. Brosens I, et al. *Am J Obstet Gynecol.* 2011;204(3):193–201.
21. Marquardt RM, et al. *Int J Mol Sci.* 2019;20(15):3822.
22. Salker MS, et al. *PLoS One.* 2012;7(12):e52252.
23. Weimar CH, et al. *PLoS One.* 2012;7(7):e41424.
24. Woods L, et al. *Nat Commun.* 2017;8(1):352.
25. Dunk C, et al. *Front Endocrinol (Lausanne).* 2019;10:160.
26. Garrido-Gomez T, et al. *Proc Natl Acad Sci U S A.* 2017;114(40):E8468–E77.
27. Gellersen B, et al. *Hum Reprod.* 2010;25(4):862–73.
28. Sampson JA. *Am J Obstet Gynecol.* 1927:422–69.
29. Aghajanova L, et al. *Biol Reprod.* 2011;84(4):801–15.
30. Lucas ES, et al. *Commun Biol.* 2020;3(1):37.
31. Gupta MB, et al. *J Clin Endocrinol Metab.* 2019;104(2):408–22.
32. Rana S, et al. *Circ Res.* 2019;124(7):1094–112.
33. Garrido-Gomez T, et al. *Am J Obstet Gynecol.* 2020;29:S0002-9378(20)31130-3.
34. Magann EF, et al. *Obstet Gynecol.* 1993;81(4):502–6.
35. Ragab A, et al. *Arch Gynecol Obstet.* 2013;288(5):1035–8.
36. Fisher SJ. *Am J Obstet Gynecol.* 2015;213(4 Suppl):S115–22.
37. Zhou Y, et al. *J Clin Invest.* 1997;99(9):2152–64.
38. Kaufmann P, et al. *Biol Reprod.* 2003;69(1):1–7.
39. Ono M, et al. *Stem Cells Cloning.* 2008;1:23–9.
40. Roberts VHJ, et al. *Hum Reprod.* 2017;32(12):2382–93.
41. Brosens I, et al. *Am J Obstet Gynecol.* 2019;221(5):437–56.
42. Zhou Y, et al. *J Clin Invest.* 2013;123(7):2862–72.
43. Alkan A, et al. *Clin Exp Obstet Gynecol.* 2006;33(1):55–8.
44. Brosens JJ, et al. *Am J Obstet Gynecol.* 2002;187(5):1416–23.
45. Conrad KP, et al. *Placenta.* 2017;60:119–29.
46. Løset M, et al. *Am J Obstet Gynecol.* 2011;204(1):84.e1–27.
47. Rabaglino MB, et al. *Hypertension.* 2015;65(2):421–9.
48. Garrido-Gómez T, et al. *FASEB J.* 2012;26(9):3715–27.
49. Xin H, et al. *Thromb Res.* 2012;129(5):563–7.
50. Garrido-Gomez T, et al. *Am J Obstet Gynecol.* 2020;222(4):376.e1–e17.
51. Wang W, et al. *Nat Med.* 2020;26(10):1644–53.
52. Munchel S, et al. *Sci Transl Med.* 2020;12(550):eaaz0131.

Advanced Cell Therapy for Asherman's Syndrome

24

Jordi Ventura and Xavier Santamaria

INTRODUCTION

The endometrium is the inner layer of the uterus, which is composed of the basal layer and the functional layer. The functional layer is the one which thickens and sheds during the menstrual cycle, whereas the basal layer maintains its size and works as the starting point for the regeneration of the functional layer. The endometrium is composed mostly of stromal and epithelial cells, but recent evidence suggests that stem progenitor cells, mostly present in the basalis layer, participate in regenerating the tissue in each menstrual cycle (1–3). At first, the principal candidates for this role were the mesenchymal stromal progenitor cells and the epithelial progenitor cells (4–6), but some groups have reported that there are bone marrow progenitor cells which collaborate in the regeneration of the endometrium, which could be the bone marrow-derived mesenchymal stem cells or the endothelial progenitor cells (EPCs) (7–10). Although this is a controversial issue, an adequate thickness of the endometrium seems to be necessary for the embryo to implant, which is usually achieved between days 20 and 24 of the natural menstrual cycle (11–13). During this period, the endometrium undergoes pronounced structural and functional changes induced by ovarian steroidal hormones, mainly estrogen and progesterone, to allow the implantation of the embryo (14, 15). Alteration in endometrial receptivity is one of the most common causes of infertility (16, 17).

Asherman's syndrome (AS) is an acquired uterine condition characterized by the presence of intrauterine adhesions (IUAs), where, in many cases, the walls of the uterus stick to one another. The loss of stroma in this condition leads to fibrous tissue formation, due to fibrin production, which leads to the creation of tissue bridges between the walls of the uterus (18). The frequency of AS in society is difficult to estimate, but the European Medicines Agency established a frequency in the European Union of four in every 10.000 people (COMP, 2017). Generally, AS is caused following a surgical intervention, such as pregnancy-related curettage, causing trauma to the basal layer of the endometrium (19, 20), although AS may develop even in the absence of any surgical procedure due to infection (endometritis) or chronic inflammation (21).

Histologically, AS causes endometrial fibrosis in which the stroma is largely replaced with fibrous tissue and the glands are replaced by inactive cubocolumnar endometrial epithelium. The functional and basal

DOI: 10.1201/9781003024941-24

layers are indistinguishable, with the functional layer being replaced by an epithelial monolayer unresponsive to hormonal stimulation, and fibrotic synechiae forming across the cavity. Sometimes, different layers of the endometrium, myometrium, or connective tissue may be involved in the resulting intrauterine adhesions. In the most severe cases, adhesions may be composed of collagen bundles, fibrous strips, or muscle with the same characteristics as normal myometrium (22). Biopsies from patients with IUAs contain 50% to 80% of fibrous tissue in the uterine wall, limiting uterine myometrial activity and reducing the perfusion of steroidal sex hormones, resulting in atrophy, compared with 13% to 20% in patients without IUAs (23).

The clinical symptoms of AS are usually a range of menstrual disorders, such as absence of bleeding or shortened days of flow. However, the most invalidating symptoms are infertility due to problems of embryo implantation in the endometrium or pregnancy complications, such as the increased likelihood of miscarriage or placenta accreta (24).

Dilation and curettage were widely used for the treatment of IUAs before the widespread use of hysteroscopy (25, 26) as the gold standard to diagnose and treat IUA. Hysteroscopy allows not only direct visualization of the intrauterine cavity, determining the location, extent, and type of adhesions (27), but also the treatment of IUA with hysteroscopic scissors at the same time, minimizing the destruction of healthy endometrium (24) and therefore reducing the need for a second surgery (28, 29). Nevertheless, fertility restoration after a hysteroscopic procedure is not always achieved, frequently requiring more surgeries in moderate and severe AS cases (30).

Moreover, other strategies, such as estrogen-based hormonal therapies, hyaluronic acid treatment, and intra-uterine device placement among others, have been proposed to prevent the recurrent formation of IUAs (31) after hysteroscopic removal. However, evidence is not clear for estrogen-based therapies and several points, such as optimal administration route and dose hormonal therapies, need to be addressed.

STEM CELLS FOR AS TREATMENT

Stem cells have the ability to divide asymmetrically and to differentiate into various cell types, while maintaining the stem cell lineage. Many kinds of stem cells are present in adult tissues that actively participate in the maintenance and regulation of the tissues thanks to their niche, which is the local microenvironment that supports the undifferentiated and multipotent state by providing adequate paracrine signals. Due to the role of stem cells in the regeneration of tissues when damage occurs, interest has increased in their use for the treatment of different conditions, such as heart disease or neurological disorders (32, 33).

As previously mentioned, endometrial stem cells are present on the basal layer and play a fundamental role in regenerating the functional layer at each menstrual cycle (34). However, their diminished number and function in AS result in a thin and afunctional endometrium, impairing embryo implantation (2, 26, 35). That is the reason why the use of stem cells for the treatment of endometrial diseases such as AS has been proposed by various researchers, since these delivered stem cells seem to be able to replace damaged cells in the endometrium (2, 9, 36–39).

One of the most widely used techniques to rapidly identify stem cells is the use of the fluorescent DNA-binding dye Hoechst 33342 in the cells, and the subsequent evaluation of the Side Population (SP) by flow cytometry. This is possible because stem cells can efficiently efflux the dye, due to a greater quantity of ATP-binding cassette (ABC) transporters than are present in cells without stem cell properties (40, 41). This technique has been widely used to identify stem cells from the endometrium, with the presence of endothelial, stromal, and epithelial cells, but with a clear predominance of endothelial cells (42–45).

Clinical treatments of AS have been proposed to use stem cells from different sources, such as the use of adipose-derived mesenchymal stem cells in a mouse model or amniotic mesenchymal stromal cells in a rat model (46, 47).

The use of bone marrow-derived stem cells (BMDSCs) to repopulate the endometrium was first demonstrated by Taylor's group (8). Endometrial biopsies of women who had undergone bone marrow transplantation from an HLA (human leukocyte antigen)-mismatched donor, demonstrating for the first time the presence of epithelial and stromal cells from the donor in the endometrial tissue of recipient women.

In a later study by this research group, AS in a mouse was caused by traumatizing the uterus in a reproducible manner (48). Then, BMDSCs from male mice were transplanted in a mouse AS model. Y+ cells were detected in the endometrium of the treated group, improving fertility after uterine damage, with nine out of ten mice conceiving in the treated group, compared with only three out of ten in the non-treated group.

According to another study, systemic administration of uterine-derived cells (UDCs) or BMDSCs resulted in better recruitment to the uterus than was achieved by local injection, with BMDSC engraftment being greater than that with UDCs (49). In that study, a mouse AS model achieved by local injury to a single uterine horn was created. Then, BMDSCs or UDCs expressing green fluorescent protein (GFP) were injected locally or systemically (i.e., intravenously). The fluorescence-activated cell sortin (FACS) analysis of the uterine tissues two or three weeks after administration revealed a greater engraftment following systemic administration of BMDSCs, compared with local administration of UCDs.

On the other hand, Zhao's group established a thin endometrium rat model by infusing ethanol into the uterine cavity (50). The BMDSCs used were mesenchymal stem cells transplanted by uterine perfusion. The uterine thickness of the treated group was improved, while the level of proinflammatory cytokines was decreased in the treated group, compared with the control group, suggesting a role for the immunomodulatory properties of BMDSCs in the treatment of endometrial diseases.

On the whole, all these findings suggest that stem cell therapies may be useful in treating patients with refractory AS or endometrial atrophy (EA), suffering infertility. However, no clear type of cells or optimal dosage has been clearly determined.

CD133+ BMDSCS FOR AS TREATMENT

Hematopoietic stem cells (HSCs) are a group of multipotent cells that reside in the bone marrow and are able to differentiate into all blood cell types. They can also differentiate into different cell lineages, such as hepatocytes (51). HSCs maintain their undifferentiated state thanks to their niche, with distinct hematopoietic progenitors living in distinct niches (52–54). One group of these HSCs is the endothelial progenitor cells (EPCs), which can originate from hemangioblast or myeloid progenitor lineages (52, 55, 56).

At first it was thought that new blood vessel formation in adults was exclusively due to angiogenesis, a process through which new blood vessels form from pre-existing vessels. However, the discovery of EPCs in the circulation changed this paradigm (57), since these cells can be mobilized from the bone marrow and form new blood vessels through a process called vasculogenesis (55, 58). Furthermore, these cells can replace damaged endothelial cells in the arteries and veins, helping to repair the endothelial monolayer in a process called re-endothelialization (59).

Cells derived from the bone marrow express different markers that allow us to distinguish some cell lines from others. A subpopulation of cells composed mostly of primitive HSCs and EPCs can be isolated by focusing on the CD133+ and VEGF-2 markers with magnet activated cell sorting (MACS) and FACS technologies (60, 61). Autologous CD133+ cells can be transplanted to patients in order to promote wound healing and revascularization in damaged tissues. Moreover, HSCs can transdifferentiate into functional cardiomyocytes when co-cultivated with rat cardiomyocytes (62, 63) to treat certain cardiovascular diseases (64). In fact, CD133+ cells can even promote the repair of injured spinal cord in animal models, enhancing angiogenesis, astrogliosis, axon growth, and functional recovery. Meanwhile, CD133- cell controls failed to promote axon growth and did not achieve functional recovery (65).

CD133+ cells usually exert effects through different mechanisms of action (MoA): paracrine activity, endometrial stem cell differentiation/cell fusion, endometrial resident stem cell recruitment/activation,

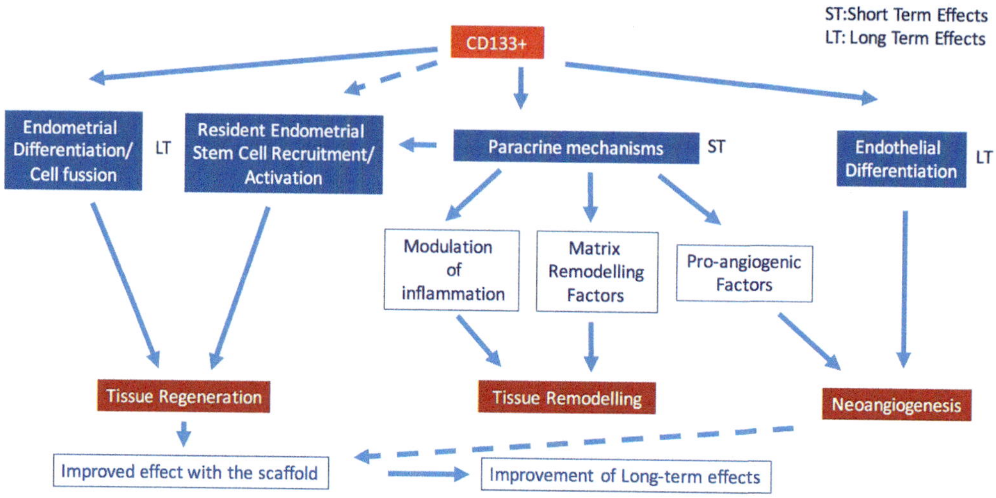

FIGURE 24.1 Different MoA of CD133+ cells consisting of tissue regeneration, tissue remodeling, or neoangiogenesis. Some mechanisms act in the short term whereas others have long-term effects.

and endothelial differentiation. Paracrine activity and endometrial resident stem cell recruitment/activation cause an immediate short-term effect and seem to work at a larger scale than endometrial and endothelial differentiation, which has a more stable long-term effect seems to be more residual. Interestingly, the paracrine effects can contribute to endothelial regeneration as well as to endometrial resident stem cell recruitment/activation as shown in Figure 24.1.

In a preclinical study, human CD133+ cells were incubated with superparamagnetic iron oxide nanoparticles and injected into a rodent AS model. The engraftment and location of the labeled cells were analyzed by Prussian blue staining, while the endometrial proliferation was analyzed by Ki67 immunohistochemistry. The accumulation of iron deposits, detected by Prussian blue, revealed that the CD133+ cells were recruited mostly around small endometrial vessels of damaged horns, whereas there was no accumulation of CD133+ cells in other organs. The Ki67 staining evaluation showed an increase of Ki67 antigen in the epithelial glands of treated horns in comparison with non-treated horns. The proliferation rate increased from a mean of 14% with a standard deviation of 10.37% to 23.15% ± 10.89% ($P < 0.01$) after intrauterine injection, whereas it increased from 6.92% ± 7.03% to 20.55% ± 10.89% ($P < 0.05$) after tail vein injection, suggesting than CD133+ cells induce the proliferation of nearby endometrial cells. By qPCR analysis, we saw that thrombospondin 1 expression was significantly upregulated (2.065 vs. 0.752 fold-change; $P < 0.05$), while IGF-1 expression was downregulated, which indicates that CD133+ cells may induce endometrial cell proliferation through paracrine effects.

On the clinical side, our group designed a prospective, experimental, no–n-controlled, open-label study involving 18 patients aged 30–45 years with refractory AS or EA, of which 16 completed the study (66). The end-point was assessing the safety and efficiency of the CD133+ BMDSCs as a potential therapy for refractory cases of AS and EA. Prior to the enrolment, eight out of the 11 patients diagnosed with AS and all the patients diagnosed with EA had previously undergone a mean of 2 (range 1–9) hysteroscopies and two subsequent cycles of hormone-replacement therapy (HRT) after each hysteroscopy with neither clinical nor sonographic improvements (Figure 24.2).

Briefly, uterine cavities were assessed following the American Fertility Society Classification. Once assessed, adhesions were removed in AS cases with sharp and blunt scissors. All patients received HRT immediately after the surgery for two months and BMDSCs were mobilized by granulocyte-colony stimulating factor (G-CSF) injection (10 μg/kg/day) for five days. After these five days, peripheral mononuclear blood cells were isolated by apheresis through peripheral venous access, and CD133+ cell selection and isolation were carried out using MACS. Between two to three volemias of blood from each patient were processed to obtain at

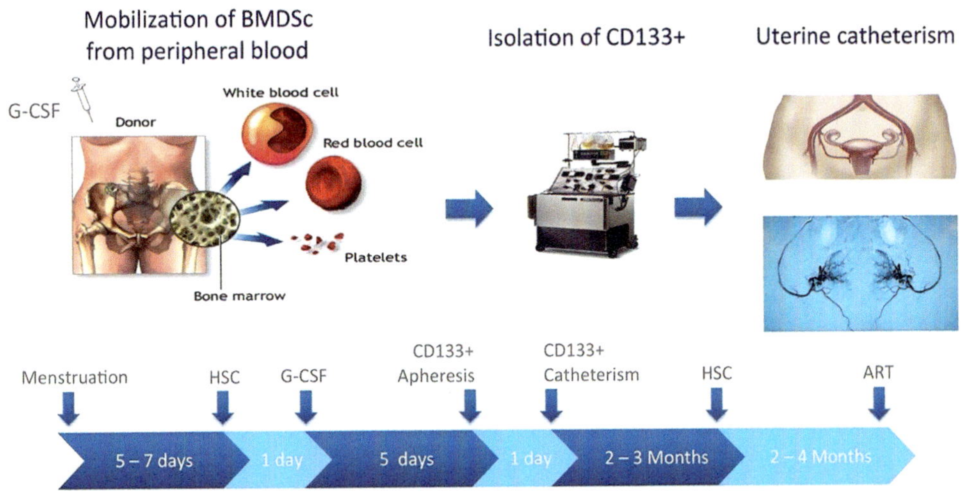

FIGURE 24.2 Summary of the study. Once patients were included in the study, an initial ecographical and hysteroscopic diagnosis was performed in the proliferative phase. Then, mobilization of BMDSCs was induced in all of the enrolled patients by pharmacological administration of recombinant human granulocyte colony-stimulating factor (G-CSF). After a second-look hysteroscopy, assisted reproductive technology (ART) was performed and reproductive outcomes, as well as endometrial thickness, were assessed.

least 50 million CD133+ cells in each case. CD133+ cells were delivered into the endometrial stem cell niche by intra-arterial catheterization, using a technique routinely performed for embolization of uterine fibroids (67) and 15 mL of a saline suspension with the CD133+ cells (containing 42–200 × 10^6 cells, mean 123.56 × 10^6) was injected into the spiral arterioles next to the endometrial stem cell niche (68) (Figure 24.3).

Three months after the stem cell instillation, a second-look hysteroscopy was performed. An improvement in the scores and stages was observed in all the patients, although the cavity was not completely normalized. Specifically, all patients diagnosed with stage III AS improved to stage I, whereas one of the two patients diagnosed with stage II showed a completely normalized endometrial cavity, with the other improving to stage I. Regarding endometrial thickness, a significant improvement was observed in the AS patients from a mean of 4.3 mm (range from 2.7 to 5.0 mm) to 6.7 mm (range 3.1–12 mm) and from a mean of 4.2 mm (range 2.7 to 5.0 mm) to 5.7 mm (range 5–12 mm) in the EA group.

After the first month of CD133+ cell treatment, the menstrual cycles resumed in 15 of the 16 patients. However, the duration and volume of menstruation decreased progressively after CD133+ cell treatment, with a mean of 5.06 days and 2.69 pads per day in the first month, to a mean of 3.25 days and 1.75 pads per day in the third month.

Angiogenesis in endometrial tissue was determined by measuring the total number of blood vessels formed after treatment with CD133+ cells and α-SMA immunofluorescence analysis in paraffin-embedded sections of the endometrium (69) before and three and six months after CD133+ cell therapy. This analysis revealed a statistically significant increase in the amount of blood vessels three months after CD133+ cell injection, together with a visible improvement of the stromal and epithelial layers of the endometrium, which progressively decreased after six months in AS and EA patients (Figure 24.4).

Finally, functionality of the endometrium was measured by assessing reproductive outcomes of the treated patients. After the cell treatment, three patients became pregnant spontaneously, resulting in two healthy babies and a miscarriage due to a premature rupture of membranes. Additionally, after 14 embryo transfers, seven positive pregnancies were obtained, resulting in three biochemical pregnancies, one miscarriage at the ninth week, one ectopic pregnancy, and three healthy new-borns out of two patients.

In summary, this study demonstrated that the use of CD133+ BMDSCs can be useful in the treatment of AS and other endometrial diseases. Indeed, the European Medicines Agency (EMA) as well as the US Food and Drug Administration (FDA) have acknowledged these results and have issued a positive

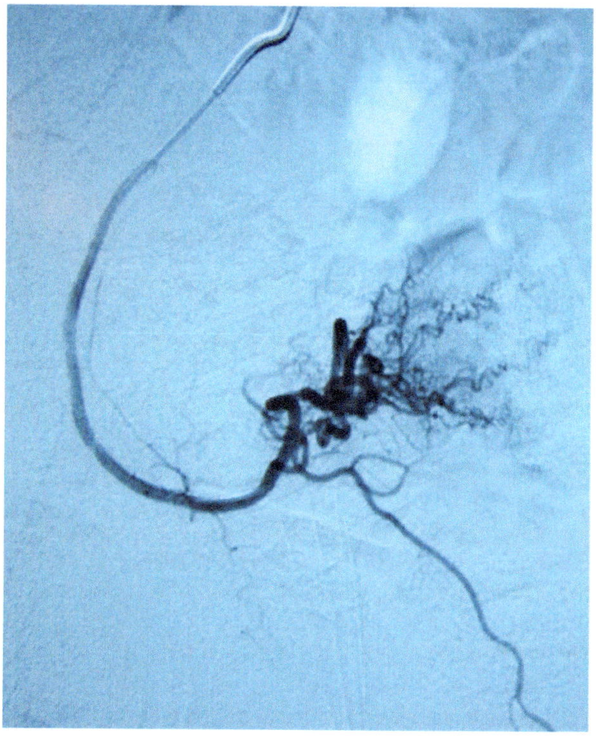

FIGURE 24.3 Spiral arteries that vascularize the basalis layer of the endometrium are clearly visible when instillation is performed in the distal part of the uterine artery.

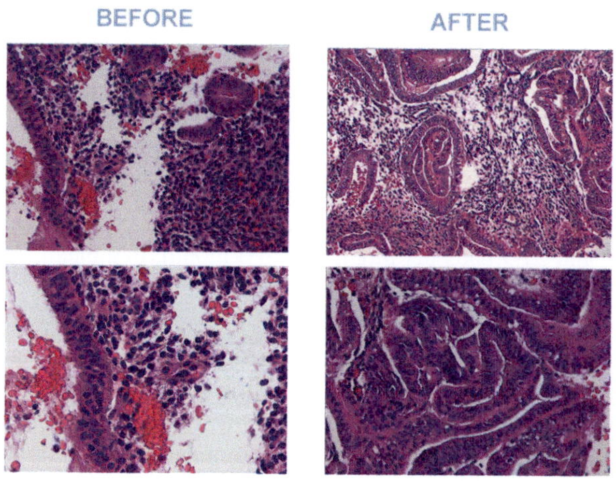

FIGURE 24.4 The endometrium has a unique capacity to reconstitute the stromal and epithelial compartments during the menstrual cycle. Intrauterine adhesions can be formed by endometrial tissue, connective tissue, or muscular tissue. Hysteroscopy is considered to be the gold standard by which to diagnose intrauterine adhesions. In severe cases, the cavity can practically be occluded by fibrosis and extensive lesions. This image shows histological improvement in a patient after stem cell therapy.

opinion to consider CD133+ cells as the first Orphan Drug Designed (ODD) therapy for the treatment of AS, categorizing these cells as Advanced Therapy Medicinal Product (ATMP); supervised phase I/II and III trials are currently being performed. Obviously, further developments need to be performed to determine optimal dosage, to achieve a long-term follow-up, and to carry out a randomized trial, but this can be considered to be the first focused proof-of-principle study. Currently, a phase I/II trial is being conducted in order to assess long-term safety and optimal dosage as well as to determine different MoAs.

FUTURE PERSPECTIVES IN STEM CELL THERAPY AND REGENERATIVE MEDICINE

In recent years, stem cells have been proposed to treat certain endometrial diseases, such as AS, with relative success, as described above. However, the use of stem cell therapies alone may not be enough in some cases of severe AS (70).

In human tissues, the cells are surrounded by the extracellular matrix (ECM), which directs cell function by a complex group of biomacromolecular interactions between the cells and the ECM. Therefore, ECM remodeling strategies, together with stem cells, seem to play a critical role in re-establishing organ functionality.

In this sense, regenerative medicine is a new strategy to treat certain pathological conditions by creating new tissues with which to restore organ function (71). Therefore, emerging regenerative medicine therapies are starting to include cell or stem cell techniques with tissue engineering, consisting of the use of scaffolds or biomaterials to enhance cell growth in three dimensions (3D), as well as ECM to represent a more accurate structural and functional complexity of the tissue (72, 73).

To achieve adequate tissue engineering, biomaterials which mimic native endometrial tissues and provide structural and mechanical support with which cells can interact are appropriate. Some biomaterials can release some drugs, growth factors, or other bioactive compounds in a controlled manner (74–76), contributing to the enhancement of uterus regeneration by providing some molecular factors, facilitating cellular signaling, stem cell recruitment, proliferation, or differentiation (77). Some research groups have used this approach for endometrial regeneration, such as Zhang's group, which used a heparin-poloxamer hydrogel with encapsulated 17β-estradiol (77), whereas Lin's group used a collagen scaffold loaded with vascular endothelial growth factor (VEGF) (78), and Cai's group, which used a new, porous scaffold which combined GelMA and Na-alginate characteristics and released bFGF (79).

Another interesting approach in tissue engineering is the use of decellularized scaffolds with high biocompatibility and low rejection effects compared with organ transplantation (80). The use of decellularized uterus and posterior regeneration by autologous cells has already been successfully used for uterus regeneration in different rat (81–83) and pig models (84). However, this approach has its limitations because of the limited availability of uterus from human donors for decellularization and the presence of some cell residues after decellularization (85).

Finally, another interesting strategy for organ regeneration consists of using 3D bioprinting technologies, which could improve the functionality of tissue-engineered scaffolds by using an appropriate bio-ink and cell source (86–88).

BIBLIOGRAPHY

1. Gargett CE, et al. *Human Reproduction Update*. 2016;22:137–63.
2. Deane JA, et al. *Current Opinion in Obstetrics and Gynecology*. 2013;25:193–200.

3. Gargett CE. *Human Reproduction Update*. 2007;13:87–101.
4. Ulrich D, et al. *Human Reproduction*. 2014;29(9):1895–905.
5. Schildberg FA, et al. *Cytometry Part A*. 2018;93:871–5.
6. Bazoobandi S, et al. *Stem Cells International*. 2020;2020:8849537.
7. Cervelló I, et al. *PLoS ONE*. 2012;7(1):e30260.
8. Taylor HS. *JAMA*. 2004;292:81–5.
9. Du H, et al. *Current Opinion in Obstetrics and Gynecology*. 2010;22:235–41.
10. Aghajanova L, et al. *Biology of Reproduction*. 2010;82(6):1076–87.
11. Achache H, et al. *Human Reproduction Update*. 2006;12:731–46.
12. di Guardo F, et al. *Medical Hypotheses*. 2020;134:109521.
13. Senturk LM. *Current Opinion in Obstetrics and Gynecology*. 2008;20(3):221–8.
14. Kumar S, et al. *Journal of Clinical Endocrinology and Metabolism*. 1998;83:4443–50.
15. Jabbour HN, et al. *Endocrine Reviews*. 2006;27:17–46.
16. Bellver J, et al. *Current Opinion in Obstetrics and Gynecology*. 2018;30:229–36.
17. Lessey BA, et al. *Fertility and Sterility*. 2019;111:611–7.
18. Evans-Hoeker EA, et al. *Seminars in Reproductive Medicine*. 2014;32(5):392–401.
19. Dreisler E, et al. *International Journal of Women's Health*. 2019;11:191–8.
20. Berman JM. *Seminars in Reproductive Medicine*. 2008;26:349–55.
21. Santamaria X, et al. *Human Reproduction*. 2018;33(8):1374–80.
22. Dmowski WP, et al. *Obstetrics and Gynecology*. 1969;34(2):288–99.
23. Yaffe H, et al. *American Journal of Obstetrics and Gynecology*. 1978;130(5):599–601.
24. Khan Z, et al. *Journal of Minimally Invasive Gynecology*. 2018;25:218–28.
25. AAGL Practice Report. *Journal of Minimally Invasive Gynecology*. 2010;17(1):1–7.
26. Yu D, et al. *Fertility and Sterility*. 2008;89:759–79.
27. Salazar CA, et al. *Current Opinion in Obstetrics and Gynecology*. 2017;29:249–56.
28. Cholkeri-Singh A, et al. *Current Opinion in Obstetrics and Gynecology*. 2016;28:250–4.
29. Hysteroscopy. *AORN Journal*. 2019;110:17–9.
30. Capella-Allouc S, et al. *Human Reproduction*. 1999;14:1230–3.
31. Johary J, et al. *Journal of Minimally Invasive Gynecology*. 2014;21:44–54.
32. Alessandrini M, et al. *South African Medical Journal*. 2019;109(8b):S71–8.
33. Müller P, et al. *Cellular Physiology and Biochemistry*. 2018;48:2607–55.
34. Gargett CE, et al. *Reviews in Endocrine and Metabolic Disorders*. 2012;13(4):235–51.
35. Gargett CE, et al. *Fertility and Sterility*. 2012;98:11–20.
36. Santamaria X, et al. *Human Reproduction Update*. 2018;24(6):673–93.
37. Kilic S, et al. *Journal of Assisted Reproduction and Genetics*. 2014;31(8):975–82.
38. Nagori C, et al. *Journal of Human Reproductive Sciences*. 2011;4:43–8.
39. Azizi R, et al. *Biomedicine and Pharmacotherapy*. 2018;102:333–43.
40. Challen GA, et al. *Stem Cells*. 2006;24(1):3–12.
41. Shimoda M, et al. In: *Methods in Molecular Biology*, edited by Biostatistical Methods. Humana Press Inc.; 2018:49–59.
42. Kato K, et al. *Human Reproduction*. 2007;22(5):1214–23.
43. Masuda H, et al. *PLoS ONE*. 2010;5(4):e10387.
44. Tsuji S, et al. *Fertility and Sterility*. 2008;90:1528–37.
45. Cervelló I, et al. *PLoS ONE*. 2010;5(6):e10964.
46. Gan L, et al. *Cytotherapy*. 2017;19(5):603–16.
47. Çil N, et al. *Molecular Biology Reports*. 2020;47(6):4541–52.
48. Alawadhi F, et al. *PLoS ONE*. 2014;9(5).
49. Liu Y, et al. *Journal of Cellular and Molecular Medicine*. 2018;22(1):67–76.
50. Zhao J, et al. *Reproductive Sciences*. 2015;22(2):181–8.
51. Lagasse E, et al. *Purified Hematopoietic Stem Cells can Differentiate into Hepatocytes in Vivo*. 2000.
52. Wilson A, et al. *Nature Reviews Immunology*. 2006;6:93–106.
53. Morrison SJ, et al. *Nature*. 2014;505:327–34.
54. Lucas D. In: *Advances in Experimental Medicine and Biology*. Springer New York LLC; 2017:5–18.
55. Ceradini DJ, et al. *Trends in Cardiovascular Medicine*. 2005;15:57–63.
56. Rehman J, et al. *Circulation*. 2003;107(8):1164–9.
57. Asahara T, et al. *Science*. 1997;275(5302):964–7.
58. Asahara T, et al. 1999. Available from: http://www.circresaha.org.
59. Urbich C, et al. *Circulation Research*. 2004;95:343–53.

60. Pelagiadis I, et al. *Cytotherapy.* 2012;14(6):701–6.
61. Rountree CB, et al. *Journal of Visualized Experiments.* 2011;56:e3183.
62. Badorff C, et al. *Circulation.* 2003;107(7):1024–32.
63. Rupp S, et al. *Basic Research in Cardiology.* 2004;99(1):61–8.
64. Rafii S, et al. 2003. Available from: http://www.nature.com/naturemedicine.
65. Kamei N, et al. *Journal of the Neurological Sciences.* 2013;328(1–2):41–50.
66. Santamaria X, et al. *Human Reproduction.* 2016;31(5):1087–96.
67. Goodwin SC, et al. *Obstetrics & Gynecology.* 2008;111.
68. Murakami K, et al. *Endocrinology.* 2014;155(11):4542–53.
69. Gee MS, et al. *American Journal of Pathology.* 2003;163:183–93.
70. Liu F, et al. *Advanced Healthcare Materials.* 2020;8(14).
71. Cossu G, et al. *The Lancet.* 2018;391:883–910.
72. Harrison RH, et al. *Tissue Engineering - Part B: Reviews.* 2014;20(1):1–16.
73. Shadish JA, et al. *Nature Materials.* 2019;18(9):1005–14.
74. Han Q, et al. *Frontiers in Bioengineering and Biotechnology.* 2020;8(February):1–9.
75. Li C, et al. *Advanced Materials.* 2019;31(17):1900291.
76. Li X, et al. *Biomaterials.* 2011;32(32):8172–81.
77. Zhang SS, et al. *FASEB Journal.* 2020;34(1):446–57.
78. Lin N, et al. *Biomaterials.* 2012;33(6):1801–7.
79. Cai Y, et al. *Acta Biomaterialia.* 2019;84:222–30.
80. Brännström M, et al. *The Lancet.* 2015;385(9968):607–16.
81. Santoso EG, et al. *PLoS ONE.* 2014;9(7):e103201.
82. Hellström M, et al. *Fertility and Sterility.* 2016;106(2):487-496.e1.
83. Miyazaki K, et al. *Biomaterials.* 2014;35(31):8791–800.
84. Campo H, et al. *Facts, Views & Vision in ObGyn.* 2019;11(1):87–99.
85. Caralt M, et al. *American Journal of Transplantation.* 2015;15(1):64–75.
86. Kesti M, et al. *Acta Biomaterialia.* 2015;11(1):162–72.
87. Hu S, et al. *Current Opinion in Cell Biology.* 2018;54:37–42.
88. Levato R, et al. *Advanced Materials.* 2020;32(12).

25 Differential Genetic Diagnosis between Leiomyoma and Leiomyosarcoma

Alba Machado-Lopez and Aymara Mas

INTRODUCTION

Uterine leiomyomas (LM, or fibroids) and leiomyosarcomas (LMS, or uterine sarcomas) are myometrial tumors with common morphological characteristics, yet cytogenetic and molecular differences indicate they are biologically unrelated tumors. These tumors are typically treated *via* surgical (laparoscopic) removal, but the absence of standardized and objective criteria by which to differentiate LM and LMS before and even after surgery represents a current diagnostic challenge (1). Indeed, this gap leads to unnecessary, invasive procedures and additional costs to patients and the health care system (2).

Whereas LM represent a frequent (~75% prevalence) and benign condition, LMS are aggressive tumors, characterized by early metastasis, poor prognosis, and high rates of recurrence, with limited therapeutic efficacy (3–5). Due to the risk of hidden LMS, the use of laparoscopic myomectomy for LM has been under debate since the FDA released a warning in 2014 after estimating the frequency of LMS as one in 350 women undergoing hysterectomy or myomectomy for supposed benign LM (2, 6, 7).

Therefore, laparoscopic myomectomy has decreased in frequency as a first-line treatment for myometrial tumors, whereas laparotomy-based procedures have become more common, along with a concomitant rise in morbidity and mortality due to the invasiveness of these procedures (8–10).

In this sense, standardized clinical management of these myometrial tumors is needed to classify patients based on the possible risk of hidden LMS, not only to establish safe diagnostic algorithms but also to reach a consensus to support treatment recommendations (11). Identification of single-nucleotide variants (SNVs), insertions/deletions (indels), and copy number variations (CNVs), as well as differential gene expression, could also provide a preliminary panel of targets and biomarkers that could be exploited as a first step towards differential tumor diagnosis (2).

DOI: 10.1201/9781003024941-25

STATE OF THE ART: WHAT DO WE KNOW TODAY?

Clinical features, imaging findings, or differences in serum levels among tumor patients and healthy individuals remain the only operational diagnostic options for LM and LMS, although these measures can lack specificity and thus complicate diagnoses (2). In this sense, the main issue to consider is the general condition of the patient, bearing in mind the clinical symptoms, not only in terms of disease but also in terms of discomfort. Clinical presentation of myometrial tumors can vary from being asymptomatic to presenting with abnormal uterine bleeding, pain, rapid abdominal tumor growth, and complications during pregnancy (12, 13). In the worst scenario of local and regional spread, myometrial tumors may manifest with gastrointestinal or urinary tract symptoms; in the case of hematogenous spread, myometrial tumors may present with pulmonary symptoms (14).

Imaging modalities, such as MRI and/or CT, can reveal specific features of LM and LMS tumors (Figure 25.1), although they have mostly been used to identify extrauterine extension into the pelvis, and features of malignancy are indistinguishable, especially for small tumors (4, 15, 16).

Other studies have reported elevated levels of the serum markers CA125 (17) and lactate dehydrogenase (LDH), specifically LDH-A and LDH-D, in patients with LMS compared to LM (18). However, these markers show overlapping expression in preoperative serum levels between LM and early-stage LMS, limiting their clinical utility.

Diagnosis of uterine tumors may also come from pathological analysis after surgical procedures. Based upon microscopic morphological features, such as coagulative tumor necrosis, atypical cytology, cellularity, and mitotic index, tumors can be categorized as benign variants of LM (mitotically active, cellular, and atypical-bizarre), smooth muscle tumors of uncertain malignant potential (STUMP), LMS, endometrial stromal sarcomas, or undifferentiated uterine sarcomas (Figure 25.1) (12). Furthermore, significantly higher levels of Ki67, the cellular tumor antigen (p53), and cyclin-dependent kinase inhibitors

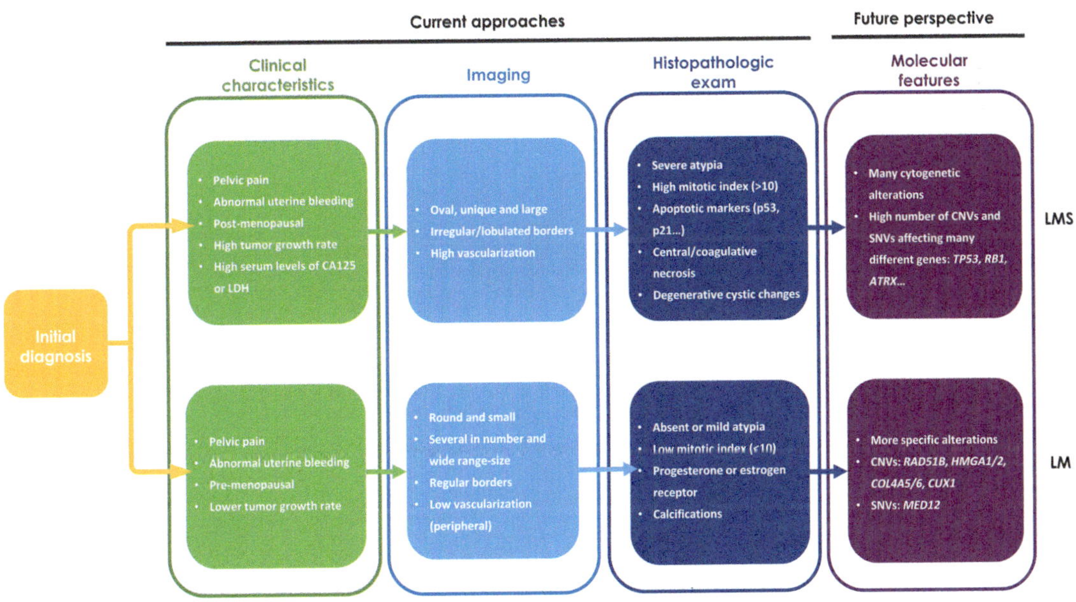

FIGURE 25.1 Current approaches for differential diagnosis of LMS and LM, and molecular features of both tumor types as the basis for molecular differentiation.

(p16, p21, p27), as well as other apoptosis regulators (BAX, BCL-2) have been detected in LMS compared to benign tumors (19, 20). Unfortunately, these features are not readily evaluated from small-volume core biopsies or needle aspirations to differentiate malignant from non-malignant myometrial tissue.

In this sense, biotechnological advances mainly focused on molecular differences may offer new promise for difficult-to-diagnose tumors such as LMS and LM (21–23).

NGS TECHNOLOGIES FOR DIFFERENTIAL DIAGNOSIS OF MYOMETRIAL TUMORS

Next-generation sequencing (NGS) technologies have been applied across various scientific disciplines, from basic research to translational medicine. Specifically, The Cancer Genome Atlas (TCGA) is one of the most relevant projects, which aims to characterize the molecular signatures of human cancers. Accordingly, genetic risk assessment, prognostic information, identification of molecular subtypes, or development of therapeutic strategies are some of the possible approaches currently being applied to breast, ovarian, endometrial vulvar, and cervical cancers.

Most "second-generation" sequencing methods used currently in research on myometrial tumors are based on Sanger sequencing but incorporate several modifications (Figure 25.2). Thermo Fisher and Illumina are the major companies developing such methodologies, which mainly differ in their chemistry. Thermo Fisher's approach is based on the pH change when a nucleotide is incorporated into a DNA chain and a hydrogen ion is released; Illumina's approach is based on "bridge amplification" and emission of a fluorescent signal when a nucleotide is incorporated to a DNA chain (24).

DNA sequencing technologies to study the genomic profile of LM and LMS include whole-genome sequencing (25, 26), whole-exome sequencing, and/or targeted sequencing (27), as well as genome-wide association studies (28). Further clinical applications of NGS include sequencing of cell-free DNA fragments circulating in a patient's bloodstream (29).

Similarly, there are many approaches based on RNA sequencing, from whole-transcriptome to whole-exome, as well as targeted sequencing in LM and LMS. The most important application of these

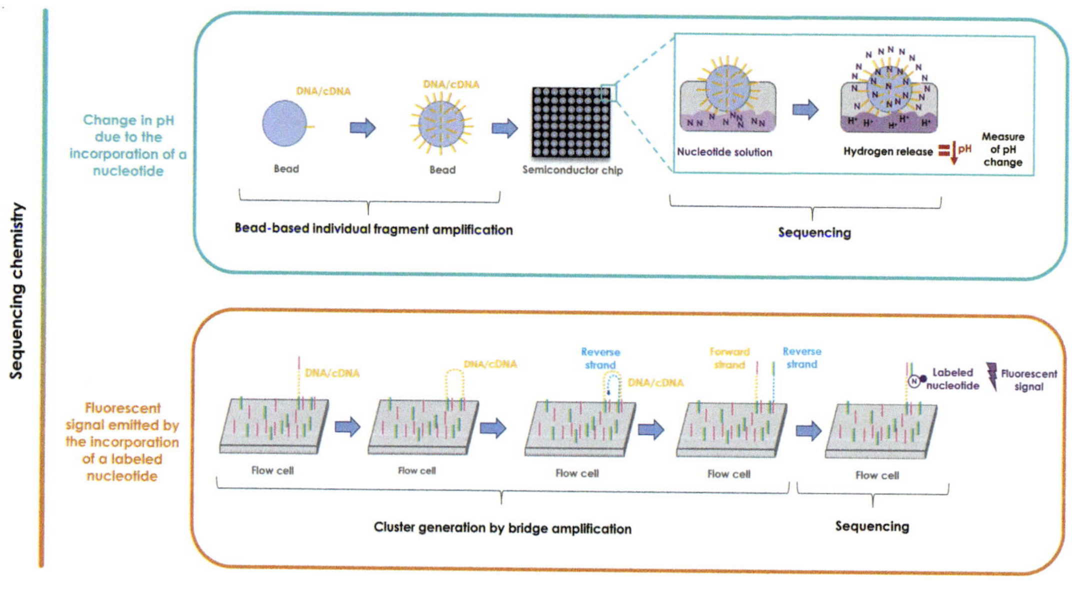

FIGURE 25.2 NGS methodology—second-generation sequencing chemistries.

technologies is related to expression analysis, such as differential gene expression (30, 31), as in the form of expression quantitative trait loci (eQTL) studies (32), but other applications include alternative splicing analysis, detection of gene fusions (23), non-coding RNA sequencing (33, 34), and cell-free RNA sequencing (35).

MOLECULAR FEATURES OF LEIOMYOMAS AND LEIOMYOSARCOMAS AS CLINICAL DIAGNOSTIC TOOLS

At the molecular level, myometrial tumors show unbalanced karyotypes as well as non-specific and complex alterations, such as SNVs, small indels, amplifications, and gene fusions (36, 37). Specifically, up to 50% of LM have cytogenetic abnormalities, mainly affecting 6p21, 7q, and 12q15 chromosome regions as the main drivers, whereas point mutations in gene *MED12* are detected in up to 70% of tumors (38). Chromosomal rearrangements between *HMGA1/HMGA2* and *RAD51B*, or between *COL4A5* and *COL4A6* are also frequent in LM, mainly resulting in overexpression of these genes or reduced expression of *CUX1* or *CUL1* due to 7q deletions (25). Furthermore, heterozygous mutations in the *FH* (fumarate hydratase) gene can cause Reed's syndrome, also known as hereditary leiomyomatosis, and renal cell cancer syndrome, which is characterized by multiple cutaneous and uterine leiomyomas (Figure 25.3).

On the other hand, LMS has a higher incidence of unbalanced karyotypes and complex alterations, accumulating frequent abnormalities across almost all chromosomes. CNVs have also been detected in LMS—losses are more frequent than gains, with losses affecting chromosomes 2, 4, 11, and 12, and gains affecting chromosomes 1, 6, and 7 (39, 40). These losses may help explain the pathogenesis of LMS, since they most frequently affect genes that may act as tumor suppressors, including *TP53*, *RB1*,

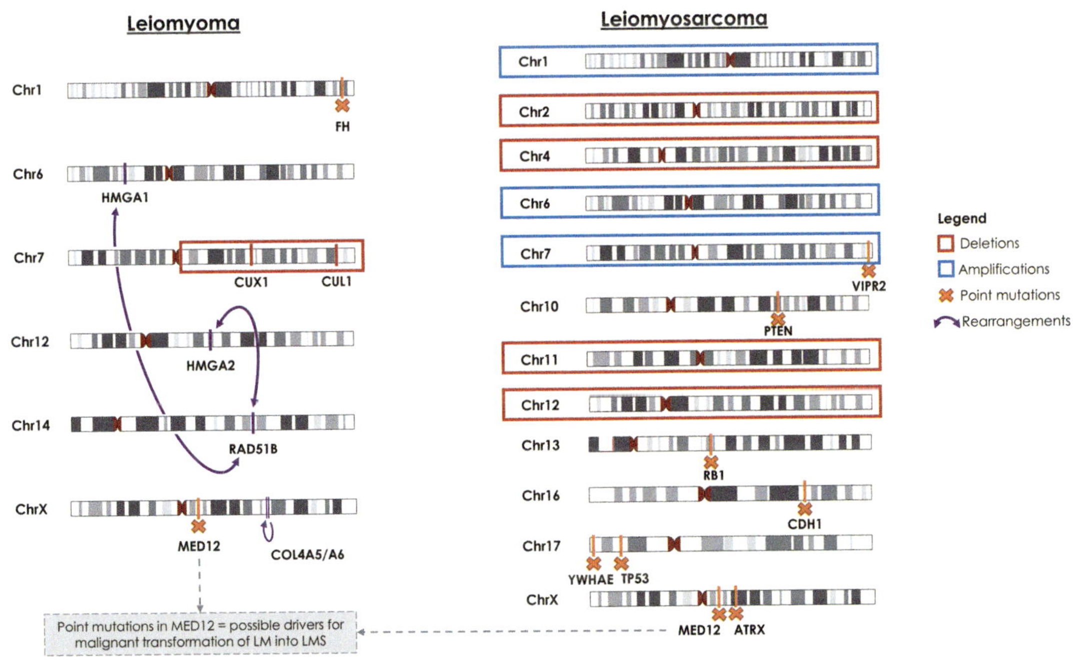

FIGURE 25.3 Most frequent genetic alterations in LM and LMS.

PTEN, and *CDH1* (26, 41). Other frequently mutated genes in LMS are *VIPR2*, *YWHAE*, and *ATRX* (42), with mutated *ATRX* associated with worse clinical outcome due to loss of expression, resulting in the alternative "lengthening of telomeres" phenotype (43, 44). These and other mutated genes, such as *MED12*, are considered to be "drivers," suggesting that LMS arise from LM or that these mutations favor malignant growth in LMS (Figure 25.3) (45).

At the RNA level, several genes, including *CDC20, CCNA2, CCNB1*, and *CCNB2*, show significantly different expression levels between LMS and LM (46). Although the number and type of such genes vary between studies, depending on the experimental approach or the sample size, most studies have detected different molecular patterns that discriminate between LMS and LM, usually reflecting altered pathways related to transcription regulation, DNA repair, or cell cycle (23, 31, 40, 47, 48). Furthermore, some non-coding RNAs, such as micro RNAs (miRNAs), show differential expression between LMS and LM, allowing molecular classification and even suggesting new treatment strategies (49).

Epigenomic features also differ between the myometrial tumors. For example, methylation profiles show not only that LMS are globally hypomethylated in comparison to LM, but also that specific loci could be used as differential markers (40, 50).

In summary, LM and LMS have different genomic/transcriptomic features that can explain their different etiopathologies, opening up new directions for investigation of these myometrial tumors. In addition, such features may help us develop clinically useful tests for early evaluation of malignancy risk and possible treatment of these pathologies.

STRENGTHS AND LIMITATIONS

In the past decade, diagnosis of patients with uterine LM or LMS has been essentially based on pathological evaluation of the tumor sample, ideally a biopsy or surgical specimen. Consequently, any formalin-fixed paraffin-embedded (FFPE) tissues remaining stored in pathology archives after diagnosis represent an immense resource for molecular analysis in both retrospective and prospective clinical studies (51).

Unfortunately, the available "omics" profiles for LMS have been limited due to its rare incidence of 1 in 350 women undergoing hysterectomy or myomectomy for presumed benign LM (23, 52–55). Moreover, nucleic acids are difficult to extract from FFPE tissue because of the need to remove paraffin and neutralize the covalent protein–DNA interactions resulting from the fixation process. In addition, tissue preparation (i.e., fixation, paraffin embedding, and archival storage) contributes to fragmentation, cross-linking, and chemical modification of FFPE tissue-derived nucleic acids. Consequently, these changes may result in identification of false positive SNVs or indels due to deamination and depurination processes, which also affect gene expression (56).

On the other hand, the power of NGS to achieve in-depth analysis of large numbers of short sequences makes this an ideal technology to apply to the usually fragmented nucleic acids extracted from FFPE specimens, although standard methods of purification and quantification are essential. Likewise, it is important to consider that damage caused by FFPE preparation is randomly distributed across all DNA fragments but can be corrected for by sequencing depth with coverage levels of $\geq 80\times$ (23, 56).

Finally, it is important to note that, although genomic/transcriptomic analyses have provided insights into the roles of CNVs and differential gene expression in disease, they have been based on analysis of the entire tumor, which implies a technical limitation due to intratumoral heterogeneity. Heterogeneity between malignant and non-malignant cells within a tumor and their interactions within the tumor microenvironment are fundamental for a better understanding of tumor biology (57, 58). This implies the importance of sampling of tumor tissues for clinical diagnosis, since localized samples may not capture the whole spectrum of genetic changes across a tumor (59).

BARRIERS TO ACCESS

Over recent decades, classical diagnosis of tumor pathology, based on morphological classification, grading, and staging, has been replaced by a growing demand for molecular tumor subclassification, information on prognosis, and response to therapeutic targets (60). To address these demands, an extended spectrum of methods is required. These encompass molecular biology techniques, such as polymerase chain reaction (PCR), fluorescence *in situ* hybridization, gene array assays, and DNA/RNA sequencing.

Recent advances in genomic/transcriptomic knowledge through NGS technologies could provide further information about myometrial tumor status based on the identification of oncogenes and tumor suppressor genes—thus benefiting not only patients but also gynecologists, pathologists, oncologists, and other doctors. As a consequence, the technical advances outlined above might shift diagnostic tumor pathology from an interpretive, subjective science to a more objective, evidence-based practice for routine patient management.

Overall, we consider that the development of molecular-driven tests linked to functional histology could provide a framework to further advance personalized medicine and selective therapy design. The challenge, however, for both the pathologist and the gynecologist, is how to effectively integrate this morphological and molecular information into a comprehensive treatment plan that includes patient-specific risk assessment (61).

CONCLUSIONS AND FUTURE PERSPECTIVES

Currently, there are few non-surgical, medical therapies approved to treat myometrial tumors, and most only provide partial or temporary relief. This may be because the mechanisms involved in the tumors' appearance, evolution, and recurrence remain largely unknown, contributing to the slow progress in developing effective therapeutic options (2). Overcoming these problems requires novel methods or strategies for rapid, cost-effective, and non-invasive identification of biomarkers, not only at the time of diagnosis, but also during defined intervals throughout treatment.

Recently, the concept of a "liquid biopsy" has emerged as a minimally invasive alternative to surgical biopsies for solid tumors that harbor highly recurrent mutations, avoiding tumor-tissue sampling before and after treatment (62). Liquid biopsy is typically accomplished through withdrawal of a blood sample or other body fluids to provide tumor-specific information from circulating tumor cells (CTCs) and/or circulating cell-free DNA (cfDNA) or RNA (cfRNA) (63–65). All these factors are present in peripheral blood, escaping from apoptotic or necrotic cells of primary tumors and releasing DNA/RNA fragments into the circulation in different forms, including mRNA, miRNA, long non-coding RNA (lncRNA) (66), and circular RNA (circRNA) (67, 68). Similarly, tumor-educated platelets—blood platelets of cancer patients that contain RNA biomarkers derived from the tumor—have been studied in many different cancer types (e.g., breast, lung, ovary) and show advantages over other liquid biopsy sources due to their abundance and ease of isolation, among other features (69) (Figure 25.4).

Considering that cfDNA/cfRNA analysis provides a global picture of the genetic status of a disease, this approach could be used to monitor tumor patients during treatment or remission. Furthermore, CTC analysis provides molecular insights into the tumor at the single-cell level, which could be used to guide drug therapy after treatment failure or disease recurrence (64). In addition to cfDNA/cfRNA and CTCs, exosomes have recently received special consideration because they also represent tumor-derived material that can be found in multiple body fluids of patients (70, 71).

All these sources, however, present technical limitations, such as low abundance or difficulties in their normalization, that will eventually be overcome with development of sequencing methods. Yet, the

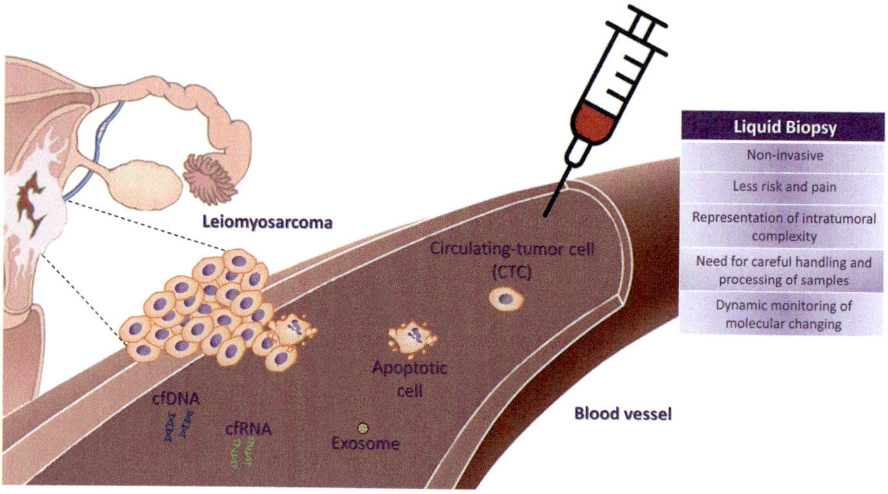

FIGURE 25.4 Liquid biopsy to identify and monitor tumor-specific mutations.

challenge is not isolation of nucleic acids from the tumor but rather detection of genetic variation like somatic mutations, insertions, deletions, or even amplifications and translocations in such a small fraction of total DNA/RNA in the bloodstream. In addition, these specific mutations may also be present because of mosaicism or benign/precancerous lesions.

Consequently, despite all the efforts to achieve comprehensive molecular analyses of cfDNA/cfRNA or CTCs for uterine tumors, the science of this approach continues to evolve, and important questions remain unanswered. Nevertheless, based on promising data recently published in this field, we anticipate that liquid biopsy, including circulating tumor DNA/RNA (ctDNA/ctRNA) or CTC analysis, might transform detection and management of these myometrial tumors soon, not only offering a better representation of intratumor complexity, but also aiding early tumor diagnosis, monitoring of tumor evolution, and assessment of patient prognosis.

REFERENCES

1. Amant F, et al. *Lancet Oncol.* 2015;16(15):1454–6.
2. Mas A, et al. *Biol Reprod.* 2019;101(6):1115–23.
3. Giuntoli RL, et al. *Gynecol Oncol.* 2003;89(3):460–9.
4. Kobayashi H, et al. *Mol Clin Oncol.* 2013;1(4):599–609.
5. Lusby K, et al. *Ann Surg Oncol.* 2013;20(7):2364–72.
6. Parker W, et al. *J Minim Invasive Gynecol.* 2016;23(3):303–8.
7. Halaska MJ, et al. *Int J Gynecol Cancer.* 2017;27(1):189–92.
8. Munro MG. *BJOG.* 2016;123(13):2188.
9. Siedhoff M, et al. *Am J Obstet Gynecol.* 2017;216(3):259.e1–e6.
10. Graebe K, et al. *Gynecol Oncol.* 2015;136(2):274–7.
11. Brölmann H, et al. *Gynecol Surg.* 2015;12(1):3–15.
12. Ip PP, et al. *Best Pract Res Clin Obstet Gynaecol.* 2011;25(6):691–704.
13. Santos P, et al. *Diagn Interv Radiol.* 2015;21(1):4–9.
14. Tanos V, et al. *Int J Surg.* 2017;43:7–16.
15. Sun S, et al. *Diagn Interv Imaging.* 2019;100(10):619–34.
16. Shwayder J, et al. *J Minim Invasive Gynecol.* 2014;21(3):362–76.
17. Juang CM, et al. *Eur J Gynaecol Oncol.* 2006;27(4):370–4.
18. Song KJ, et al. *Medicine (Baltimore).* 2018;97(14):e0268.

19. Leiser AL, et al. *Gynecol Oncol.* 2006;101(1):86–91.
20. Hwang H, et al. *J Clin Pathol.* 2015;68(9):710–7.
21. Ravegnini G, et al. *Mod Pathol.* 2013;26(5):743–9.
22. Bertsch E, et al. *Mod Pathol.* 2014;27(8):1144–53.
23. Mas A, et al. *Am J Obstet Gynecol.* 2019;221(4):320.e1–e23.
24. Kumar KR, et al. *Semin Thromb Hemost.* 2019;45(7):661–73.
25. Mehine M, et al. *N Engl J Med.* 2013;369(1):43–53.
26. Chudasama P, et al. *Nat Commun.* 2018;9(1):144.
27. Rao U, et al. *J Clin Pathol.* 2019;72(12):810–6.
28. Yatsenko SA, et al. *Fertil Steril.* 2017;107(2):457–66.e9.
29. Hemming ML, et al. *JCO Precis Oncol.* 2019;2019:PO.18.00235.
30. Anjum S, et al. *Anticancer Agents Med Chem.* 2019;19(14):1703–18.
31. Skubitz KM, et al. *Cancer.* 2003;98(5):1029–38.
32. Liu B, et al. *Sci Rep.* 2018;8(1):7401.
33. Georgieva B, et al. *Genomics.* 2012;99(5):275–81.
34. Yokoi A, et al. *Cancer Sci.* 2019;110(12):3718–26.
35. Dvorská D, et al. *Int J Mol Sci.* 2019;20(15):3825.
36. Walker CL, et al. *Science.* 2005;308(5728):1589–92.
37. Islam MS, et al. *Fertil Steril.* 2013;100(1):178–93.
38. Mehine M, et al. *N Engl J Med.* 2013;369(1):43–53.
39. Raish M, et al. *J Cancer Res Clin Oncol.* 2012;138(7):1173–86.
40. Miyata T, et al. *Sarcoma.* 2015;2015:412068.
41. Agaram NP, et al. *Genes Chromosomes Cancer.* 2016;55(2):124–30.
42. Jour G, et al. *Hum Pathol.* 2014;45(8):1563–71.
43. Liau JY, et al. *Am J Surg Pathol.* 2015;39(2):236–44.
44. Hensley ML, et al. *Clin Cancer Res.* 2020;26(14):3881–8.
45. Makinen N, et al. *Mol Cancer.* 2017;16(1):101.
46. Barlin JN, et al. *Neoplasia.* 2015;17(2):183–9.
47. Quade BJ, et al. *Genes Chromosomes Cancer.* 2004;40(2):97–108.
48. Adams CL, et al. *Exp Mol Pathol.* 2019;110:104284.
49. de Almeida B, et al. *Int J Mol Sci.* 2017;19(1):52.
50. Kommoss FKF, et al. *J Cancer Res Clin Oncol.* 2020;146(1):97–104.
51. Kerick M, et al. *BMC Med Genomics.* 2011;4:68.
52. Chudasama P, et al. *Nat Commun.* 2018;9(1):144.
53. Cuppens T, et al. *Int J Cancer.* 2018;142(6):1230–43.
54. Croce S, et al. *Clin Cancer Res.* 2020;26(4):855–61.
55. ACOG Committee. *Obstet Gynecol.* 2019;133(3):e238–e48.
56. Hedegaard J, et al. *PLoS One.* 2014;9(5):e98187.
57. Hanahan D, et al. *Cancer Cell.* 2012;21(3):309–22.
58. Commandeur AE, et al. *Hum Reprod Update.* 2015;21(5):593–615.
59. Marusyk A, et al. *Nat Rev Cancer.* 2012;12(5):323–34.
60. Müller B, et al. *Hum Genet.* 2016;135(3):259–72.
61. Donovan MJ, et al. *Cancer.* 2009;115:3078–84.
62. Diaz LA, et al. *J Clin Oncol.* 2014;32(6):579–86.
63. Zhang Z, et al. *EBioMedicine.* 2018;34:267–74.
64. Mansilla C, et al. *Crit Rev Oncol Hematol.* 2018;126:129–34.
65. Kozminsky M, et al. 2019;6(2):1801254.
66. Qi P, et al. *Mol Cancer.* 2016;15(1):39.
67. Sole C, et al. *Semin Cancer Biol.* 2019;58:100–8.
68. Valihrach L, et al. *Mol Aspects Med.* 2020;72:100825.
69. In 't Veld S, et al. *Blood.* 2019;133(22):2359–64.
70. Kim KM, et al. *Wiley Interdiscip Rev RNA.* 2017;8(4):10.1002.
71. Castro-Giner F, et al. *Diagnostics (Basel).* 2018;8(2):31.

Non-Invasive Prenatal Testing (NIPT) 26

Nuria Balaguer, Emilia Mateu-Brull, and Miguel Milán

INTRODUCTION

Non-invasive prenatal testing (NIPT) is a class of screening for fetal chromosomal or genetic conditions utilizing cell-free fetal DNA (cffDNA) in maternal plasma or serum (1). Until the development of this tool, the diagnosis of fetal chromosomal or genetic disorders required invasive sampling of fetal or placental tissues by amniocentesis or chorionic villus sampling (CVS). Although accurate, invasive sampling procedures involve a small but definite risk of fetal loss and maternal morbidity (2). Therefore, non-invasive methods of prenatal testing, like NIPT, are highly favored.

The presence of cffDNA was first reported by Lo et al. in 1997 (3). On average, 10–20% of cell-free DNA (cfDNA) in maternal plasma is of fetal origin (4). This percentage, commonly known as the fetal fraction (FF), varies between individuals and generally increases with gestational age (4, 5). After delivery, cffDNA is completely cleared from the maternal circulation within hours (6). It is therefore assumed that cffDNA does not persist from one pregnancy to the next, so there is no risk of contamination in later NIPT analyses.

cffDNA originates in the placenta, mainly from apoptotic trophoblasts (7, 8), and can be detected as early as five weeks gestation (9). Unlike intact genomic DNA, cfDNA in maternal plasma is fragmented, with fetal fragments shorter than cell-free maternal DNA (10). Thus, the peak size of cell-free maternal DNA appears to be 166 base pairs (bp), whereas the highest fetal DNA signal is detected at 143 bp. This distribution of fetal and maternal DNA across the entire genome is fairly even (11), allowing relatively precise estimation of the FF. The minimum FF threshold for adequate NIPT performance varies by assay but is typically between 2% and 4% (12). As the FF increases, a greater statistical separation between aneuploid and euploid pregnancies provides greater confidence in the NIPT result.

NIPT is more accurate than the traditional combined screening test (13), with sensitivity and specificity reported above 99% for trisomy (T) 21 and a false-positive rate (FPR) under 1% (14). This accuracy varies among genetic conditions, with less accuracy for aneuploidies other than T13, T18, and T21. Test performance is influenced by several factors, including placental mosaicism, maternal body mass index (BMI), FF, maternal chromosomal abnormalities or malignancies, the presence of a vanishing twin (VT), and singleton *versus* multiple pregnancies.

NIPT is currently offered as early as 9–10 weeks gestation and harnesses the ease of a serum screen to provide accurate and informative fetal chromosomal data. However, NIPT is a screening tool, and women with a positive result require invasive testing to confirm NIPT findings (15). Understanding the

DOI: 10.1201/9781003024941-26

basics of NIPT and the potential causes of false- positive/negative results is essential for comprehensive counseling of patients before and after testing. Proper counseling ensures the responsible application of NIPT and facilitates evidence-based decision making for the prospective parents (16).

STATE OF THE ART: WHAT IS ALREADY KNOWN?

After the first large-scale clinical trials demonstrating the robustness of NIPT technology in 2011 (17), the tool was rapidly introduced into clinical practice in dozens of countries worldwide. Due to constant improvements in next-generation sequencing (NGS), the scope of NIPT has expanded to include sex chromosome aneuploidies (SCAs), panels of specific microdeletions, and, most recently, genome-wide copy number variants (CNVs) and rare autosomal trisomies (RATs) (Figure 26.1).

This wealth of options drove the commercial sector to update its product portfolio to offer basic panels (T13, T18, T21, and SCAs) along with more complex screening options. Genome-wide detection of fetal *de novo* mutations and the study of maternal inheritance are also likely to become available. The cfDNA test for single-gene disorders, termed non-invasive prenatal diagnosis (NIPD), encompasses an increasing array of conditions but is less widely available (18). Additional emerging areas of investigation

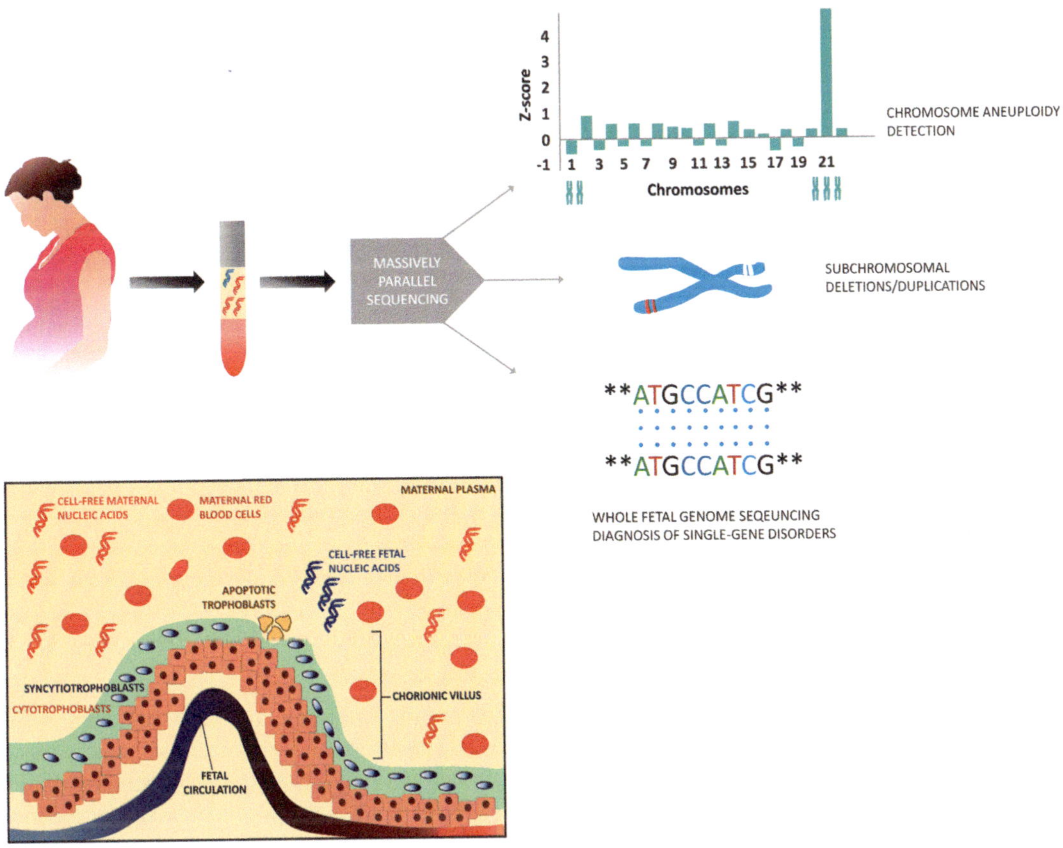

FIGURE 26.1 An overview of the currently available applications of massive sequencing (MPS)-based NIPT. Adapted from (1).

include determining the relationship between circulating DNA and nucleosomal structure (19, 20), the tissues of origin of circulating DNA (21, 22), and whether cffDNA has any biological or pathogenic function.

This chapter presents an overview of the technologies used for NIPT screening and their clinical applications. We also examine some of the ethical, legal, and social issues associated with NIPT implementation in a clinical setting and discuss emerging concerns associated with the inclusion of a broader range of medical conditions.

METHODOLOGY

There are many approaches to prenatal cfDNA testing. The most commonly used are whole-genome sequencing (WGS), targeted genome sequencing, and single-nucleotide polymorphism (SNP)-based sequencing. These techniques can determine both the FF and the likelihood of aneuploidy. The fundamentals of each cfDNA testing platform are discussed below.

WGS-Based NIPT

WGS-based NIPT was the first clinical NIPT method. In WGS-based NIPT, cfDNA is sequenced without differentiating between maternal and placental DNA. Sequenced fragments (reads) are aligned to a reference genome to map their chromosomal origin. Finally, the number of reads corresponding to each chromosome of interest is compared to a euploid reference sequencing run, and the fetal chromosomal copy number is determined (23, 24). Read density in WGS-based NIPT algorithms is usually assessed by tiling each chromosome with non-overlapping regions of equal size, counting the number of reads per region, and averaging the reads per bin over a region of interest (e.g., microdeletion or whole chromosome) (23–25). In turn, although there are multiple sources of bias in WGS data (e.g., GC bias, non-unique sequence), bias can be corrected based on origin (26).

The number of reads obtained with random sequencing ranges from 10 to 30 million per sample depending on the sequencing platform (23, 27, 28). Usually, all the sequencing information is used to normalize the data and make samples comparable. In addition to aneuploidies, this methodology allows the detection of the most prevalent microdeletions (29). This method is advantageous because the higher the amount of cffDNA sequenced and analyzed, the greater the statistical power to correctly identify aneuploidies in low-FF scenarios. On the other hand, a disadvantage to this approach is the interference from maternal karyotype abnormalities, especially with technologies that use normalized chromosome values (NCV) for analyses.

Targeted Genome Sequencing

Targeted sequencing uses targeted PCR amplification, sequencing, counting, and sequence comparison with a reference genome to determine fetal aneuploidies (or abnormalities). Targeted sequencing uses non-polymorphic loci to determine chromosome copy number and polymorphic loci to estimate the FF (28, 30). One of the advantages of targeted genome sequencing is that it is not costly and requires the lowest number of sequencing reads among the three methodologies described here (the number of reads per sample ranges from 420,000 to 1 million). Targeted sequencing shares one drawback with the WGS-based NIPT method, which is that, as a shotgun-based approach, an abnormal maternal karyotype may interfere with the analysis.

SNP-Based NIPT

SNP-based NIPT is based on specific amplification and sequencing of single-nucleotide polymorphisms (SNPs). SNPs are the most common genetic variation in the human genome, and each SNP represents a change in one nucleotide (31). Informative SNPs are those for which the mother is homozygous and the fetus is heterozygous, so parental genomic DNA is needed for this method. The required maternal genomic DNA is obtained from the leukocyte layer formed after centrifugation of whole blood samples. Paternal genomic DNA is optional, and can be obtained from a blood or a buccal swab sample. All samples undergo PCR amplification specific to selected SNPs, and PCR products are then sequenced. A specific algorithm subtracts parental genomic information to deduce placental DNA data and determine the copy number of each interrogated chromosome (32, 33). The number of sequencing reads obtained in SNP-based NIPT is 6 million reads per sample on average (32, 34). This approach also enables the reporting of deletions and microdeletions (35). Among NIPT, the SNP-based method demonstrates the greatest accuracy in SCA detection. It is also the only method capable of detecting triploidy. Furthermore, SNP-based NIPT does not detect off-target abnormalities.

CLINICAL RESULTS

Analysis of Common Aneuploidies

cfDNA screening is the best screening method for detecting T21 and has the highest detection rate (DR) with the lowest FPR, compared with other conventional screenings. An updated meta-analysis that included studies between January 2011 and January 2015 (14) found a 99.7% DR for T21, with an FPR of 0.04%. These figures favorably exceed those obtained with the first-trimester combined screening (FCT), where an approximately 85–90% DR and an FPR of 5% were obtained. In the same meta-analysis, T18 and T13 were found to have a DR of 97.9% and 99.0%, respectively, with an FPR of 0.04% for each aneuploidy.

In addition to studies intended to demonstrate NIPT's high sensitivity in both high-and general-risk populations, other initiatives, such as the TRIDENT-1 study in the Netherlands, were undertaken to evaluate the clinical utility of genome-wide NIPT. For example, the TRIDENT-1 study conducted with women at increased risk of common trisomies (risk ≥ 1/200) (based on FCT or medical history) demonstrated that 80% of the findings from the fetus or the placenta were of clinical relevance for pregnancy management (36). Follow-up testing in the TRIDENT-2 study confirmed high sensitivity for T21, T18, and T13 screening of 98%, 91%, and 100%, respectively (37).

SCAs

The frequency of SCAs is around 1.88/10,000 births (38). Testing for SCAs was first incorporated as a screening option in 2012 for monosomy X (39). This was promptly followed by testing for other SCAs, including 47, XXX; 47, XXY; and 47, XYY; which are frequently mild or asymptomatic and do not meet the classical criteria for population-based screening. However, the NIPT performance for the X and Y chromosomes is not as accurate as it is for chromosome 21 (40). The most recent meta-analysis estimated the sensitivity for monosomy X detection at 95.8% with an FPR of 0.14%. For the other SCAs, the DR was estimated at 100% and FPR at 0.004% (14) (Table 26.1).

Usually, the positive predictive value (PPV) linked to the detection of SCAs is lower than for autosomal aneuploidy, mainly because of confined placental mosaicism (CPM) or maternal SCAs (41).

TABLE 26.1 Performance of cfDNA for Aneuploidies

ANEUPLOIDY	*DR (95% CI)[A]*	*FPR (95% CI)[A]*
T21	99.7% (99.1–99.9%)	0.04% (0.02–0.08%)
T18	98.2% (95.5–99.2%)	0.05% (0.03–0.07%)
T13	99.0% (65–100%)	0.04% (0.02–0.07%)
Monosomy X	95.8% (70.3–99.5)	0.14% (0.05–0.38%)
Other SCAs	100% (83.6%–100%)	100% (83.6%–100%)

Abbreviations: DR: detection rate; FPR: false-positive rate; CI: confidence interval; SCAs: sex chromosome aneuploidies.

Considering these factors, most guidelines advise careful pre-test counseling for patients contemplating SCA screening.

RATs

There is little literature regarding RATs identified by NIPT. With a current registered frequency of 0.3–0.8% (38), RATs are usually associated with either placental/fetal mosaicism or uniparental disomy (UPD), which both involve a fetus that inherits both chromosomes from one parent (42). In 2014, Lau et al. described one of the first clinical studies employing a genome-wide cfDNA prenatal screening approach, reporting six out of 1,982 (0.3%) pregnancies with a RAT. Interestingly, all resulted in a normal outcome, but two pregnancies were affected by intrauterine fetal growth restriction (IUFGR) (43).

Fiorentino et al. used a genome-wide approach and reported 17 trisomies from 12,078 patients (0.14%). Trisomies 22, 7, and 14 were the most common. All seven miscarriages with RATs were confirmed on follow-up testing (four cases of T22 and three cases of T15). The only viable pregnancy with T15 was disomic on amniocentesis but had maternal UPD 15 causing Prader–Willi syndrome. True-fetal mosaicism (TFM) was reported for single cases of trisomies 7, 9, and 22 (44).

In a large retrospective study, Pertile et al. described a US cohort of 72,932 cases with 246 RATs (0.3%) and an Australian cohort of 16,885 NIPT samples with 60 RATs (0.4%). The most common RATs in these cohorts were trisomies of chromosomes 15, 7, 6, 22, and 2. From the 52 samples with clinical outcomes, they reported 22 miscarriages (the majority involving T15), five cases of TFM, and five with UPD. Only 14 of the 52 cases (27%) with outcome data resulted in normal births (45). Scott et al. identified 28 RATs (0.1%) from a cohort of 23,388 NIPT samples. The most common trisomies involved chromosome 7 (n = 6), followed by chromosome 16 (n = 4) and chromosomes 4 and 22 (each n = 3). Of the 28 RAT cases, six miscarried, two cases had TFM confirmed on amniocentesis (one had structural abnormalities and the other had both a trisomy and UPD 15 with no structural abnormalities seen on ultrasound). One case with mosaic T10 on CVS and structural abnormalities seen on ultrasound had a likely, but unproven, fetal mosaicism. There were 18 live births (all with IUFGR) and 15 were phenotypically normal. Surprisingly, three babies (16.6%) were born with structural abnormalities despite a normal microarray result following amniocentesis (46).

Finally, Van Opstal et al. reported 27 whole-chromosome RATs (plus 14 other chromosomal aberrations involving deletions or duplications by WGS) in 2,527 reported NIPT samples (1.1%) from the TRIDENT-1 study. Two-thirds (18 out of 27; 66%) of the whole-chromosome RAT cases resulted in phenotypically normal babies. However, five of the 22 cases thought to have CPM displayed multiple congenital abnormalities (36). In contrast, in the TRIDENT-2 study, NIPT indicated a high-risk for 101 (0.18%) RATs and 95 (0.16%) subchromosomal aneuploidies (SAs). T7 was the aneuploidy most frequently detected (n = 32), followed by T16 (n = 14), T8 (n = 13), and T20 (n = 11). Diagnostic follow-up was available for 97 of the 101 cases. In six of 97 cases (65), a trisomy or a UPD was confirmed in the fetus. This

TABLE 26.2 Summary of RAT Detection Studies

STUDY (REFERENCE)	*POPULATION SCREENED*	*STUDY SIZE*	*RARE TRISOMY NO. (%)*	*MOST COMMON RATS*
Lau et al. (2014) (43)	General population	1,982	7 (0.35)	22
Brady et al. (2016) (47)	Increased-risk	4,000	11 (0.28)	7
Fiorentino et al. (2017) (44)	Increased-risk	12,078	17 (0.14)	22, 7, 15
Pertile et al. (2017) (45)	General population	89,817	306 (0.34)	7, 15, 16, 22
Ehrich et al. (2017) (48)	Increased-risk	10,000	78 (0.78)	16, 7, 3
Scott et al. (2018) (46)	General population	23,388	28 (0.12)	7, 16, 22, 4
Brison et al. (2018) (49)	General population	19,735	58 (0.29)	7, 16, 22
Van Opstal et al. (2018) (36)	Increased-risk	2,527	29 (0.91)	16, 7, 9
Chatron et al. (2019) (50)	Increased-risk	1,617	10 (0.62)	16
Van der Meij et al. (2019) (37)	General population	56,818	101 (0.18)	7, 16, 8, 20

included two cases of mosaic T16 and one case of mosaic T22. Furthermore, three cases of fetal UPD were confirmed on chromosomes 9, 12, and 15 after NIPT indicated a trisomy for these chromosomes. In four cases (4%), the trisomy was confirmed with CVS (three T8 and one T13 + 20), but follow-up fetal investigations on amniotic fluid (AF) or cord blood were normal, confirming the CPM diagnosis. In 87 of 97 cases (90%), diagnostic follow-up, mostly in AF during pregnancy or in cord blood after birth, showed no trisomy in the fetus. However, these tests do not exclude the presence of CPM (37) (Table 26.2, Table 26.3, Figure 26.2).

ATs are associated with a range of pregnancy complications, including miscarriage, intrauterine fetal death (IUFD), TFM, and UPD, as well as IUFGR related to placental insufficiency. Even after normal amniocentesis, there is a possibility of a fetus with structural anomalies or a congenital defect. However, many normal pregnancy outcomes are also recorded. Large cohorts are necessary to estimate the real incidence of RATs in the general population; this may be helpful in quantifying the risk for adverse pregnancy outcomes based on a positive result on some or all RATs using cfDNA screening. Further cytogenetic and pregnancy outcome data are needed from different centers (48) before clinical recommendations can be made.

SAs

SAs may result in physical and/or intellectual impairments that can be highly variable and dependent on the specific chromosomal region affected and the amount of duplicated or deleted genetic material (51). SAs in general are common and do not increase with maternal age, unlike the common trisomies. Pathogenic CNVs can occur across the genome, but around 25% are recurrent (52). The most common is the 22q11.2 deletion, which causes DiGeorge syndrome, with a frequency of 1 in 992 in a low-risk obstetric population (Grati et al., 2015) (52). Although NIPT has the potential to include CNV screening, most commercial platforms only report the detection of the most common recurrent microdeletion syndromes, with few detecting all CNVs that are 7 Mb or greater (18): 5p− (cri-du-chat syndrome), 22q11.2− (DiGeorge syndrome), 15q− (Prader–Willi/Angelman syndrome), 4p− (Wolf–Hirschhorn syndrome), and 1p36 (53) (Table 26.4).

Yin et al. developed a method to identify 71.8% of CNVs using 3.5 million reads but the performance dropped to 41.2% when CNVs were below 5 Mb (54). However, Srinvasan et al. detected fetal CNVs as small as 300 kb using an extremely high depth of sequencing of 600–1,300 million reads (29). Straver et al. reported the detection of large CNVs (over 20 Mb) with low sequencing depth, which had limited clinical value (55). Lo et al. reported 64.5% (20/31) accuracy when 4–6 million reads were used to analyze samples with 3-Mb to 42-Mb CNVs. However, if CNVs were smaller than 6 Mb, only 5 in 13 cases

TABLE 26.3 RAT Cases from Studies Documenting Outcomes

Chr.	Lau *et al.* (2014)	Brady *et al.* (2016)	Fiorentino *et al* (2017)	Pertile *et al.* (2017)	Ehrich *et al.* (2017)	Scott *et al.* (2018)	Brison *et al.* (2018)	Van Opstal *et al.* (2018)	Chatron *et al.* (2019)	Van der Meij *et al.* (2019)
1		**1**			**1**					
2				**5**	**3**	**1**	**4**	**1**		
3	**1**			**1**	**10**	**1**	**1**	**1**		**3**
4				**3**	**2**	**3**				**1**
5				**1**		**1**				**2**
6	**1**									
7	**1**	**3**	**4**	**9**	**11**	**6**	**13**	**6**	**1**	**32**
8	**1**	**1**		**3**	**5**	**2**	**4**	**3**	**1**	**13**
9		**1**	**1**	**4**	**3**	**1**	**2**	**3**		**4**
10		**1**		**3**	**2**	**1**	**2**			**1**
11					**1**		**1**			**1**
12			**1**							**4**
14			**1**	**1**	**6**	**2**				**3**
15		**1**	**4**	**14**	**7**	**3**	**4**	**1**	**1**	**4**
16		**1**	**1**	**7**	**15**	**4**	**13**	**9**	**6**	**14**
17							**1**			**1**
19										
20		**1**		**3**	**5**	**1**	**4**	**1**		**11**
22	**2**	**1**	**5**	**4**	**7**	**3**	**5**	**3**	**1**	**5**
TOTAL	6	11	17	58	78	29	54	28	10	99

Abbreviations: RAT: rare aneuplidy trisomy; MC: miscarriage; FSA: fetal structural abnormalities; CPM: confined placental mosaicism; TFM: true fetal mosaicism; UPD: uniparental disomy; IUFD: intrauterine fetal death; FGR: fetal growth restriction; LB: live birth; NAD: no abnormality detected.

* Percentages calculated with respect to the 235 cases with follow-up.

were identified. Nevertheless, although the majority of CNVs larger than 6 Mb could be identified, the authors concluded that the test had limited clinical utility, in part because FF could not be measured for all samples (56). Chen et al. developed their own algorithm of analysis known as fetal copy number analysis through maternal plasma sequencing (FCAPS), which detected over 90% of CNVs greater than 10 Mb at 10% FF (57). Later, they evaluated the performance of FCAPS in detecting CNVs using selected clinical samples with known CNVs. In this study, 33 samples with deletions/duplications ranging from 1 to 129 Mb were detected with consistent CNV size and location to karyotyping/microarray results. The sensitivity and specificity of detection of deletions/duplications were 84.21% and 98.42%, respectively (58).

Lefkowitz et al. conducted a retrospective, blinded study of approximately 1,200 subjects with pregnancies at increased risk for fetal chromosomal abnormalities. Clinical sensitivity and specificity for CNVs > 7 Mb were determined to be 97.7% and 99.9%, respectively (59). This group later reported on a more extensive series of 10,000 clinical samples, in which 154 results, representing 27.8% of all abnormal calls, were only detectable by genome-wide analysis. Because no outcome data were available, they could not provide conclusive figures for the sensitivity and specificity calculations.

Fiorentino et al. detected 30 pregnancies (out of 12,114 patients) with positive results for a variety of CNV aberrations other than common aneuploidies, including both segmental chromosome imbalances and RATs. Clinically relevant segmental chromosomal imbalances were detected in 13 pregnancies and confirmed by invasive prenatal diagnosis in eight cases. Interestingly, a fetus with unbalanced translocations was identified, which was subsequently found to be inherited from the mother who was unaware of

Total (all studies)	Lost to follow up (%)	MC (n,[%])	FSA (n,[%])	CPM (n,[%])	TFM (n,[%])	UPD (n,[%])	IUFD (n,[%])	FGR (n,[%])	LB,NAD (n,[%])
2	2 (100)	0 (0)	0 (0)	0 (0)	0 (0)	0 (0)	0 (0)	0 (0)	0 (0)
14	3 (21.4)	1 (7.1)	0 (0)	2 (14.3)	4 (28.6)	0 (0)	1 (7.1)	2 (14.3)	2 (14.3)
18	12 (66.6)	0 (0)	0 (0)	1 (5.6)	0 (0)	0 (0)	0 (0)	1 (5.6)	4 (22.2)
9	2 (22.2)	0 (0)	0 (0)	0 (0)	0 (0)	2 (22.2)	1 (11.1)	2 (22.2)	3 (33.3)
4	3 (75)	0 (0)	1 (25)	0 (0)	0 (0)	0 (0)	0 (0)	0 (0)	0 (0)
1	0 (0)	0 (0)	0 (0)	0 (0)	1 (100)	0 (0)	0 (0)	1 (100)	0 (0)
86	40 (46.5)	0 (0)	2 (2.3)	7 (8.1)	2 (2.3)	0 (0)	0 (0)	4 (4.7)	31 (36.0)
33	20 (60.6)	0 (0)	0 (0)	5 (15.6)	0 (0)	0 (0)	0 (0)	0 (0)	8 (25)
19	5 (25)	3 (15)	4 (20)	0 (0)	3 (15)	1 (5)	0 (0)	2 (10)	5 (25)
10	6 (60)	1 (10)	1 (10)	0 (0)	0 (0)	0 (0)	0 (0)	0 (0)	2 (20)
3	2 (66.7)	0	0 (0)	0 (0)	0 (0)	0 (0)	0 (0)	0 (0)	1 (33.3)
5	2 (40)	1 (20)	0 (0)	0 (0)	0 (0)	1 (20)	0 (0)	0 (0)	1 (20)
13	8 (61.5)	2 (15.4)	1 (7.7)	0 (0)	0 (0)	0 (0)	0 (0)	1 (7.7)	1 (7.7)
39	12 (30.8)	16 (41.0)	0 (0)	0 (0)	3 (7.7)	6 (15.4)	0 (0)	0 (0)	2 (5.1)
70	16 (22.9)	4 (5.7)	4 (5.8)	3 (4.3)	7 (10)	2 (2.9)	0 (0)	14 (20)	20 (28.6)
2	0 (0)	0 (0)	0 (0)	0 (0)	0 (0)	0 (0)	0 (0)	0 (0)	0 (0)
0	0 (0)	0 (0)	0 (0)	0 (0)	0 (0)	0 (0)	0 (0)	0 (0)	0 (0)
26	19 (73.1)	1 (3.8)	0 (0)	2 (7.7)	0 (0)	0 (0)	0 (0)	1 (3.8)	3 (11.5)
36	10 (27.8)	11 (30.6)	1 (2.8)	3 (8.3)	3 (8.3)	0 (0)	0 (0)	0 (0)	10 (27.8)
390		40 (17.0*)	14 (6.0*)	23 (9.8*)	23 (9.8*)	12 (5.1*)	2 (0.9*)	28 (11.9*)	93 (39.6*)

being a balanced translocation carrier. The sensitivity for segmental imbalances was 100%, and specificity was 99.6% (44).

Pertile et al. employed the WISECONDOR (**WI**thin **S**ampl**E** **CO**py **N**umber aberration **D**etect**OR**) algorithm to detect copy number imbalances of 10–20 Mb at FF above 5% using low-coverage data. They prospectively screened 15,600 samples from high- and average-risk women and identified 18 pregnancies at increased risk of segmental aneuploidy (0.12%), of which 11 cases were confirmed after prenatal diagnosis. Performance statistics for segmental aneuploidy in this validation study were a test sensitivity of 78.6%, specificity of 99.9%, PPV of 61.1%, and a negative predictive value (NPV) > 99.9% (45). Van Opstal et al. used WISECONDOR as well, reporting 12 segmental aneuploidies among 2,553 primarily high-risk FCT referrals (risk > 1/200), in which six cases were confirmed after diagnostic testing (PPV 50%) (36). In the TRIDENT-2 study, 95 cases were identified with an SA, with one case of IUFD without diagnostic testing and three that were lost to follow-up. Of the remaining 91 cases, 29 (32%) were confirmed in the fetus after diagnostic testing. From the findings derived from their study, they concluded that women with high-risk NIPT results for SAs were significantly older than women with low-risk results. This suggests that, similar to the risk of common trisomies, the risk of a high-risk SA result increases with maternal age. Interestingly, this unexpected relationship was not found for the RATs (37).

Each of these studies suggests that, although PPVs are less common than trisomies, they still represent a high-risk category that is useful for prenatal diagnosis. Practitioners offering extended NIPT, including CNVs, should provide comprehensive counseling before and after testing, disclosing the possibility of no findings and the need for confirmation of positive results with invasive testing (60). Many CNV syndromes present with variable expression, and accurate prediction of phenotypic severity in the absence of ultrasound findings is not possible. Because of this, most national and professional society guidelines do not currently recommend the use of genome-wide cfDNA-based NIPT for either CNVs

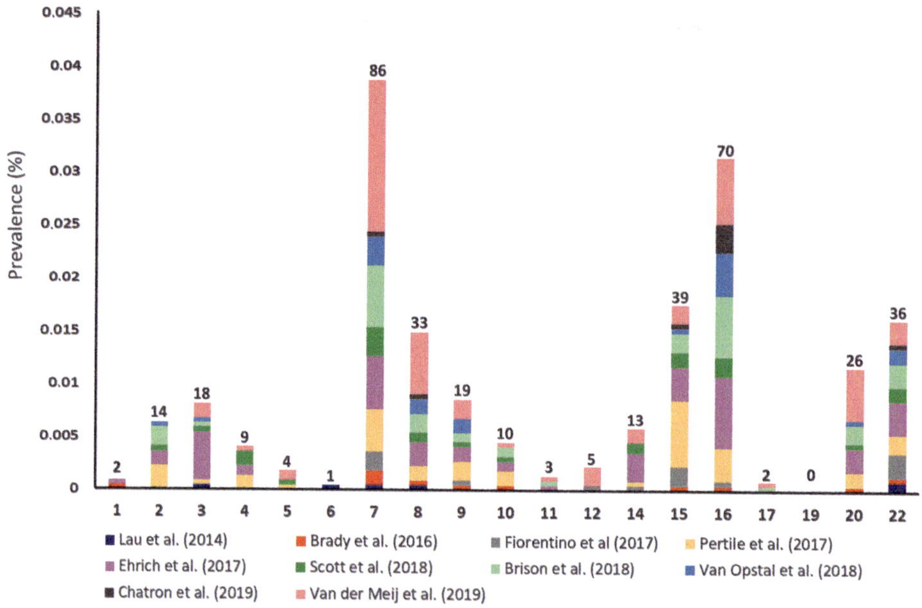

FIGURE 26.2 The reported frequency of each RAT from the studies shown in Table 26.2 as a proportion of the total cases tested (n = 221,962). The absolute numbers of each RAT are indicated above each column.

TABLE 26.4 Recurrent Chromosomal Deletions Causing Genetic Syndromes Commonly Offered in Expanded NIPT by Commercial Providers

CHROMOSOMAL DISORDER	*CHROMOSOMAL LOCUS*	*INCIDENCE*	*DR*	*SPECIFICITY*
DiGeorge syndrome	22q11.2	0.736	> 97%	> 99%
Prader–Willi syndrome	15q11.2–q13	1:10,000–30,000	> 97%	> 99%
Angelman syndrome	15q11.2–q13	1:12,000–20,000	> 97%	> 99%
Cri-du-chat syndrome	5p	1:15,000–50,000	> 97%	> 99%
1p36 deletion	1p36	1:5,000	> 97%	> 99%

Source: Adapted from (35).

or RATs (18). However, due to the value of their detection, several commercial, academic, and national public health services have begun to incorporate or consider CNV screening.

NIPD of Monogenic Disorders

Another application of cfDNA prenatal testing is the NIPD of monogenic disorders. Pathogenic mutations causing monogenic disorders include single-base changes, small insertions, and deletions of a few bases, much larger duplications and deletions, and more complex rearrangements, such as inversions (61).

Currently, there are two main analytical approaches for non-invasive diagnosis of autosomal recessive diseases and X-linked diseases: the relative mutation dosage (RMD) method and the relative haplotype dosage (RHDO) method (11, 62, 63). These approaches determine which paternal allele the fetus has inherited despite an excess of maternal DNA among the cfDNA in the maternal plasma (64). Table 26.5 reflects the clinical applications of RMD and RHDO for NIPD (64).

RMD requires precise quantification of mutant and wild-type (WT) alleles in cfDNA and the application of statistical methods to clarify that measured imbalances reflect fetal genotype signal rather than

TABLE 26.5 Applications and Limitations of RMD and RHDO for NIPD

APPLICATIONS/ LIMITATIONS	*RMD*	*RHDO*
Clinical applications	• To detect SNPs and small deletions.	• To detect SNP mutations, including large structural variants and mutations in regions with similar sequence context.
	• With a predesigned assay, it is more cost-effective to detect single mutations with simple instrumentation.	• More efficient to analyze a panel of mutations in one or multiple genes.
	• Suitable for detecting common mutations in the population.	• Can multiplex multiple patient samples with different mutations in one test.
Limitations	• Cannot detect large structural variants and mutations in regions with similar sequence context.	• Need to phase parental haplotypes.
	• Need to design mutation-specific assay.	• Not suitable for *de novo* mutations.

Abbreviations: RMD: relative mutation dosage; RHDO: relative haplotype dosage; NIPD: non-invasive prenatal diagnosis.

technical noise. Unfortunately, standard NGS protocols are not sensitive enough for these applications since amplification bias between mutant and WT alleles can lead to inaccurate allelic fractions. Possible solutions to these limitations include the use of nested PCR or digital PCR (dPCR) (65, 66). Proof-of-principle studies for NIPD using dPCR were reported for several recessive and X-linked conditions, including β-thalassemia (62, 67), sickle-cell disease (68), hemophilia (69, 70), and recessive forms of deafness (71). However, dPCR is highly sensitive and has a limited capacity for multiplexing compared to NGS, which confines the number of mutations that can be assayed in one test (18).

RHDO analysis differentiates and quantifies haplotypes linked to a gene mutation site. RHDO processes a large amount of parental SNP data to reveal its haplotype and quantify the allele ratio in maternal blood. Notably, RHDO determines the inheritance of complex genomic variants under the resolution of cfDNA fragmentation, such as exonic deletions and the intron-22-related inversions within the *F8* gene, which cause hemophilia A (70). RHDO can also be applied to genes with homologous pseudogenes that make mutation detection difficult, particularly *CYP21A*-related congenital adrenal hyperplasia (63). Contrary to RMD approaches, RHDO has been implemented clinically, and testing for Duchenne muscular dystrophy (72), spinal muscular atrophy (73), and cystic fibrosis (74) are available.

It is important to note that exonic variants cause the majority of monogenic disorders. Another approach for NIPD is whole-exome sequencing (WES) or WGS (11, 75–77). Once WES or WGS of cfDNA from maternal plasma is performed, samples undergo a tailored bioinformatic analysis of a specific variant or panel of variants related to the disorder of interest or clinical presentation (61). Invasive prenatal WES/WGS, performed on amniocentesis samples, enables the diagnosis of monogenic diseases in the earliest stages of life, affording early treatment or pregnancy termination (78, 79). Also, accurate genome-wide non-invasive genotyping depends on the FF and the sequencing depth; these can both become limitations, even with NGS-based approaches (80). Therefore, improving the computational and algorithmic aspects of these approaches could help overcome such problems.

STRENGTHS AND LIMITATIONS

NIPT is currently considered to be at the vanguard of genomic medicine due to its high specificities and sensitivities for the screening of common aneuploidies. These figures have progressively increased to exceed those of the FCTs, increasing the popularity of NIPT. Despite the high accuracy reported, NIPT

TABLE 26.6 Causes of Aberrant NIPT Results

CATEGORY	*INCIDENCE (%)*	*CAUSE*
FP	0.3	CPM
		Maternal mosaicism
		Statistical chance as the cut-off for a positive test
		Maternal transfusions or heterologous organ transplantations
		VT
FN	0.01	CPM
		Maternal transfusions
		Insufficient amount of total and/or fetal/placental DNA
Non-reportable	0.3–5.4	Insufficient amount of total and/or fetal/placental DNA
		Autoimmune disease
		Heparin administration
		VT
Incidental findings	0.1	Maternal neoplasia
		Fetal chromosomal abnormalities not intended for testing

Source: Adapted from (81).

cannot be considered diagnostic due to the placental-origin cfDNA used to perform the screening. False-positive (FP) and false-negative (FN) results have important implications for pregnancy management and parental decisions (Table 26.6).

Causes of FP NIPT Results

Confined placental mosaicism

It is well documented from conventional cytogenetic examination of CVSs that placental chromosome constitution can differ from that of the fetus (i.e., confined placental mosaicism, CPM) (82). CPM occurs when two (or more) different cell lines with different chromosomal notation exist in a fetoplacental unit derived from a single zygote (83, 84). In most cases, mosaicism is found only in the placenta and will lead to a normal fetal outcome. However, CPM can alter placental function and lead to intrauterine growth restriction, pregnancy loss, or perinatal death. Experience gained from CVS indicates that this may occur in 1–2% of pregnancies with a high risk of aneuploidy, and is more likely with monosomy X and T13 than with T21 or T18 (47). Depending on the affected tissue types, fetoplacental discrepancies can be characterized into three subtypes of CPM when the karyotype in amniocytes is normal and three subtypes of TFM when the karyotype in amniocytes is abnormal (85) (Figure 26.3). Therefore, a positive NIPT result should always be confirmed with invasive testing before any irreversible procedure is performed, because the result may not correlate with the true fetal genotype.

Vanishing twins

The second cause of FP results is a vanishing twin (VT). VT is described as the disappearance of an embryo and/or gestational sac after registration of fetal activity in a twin gestation (86). The presence of a VT can cause an FP NIPT result when the demised twin is aneuploid. This is because the placenta from the deceased twin is still present at the time of testing and continues to shed DNA weeks after its death.

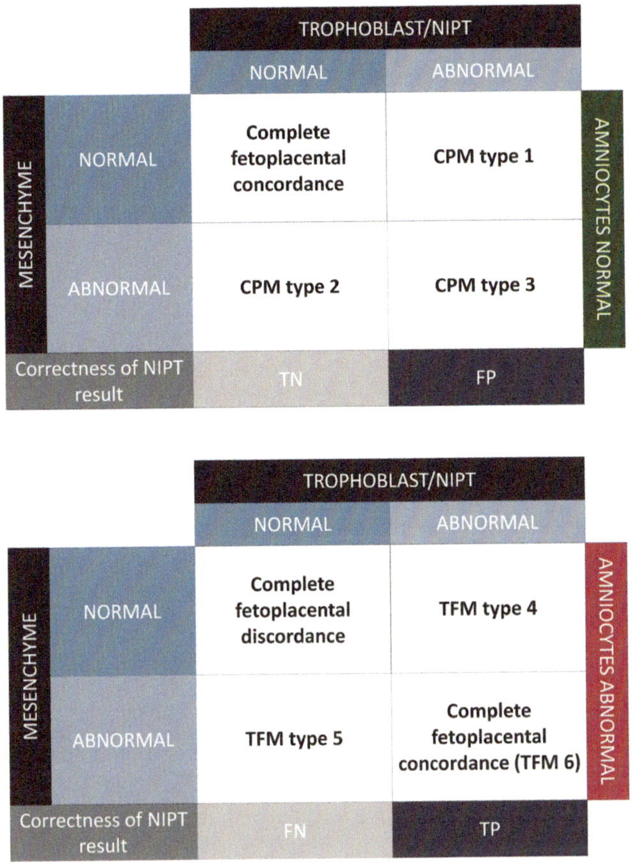

FIGURE 26.3 Distribution of placental cell lines in CPM and TFM. (a) Types of CPM (amniocytes with a normal karyotype). (b) Types of TFM (amniocytes with an abnormal karyotype). FP: false positive, FN: false negative TN: true negative, TP: true positive. Adapted from (16).

There is little information regarding how long after fetal death the placenta of the dead fetus contributes to the fetal DNA pool in the maternal plasma (87, 88). In any case, this is a special obstetric population and there are a series of considerations that must be taken into account before performing NIPT (89):

- An ultrasound scan is necessary before cfDNA testing to determine gestational age, number of gestational sacs, and the number and viability of fetuses.
- The laboratory should be informed about the presence of a VT before testing in order to interpret cfDNA results correctly.
- A patient with VT must receive appropriate pre-test counseling and be informed about the increased risk of FP cases and sex discrepancy.
- cfDNA analysis in the VT population should be requested after the 14th week of pregnancy to reduce the chances of obtaining an FP result.
- Fetal sex should not be reported in VT. Sex determination should be made by sonographic examinations at or beyond week 16 of gestation.

Maternal factors

Maternal acquired or constitutional genomic alterations may be responsible for false prenatal cfDNA-based testing results. Thus, since the majority of cfDNA is maternal in origin, there may be serious difficulties in the interpretation of NIPT results. The counting statistics of conventional Z-score chromosome-wide analysis methods can be affected by maternal CNVs leading to FP and FN NIPT results. Maternal mosaicism can also be a source of FP results. Several reports have demonstrated that loss of an X chromosome occurs with increasing frequency as maternal age increases. This can cause an FP NIPT result for monosomy X (90, 91). Additionally, maternal malignancies can result in multiple whole-chromosome or segmental gains or losses, which are absent from the fetus (92). Genome-wide profiling has the potential to identify "tumor-like" genome-wide aneuploidy profiles and identify presymptomatic cancers in pregnant women. Finally, autoimmune diseases, such as lupus erythematosus (93) or thrombocytopenic purpura (94, 95), have been described in association with abnormal cfDNA profiles.

False-Negative Results

One of the most important causes of FN results is the issuance of NIPT results with incorrect FF estimations, which are placed lower than the threshold of the technique used to perform sequencing. Hence, causes of low FF are high maternal body mass index, early gestational age, hemolysis in the sample due to storage and shipping conditions, or maternal therapy such as anticoagulation medication (96–98). Multiple studies have found an association between increased maternal weight and decreased FF. The comparatively lower FF in obese women may be explained by greater cell turnover, apoptosis of adipose tissue, or a dilution effect due to an increase in maternal blood volume (99, 100).

Chromosomal mosaicism may also lead to FN NIPT results (16). Two types of mosaicism are associated with a normal karyotype in the trophoblast while the fetus itself has a chromosomal aberration. The most common FN is TFM type V, characterized by the presence of a chromosome alteration in the fetus and the mesenchymal core of the placenta, with a chromosomally normal cytotrophoblast. TFM type V results in confined fetal mosaicism with a normal karyotype in both short- (STC) and long-term cultured villi (LTC) and abnormal cytogenetic results in the fetus (101).

Barriers to Access

Cost-effectiveness is important for any medical screening (102, 103). The cost of cfDNA screening may be one of the most considerable challenges to widespread clinical use. In the USA, cfDNA screening cost is high, ranging from approximately $700 to over $2,700 per test (104, 105). The actual price to the patient depends on the laboratory used, the conditions screened for, whether the patient has insurance, and, if so, the extent to which insurance covers the cost (106). Patients with financial difficulties may be unable to afford NIPT, especially in countries without public healthcare. So far, implementation of NIPT is not equivalent across countries; for example, in the Netherlands, access to NIPT is included in the health care system, so the cost to the patient is low. Differences in the co-payment policy disproportionately affect pregnant women of lower socio-economic status and lead to unequal access to first-trimester prenatal screening. This may imply that women who have the greatest interest in prenatal screening (for whom taking care of a child with a chromosomal abnormality might be hardest due to lack of resources) are least likely to access the screening.

There are also increasing fears that cfDNA screening may reduce the number of care facilities for children with disabilities such as Down syndrome (107, 108). This could mean that women would not have a choice in whether they receive prenatal screening. Therefore, an essential challenge for the implementation of cfDNA screening is to safeguard women's autonomy and ensure informed consent. Informed decision making for prenatal screening is already a challenge; the relative ease of cfDNA testing and its expanding scope may reinforce this challenge and lead to inappropriate use of the test in practice. Continuing education and training of health care professionals is vital to support the decision-making process in a way that is valued by pregnant women and their partners (109).

CONCLUSIONS

- NIPT provides a cost-effective, high-sensitivity screening test for common aneuploidies (T13, T18, and T21).
- Global implementation of NIPT has dramatically reduced the number of invasive prenatal procedures performed. Hence, the number of procedure-related miscarriages and adverse events for the mother has been minimized.
- Advances in genome-wide sequencing methodologies have enabled commercial providers to report on a broader range of fetal genetic abnormalities, including SCAs, RATs, and CNV syndromes. However, the low PPVs generate controversy regarding the clinical utility of screening, limiting the support from the leading international societies.
- More clinical validation studies are required to investigate the sensitivity and FPR of the tests intended to detect subchromosomal deletions/duplications. Thus, whether it is clinically meaningful to screen for these conditions depends on disease severity, frequency, prognosis, and other factors.
- There is an increasing commercial interest in providing NIPD to screen for monogenic conditions in low-risk pregnancies.
- The continuing education of physicians and patients about pre-test and post-test counseling is crucial to ensure NIPT-related tests are implemented appropriately to provide maximal benefits for families. Thus, good pre-test counseling should include a discussion of the following (110):
  - The scope and nature of conditions being tested for.
  - The performance capabilities of the test for each condition, including DR, FPR, and no-call rate.
  - An explanation highlighting that FP results may be more common than true-positive results for rare conditions.
  - The need to confirm results through additional testing.
  - The potential to detect maternal chromosome abnormalities and constitutional changes associated with malignancy.
  - Uncertainties associated with mosaicism and unexpected findings.

FUTURE PERSPECTIVES

Widespread adoption of NIPT is likely to continue, leading to a dramatic reduction in invasive sampling for prenatal genetic diagnostic tests. From a technical point of view, non-invasive deep sequencing of the fetal genome may serve as a universal screening or even diagnostic test to detect genetic diseases in a fetus (1). However, whether it should be routinely offered to pregnant women without a known risk is a complex issue requiring ethical and socio-economical discussions. Regardless, genome-wide profiling will improve overall pregnancy management. A workflow for the follow-up of NIPT results is proposed in Figure 26.4 (47). Despite technological advances in NIPT for monogenic diseases, widespread clinical uptake will remain hindered by the rarity of individual genetic diseases and the subsequent need for individual, customized development.

In summary, cfDNA is part of the changing landscape of reproductive medicine. The development of genomic screening techniques will improve the capacity to detect and eventually treat inherited disorders. Test developers and providers, health care professionals, and women and couples engaging in the debate about the most responsible use of prenatal genetic testing will drive the figure of NIPT technology.

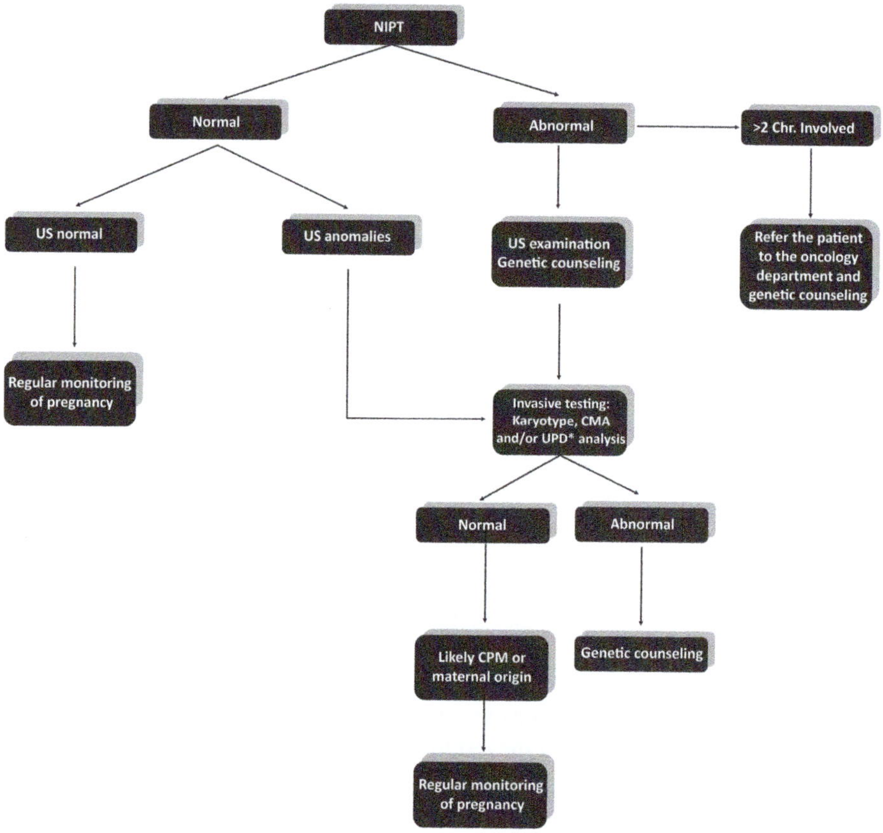

FIGURE 26.4 Proposed decision tree for NIPT results management. In cases where NIPT indicates a low risk for the aneuploidies tested for and in the absence of ultrasound anomalies, regular pregnancy monitoring should be recommended. If the NIPT result is high-risk for common aneuploidies (for chromosomes 13, 18, and 21), invasive testing is needed to confirm the findings. If the results from the two testing types are discordant, CPM should be suspected. In such cases, a maternal karyotype or microarray should be performed to exclude a maternal cause (maternal CNV, maternal mosaicism, or a maternal tumor). Also, the analysis of placental tissue at birth could provide a definite diagnosis. If autosomal aneuploidies other than 13, 18, or 21 are detected, ultrasound examination and genetic counseling should be provided to inform evidence-based decision making (invasive testing or continue with the pregnancy). Likewise, if CPM is suspected, there is a risk for UPD because of trisomy rescue; therefore, it is worthwhile to perform UPD analysis, particularly when chromosomes 6, 7, 11, 14, 15, or 20 are involved due to the presence of imprinting disorders. Finally, in the case of (segmental) aneuploidies on > 2 chromosomes and identical genome representation profiles after repeating the test with a new blood specimen, a tumor may be suspected as a possible biological cause. In these cases, referral to oncology may be warranted. Adapted from (47).

REFERENCES

1. Wong FC, et al. *Annu Rev Med.* 2016;67:419–32.
2. Mujezinovic F, et al. *Obstet Gynecol.* 2007;110(3):687–94.
3. Lo YM, et al. *Lancet.* 1997;350(9076):485–7.
4. Lun FM, et al. *Clin Chem.* 2008;54(10):1664–72.
5. Lo YM, et al. *Am J Hum Genet.* 1998;62(4):768–75.
6. Lo YM, et al. *Am J Hum Genet.* 1999;64(1):218–24.
7. Masuzaki H, et al. *J Med Genet.* 2004;41(4):289–92.

8. Alberry M, et al. *Prenat Diagn.* 2007;27(5):415–8.
9. Rijnders RJ, et al. *Prenat Diagn.* 2003;23(13):1042–4.
10. Chan KC, et al. *Clin Chem.* 2004;50(1):88–92.
11. Lo YM, et al. *Sci Transl Med.* 2010;2(61):61ra91.
12. Hui L, et al. *Prenat Diagn.* 2020;40(2):155–63.
13. Dondorp W, et al. *Eur J Hum Genet.* 2015;23(11):1438–50.
14. Gil MM, et al. *Ultrasound Obstet Gynecol.* 2017;50(3):302–14.
15. Sacco A, et al. In: Page-Christiaens L, Klein H-G, editors. *Noninvasive Prenatal Testing (NIPT).* Academic Press; 2018:51–66.
16. Harasim T, et al. In: Page-Christiaens L, Klein H-G, editors. *Noninvasive Prenatal Testing (NIPT).* Academic Press; 2018:67–82.
17. Ehrich M, et al. *Am J Obstet Gynecol.* 2011;204(3):205.e1–11.
18. Shaw J, et al. *Reproduction.* 2020;160(5):A1–A11.
19. Snyder MW, et al. *Cell.* 2016;164(1–2):57–68.
20. Straver R, et al. *Prenat Diagn.* 2016;36(7):614–21.
21. Sun K, et al. *Proc Natl Acad Sci U S A.* 2015;112(40):E5503–12.
22. Guo S, et al. *Nat Genet.* 2017;49(4):635–42.
23. Chiu RW, et al. *Proc Natl Acad Sci U S A.* 2008;105(51):20458–63.
24. Fan HC, et al. *Proc Natl Acad Sci U S A.* 2008;105(42):16266–71.
25. Fan HC, et al. *PLoS One.* 2010;5(5):e10439.
26. Muzzey D. In: Page-Christiaens L, Klein H-G, editors. *Noninvasive Prenatal Testing (NIPT).* Academic Press; 2018:25–47.
27. Stumm M, et al. *Prenat Diagn.* 2014;34(2):185–91.
28. Sparks AB, et al. *Am J Obstet Gynecol.* 2012;206(4):319.e1–9.
29. Srinivasan A, et al. *Am J Hum Genet.* 2013;92(2):167–76.
30. Sparks AB, et al. *Prenat Diagn.* 2012;32(1):3–9.
31. Dar P, et al. *Clin Lab Med.* 2016;36(2):199–211.
32. Zimmermann B, et al. *Prenat Diagn.* 2012;32(13):1233–41.
33. Samango-Sprouse C, et al. *Prenat Diagn.* 2013;33(7):643–9.
34. Norwitz ER, et al. *Rev Obstet Gynecol.* 2013;6(2):48–62.
35. Wapner RJ, et al. *Am J Obstet Gynecol.* 2015;212(3):332.e1–9.
36. Van Opstal D, et al. *Genet Med.* 2018;20(5):480–5.
37. van der Meij KRM, et al. *Am J Hum Genet.* 2019;105(6):1091–101.
38. Suciu ID, et al. *J Med Life.* 2019;12(3):221–4.
39. Bianchi DW, et al. *Obstet Gynecol.* 2013;121(5):1057–62.
40. Hui L, et al. *Annu Rev Med.* 2017;68:459–72.
41. Zhang B, et al. *Prenat Diagn.* 2020;40(4):463–9.
42. Benn P. *Clin Genet.* 2016;90(6):477–85.
43. Lau TK, et al. *Ultrasound Obstet Gynecol.* 2014;43(3):254–64.
44. Fiorentino F, et al. *Prenat Diagn.* 2017;37(6):593–601.
45. Pertile MD, et al. *Sci Transl Med.* 2017;30;9(405):eaan1240:1–11.
46. Scott F, et al. *Prenat Diagn.* 2018;38(10):765–71.
47. Brady P, et al. *Clin Genet.* 2016;89(5):523–30.
48. Ehrich M, et al. *Genet Med.* 2017;19(12):1332–7.
49. Brison N, et al. *Prenat Diagn.* 2018;38(4):258–66.
50. Chatron N, et al. *Ultrasound Obstet Gynecol.* 2019;53(1):129–30.
51. Hillman SC, et al. *Ultrasound Obstet Gynecol.* 2011;37(1):6–14.
52. Grati FR, et al. *Prenat Diagn.* 2015;35(8):801–9.
53. Weise A, et al. *J Histochem Cytochem.* 2012;60(5):346–58.
54. Yin AH, et al. *Proc Natl Acad Sci U S A.* 2015;112(47):14670–5.
55. Straver R, et al. *Nucleic Acids Res.* 2014;42(5):e31.
56. Lo KK, et al. *Am J Hum Genet.* 2016;98(1):34–44.
57. Chen S, et al. *Prenat Diagn.* 2013;33(6):584–90.
58. Liu H, et al. *PLoS One.* 2016;11(7):e0159233.
59. Lefkowitz RB, et al. *Am J Obstet Gynecol.* 2016;215(2):227.e1–e16.
60. Grati FR, et al. *Prenat Diagn.* 2019;39(2):70–80.
61. Allen S, et al. In: Page-Christiaens L, Klein H-G, editors. *Noninvasive Prenatal Testing (NIPT).* Academic Press; 2018:157–77.

62. Lun FM, et al. *Proc Natl Acad Sci U S A*. 2008;105(50):19920–5.
63. New MI, et al. *J Clin Endocrinol Metab*. 2014;99(6):E1022–30.
64. Chiu EKL, et al. *Prenat Diagn*. 2018;38(1):52–8.
65. Xiong L, et al. *BJOG*. 2018;125(4):461–8.
66. Cutts A, et al. *Blood*. 2019;134(14):1190–3.
67. Camunas-Soler J, et al. *Clin Chem*. 2018;64(2):336–45.
68. Barrett AN, et al. *Clin Chem*. 2012;58(6):1026–32.
69. Tsui NB, et al. *Blood*. 2011;117(13):3684–91.
70. Hudecova I, et al. *Blood*. 2017;130(3):340–7.
71. Chang MY, et al. *Sci Rep*. 2016;6:37153.
72. Parks M, et al. *Prenat Diagn*. 2016;36(4):312–20.
73. Parks M, et al. *Eur J Hum Genet*. 2017;25(4):416–22.
74. Chandler NJ, et al. *Clin Chem*. 2020;66(1):207–16.
75. Fan HC, et al. *Nature*. 2012;487(7407):320–4.
76. Kitzman JO, et al. *Sci Transl Med*. 2012;4(137):137ra76.
77. Best S, et al. *Prenat Diagn*. 2018;38(1):10–9.
78. Lord J, et al. *Lancet*. 2019;393(10173):747–57.
79. Petrovski S, et al. *Lancet*. 2019;393(10173):758–67.
80. Rabinowitz T, et al. *Comput Struct Biotechnol J*. 2020;18:2463–70.
81. Samura O, et al. *Taiwan J Obstet Gynecol*. 2020;59(1):16–20.
82. Kalousek DK, et al. *Science*. 1983;221(4611):665–7.
83. Kalousek DK, et al. *J Med Genet*. 1996;33(7):529–33.
84. Lestou VS, et al. *Arch Dis Child Fetal Neonatal Ed*. 1998;79(3):F223–6.
85. Grati FR, et al. *Genet Med*. 2014;16(8):620–4.
86. Levi S. *J Clin Ultrasound*. 1976;4(1):3–5.
87. Curnow KJ, et al. *Am J Obstet Gynecol*. 2015;212(1):79.e1–9.
88. Niles KM, et al. *Ultrasound Obstet Gynecol*. 2018;52(4):547–8.
89. Balaguer N, et al. *Prenat Diagn*. 2021;41(10):1241–8.
90. Russell LM, et al. *Cytogenet Genome Res*. 2007;116(3):181–5.
91. Guttenbach M, et al. *Am J Hum Genet*. 1995;57(5):1143–50.
92. Bianchi DW, et al. *JAMA*. 2015;314(2):162–9.
93. Chan RW, et al. *Proc Natl Acad Sci U S A*. 2014;111(49):E5302–11.
94. Hui L, et al. *Ultrasound Obstet Gynecol*. 2014;44(2):242–3.
95. Hui CY, et al. *BMJ Case Rep*. 2016;2016:bcr2016216593:1–3.
96. Barrett AN, et al. *PLoS One*. 2011;6(10):e25202.
97. Burns W, et al. *Prenat Diagn*. 2017;37(11):1125–9.
98. Ma GC, et al. *Ultrasound Obstet Gynecol*. 2018;51(2):276–7.
99. Juul LA, et al. *Acta Obstet Gynecol Scand*. 2020;99(6):744–50.
100. Haghiac M, et al. *Obesity (Silver Spring)*. 2012;20(11):2213–9.
101. Mardy A, et al. *Am J Med Genet C Semin Med Genet*. 2016;172(2):118–22.
102. Grimes DA, et al. *Lancet*. 2002;359(9309):881–4.
103. Bunnik EM, et al. *J Med Ethics*. 2020;46(3):194–8.
104. Lo JO, et al. *Obstet Gynecol Surv*. 2014;69(2):89–99.
105. Benn P, et al. *Ultrasound Obstet Gynecol*. 2013;42(1):15–33.
106. Grace MR, et al. *Obstet Gynecol Surv*. 2016;71(8):477–87.
107. Kellogg G, et al. *J Genet Couns*. 2014;23(5):805–13.
108. van Schendel RV, et al. *J Genet Couns*. 2017;26(3):522–31.
109. Kuppermann M, et al. *JAMA*. 2014;312(12):1210–7.
110. Di Renzo GC, et al. *Am J Obstet Gynecol*. 2019;220(6):537–42.

Advances in Non-Invasive Diagnosis of Single-Gene Disorders and Fetal Exome Sequencing

27

Liesbeth Vossaert, Roni Zemet, and Ignatia B. Van den Veyver

INTRODUCTION

Congenital structural anomalies affect approximately 2–4% of pregnancies, and are caused by environmental factors, genetic defects, or are multifactorial.[1] Recent data support the observation that, depending on the type of defect and organ system involved, about 30% of congenital anomalies are caused by aneuploidy. Chromosomal microarray analysis (CMA) provides an overall additional diagnostic yield of 4–7% when there are fetal congenital anomalies. This number increase to > 10% when there are multiple fetal congenital anomalies, but still leaves 60% of cases undiagnosed. This is in part because this strategy cannot diagnose the more than 7,000 known monogenic disorders, which together affect approximately 1% of all live births. Not surprisingly, prenatal gene panels or exome sequencing on amniotic fluid samples obtained from pregnancies with fetuses with congenital anomalies have revealed that *de novo* and inherited pathogenic variants in single genes are an important cause of prenatally detected congenital anomalies. In addition, expanded carrier screening now identifies more carrier parents at risk of having children with recessive monogenic disorders, increasing the need for prenatal diagnostic testing for these disorders.[4] Despite this, most testing for monogenic disorders is only carried out in high-risk pregnancies on fetal samples obtained through an invasive diagnostic procedure. Monogenic disorders are not yet routinely included in non-invasive prenatal screening and testing paradigms, but extensive research is ongoing on the analysis of circulating fetal cell-free DNA (cfDNA) or circulating fetal cells for non-invasive prenatal diagnosis (NIPD) and screening (NIPS) assays for single-gene disorders and some have already been introduced clinically.

DOI: 10.1201/9781003024941-27

STATE OF THE ART

To date, testing for pathogenic variants that cause single-gene disorders is mainly performed on high-risk pregnancies and requires amniocentesis or chorionic villus sampling (CVS), such as when a fetus is diagnosed with anomalies on ultrasound, and the recommended first-line genetic test is a CMA with or without a karyotype. When these do not clarify the cause of the anomalies, the next level of testing offered is often a gene panel that includes genes relevant to the observed phenotype, and, increasingly common, whole-exome sequencing. Alternatively, when a pregnancy is at high risk of a single-gene disorder because of a significant family history or results of carrier screening, diagnostic testing on amniotic fluid or CVS for the specific variant is done. Non-invasive fetal cfDNA analysis for monogenic conditions is already offered on a small scale, and panels for non-invasive screening for selected dominant monogenic disorders have been introduced on a limited scale. These can be considered when a patient declines a diagnostic procedure, but many of these tests are not yet fully clinically validated. This chapter reviews the current progress with non-invasive and invasive prenatal testing approaches for single-gene disorders (overview in Figure 27.1).

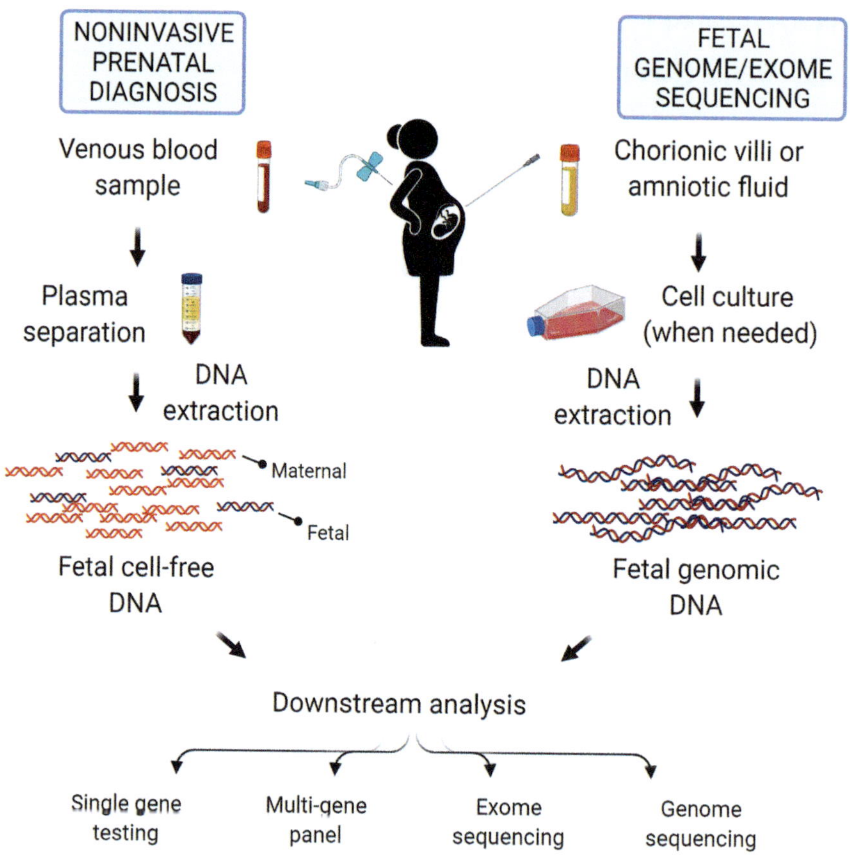

FIGURE 27.1 Monogenic disorder testing approaches: non-invasive prenatal diagnosis and fetal exome/genome sequencing after invasive sampling. Created with BioRender.com.

NON-INVASIVE DIAGNOSIS OF SINGLE-GENE DISORDERS

Methodology

To date, most non-invasive prenatal testing development has focused on analyzing circulating fetal DNA (cfDNA) in maternal plasma to screen for fetal aneuploidies (Chapter 26) and specific deletion and duplication syndromes caused by subchromosomal copy number abnormalities. Analysis of cfDNA is also the primary approach for non-invasive prenatal test development for single-nucleotide variants (SNVs) that cause monogenic disorders. Because cfDNA in maternal plasma is predominantly maternal in origin, with a much smaller amount being fetal cfDNA derived from apoptotic trophoblasts, an important parameter for cfDNA testing is the fetal fraction (FF), or the proportion of all cfDNA that is of fetal (trophoblast) origin. The FF can vary widely between samples, but is usually between 5% and 20%. For pregnancies with male fetuses, the FF can be estimated by quantifying Y chromosome-derived cfDNA fragments. For pregnancies with female fetuses, other measurements are needed, such as precise quantification of fetal alleles at informative polymorphisms by polymerase chain reaction (PCR) or next-generation sequencing (NGS) methods,[5,6] or, more rarely, assessment of differential DNA methylation at the *RASSF1A* promotor between trophoblast-derived cfDNA and maternal cfDNA.[7] Generally, a FF of ≥ 4% is considered acceptable, but some assays may work at lower FF. Depending on the type of fetal SNV of interest, different strategies are used for downstream analysis of circulating cfDNA (Figure 27.2).

Detecting paternally inherited and de novo variants in cfDNA

NIPD assays for autosomal dominant conditions caused by paternally inherited or *de novo* variants aim to detect a unique fetal variant present at a concentration that is approximately half of the FF against a high background of maternal wild-type (WT) alleles (Figure 27.2). The earliest application of cfDNA-based NIPD was PCR-based assays targeting Y chromosome-specific sequences to determine fetal sex[8] or *RHD* gene-specific sequences to assess the fetal Rhesus D status for Rh-negative women.[9] Subsequently, PCR-based assays were developed for paternally inherited pathogenic variants or for suspected *de novo* dominant genetic disorders based on ultrasound findings, such as a PCR combined with restriction enzyme digestion for *FGFR3* variants that cause achondroplasia and thanatophoric dysplasia.[10,11] An approach with greater sensitivity that uses digital droplet PCR to detect these variants has also been developed.[5,12–14] More recently, NGS-based assays have been developed and some introduced into clinics, that can detect multiple potential disease-causing variants in specific genes or panels of genes, ranging from a diagnostic assay for *FGFR3*-related disorders and other skeletal dysplasias to a broader screening panel that includes 30 genes.[15–17]

Detecting maternally inherited variants in cfDNA

Testing for a maternally inherited variant is more complicated because the variant is also present in the maternal cfDNA fraction. In cases where both parents are heterozygous carriers for different pathogenic variants of a gene associated with an autosomal recessive disorder, PCR or NGS analysis can be done to detect or exclude the presence of the paternal pathogenic variant in the fetal cfDNA. With this strategy, invasive testing to determine the status of the maternal variant is only offered for those pregnancies where the paternal variant is detected non-invasively, potentially reducing the number of diagnostic procedures for these circumstances by 50%.

More recently, relative mutation dosage (RMD) and relative haplotype dosage (RHDO) methods have been developed. These are dosage-based methods that detect small differences in allele ratios that accompany the presence or absence of a maternally inherited variant in the fetal fraction. RMD directly assesses

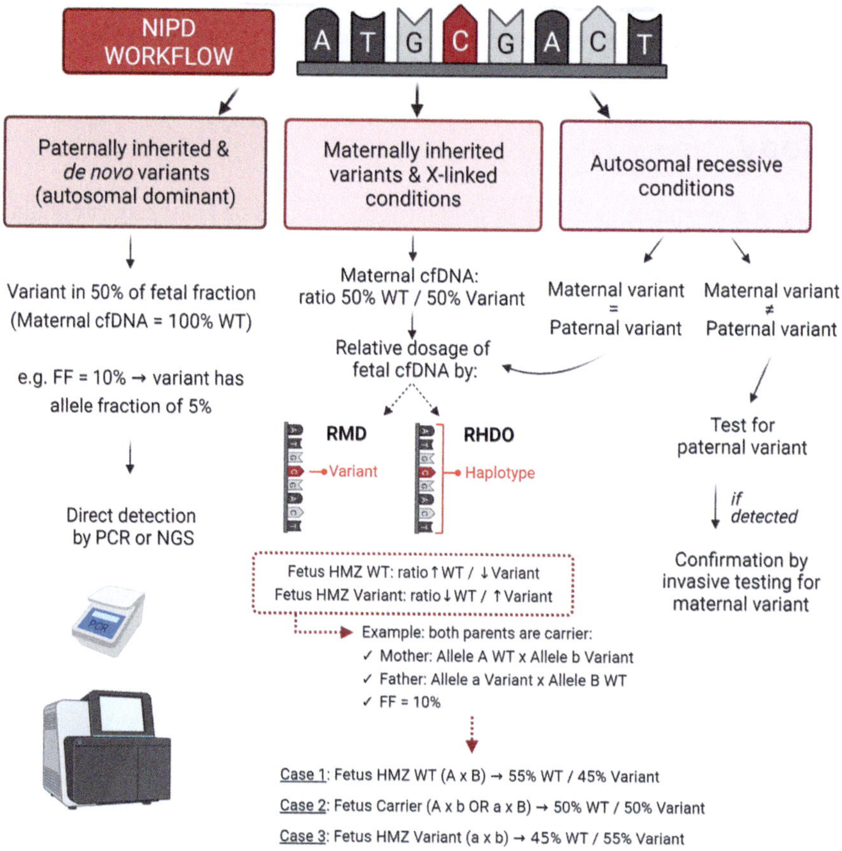

FIGURE 27.2 NIPD workflow. The non-invasive prenatal diagnosis set-up depends on the gene and variant of interest. Abbreviations: cfDNA: cell-free DNA, FF: fetal fraction, HMZ: homozygous, RHDO: relative haplotype dosage, RMD: relative mutation dosage, WT: wild-type. Created with BioRender.com.

the relative proportion of the WT and the pathogenic variant. That proportion is 50:50 in the maternal fraction of the cfDNA of a heterozygous mother, but becomes slightly skewed toward the WT variant if the fetus does not inherit that variant, remains 50:50 if the fetus is heterozygous or skews toward the pathogenic variant if the fetus is homozygous for that variant, or towards the pathogenic variant if that variant is inherited by the fetus. Digital PCR and NGS-based RMD approaches have been developed for a range of conditions.[18–28] In RHDO, the dosage of multiple informative single-nucleotide polymorphisms (SNPs), either by themselves or along with the variant of interest, is determined to get a more accurate result.[29–33] This assay can be used for genes or variants that are difficult to target directly, e.g., because of the type of variant, the presence of highly homologous pseudogenes,[34] or where there are more complex genomic rearrangements.[35] Thus, RHDO is not limited to analyzing one specific pathogenic variant, and could be applied for multiple variants and genes simultaneously, provided there is sufficient coverage. To set up and interpret RHDO, both parental samples and a sample of an unaffected or affected sibling are typically needed.

Whole-genome cfDNA analysis

The potential of genome sequencing of cell-free fetal DNA (cffDNA) was first reported[36–38] about a decade ago, but the high sequencing cost has hampered clinical implementation. Che et al. recently described a newer genome-wide NGS haplotyping approach.[39] Their technique, named cffDNA haplarithmisis, uses targeted capture of approximately 250,000 SNPs, which include some 250 disease-associated regions, allowing detection of pathogenic variants associated with monogenic disorders along with fetal aneuploidies, but samples from both parents and possibly from other family members are required to establish the full haplotype. Chan and colleagues published a genome-wide sequencing-based relative allelic dosage analysis method to resolve the maternal inheritance for thousands of SNPs across the genome and detect fetal *de novo* mutations.[40]

Clinical Results and Applications

The use of non-invasive testing for monogenic disorders is far less widespread than NIPS for aneuploidy screening. NIPD for monogenic disorders is mostly still performed in a research setting, except for the UK, where it is clinically offered for *FGFR2*- and *FGFR3*-related disorders, skeletal dysplasias caused by variants in collagen genes, cystic fibrosis, spinal muscular atrophy, Duchenne/Becker muscular dystrophy (DMD/BMD), and for known familial mutations in a range of rare single-gene conditions.[41,42] Other panels for non-invasive screening for single-gene disorders, including genes associated with Noonan spectrum disorders, skeletal dysplasias, craniosynostosis disorders, and other conditions have been clinically implemented in the United States.[15,26]

Fetal sex and RhD status determination were among the initial NIPD applications by PCR, and, more recently, a study using an NGS-based method for non-invasive ABO blood group typing was published.[43] Determining the fetal RhD status is valuable for the management of pregnancies at risk for hemolytic disease of the fetus and newborn due to RhD incompatibility. Meta-analyses of published *RHD*-NIPD studies note a very high test efficiency and utility: a review from 2006[44] that included 37 studies indicated over 94% accuracy, with 16 studies reaching 100%, whereas a more recent analysis found sensitivity and specificity levels over 99% and a performance equivalent to standard newborn serology testing.[45] This test is now integrated into prenatal care of RhD-negative women in some countries, but is not currently widely used in the US.

Non-invasive fetal sex determination and, subsequently, specific detection of pathogenic variants on the X chromosome, have been beneficial for families with X-linked disorders. Non-invasive fetal sex determination reduces the need for CVS or amniocentesis for X-linked disorders that do not affect female carriers,[46] and has value when there is fetal ambiguous genitalia or for suspected congenital adrenal hyperplasia to guide the decision on prenatal treatment with maternal steroid hormone administration.[34,47] Despite these benefits, concerns have been raised about its use for sex-selective termination of pregnancy.[48] Different groups have reported high accuracy with specific non-invasive prenatal DMD diagnosis, including a recent research study by Chen et al.[49] who developed a cfDNA test for pregnant DMD carrier mothers that avoided the need for proband or other family member samples.

One of the clinically validated NIPD tests for autosomal dominant disorders is for variants in genes that cause, primarily *de novo*, dominant skeletal dysplasias, such as specific pathogenic variants in *FGFR3*, which cause achondroplasia and thanatophoric dysplasia, and skeletal dysplasias with very different severity and prognosis. Other examples include variants in *COL1A1* and *COL1A2*, which cause osteogenesis imperfecta that can present as a severe lethal form or milder form. Thus, early exact molecular diagnosis is important for pregnancy management decisions and counseling, and cfDNA-based NIPD for *FGFR3* variants has now been reported with sensitivities and specificities up to 100% for common hotspot variants,[17,50] and for *COL1A1* and *COL1A2* variants.[15,51]

Autosomal recessive disorders for which cfDNA-based NIPD assays have been developed are those with greater prevalence in the population. These include β-thalassemia and sickle-cell disease, hemoglobinopathies caused by variants in the *HBB* gene,[52] with reported sensitivities and specificities ranging from 75% up to 100%, depending on factors such as the FF and the testing methodology applied, with digital PCR and NGS yielding a greater accuracy than older PCR-based methods. Additionally, cfDNA-based NIPD for several other autosomal recessive conditions, such as cystic fibrosis, congenital adrenal hyperplasia, autosomal recessive polycystic kidney disease, and recessive forms of skeletal dysplasias, have been explored as well.[22,34,42,53]

Strengths and Limitations

The primary advantage of NIPD is its non-invasive nature, thereby avoiding the risks associated with CVS or amniocentesis. In addition, it can be performed as early as 9–10 weeks' gestation, permitting earlier pregnancy management decisions.[54]

Because cffDNA is derived from the placenta, confined placental mosaicism (CPM) is a potential limitation. However, unlike with NIPS for fetal aneuploidies, CPM is not a significant concern for monogenic conditions. The analysis of cffDNA is technically challenging and the number of variants that can be detected remains limited with PCR and panel approaches. Although numerous assay development studies and some validation studies have been published, clinical availability of NIPD is still limited, and certain variants, such as small copy number abnormalities and structural variants, cannot be detected. More efforts will be needed to further expand the scope of NIPD for monogenic disorders. One promising report from 2019[55] described a comprehensive cfDNA analysis customized target capture sequencing method, applicable to aneuploidies (targeting selected regions for chromosomes 13, 18, 21, X, and Y), microdeletions (including 22q11.2, 1p36, Wolf–Hirschhorn syndrome, and Smith–Magenis syndrome), and known pathogenic variants for 50 autosomal recessive disorders. Another study by Che et al.[39] also describes the concurrent genome-wide analysis for aneuploidies and monogenic disorders. Future technological improvements should help the development of more comprehensive, non-invasive, prenatal tests for clinical use.

FETAL EXOME SEQUENCING

Methodology

Principles of next-generation sequencing

Genome and exome sequencing are technologies that examine the genome at a nucleotide level. NGS is achieved by sequencing many distributed, overlapping sites throughout the genome in a massively parallel manner.[56,57] Briefly, for NGS library preparation, genomic DNA is sheared into 50–400 nt fragments, ligated to adapters, and purified. For panel or exome sequencing, an enrichment step is included to capture regions of interest using "DNA baits." Massively parallel sequencing of the NGS library then rapidly and accurately derives the nucleotide sequences of each fragment, which are then aligned to a reference genome to detect potential sequence variants[56–58] (Figure 27.3). Two quality parameters for NGS accuracy are the sequencing depth, which refers to the number of overlapping reads for each base-pair, and the breadth of sequence coverage, which is the fraction of the reference sequence that is covered at sufficient depth. The American College of Medical Genetics and Genomics (ACMG) recommends an average depth of $\geq$ 100-fold with 90–95% of the sequence covered at least 10-fold for diagnostic exome sequencing (ES).[59] With recent advances in NGS technology, clinical laboratories can offer increasingly better sequencing depth and shorter turnaround time (TAT) to results, which is crucial for fetal ES.[60,61]

Analysis and interpretation of ES

In ES, the DNA baits are designed to capture coding exons along with exon-intron junctions, which comprise 1–2% of the human genome, but contain 85% of the disease-causing variants.[59] Given the high level of variability of the human genome, 20,000 to 50,000 variants are present per sequenced exome, which are mostly benign polymorphisms, making the identification of the causative variant for a phenotype or disease a cumbersome process.[62,63] ACMG has published variant interpretation and reporting guidelines for laboratories to promote a standardized variant classification[64] into pathogenic, likely pathogenic,

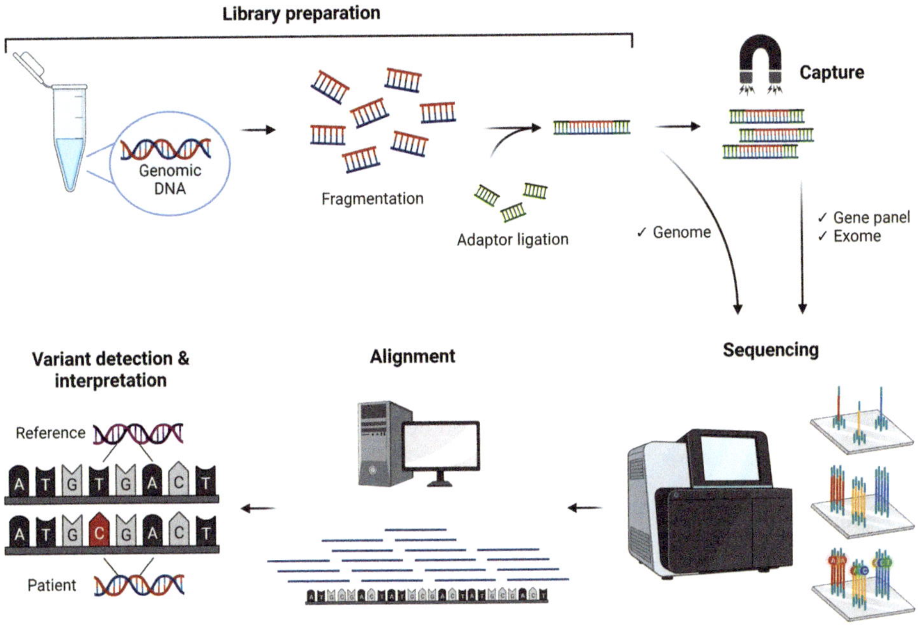

FIGURE 27.3 Next-generation sequencing workflow. Created with BioRender.com.

variant of uncertain significance (VUS), likely benign, or benign. The variants are further sorted into genes related to the patient's phenotype and indication for sequencing or to incidental findings. Clinically important variants are then prioritized and interpreted based on variant frequency in a healthy population, their inheritance and segregation pattern within families, information in clinical variant databases, and available patient reports, keeping in mind exceptions due to incomplete penetrance or variable expressivity.[56,65–69] Nonsense, frameshift, splice site variants, and deletions that result in the absence of a gene product are more likely to be pathogenic, whereas missense variants can have variable effects and interpretation is informed by additional computational predictions and data from animal models or functional assays. Finally, clinical correlation is critical, and requires collaboration and data sharing between the clinician, the diagnostic laboratories, and other scientists.[70] This is more challenging prenatally, because phenotypic information is limited to features seen on prenatal imaging, which can evolve over time in the developing fetus. In addition, the prenatal phenotypes of many genetic conditions are not well described in existing databases. Therefore, initiatives are underway to improve prenatal phenotypic information in the current clinical databases and develop an international registry of fetal phenotypes. Of note, reported variants are validated by an independent method, such as Sanger sequencing.

Considering that new gene-disease associations are regularly discovered, reanalysis of ES data could uncover a genetic diagnosis for patients with an initially unexplainable ES result.[71] For non-diagnostic prenatal ES results, reanalysis should be considered if a new phenotype develops over time, when a future pregnancy is planned, or once a considerable amount of time has passed since the initial testing. Laboratories should have a defined policy regarding ES reanalysis and the release of an updated report.[1,72,73]

Technical limitations of ES

NGS cannot cover all sequence variants.[57,65,74] It performs well for single-nucleotide variants and small insertion/deletions, but regions with high CG content are more difficult to capture. Because the consensus sequence is built from the alignment of overlapping short fragments, regions with high homology to other sequences within the genome are challenging. These include duplicated genes or exons, repetitive sequences, short repeat expansions, pseudogenes, and highly homologous gene families. Structural chromosomal abnormalities or aneuploidy can be detected by genome sequencing (GS), but not currently as effectively by ES. Low-level mosaic variants are also challenging, but can be identified provided that the sequencing depth is adequate. Haplotype information can aid in detecting uniparental disomy.

Results

Clinical use of ES in prenatal genetic diagnosis

Rapid high-throughput NGS technology has revolutionized clinical and particularly prenatal genetic diagnosis because of its increased speed at a relatively low cost. Currently, ES and GS analysis can be performed within days, and is widely used in the pediatric and adult population with a diagnostic yield of 25–50%.[56,75–77] Their introduction into reproductive and prenatal medicine is more novel,[72,74,78] and, until recently, prenatal testing for single-gene disorders had been limited to rare cases with an increased risk for a specific genetic disorder. More commonly, the fetus has a *de novo* presentation or genetic condition, in which case using a single-gene test or a multigene panel is ineffective because the prenatal phenotype is not always well defined. There can also be genetic heterogeneity, or the causative genes for the observed fetal anomalies may not be known, and some conditions are multifactorial with both genetic and environmental causes. Therefore, a more comprehensive approach of ES is required.

Proband vs. trio sequencing in prenatal ES

Prenatal ES is best performed as a trio, where fetal, maternal, and paternal DNA are sequenced and analyzed together.[72] Integrated analysis of all three samples allows for faster determination of whether a variant is *de novo* or inherited, the parental origin, the phase of biallelic variants, and parental carrier status, to improve interpretation of a variant's pathogenicity. Trio analysis has been shown to have a higher diagnostic yield compared with proband analysis (24–31% vs. 14–23.6%).[79,80] For proband-only sequencing, the potential pathogenicity of identified variants needs to be confirmed by other means, including variant segregation analysis within the family.[72]

Current clinical experience

Early experience with prenatal ES consisted of case reports and smaller series, sometimes embedded within a more extensive postnatal series of diagnostic ES.[75,76,81,82] In 2017, a review of available studies with more than five cases found that the reported ES diagnostic rate varied between 6.2% and 80% for fetal anomalies.[74] More recent, larger studies of prenatal ES with > 50 cases (Table 27.1)[83–89] reported an incremental diagnostic yield of 8.5–31.5% in structurally abnormal fetuses following non-diagnostic karyotype and CMA results, with a higher rate for more complex fetal anomalies and a lower rate for studies that included a less-selected cohort.[85,86]

A few studies have examined the advantage of ES for prenatally diagnosed anomalies in specific organ systems, emphasizing the importance of case selection. In a study focusing on fetuses with congenital anomalies of the kidneys and urinary tract (CAKUT), pathogenic variants were detected in 4/30 (14%) cases after a negative CMA analysis.[90] For skeletal dysplasias, Chandler et al. reported an 81% (13/16) diagnostic rate for whole-exome sequencing (WES),[91] and, for nonimmune hydrops fetalis, Sparks et al. reported a 29% (37/127) diagnostic rate. ES has also been a useful testing tool in 6.1% of cases of intrauterine fetal death (15/246).[84,92]

The impact of prenatal ES on medical management

Finding the cause of fetal anomalies is important to predict prognosis, establish recurrence risk, and guide clinical management. As a single comprehensive test, ES can prevent the "diagnostic odyssey," reducing

TABLE 27.1 Overview of Recent Larger Prenatal ES Studies

STUDY	*NO. CASES*	*STUDY DESIGN*	*DIAGNOSTIC RATE*	*POTENTIALLY PATHOGENIC VARIANTS*
Sparks et al. (2020)[83]	127	Trio/quad-WES (117), duo-WES (9), and proband-only WES (1); fetuses with unexplained nonimmune hydrops fetalis; normal karyotype/CMA results	**37/127 (29.1%)**	12/127 (9.4%)
Lord et al. (2019)[85]	610	Trio-WES (596) and duo WES (14); fetuses with structural anomalies and normal karyotype/CMA results	**52/610 (8.5%)** 22/143 (25.4%) with multisystem anomalies	24/610 (3.9%)
Petrovski et al. (2019)[86]	234	Trio-WES; Fetuses with structural anomalies and normal karyotype/CMA results	**24/234 (10.3%)**	46/234 (19.7%)
Daum et al. (2018)[106]	77	Proband-only WES (44) and Trio-WES (33); Fetal DNA or POC, sonographic malformation(s) or isolated NT; normal CMA results	**16/77 (20.8%)** - 11/44 (25%) proband-only - 5/33 (15%) trio	Not stated
Normand et al. (2018)[87]	146	Proband-only WES (51), Trio-WES (33) or Prenatal Trio-WES (62); Fetuses or POC, with structural anomalies and a suspected Mendelian disorder	**46/146 (31.5%)**	Not stated
Fu el al. (2018)[88]	196	Proband-only WES (147), Trio-WES (49); Fetal structural anomalies; normal karyotype/CMA results	**47/196 (24%)**	25/196 (12.8%)
Yates et al. (2017)[89]	84	Trio/Quad WES (52) and proband-only or duo WES (32); Fetal demise or termination of pregnancy (following structural anomalies)	**17/84 (20.2%)**	38/84 (45.2%)

Abbreviations: POC, product of conception; WES, whole-exome sequencing.

the time to diagnosis and the cost of serial testing. Knowing a genetic diagnosis prenatally can help guide delivery plans and decisions regarding perinatal intervention and treatment. For example, prenatal diagnosis of metabolic disorders, such as urea cycle disorders and pyruvate dehydrogenase deficiency, facilitates earlier neonatal treatment with improved long-term outcomes.[93] *In utero* or postnatal mesenchymal stem cell transplantation for fetuses or children with osteogenesis imperfecta, and stem cell transplantation for α-thalassemia, are currently in clinical trials,[94,95] and, with the development of gene therapies, *in utero* treatments may become an option for more conditions in the future. Prenatal genetic diagnosis can also improve candidate selection and outcome prediction of *in utero* fetal interventions, such as treatment with fetoscopic endotracheal occlusion for congenital diaphragmatic hernia or prenatal cardiac intervention for hypoplastic left heart syndrome. Diagnosis of a lethal genetic condition allows parents to make choices regarding pregnancy management, including termination within legal limits, palliative care, and mode and place of delivery. Establishing a diagnosis and mode of inheritance also guides recurrence risk counseling and future reproductive choices, including preimplantation genetic testing, targeted prenatal testing, or the use of donor gametes for subsequent pregnancies.

In addition, prenatal ES data have already contributed new information on the prenatal phenotype of known Mendelian diseases and accelerated the discovery of candidate genes for disorders with previously unknown genetic etiology. ES has also revealed new genetic etiologies of uniquely prenatal phenotypes, such as nonimmune hydrops or cystic hygroma[83] and prenatally lethal disorders.[84,96]

However, ES also increases the likelihood of identifying VUS, incidental findings unrelated to phenotype, variants in genes on the ACMG secondary finding list, and, in the case of trio-exome, incidental parental findings, misattributed parentage, and undisclosed consanguinity.[97]

Limitations

Technical challenges

Timely return of results is critical for prenatal (trio) ES, and laboratories should implement procedures to assure rapid turn-around time (TAT) by optimizing sequencing equipment, reagents, protocols, and workflows to speed up interpretation. Cell culture may be required due to a limited amount of available DNA when prepared directly from amniotic fluid or CVS samples, prolonging the TAT. Furthermore, maternal cell contamination may interfere with the fetal result interpretation and must be excluded. Finally, ES is challenged by the incomplete coverage of some relevant genes. Prenatally, the limited timeframe sometimes precludes extending coverage of candidate genes by alternative sequencing methods.

Prenatal results interpretation

The final interpretation of identified variants considers correlation with phenotypic information,[1] but this is particularly challenging prenatally because of limited phenotypic information from fetal imaging. Accurate identification of fetal anomalies depends on the sophistication of imaging equipment, maternal body habitus, fetal position, gestational age, and the provider's experience; both over- and under-diagnosis of fetal structural anomalies have been documented.[74] Prenatally, phenotypes may evolve as development progresses or be incompletely penetrant, and subtle dysmorphic features can be missed. Certain postnatal phenotypes, such as intellectual disability and metabolic disorders, cannot be assessed prenatally.

Challenges related to how pathogenicity is interpreted and reported are amplified in prenatal diagnosis and careful pre- and post-test counseling are essential.[1] Laboratories may have different reporting policies for VUS which can cause significant anxiety and uncertainty and influence prenatal decision making. Usually, those reported are only in genes that fit the fetal phenotype, such as in autosomal recessive conditions if the VUS is predicted to be damaging and in *trans* with a (likely) pathogenic variant.

Incidental and secondary findings

ES can result in identifying incidental and secondary findings in the fetus or the parents. Fetal incidental findings include clinically significant variants in genes associated with neurodevelopmental disorders, intellectual disability, or metabolic conditions that do not present with a prenatally detectable anomaly. ACMG recommends that highly penetrant pathogenic variants in genes known to cause moderate to severe childhood-onset disorders are reported, but not carrier status for autosomal recessive disorders and, in female fetuses, for X-linked disorders.[1] ACMG also recommends that the possibility of identifying secondary findings in the fetal or parental exomes should be discussed as part of the pre-test counseling, with the option of opting out of reporting of these variants. Another important point in pre-test counseling is the possible identification of misattributed parentage and undisclosed consanguinity, which is preferentially discussed with the mother privately, allowing her to be fully informed in a confidential manner before providing consent. If consanguinity is suggestive of a relationship between first- or second-degree relatives, the laboratory is obliged to communicate the results to the referring clinician, who is responsible for assessing clinical correlation and investigating concerns for abuse.[97]

BARRIERS TO ACCESS

Evaluating the benefits from a new medical technology must balance the ethical principles of autonomy, beneficence, nonmaleficence, and justice. Respect for autonomy validates the parents' role as decision makers for their future children, but to exert it they must be informed of the test's implications, through objective, comprehensive, and understandable pre-test counseling. Parental autonomy must be weighed against that of their future child,[98] as they may choose to terminate the pregnancy based on genetic data or may raise their children differently, which is particularly relevant for adult-onset disorders. Autonomy should also be balanced against the provider's responsibility of beneficence and nonmaleficence with reporting of VUS, incidental, and secondary findings.[99] The possible actions after discovering a pathogenic variant in an actionable gene differ between prenatal and postnatal testing, primarily because the option of pregnancy termination exists. The diagnosis of secondary findings in the fetus or parents means other family members who were not part of the decision-making process are also at risk. The justice principle considers equitable access to health care and medical resources. Prenatal ES is a costly test, not reimbursed by all private health insurers or national health systems, and is still performed on a relatively small scale outside of research settings.

Counseling pregnant women undergoing prenatal testing, particularly ES, is complex, lengthy, and needs to be conducted by professionals trained in individual risk assessment and principles of non-directive counseling, who can explain complicated genetic concepts in plain language (Table 27.2)[100] during stressful times when parents are faced with recently discovered fetal anomalies.

The cost-effectiveness of diagnostic prenatal ES and GS has not yet been comprehensively evaluated. Important considerations, in addition to the sequencing cost itself, include the cost of follow-up testing, pre- and post-test counseling, variant interpretation, and data storage and reinterpretation. This has to be weighed against cost savings to the healthcare system from avoiding a lengthy and expensive "diagnostic odyssey," adjusting management plans for the pregnancy and perinatal care, more informed decisions about time, mode, and place of delivery, reducing metabolic risk factors to avoid a neonatal metabolic crisis, or initiating palliative care plans for lethal conditions.

PROFESSIONAL GUIDANCE

In a committee opinion in 2016, the American College of Obstetricians and Gynecologists (ACOG) stated that the routine use of ES and GS for prenatal diagnosis is not yet recommended outside of the context of clinical trials.[101] In 2018, the International Society for Prenatal Diagnosis, the Society for Maternal-Fetal Medicine, and the Perinatal Quality Foundation issued a joint position statement about fetal genome sequencing,[72] that indicated that prenatal genome-wide sequencing is best performed in a research setting, though it may be acceptable for selected cases where routine testing was non-diagnostic if coordinated by multidisciplinary teams with special expertise. They concluded that genome-wide sequencing as a routine prenatal diagnostic test cannot be supported due to insufficient validation data and knowledge about its benefits and pitfalls. In January 2020, the ACMG stated that ES may be considered where a diagnosis cannot be obtained using routine prenatal tests (karyotype and CMA) for a fetus with one or more significant anomalies and that trio analysis is preferred to proband or duo testing.[1] They stated that, upon suspicion of a specific diagnosis, molecular testing for that condition with a single-gene or panel test should be the first tier. This can be difficult as prenatal phenotypes can be limited or atypical. Finally, they state that clinical use of ES for sonographic markers suggestive of aneuploidy or a history of recurrent unexplained pregnancy loss is currently not supported.

TABLE 27.2 Elements of Genetic Counseling for Prenatal WES

Pre-Test Counseling

1. Testing is optional
2. Associated risks of invasive procedure (amniocentesis or CVS), and informed consent
3. Both parents included (trio sequencing), and both parents sign informed consent
4. Independent consent option for mother, to address possible nonpaternity detection and to allow her to decline the test
5. Pregnancies conceived using known donors—discuss informing the donors and including them in the consent form
6. Turn-around time for results and possibility of late results, including after birth
7. Realistic expectations about diagnostic rates, and limited clinical utility data (PPV and NPV)
8. Categories of results, including negative report
9. Possibility for variants of uncertain significance and reporting policy
10. Possibility for incidental and secondary findings in fetus and parents, including adult-onset disorders, and carrier status, reporting policy, and option to opt-out
11. Implications of results for parents and other family members
12. Potential for the discovery of misattributed parentage and consanguinity
13. Importance of data sharing and future reinterpretation, especially before future pregnancy
14. Cost
15. Possible effects on future insurability

Post-Test Counseling

1. Confirm choices discussed in pre-test counseling
2. Disclose results and implications for the fetus and his/her family (prognosis and potential therapy)
3. Address confirmatory testing when needed
4. Address decisions about pregnancy, delivery, and neonatal management
5. Address implications of results for parents and other family members
6. Discuss reproductive decisions
7. Facilitate referrals and care
8. Provide support

Abbreviations: CVS, chorionic villus sampling; NPV, negative predictive value; PPV, positive predictive value.

CONCLUSIONS AND FUTURE PERSPECTIVES

Having a genetic diagnosis where fetal anomalies are detected guides prenatal and perinatal management, and future reproductive choices. The standard genetic workup still leaves a considerable portion of cases undiagnosed, and adequate testing for monogenic disorders can help close this gap. Current estimates are that ES finds a genetic diagnosis in 8.5–31% of pregnancies with fetal anomalies, but more research is needed to address challenges related to clinical utility, case selection, variant interpretation, cost, and access. To date, the depth and coverage required to survey the entire exome or genome for single-gene disorders requires fetal DNA obtained through an invasive procedure and a trio-sequencing approach is preferred to obtain results in a timely manner. Non-invasive prenatal sequencing of cffDNA is still limited to disease-specific assays or smaller panels, but the technical feasibility of non-invasive sequencing-based testing of the fetal genome from cffDNA has been reported.[15,37,39,42] Finally, ongoing work on the isolation of intact fetal cells and trophoblasts from maternal blood and from the endocervical canal holds great promise as a non-invasive source of fetal DNA[102–105]. Both time and cost currently preclude these from being applied clinically, but this can become a reality in the future as the technology continues to develop.

REFERENCES

1. Monaghan KG, et al. *Genet Med*. 2020;22(4):675–80.
2. Hillman SC, et al. *Ultrasound Obstet Gynecol*. 2013;41:610–20.
3. Wapner RJ, et al. *N Engl J Med*. 2012;367(23):2175–84.
4. Gregg AR, et al. *Semin Perinatol*. 2018;42:303–6.
5. Barrett AN, et al. *Methods Mol Biol*. 2014;1160:215–28.
6. Barrett AN, et al. *PLoS One*. 2017;12(10):e0186771:1–16.
7. Chan KCA, et al. *Clin Chem*. 2006;52(12):2211–8.
8. Hill M, et al. *Clin Genet*. 2011;80(1):68–75.
9. Banch Clausen F. *Prenat Diagn*. 2014;34:409–15.
10. Chitty LS, et al. *Ultrasound Obstet Gynecol*. 2011;37(3):283–9.
11. Chitty LS, et al. *Prenat Diagn*. 2013;33(5):416–23.
12. Gruber A et al. *Clin Chem Lab Med*. 2018;56(5):728–38.
13. Perlado S, et al. *PLoS One*. 2016;11(4):e0153258:1–12.
14. Orhant L, et al. *Prenat Diagn*. 2016;36(5):397–406.
15. Zhang J, et al. *Nat Med*. 2019;25(3):439–47.
16. Dello Russo C, et al. *J Mol Diagn*. 2019;21(4):572–9.
17. Chitty LS, et al. *Prenat Diagn*. 2015;35(7):656–62.
18. Luna FMF, et al. *Proc Natl Acad Sci U S A*. 2008;105(50):19920–5.
19. Tsui NBY, et al. *Blood*. 2011;117(13):3684–91.
20. Barrett AN, et al. *Clin Chem*. 2012;58(6):1026–32.
21. Lv W, et al. *Clin Chem*. 2015;61(1):172–81.
22. Chang MY, et al. *Sci Rep*. 2016;6:37153:1–10.
23. Xiong L, et al. *BJOG An Int J Obstet Gynaecol*. 2018;125(4):461–8.
24. Cutts A, et al. *Blood*. 2019;134(14):1190–3.
25. Camunas-Soler J, et al. *Clin Chem*. 2018;64(2):336–45.
26. Tsao DS, et al. *Sci Rep*. 2019;9(1):14382.
27. Caswell RC, et al. *Clin Chem*. 2020;66(7):958–65.
28. van Campen J, et al. *Br J Haematol*. 2020;190(1):119–24.
29. Lam KWG, et al. *Clin Chem*. 2012;58(10):1467–75.
30. You Y, et al. *Genet Med*. 2014;16(8):594–600.
31. Chandler NJ, et al. *Clin Chem*. 2020;66(1):207–16.
32. Han L, et al. *Prenat Diagn*. 2020;40(3):324–32.
33. Young E, et al. *J Mol Diagn*. 2020;22(9):1151–61.
34. New MI, et al. *J Clin Endocrinol Metab*. 2014;99(6):E1022–36.
35. Hudecova I, et al. *Blood*. 2017;130(3):340–7.
36. Lo YMD, et al. *Sci Transl Med*. 2010;2(61):61ra91:1–14.
37. Kitzman JO, et al. *Sci Transl Med*. 2012;4(137):137ra76:1–9.
38. Christina Fan H, et al. *Nature*. 2012;487(7407):320–4.
39. Che H, et al. *Genet Med*. 2020;22(5):962–73.
40. Chan KCA, et al. *Proc Natl Acad Sci U S A*. 2016;113(50):E8159–68.
41. Jenkins LA, et al. *Prenat Diagn*. 2018;38(1):44–51.
42. Hayward J, et al. *Semin Fetal Neonatal Med*. 2018;23(2):94–101.
43. Rieneck K, et al. *Transfus Med Hemother*. 2020;47(1):45–53.
44. Geifman-Holtzman O, et al. *Expert Rev Mol Diagn*. 2008;8(6):727–51.
45. Runkel B, et al. *BMC Pregnancy and Childbirth*. 2020;20(1):83:1–10.
46. Hyett JA, et al. *Prenat Diagn*. 2005;25(12):1111–6.
47. Tardy-Guidollet V, et al. *J Clin Endocrinol Metab*. 2014;99(4):1180–8.
48. Bowman-Smart H, et al. *Prenat Diagn*. 2020;40:398–407.
49. Chen M, et al. *Prenat Diagn*. 2020;40(8):918–24.
50. Ren Y, et al. *Prenat Diagn*. 2018;38(11):821–8.
51. Yan H, et al. *Prenat Diagn*. 2020;40(11):1459–65.
52. Hudecova I, et al. *Best Pract Res Clin Obstet Gynaecol*. 2017;39:63–73.
53. Hill M, et al. *Prenat Diagn*. 2015;35(10):950–8.

54. Scotchman E, et al. *Eur J Obstet Gynecol Reprod Biol.* 2020;253:320–7.
55. Koumbaris G, et al. *Mol Cytogenet.* 2019;12(1):48–56.
56. Biesecker LG, et al. *N Engl J Med.* 2014;370(25):2418–25.
57. Normand EA, et al. *Fertil Steril.* 2018;109(2):213–20.
58. Bamshad MJ, et al. *Nat Rev Genet.* 2011;12:745–55.
59. Rehm HL, et al. *Genet Med.* 2013;15(9):733–47.
60. Willig LK, et al. *Lancet Respir Med.* 2015;3(5):377–87.
61. Drury S, et al. *Prenat Diagn.* 2015;35(10):1010–7.
62. Gonzaga-Jauregui C, et al. *Annu Rev Med.* 2012;63:35–61.
63. Gilissen C, et al. *Eur J Hum Genet.* 2012;20:490–7.
64. Richards S, et al. *Genet Med.* 2015;17(5):405–23.
65. Xue Y, et al. *Genet Med.* 2015;17:444–51.
66. Firth H V., et al. *Am J Hum Genet.* 2009;84(4):524–33.
67. Lek M, et al. *Nature.* 2016;536(7616):285–91.
68. Landrum MJ, et al. *Nucleic Acids Res.* 2016;44(D1):D862–8.
69. Stenson PD, et al. *Hum Genet.* 2017;136:665–77.
70. Watson MS. *Genet Med.* 2017;19(7):721–2.
71. Ewans LJ, et al. *Genet Med.* 2018;20(12):1564–74.
72. International Society for Prenatal Diagnosis; Society for Maternal and Fetal Medicine; Perinatal Quality Foundation. *Prenat Diagn.* 2018;38(1):6–9.
73. Deignan JL, et al. *Genet Med.* 2019;21(6):1267–70.
74. Best S, et al. *Prenat Diagn.* 2018;38(1):10–9.
75. Yang Y, et al. *N Engl J Med.* 2013;369(16):1502–11.
76. Yang Y, et al. *JAMA.* 2014;312(18):1870.
77. Meng L, et al. *JAMA Pediatr.* 2017;171(12):e173438:1–10.
78. Hillman SC, et al. *Ultrasound Obstet Gynecol.* 2015;45:4–9.
79. Retterer K, et al. *Genet Med.* 2016;18(7):696–704.
80. Wright CF, et al. *Lancet.* 2015;385(9975):1305–14.
81. Filges I, et al. *Clin Genet.* 2014;86(3):220–8.
82. Talkowski ME, et al. *N Engl J Med.* 2012;367(23):2226–32.
83. Sparks TN, et al. *N Engl J Med.* 2020;383(18):1746–56.
84. Stanley KE, et al. *N Engl J Med.* 2020;383(12):1107–16.
85. Lord J, et al. *Lancet.* 2019;393(10173):747–57.
86. Petrovski S, et al. *Lancet.* 2019;393(10173):758–67.
87. Normand EA, et al. *Genome Med.* 2018;10(1):74.
88. Fu F, et al. *Ultrasound Obstet Gynecol.* 2018;51(4):493–502.
89. Yates CL, et al. *Genet Med.* 2017;19(10):1171–8.
90. Lei TY, et al. *Nephrol Dial Transplant.* 2017;32(10):1665–75.
91. Chandler N, et al. *Genet Med.* 2018;20(11):1430–7.
92. Filges I, et al. *Prenat Diagn.* 2015;35:1005–9.
93. Natarajan N, et al. *Pediatr Radiol.* 2016;46:1354–7.
94. Le Blanc K, et al. *Transplantation.* 2005;79(11):1607–14.
95. Götherström C, et al. *Stem Cells Transl Med.* 2014;3(2):255–64.
96. Shamseldin HE, et al. *Genet Med.* 2018;20(4):420–7.
97. Rehder CW, et al. *Genet Med.* 2013;15(2):150–2.
98. Yurkiewicz IR. *N Engl J Med.* 2014;370(3):195–7.
99. Amor DJ, et al. *Prenat Diagn.* 2020;40(12):1508–14.
100. Westerfield LE, et al. *Prenat Diagn.* 2015;35(10):1022–9.
101. Committee Opinion No.682. *Obstet Gynecol.* 2016;128(6):e262–8.
102. Beaudet AL. *Am J Med Genet Part C Semin Med Genet.* 2016;172(2):123–7.
103. Vossaert L, et al. *Am J Hum Genet.* 2019;105(6):1262–73.
104. Drewlo S, et al. *Placenta.* 2017;60(1):S27–S31.
105. Jain C V, et al. *Sci Transl Med.* 2016;8(November):363re4.
106. Daum H, et al. *Ultrasound Obstet Gynecol.* 2019;53(1):80–6.

Chromosome Abnormalities in Human Pregnancy Loss

A Review of Cytogenetic and Molecular Analyses

28

Kathy Hardy and Terry Hassold

INTRODUCTION

Pregnancy loss occurs with astonishing frequency in our species. From studies involving couples attending assisted reproduction facilities, it is clear that a large proportion of conceptuses perish in the earliest stages of pregnancy. However, even among those pregnancies that survive to the time of clinical recognition (about 6–8 weeks gestation), at least 15%–20% terminate as spontaneous abortions (SABs). In this review, we focus on the contribution of chromosome abnormalities to SABs, summarizing methodologies used to detect them and factors contributing to their occurrence.

ANALYSIS OF SPONTANEOUS ABORTIONS: TECHNICAL CONSIDERATIONS

Collection of Tissue Samples

Tissue samples from spontaneous abortions must be of inner cell mass or fetally derived extra-embryonic origin to be of use in assessing the reason for the abortion. Indeed, as discussed in more detail below,

DOI: 10.1201/9781003024941-28

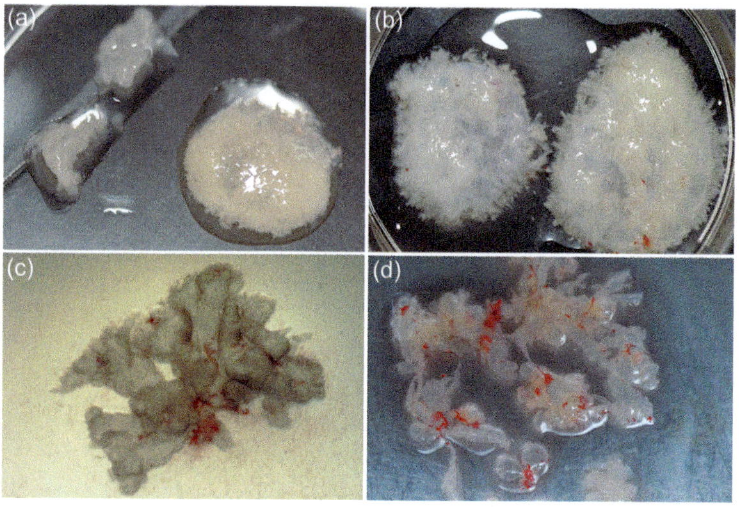

FIGURE 28.1 Tissue samples from SABs. Samples of SAB tissues forwarded for cytogenetic studies. (a) A typical sample containing maternal decidua (left) and a small conceptus (right). The maternal tissue is thick and rubbery, while the fetally derived tissue, including overlying villous material, has more delicate membranes. (b) Failed twin pregnancy, with two intact sacs that are discrepant in size. (c) Fragments of villi obtained by dissection of a sample containing no discernible fetal material. (d) Villus morphology is variable and may be related to the specific chromosomal error. This sample shows swollen, hydropic villi (with a grape-like appearance), typical of triploidy of paternal origin.

contaminating maternal tissue is one of the most common, yet underappreciated, reasons for failure to identify the cause of the pregnancy loss (Figure 28.1a). The ideal tissue sample is a complete conceptus from an early loss (Figure 28.1b). This is not feasible in many instances, and fragments of fetal membranes with attached chorionic villi or simply villus fronds provide the next best options (Figure 28.1c and d). Such fragments can be obtained from samples where tissue from the conceptus has disintegrated or never formed. Later gestational age conceptions (post-12 weeks) may be represented by tissue from the extra-embryonic membranes, umbilical cord, placenta, or fetus proper. Importantly, the time between fetal demise and sample collection will affect tissue viability differentially. The embryo/fetus seems to be the first to die, followed by the umbilical cord, placenta, and finally the extra-embryonic membranes. This last tissue is therefore the most reliable in providing a fetal result.

Methodological Approaches

Historically, SAB samples have been analyzed using conventional cytogenetic methodology. This involves tissue culture, slide preparation, banding, and microscopic analysis of metaphase spreads to obtain a karyotype. However, as molecular techniques have evolved, so too has the study of SABs. Fluorescence *in situ* hybridization (FISH) can be applied in instances in which no dividing cells are present, or when the morphology of the metaphases is too poor for identification of individual chromosomes/chromosome regions [1–3]. In its initial application, FISH also required tissue culture and slide preparation. However, as probe sets became more sophisticated and covered wider parts of the genome, comparative genomic hybridization (CGH) approaches became available [4–7]. This technique still requires the traditional steps from culture to slide preparation, although the fluorescence signals can then be measured computationally. The subsequent application of proteases to disaggregate tissues has allowed laboratories to apply fluorescence techniques to the primary sample, thus removing the need for tissue culture. More recently, molecular approaches have further advanced, with the introduction of whole-genome microarray technology [8, 9].

Each of the approaches described above has advantages and limitations (e.g., [3, 10, 11]). Traditional cytogenetics is hampered by tissue culture failure, maternal tissue contamination, and maternal overgrowth [12–15]. Indeed, successful tissue culture results have ranged from 37% [16] to 95% [17] in different published series. Speed of analysis, the ability to analyze all samples, and elimination of time-consuming tissue culture are advantages of molecular tools over classical cytogenetics. However, molecular methods are also hampered by maternal tissue contamination [3, 18–20], although the newer techniques may utilize SNPs to differentiate maternal from fetal tissue [9, 19]. In addition, array-based techniques may be hampered by poor-quality DNA samples [10, 20] and they rely on the premise that the fetal material contains an unbalanced genome. Thus, conditions such as triploidy, tetraploidy, and balanced structural rearrangements pose problems for array-based approaches [18, 19, 21], suggesting the utility of a combined approach that utilizes both array-based methods and either FISH or flow cytometry [20, 22].

THE CONTRIBUTION OF CHROMOSOME ABNORMALITIES TO SPONTANEOUS ABORTIONS: RESULTS OF 50 YEARS OF STUDIES

Early studies conducted in the 1960s and 1970s demonstrated that a small, but clinically important, proportion of newborn individuals had recognizable numerical or structural chromosome abnormalities. Indeed, in a review of studies of more than 50,000 consecutive newborns, Hook and Hamerton [23] reported abnormalities in approximately 1/200 individuals, with sex chromosome trisomies (47,XXX; 47,XXY; and 47,XYY) and trisomy 21 being the most common specific abnormalities. Similar studies of late fetal wastage (i.e., stillbirths) indicated a much higher level of chromosome abnormality but, in general, the types of abnormalities were in keeping with those identified in the newborn series (e.g., [24]).

These results prompted investigators to ask whether the other common category of clinically recognized pregnancy—spontaneous abortions (i.e., fetal losses occurring between about 6–8 to 20 weeks gestation)—might also include cases of chromosome abnormalities. Spearheaded by initial studies of Carr and colleagues (e.g., [25]), it soon became clear that a large proportion of SABs were attributable to chromosome abnormalities, and that the types of abnormalities were much more varied than those identified in stillbirths or newborn individuals. Tables 28.1 through 28.4 provide a summary of representative analyses of SABs; a description of cytogenetic studies and their results are provided in Tables 28.1 and 28.2 and more recent, molecular and molecular, cytogenetic studies are reviewed in Tables 28.3 and 28.4. Several important conclusions are clear from these analyses.

Maternal Contamination Complicates Interpretation of SAB Data

For most purposes, karyotypic analysis provides an unambiguous approach to the identification of numerical or major structural abnormalities. However, in the instance of SABs, fetally derived tissue may be scant, unviable, or difficult to culture. Coupled with the presence of maternally derived placental material, artifactual 46,XX results are a common complication of SAB studies. The consequences of maternal contamination are readily apparent from the results of cytogenetic studies of SABs (Table 28.2). For example, the sex ratio among euploid samples (i.e., 46,XY; 46,XX) varies by an order of magnitude among the different studies, from nearly 3:1 to 0.3:1, presumably reflecting differences in the likelihood of including maternal tissue samples. Accordingly, the overall proportion of chromosome abnormalities is also remarkably variable among the different studies, with the lowest chromosome abnormality rates typically tracking with the lowest sex ratios.

TABLE 28.1 Summary of Methodology, Dates of Study, and Study Population Characteristics for Representative Cytogenetic Studies of SABs

STUDY	*APPROACH*	*YEARS*	*MEAN MATERNAL AGE (IN YEARS)*	*GESTATIONAL AGE RANGE (IN WEEKS)*	*NO. CASES*
Creasy et al. [26]	Cultured tissue; G- and Q-banding	1971–1974	—	8–38	941
Lauritsen [27]	Cultured tissue; aceto-orcein, Q-banding	1971–1973	—	< 16	255
Byrne et al. [28]	Cultured tissue; G-banding	1977–1981	—	—	1356
Andrews et al. [29]	Cultured tissue; G- and Q-banding	—	25.9	9–28	154
Eiben et al. [30]	Direct preparations; G-, Q-, and C-silver staining	—	30.1	6–24	140
Ohno et al. [14]	Direct preparations; G-, Q-, and C-banding	—	30.8	6–16	144
Menasha et al. [31]					
Period A	Cultured tissue; non-specified banding	1990–1997	35.6	—	717
Period B	Cultured tissue, direct preparations; non-specified banding	1998–2002	36.6	—	1203
Cheng et al. [32]	Cultured tissue; G-banding	1995–2013	32.4	—	223
Choi et al. [33]	Cultured tissue; G-banding	2000–2013	30.3	< 10–20	164
Hardy et al. [34]					
Hawaii	Cultured tissue; non-banded and Q-banding	1976–1985	28.1	2–32	2899
Emory	Cultured tissue; Q-banding	1989–1992	32	—	1365
CWRU	Cultured tissue; Q-banding	1993–1997	30.5	—	883
Perth-1	Cultured tissue; G-banding	1996–2007	34.8	< 12	1188
Perth-2	Cultured tissue; G-banding	2008–2015	35.1	< 12	1984

TABLE 28.2 Summary of Karyotypic Results for Representative Cytogenetic Studies of SABs

STUDY	*NORMAL (XY:XX)*	*SEX CHROMOSOME MONOSOMY*	*TRISOMY*	*TRIPLOIDY*	*TETRAPLOIDY*	*STRUCTURAL ABNORMALITY*	*OTHERS*	*% CHROMOSOMALLY ABNORMAL*
Creasy et al, [26]	654 (1.27)	68	152	38	12	10	7	69.5
Lauritsen [27]	115 (0.92)	40	65	14	12	4	5	54.9
Byrne et al. [28]	816	86	301	85	28	17	23	39.8
Andrews et al. [29]	125 (1.23)	8	15	3	1	1	1	18.8
Eiben et al. [30]	72 (0.76)	5	43	10	1	2	7	48.6
Ohno et al. [14]	44 (0.83)	7	69	9	1	6	8	69.4
Menasha et al. [31]:								
Period A	410 (0.33)	42	208	32	6	15	4	42.8
Period B	411 (0.71)	54	572	91	18	31	26	65.8
Cheng et al. [32]	98 (0.44)	16	73	8	6	9	13	56.1
Choi et al. [33]	81	12	53	6		6	6*	50.6
Hardy et al. [34]:								
Hawaii	1433 (0.87)	263	844	180	66	69	44	50.6
Emory	530 (0.66)	96	529	87	38	45	40	61.2
CWRU	499 (0.66)	40	239	59	13	21	12	43.5
Perth-1	330 (1.60)	89	584	80	30	33	42	72.2
Perth-2	413 (2.93)	157	1048	167	49	58	92	79.2

TABLE 28.3 Summary of Methodology, Dates of Study, and Characteristics of Study Populations for Representative Molecular or Molecular Cytogenetic Studies of SABs

STUDY	*APPROACH*	*YEARS*	*MEAN MATERNAL AGE*	*GESTATIONAL*	*CASES AGES*
Zhang et al. [35]	Cultured tissue; G-banding, PCR-based microsatellite genotyping, array CGH	2006–2007	—	—	115
Gao et al. [22]	Cultured tissue; G-banding, array CGH, FISH, QF-PCR	—	32	< 12	100
Jenderny [8]	Cultured tissue; G-banding, QF-PCR	2002–2013	—	7–34	398
Levy et al. [9]	DNA from fetally derived tissue; SNP-based chromosomal microarray	2010–2012	36.2	3–20	1861

Notably, the largest study to use a single-nucleotide polymorphism (SNP)-based approach (9; Table 28.4) removed from consideration 528 46,XX cases in which the genotypic information matched that of the mother, effectively raising the sex ratio from approximately 0.4 to nearly 0.9. Similarly, in a recent SNP analysis by Lathi et al. [19], in which results on specific categories of chromosome abnormalities were not detailed, 269 of 456 46,XX results were judged to reflect maternal contamination. Clearly, continued use of SNP-based analyses will eliminate much of the confusion created by inclusion of spurious 46,XX results.

For both the cytogenetic and molecular/molecular cytogenetic studies, we restricted our analysis to studies with relatively large study populations (i.e., > 100 cases). Studies were excluded from consideration if they primarily involved recurrent miscarriages or were restricted to assisted reproductive technology (ART)-derived pregnancies. For all studies, we attempted to provide information only when it was clearly reported in the manuscript; however, in some instances in which the karyotypes were unclear, we excluded the case or assigned it to the most likely karyotypic category.

Numerical Chromosome Abnormalities Are the Leading Cause of SABs

The above concerns notwithstanding, the results of both the cytogenetic (Table 28.2) and molecular/molecular cytogenetic studies (Table 28.4) make it clear that chromosome abnormalities are the primary cause of SABs. Indeed, of the overall total of 16,090 abortuses reported in Tables 28.1 through 28.4, 9,054 (56.3%) were chromosomally abnormal.

Intriguingly, by comparison with liveborn or stillborn conceptuses, SABs are characterized by a wide variety of chromosome abnormalities, the majority of which involve single missing or extra whole chromosomes (i.e., aneuploidy). The single most-common specific abnormality, sex chromosome monosomy (45,X), accounts for 6.6% of the cases reported in Tables 28.1 through 28.4. Trisomies are the most common class of abnormality, identified in 36.2% of abortuses. Trisomies for all chromosomes have been identified, although a few (e.g., trisomies 1 and 19) are extremely rare. Conversely, other trisomies are quite common, i.e., trisomy 16 is the most commonly identified trisomy and, together with trisomies 15, 21, and 22, these four conditions account for well over one-half of all single trisomies. Importantly, these results are in relatively good agreement with recent studies of human preimplantation embryos

TABLE 28.4 Summary of Karyotypic Results for Representative Molecular or Molecular Cytogenetic Studies of SABs

STUDY	*NORMAL (XY:XX)*	*SEX CHROMOSOME MONOSOMY*	*TRISOMY*	*TRIPLOIDY*	*TETRAPLOIDY*	*STRUCTURAL ABNORMALITY*	*OTHERS*	*% CHROMOSOMALLY ABNORMAL*
Zhang et al. [35]	45	7	46	4	6	2	5	60.9
Gao et al. [22]	39 (2.08)	4	49	4	0	3	1	61.0
Jenderny [8]	153 (0.94)	17	141	32	8	9	38	61.6
Levy et al. [9]	755 (0.86)	53	794	114	4	38	90	59.4

from assisted reproduction technology (ART)-derived pregnancies. For example, similar to SABs, chromosomes 15–22 are more likely to be present in an aneuploid state in preimplantation embryos than are larger chromosomes (e.g., [36, 37]). Thus, while it is clear that selection eliminates a large number of aneuploid conceptuses between the time of conception and the time of clinical recognition, it also seems likely that certain chromosomes are more likely than others to undergo meiotic mis-segregation.

In addition to aneuploid abnormalities, two classes of polyploids are common in SABs: triploids, accounting for 6.4% of cases, and tetraploids, occurring in 1.9% of SABs. Surprisingly, structural abnormalities—which are nearly as common as numerical abnormalities in the newborn series [23]—are uncommon in SABs, accounting for only 2.4% of the cases reported in Tables 28.1 through 28.4. Furthermore, the initial application of molecular analytic approaches, which should make it possible to identify subtle structural abnormalities, appears not to have substantially increased the frequency of such abnormalities in SABs (e.g., [9]).

Maternal Age Is the Primary Determinant of Chromosome Abnormality Rate in SABs

Surprisingly, there is considerable variation in the rates of chromosome abnormality among different studies of SABs, even after taking into account apparent differences in levels of maternal contamination. In general, abnormality levels appear to have increased over time, with studies conducted in the 1970s–1980s typically reporting rates of 40%–50% (e.g., [26, 28, 29, 38–42]), whereas several recent studies have reported rates of 60% or more (e.g., [9, 19, 22, 31, 33, 43–46]).

Although the reasons for these differences are not known, they might reflect real changes in the likelihood of conceiving chromosomally abnormal fetuses (e.g., because of changing environmental conditions) or they might simply be attributable to changes in population demographics over time. To discriminate between these possibilities, we recently compared results among five different cytogenetic studies of SABs in which we participated: one study was conducted in Honolulu, HI in the 1970s and 1980s (by TH), one in Atlanta, GA in the 1980s and 1990s (by TH), one in Cleveland, OH in the 1990s (by TH), and two in Perth, Australia from the 1990s to the present (by KH). The results were instructive [34, 47]. Consistent with expectations, we saw an increase in the rate of chromosome abnormality over time, with the earliest studies having abnormality rates of approximately 40%–50%, and the rates in the more recent ones exceeding 70% (Table 28.2). However, analyses of racial/ethnic background revealed no important effect, nor was among-study variation in gestational age a factor. Indeed, an examination of the types of chromosome abnormalities showed that the only class of abnormality that varied among the studies was trisomy, and that the differing levels of trisomy could be completely explained by maternal age, i.e., in the earliest study (conducted in Hawaii), the mean maternal age was 28.1 years, whereas in the most recent study (conducted in Perth), the comparable mean was 35.1 years. Taken together, our results suggest that the primary variable affecting the rate of chromosome abnormality is maternal age, and that this exerts its effect by varying the likelihood of mis-segregation during maternal meiosis. This is not to say that other factors, including environmental contaminants, have no role in mediating chromosome abnormality rates, only that the magnitude of any such effect is minimal by comparison with that exerted by maternal age.

HOW DO THE ABNORMALITIES ORIGINATE?

For many human genetic disorders, the low frequency of the abnormality complicates attempts to directly study the origin of the underlying mutation(s); instead, alternative approaches (e.g., *in vitro* analyses or generation of animal models) are needed. However, this is not the case for numerical chromosome abnormalities. For reasons that we do not understand, the rates of numerical abnormalities in humans are at least an order of magnitude higher than in other mammalian species that have been appropriately studied, meaning that ample material is available for analysis. Given the incidence and variety of these abnormalities in spontaneously aborted fetuses, SABs have been especially useful for investigating the origin of the abnormalities. These studies have relied on the inheritance of polymorphisms (Figure 28.2 [48]) to address three questions about the origin of the abnormalities:

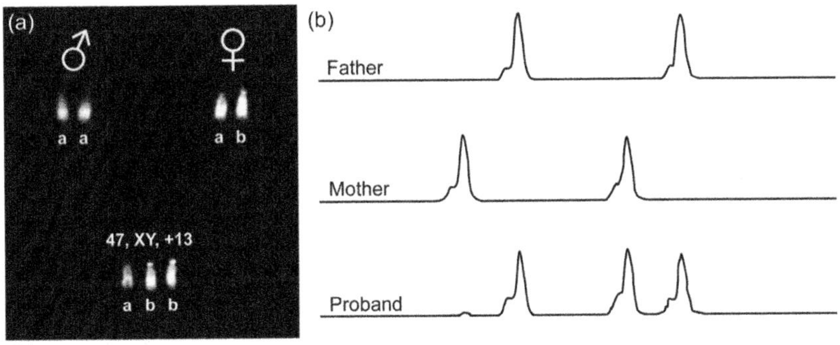

(c)

LOCUS	PROBAND	FATHER	MOTHER	MAT/PAT	R/N
Centromere					
D14S742	ABC	AC	BC		N
MYH7	ABC	AB	AC		N
D14S581	AAB	AB	AB		-
D14S615	ABC	AC	BC		N
D14S49	AAB	AB	AC		R
D14S1432	ACC	AB	AC	MAT	R
D14S587	ABC	BB	AC	MAT	N
D14S1429	BCC	AC	BC		N
D14S588	ABB	BB	AB		N
D14S43	ABC	AA	BC	MAT	N
D14S1433	ABC	AB	BC		N
D14S617	ABB	BC	AB		N
D14S611	AAB	AA	AB		N
D14S1426	BCD	AB	CD	MAT	N
D14S1007	ABC	AC	BC		N

FIGURE 28.2 Polymorphism analysis of human chromosome abnormalities. Several different approaches have been used to investigate the origin of chromosome abnormalities. Early efforts made use of blood group markers (e.g., from Sanger, R et al., *J Med Genet*, 8, 417–426, 1971) but these were gradually replaced by chromosome heteromorphisms, variable markers detectable by conventional cytogenetic methodology and occurring at pericentromeric regions of eight of the 23 autosomes. (a) An example of Q-banding chromosome heteromorphism analysis, indicating a maternal origin for a case of trisomy 13. With the demonstration of sequence-based variants, DNA polymorphism analysis (initially involving RFLPs, and subsequently, minisatellites, microsatellites, and most recently, SNPs) became the standard technique for analysis of the origin of chromosome abnormalities. (b) A microsatellite-based analysis of trisomy 14, indicating a paternally derived additional chromosome. In addition to assessing parental origin, DNA polymorphisms have been used extensively to determine the meiotic/mitotic stage of origin of trisomies, as well as the amount of recombination that occurred between the nondisjunctional homologs. (c) An example of this type of analysis, as applied to a case of trisomy 14. Multiple polymorphic sites along chromosome 14 were examined, with four providing evidence of a maternal origin. Subsequent analysis focuses on whether, at individual polymorphic sites, the trisomic conceptus received both maternal alleles (non-reduced, or "N") or two copies of the same maternal allele (reduced, or "R"). If pericentromeric sites are N, this is consistent with a meiosis I error and, if R, a meiosis II error; thus, this case is of meiosis I origin. Finally, switches between N and R at adjacent loci indicate points of crossing-over that occurred in the maternal meiosis that generated the additional chromosome; in this case, two cross-overs occurred between the chromosome 14 homologs in the trisomy-generating meiosis.

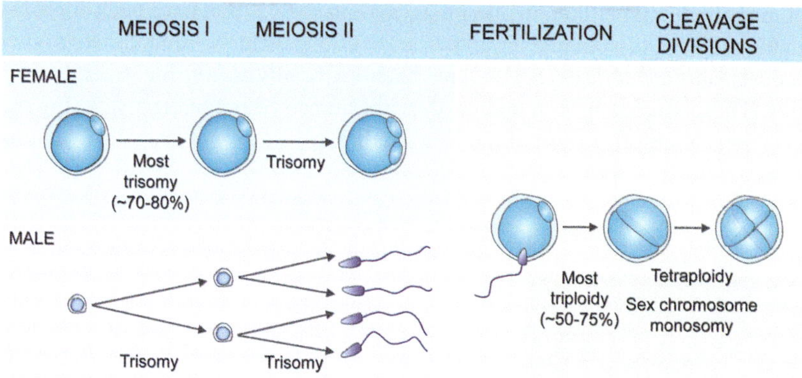

FIGURE 28.3 Summary of studies investigating the origin of chromosomally abnormal SABs. The results of analyses of numerical chromosome abnormalities indicate that errors can occur at any of the possible meiotic stages (maternal meiosis I and II, paternal meiosis I and II), as well as at the time of fertilization and during the first few cleavage divisions. However, against this background, certain timepoints appear to be especially vulnerable. Perhaps most importantly, abnormalities at maternal meiosis I are the leading cause of trisomy and arguably the most common cause of spontaneous abortion. Additionally, fertilization by two sperm appears to be the leading cause of triploidy, while errors in early mitotic cell divisions are likely causes of tetraploidy and sex chromosome monosomy.

- In which parent did the abnormality originate, i.e., is the extra or missing chromosome(s) maternal or paternal in origin?
- At what stage of development did the abnormality originate, i.e., at meiosis I or meiosis II, at the time of fertilization, or in the early cleavage divisions?
- Was abnormal recombination a contributor to the abnormality? By examining polymorphic markers along the chromosome, the recombinational "history" of a chromosome can be re-created and compared between chromosomally normal and abnormal progeny. This approach has been used extensively to investigate the role that abnormal recombination plays in the generation of trisomic SABs.

The utility of polymorphism analysis in studying the different classes of numerical abnormality is summarized below, and in Figures 28.2 and 28.3.

Sex Chromosome Monosomy

Unlike abnormalities in which additional chromosome material is present in the conceptus, the 45,X condition involves a missing chromosome, either an X or a Y chromosome. Thus, it is only possible to specify the parental origin of the X chromosome that is available and, by exclusion, the parental origin of the sex chromosome that is missing. Such studies have been conducted both in SABs and in liveborn Turner syndrome individuals, with similar conclusions (e.g., [49–51]). Specifically, among approximately 100 apparently non-mosaic 45,X cases, some 70% had a single, maternally derived X chromosome and the remainder were Xpat cases. There was no obvious difference in results between the spontaneously aborted and liveborn cases, indicating that the parental origin of the single X does not affect survival.

However, mosaicism for a second, normal, or structurally rearranged sex chromosome is much more common in liveborn Turner syndrome individuals than in aborted 45,X conceptuses, indicating an association with survival to term [52, 53]. Additionally, studies of liveborn 45,X individuals suggest phenotypic variation between 45,Xpat and 45,Xmat cases (e.g., [54]); thus, although there does not appear to be a correlation between parent-of-origin of the single X and survival to term, subtler phenotypic effects apparently are affected by the origin of the X.

Although we can obtain information on the parental origin of sex chromosome monosomy, the absence of the "offending" chromosome prevents us from studying the mechanism or stage of origin of the error. Nevertheless, the high frequency of the 45,X condition by comparison with sex chromosome trisomies, which are thought to originate from meiotic nondisjunctional errors [49], suggest that 45,Xs arise from a different mechanism. Random loss of a sex chromosome, presumably at an early mitotic division, is an attractive possibility, since this would generate an apparent bias toward Xmat cases. That is, assuming an equal likelihood of loss of the X or Y chromosome in early stage 46,XX and 46,XY embryos, we would expect to generate 45,Y conceptuses (presumably early lethals) and 45,X conceptuses, two-thirds of which would carry a single, maternally derived X chromosome.

Trisomy

Over the past 25 years, the parent and meiotic stage of origin of human trisomies have been extensively investigated. Because of its clinical importance, trisomy 21 has received the most attention, with nearly 1,000 cases having been analyzed (e.g., [55, 56]). From these studies, two aspects of chromosome 21 mis-segregation are now clear. First, the vast majority of trisomies 21 involve errors at maternal meiosis I, e.g., in the largest series of cases, Sherman et al. [57] observed that approximately 90% of cases were maternally derived, with 70% of these attributable to meiosis I errors, with the remainder ascribed to meiosis II errors [58]. Second, abnormal meiotic recombination is an important contributor to trisomy 21. Indeed, the overwhelming majority of trisomy 21 cases appears to be linked to one of three types of susceptible crossover configurations: failure of recombination with maternal meiosis I errors, extremely distal crossovers with maternal meiosis I errors, and extremely proximal crossovers with apparent maternal meiosis II errors. Since meiotic recombination occurs in the fetal ovary in mammalian females, this means that a proportion of human oocytes are predisposed to mis-segregate right from the beginning of meiosis. However, intriguingly, the relationship between these recombination risk factors and the only other known etiological agent for trisomy 21—advancing maternal age—is still unclear. Oliver et al. [59] reported an increase in telomeric exchanges in meiosis I cases involving young women, an increase in pericentromeric exchanges among meiosis II cases involving older women, and no clear linear association between maternal age and cases involving failure to recombine. Thus, the effect of recombination on trisomy is likely complex, presumably reflecting the fact that there are multiple maternal age-dependent and age-independent routes to trisomy 21.

How do the results for trisomy 21 compare with those for other trisomies? Given the complexity of errors involving chromosome 21, it is perhaps not surprising that other trisomies exhibit similarities and differences. For example, like trisomy 21, maternal meiotic errors predominate for most trisomies, and errors of recombination, especially absence of crossovers, have been observed for all trisomies that have been appropriately studied (e.g., [60]). However, there are differences in the details. For example, trisomy 18 typically involves a maternal error at maternal meiosis II, not meiosis I [61]; trisomy 16, the most common trisomy, is associated with telomeric exchanges but rarely, if ever, with failure to recombine [62]; and the 47,XXY condition is as likely to be paternal as maternal in origin [63]. Thus, while we have a long way to go in our understanding of the causes of human nondisjunction, one fact is clear: should we ever get to the point where we think about preventing the occurrence of meiotic mis-segregation, we will need many "magic bullets" to accomplish the task.

Triploidy

Triploidy can result from a diploid contribution from the father (diandry), from fertilization by either two sperm (dispermy) or by a diploid sperm; alternatively, triploidy may be maternal in origin (digyny),

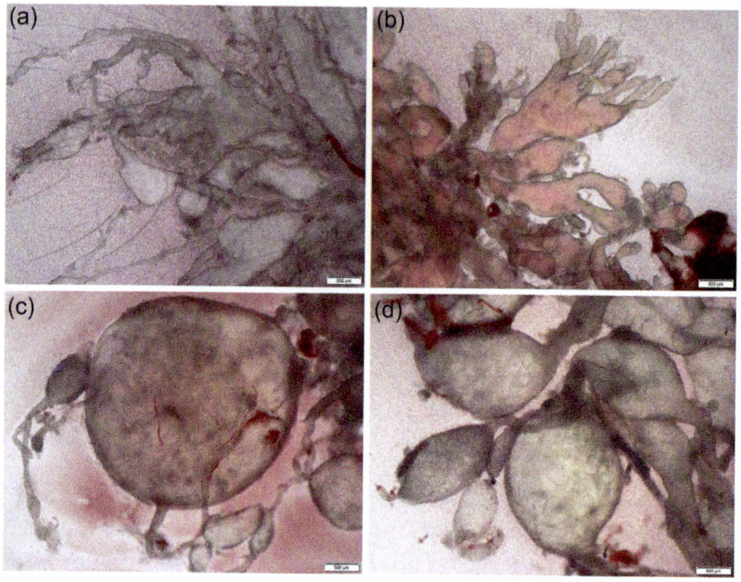

FIGURE 28.4 Images of chorionic villi, taken using an Olympus SZ61 dissecting microscope and Olympus DP22 camera and capture software, at a magnification of 2.5 × 10 (25×). Image (a) represents 47,XX,+16. Image (b) is from a 47,XY,+16. Both conceptions demonstrate similar abnormal branching of the villus fronds. Images (c) and (d) represent two different SABs with 69,XXY. Note the large hydropic villus structures, as well as smaller villi.

because of failure of either of the two meiotic divisions. DNA polymorphism analyses indicate that dispermy is the most common mode of origin, accounting for over two-thirds of cases [64, 65]. Of the remaining maternally derived cases, errors at meiosis II appear to be the most common source of the additional haploid set [66], although errors at meiosis I and, intriguingly, "dieggy" (i.e., fusion of two oocytes), have also been reported [63].

Unlike most other numerical abnormalities, the parental origin of triploidy has a profound effect on phenotype. Specifically, diandric triploids typically abort between 10 to 20 weeks of gestation, with limited development of fetal structures but good development of the extra-embryonic membranes and villi [67]; indeed, diandric triploids are frequently diagnosed as partial hydatidiform moles histologically based on villus structures and as demonstrated in Figure 28.4c and d (for discussion of moles, see [68]). In contrast, there appear to be two general categories of maternally derived triploids. The more common category aborts very early in pregnancy but a subset of cases is associated with good fetal development and with abortion late in gestation; indeed, the extremely small proportion of cases of triploidy that survive to term are thought to be maternal in origin. Thus, triploidy provides an important example of the existence of imprinted loci, although the phenotypic contribution of the specific loci is not known.

Tetraploidy

Very little effort has been directed at studying the origin of tetraploidy. However, virtually all tetraploid SABs have 92,XXYY or 92,XXXX karyotypes, consistent with doubling of 46,XY or 46,XX fertilized eggs, and early chromosome heteromorphism studies indicated they consisted of two duplicated sets of chromosomes (e.g., [69]). Thus, it seems likely that most tetraploids arise from a failure of the chromosomes to divide at the first or a very early embryonic cell division. Both hyper- and hypo-tetraploidy occur on rare occasions, consistent with this doubling event. Two recent cases in the Perth laboratory were hyper-tetraploid: 94,XXYY,+16,+16 and 94,XXYY,+22,+22. One case was hypo-tetraploid being 90,XX.

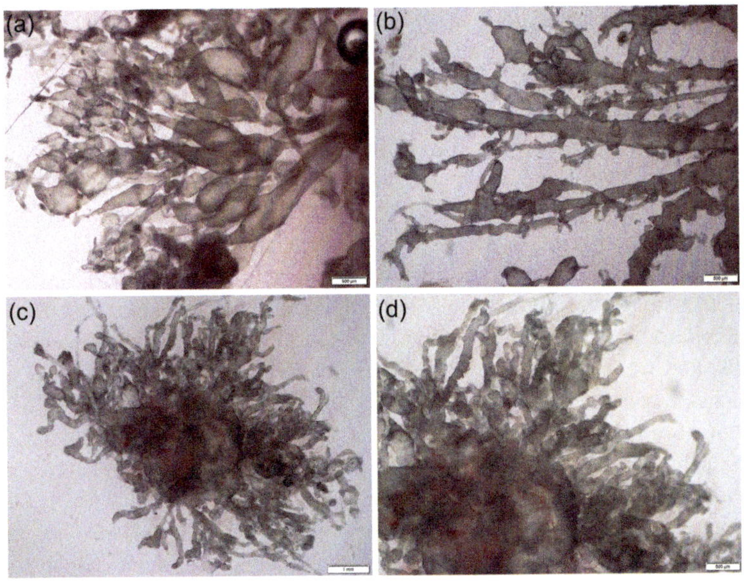

FIGURE 28.5 Villus structures of two different 46,XY conceptions (a, c, d) and a 45,X case (b). Image (a) demonstrates limited branching with many swollen terminal villi. Image (b) shows part of the long, filamentous villi typical of monosomy X conceptions. Images (c) and (d) are two different magnifications of the same normal male conception; (c) is at 10× and (d) is 25×. The villi show limited branching, typical of a 5- to 6-week conception, with little evidence of villus tree structures.

Chorionic Villus Structure

Little literature is available on the microscopic observation of 3-D chorionic villus structure from early conceptions. Utilizing laboratory observations and photography in the Perth laboratory, there appear to be correlations between the villus morphology and the chromosomal constitution. It has long been known that histological examination identifies many partial molar pregnancies due to characteristic hydropic villi mixed with more normal villus structures. Our images demonstrate these structures in three dimensions (Figure 28.4c and d). These villi can be compared with trisomy 16 (Figure 28.5a and b), where the villi show quite abnormal villus tree structures. Both normal male and monosomy X karyotypes demonstrate simple villus structures (Figure 28.5a and b, respectively) with little development of the villus tree structure required for placental development. Early development of a normal male conception at 5–6 weeks presents simple unbranching villi (Figure 28.5c and d).

OVERVIEW: WHERE DO WE GO FROM HERE?

Fifty years of studies of SABs have provided us with a wealth of knowledge about the incidence and origin of human chromosome abnormalities. It is now clear that numerical chromosome abnormalities are the leading cause of pregnancy failure in our species; indeed, it could be argued that, cumulatively, they represent the most common category of *de novo* mutation in humans. The parent and stage of origin of these abnormalities is complex, with errors identified at all possible stages of meiosis, at fertilization, and likely during the early mitotic divisions (Figure 28.3). However, against this background, it is also clear that maternal meiotic errors are especially common and represent the most important source of SAB-causing mutations. These occur in high frequencies in all human populations that have been appropriately studied and are presumably under selective pressure, although the evolutionary benefit remains obscure.

Given the extensive data available on the incidence and types of chromosome abnormalities in SABs, is there a compelling reason to continue these analyses? Obviously, the information is clinically warranted, since it may provide the reason for the demise of a wanted pregnancy, as well as often being of prognostic value. However, from a research perspective, this is clearly a field that is "mature," and it might be argued that there is little left to learn from it. We think this conclusion is incorrect, for two reasons. First, most of our information on SABs has come from conventional cytogenetics and, although this is suitable for detection of numerical abnormalities and large structural abnormalities, the level of resolution is insufficient to detect small, but potentially SAB-associated structural rearrangements. The advent of array- and sequence-based approaches now provides the opportunity to ask whether such abnormalities contribute to SABs, and we suggest that this is an important area of future clinical research. Second, SABs provide a valuable resource for examining the origin of chromosome abnormalities in natural conceptions and provide an important comparison group for ART-associated pregnancies. We have already seen the utility of such comparisons, e.g., in studies comparing the types and frequencies of different aneuploidy conditions in ART-derived preimplantation embryos compared with those associated with SABs (e.g., [37]). Recent "MeioMapping" studies [70] demonstrate that it will also be possible to compare the mechanism(s) of origin of specific types of chromosome abnormalities between these categories of pregnancies. Thus, whereas studies of SABs may no longer be innovative or ground-breaking, if properly orchestrated, they still have much to teach us.

ACKNOWLEDGMENT

Work summarized in this review that was conducted in the Hassold Laboratory was supported by Grant No. R37 HD21341.

REFERENCES

1. Horiuchi I, et al. *Mol Hum Reprod.* 1997;3:445–450.
2. Jobanputra V, et al. *Hum Reprod.* 2002;17:1166–1170.
3. Shearer BM, et al. *Genet Med.* 2011;13:545–552.
4. Bell KA, et al. *Fertil Steril.* 2001;75:374–379.
5. Danielly M, et al. *Hum Reprod.* 1998;13:805–809.
6. Danielly M, et al. *Prenat Diagn.* 1999;19:100–104.
7. Fritz B, et al. *Eur J Hum Genet.* 2001;9:539–547.
8. Jenderny J. *Mol Cytogenet.* 2014;7:38.
9. Levy B, et al. *Obstet Gynecol.* 2014;124:202–209.
10. Caramins MC, et al. *Genet Med.* 2011;13:46–51.
11. Hardy K, et al. *Transl Pediatr.* 2015;4:189–200.
12. Gardo S, et al. *Eur J Obstet Gynaecol Reprod Biol.* 1992;47:117–120.
13. Morton NE, et al. *Ann Hum Genet.* 1988;52:227–235.
14. Ohno M, et al. *Obstet Gynecol.* 1991;77:394–398.
15. Sanchez JM, et al. *Prenat Diagn.* 1999;19:601–603.
16. Risch N, et al. *Am J Hum Genet.* 1986;39:68–78.
17. Be C, et al. *Rev Med Chil.* 1997;125:317–322.
18. Bug S, et al. *Mol Cytogenet.* 2014;7:43.
19. Lathi RB, et al. *Fertil Steril.* 2014;101:178–182.

20. van den Berg MMJ, et al. *Biochim Biophys Acta.* 2012;1822:1951–1959.
21. Menten B, et al. *BMC Med Genet.* 2009;10:89.
22. Gao J, et al. *Mol Cytogenet.* 2012;5:33.
23. Hook EB, et al. In Hook EB and Porter IH, Eds. *Population Cytogenetics.* New York: Academic Press; 1977:63–79.
24. Bauld R, et al. *Arch Dis Child.* 1974;49:782–788.
25. Carr DH. *Obstet Gynecol.* 1965;26:308–326.
26. Creasy MR, et al. *Hum Genet.* 1976;31:177–196.
27. Lauritsen J. *Acta Obstet Scand Suppl.* 1976;52:1–29.
28. Byrne J, et al. *Teratology.* 1985;32:297–315.
29. Andrews T, et al. *Hum Genet.* 1984;66:77–84.
30. Eiben B, et al. *Hum Genet.* 1987;77:137–141.
31. Menasha J, et al. *Genet Med.* 2005;7:251–263.
32. Cheng HH, et al. *J Assist Reprod Genet.* 2014;31:1059–1064.
33. Choi TY, et al. *Obstet Gynecol Sci.* 2014;57:518–525.
34. Hardy K, et al. *Am J Med Genet.* 2016;170:2671–2680.
35. Zhang YX, et al. *Clin Genet.* 2009;75:133–140.
36. Fragouli E, et al. *Hum Genet.* 2013;132:1001–1013.
37. Franasiak JM, et al. *J Assist Reprod Genet.* 2014;31:1501–1509.
38. Boue J, et al. *J Med Genet.* 1975;12:265–268.
39. Geisler M, et al. *Hum Genet.* 1978;45:239–251.
40. Hassold T, et al. *Ann Hum Genet.* 1980;44:151–179.
41. Kajii T, et al. *Hum Genet.* 1980;55:87–98.
42. Lauritsen JG, et al. *Hereditas.* 1972;71:160–163.
43. Hogge WA, et al. *Am J Obstet Gynaecol.* 2003;189:397–400.
44. Philipp T, et al. *Hum Reprod.* 2003;18:1724–1732.
45. Shen J, et al. *Mol Cytogenet.* 2016;9:7.
46. Werner M, et al. *J Assist Reprod Genet.* 2012;29:817–820.
47. Hassold T, et al. *Cytogenet Cell Genet.* 1984;38:200–205.
48. Sanger R, et al. *J Med Genet.* 1971;8:417–426.
49. Hall H, et al. *Curr Opin Genet Devel.* 2006;16:323–329.
50. Hassold T, et al. *Hum Genet.* 1992;89:647–652.
51. Jacobs P, et al. *Ann Hum Genet.* 1997;61:471–483.
52. Fernández-García R, et al. *Clin Genet.* 2000;58:201–208.
53. Hook EB, et al. *Hum Genet.* 2014;133:417–424.
54. Skuse DH, et al. *Nature.* 1997;387:705–708.
55. Ghosh S, et al. *Am J Med Genet A.* 2009;149A:1415–1420.
56. Oliver TR, et al. *PLoS ONE.* 2014;9:e99560.
57. Sherman SL, et al. *Biochem Soc Trans.* 2006;34:578–580.
58. Middlebrooks CD, et al. *Hum Mol Genet.* 2014;23:408–417.
59. Oliver T, et al. *PLoS Genet.* 2008;4(3):e1000033.
60. Lamb N, et al. *Cytogenet Genome Res.* 2005;111:250–255.
61. Bugge M, et al. *Hum Mol Genet.* 1998;7:661–669.
62. Hassold T, et al. *Am J Hum Genet.* 1995;57:867–874.
63. Hassold TJ, et al. *Am J Hum Genet.* 1991;49:253–260.
64. Joergensen MW, et al. *Am J Obstet Gynecol.* 2014;211:370. e1–e19.
65. Zaragoza MV, et al. *Am J Hum Genet.* 2000;66.1807–1820.
66. McFadden D, et al. *Clin Genet.* 2000;58:192–200.
67. Pflueger SMV. In Gersen SL and Keagle MB, Eds. *Principles of Clinical Cytogenetics.* Totowa, NJ: Humana Press; 1999:317–343.
68. Szulman A, et al. *Am J Obstet Gynecol.* 1978;131:665–671.
69. Kajii T, et al. *Cytogenet Cell Genet.* 1977;18:109–125.
70. Ottolini C, et al. *Nat Genet.* 2015;47:727–735.

Products of Conception
Current Methodologies and Clinical Applications

29

Nasser Al-Asmar and Marcia Riboldi

INTRODUCTION

Miscarriage is defined as gestational loss that occurs before a fetus is viable. This includes losses during pregnancy that occur before the 20th week of gestation or the loss of a fetus whose weight is less than 500 g. An estimated 23 million miscarriages occur every year worldwide, translating to 44 pregnancy losses each minute. Despite advances in obstetric care, miscarriage is the most common complication observed during the first trimester of pregnancy, and the overall risk of miscarriage is 15.3% (95% CI 12.5–18.7%) among recognized pregnancies. The frequency of women who have had one miscarriage is 10.8% (10.3–11.4%), two miscarriages is 1.9% (1.8–2.1%), and three or more miscarriages is 0.7% (0.5–0.8%) worldwide (1).

The primary cause of early pregnancy loss is the presence of a chromosomal abnormality that is incompatible with life. Chromosomal abnormalities are present in up to 50% of first-trimester losses (2) and the causes of the remaining 50% of miscarriages and perinatal loss are heterogeneous. One-quarter to one-half of all miscarriages reported are of unexplained origin. Established causes of pregnancy loss include parental or *de novo* chromosomal aberrations (2), antiphospholipid syndrome (3), inherited thrombophilia, such as factor V Leiden and prothrombin G20210A gene mutations (4, 5), congenital or acquired uterine anomalies (6), endocrine, autoimmune, or alloimmune disturbances (7), as well as the consequences of lifestyle habits (e.g., smoking, obesity, or psychological stress) (8, 9). Once uterine malformations, endocrine pathologies, and antiphospholipid syndrome are ruled out, parental karyotype assessment should be considered, especially if the miscarriages were karyotypically abnormal (10).

Analysis of products of conception (POC) facilitates diagnosis of cytogenetic anomalies that may explain pregnancy loss. Defining POC karyotype has an impact on a couple's reproductive future because it distinguishes between non-hereditary and hereditary chromosomal abnormalities (e.g., reciprocal translocation carriers) and detects gestational trophoblastic diseases like hydatidiform mole formation to assist with post-loss diagnosis and follow-up. POC analysis also benefits women experiencing recurrent pregnancy loss (RPL). RPL is defined as two or more consecutive pregnancy losses in the first or early second trimester (2). Although the

DOI: 10.1201/9781003024941-29

overall incidence of RPL is low and is estimated to occur in less than 5% of pregnant women (11), an abnormal embryonic karyotype is the most common cause of recurrent miscarriage (12).

In summary, POC analysis is essential to determining the cause of sporadic and recurrent pregnancy loss and assists in the estimation of recurrence risk in future pregnancies, thus improving the chances of a healthy full-term pregnancy.

INDICATIONS

Although it would be useful to analyze every spontaneous abortion, miscarriage analysis would be of special interest in couples with recurrent miscarriages as well as infertile patients undergoing assisted reproductive technology (ART) treatment.

Couples with Recurrent Miscarriage

Couples with previous miscarriages or aneuploid conceptions due to abnormal sex chromosomes and/or autosomes have increased risk of additional instances of fetal aneuploidy. Women who have had a previous trisomic pregnancy, particularly those over 35 years old, are at increased risk of subsequent trisomic pregnancies. The relative risk of trisomy 21 following a previous trisomy 21 conception is greater for women older than 35 years (13, 14). Implantation and pregnancy rates also decrease, and miscarriage rates increase as the number of previous spontaneous abortions increases (15).

Infertile Couples Undergoing ART Treatments

In the general population, spontaneous abortion accounts for 15–20% of all clinically recognized pregnancies during the first trimester. The incidence of chromosomal abnormalities in spontaneous pregnancy loss is approximately 50% (2).

Nevertheless, several studies have found that patients undergoing ART present higher rates of chromosomally abnormal POC compared with age-matched controls (16). Some studies report chromosomal abnormality rates in POC from infertile patients ranging between 63 and 76%, with rates higher than 50% estimated in the general population (17).

Additional studies suggest that not only are the frequencies of chromosomal abnormalities higher in infertile populations undergoing ART, but the percentage and type of chromosomal aberrations also vary depending on the type of ART used (18). Martinez et al. reported an increased percentage of chromosomal abnormalities in miscarriages from an intrauterine insemination (IUI) cycle compared to those from an IVF cycle, with ART showing a lower chromosomal abnormality rate in POC after IVF treatment in patients in an oocyte donation program (18). They also determined that the types of chromosomal aberrations were different between the general population and infertile patients; specifically, that monosomy X presented at twice the incidence among women receiving ART compared with women who experienced a spontaneous abortion after natural conception, whereas polyploidy presented at twice the incidence in women experiencing a spontaneous abortion after natural conception compared with patients undergoing ART.

Couples with Severe Male Factor

Campos-Galindo et al. published in 2015 (19) that couples undergoing ART with own oocytes, in which the men had a sperm concentration lower than 5 million/mL, produced 75% abnormal results, whereas, when the men had a concentration higher than 5 million/mL, couples produced 51% abnormal results. Men with oligoasthenozoospermia, with low motility and/or high follicle stimulating hormone (FSH) levels, show frequent synaptic abnormalities that may underlie production of aneuploid and/or diploid spermatozoa. This may contribute to the observed increased risk of recurrent abortion resulting from the presence of chromosome abnormalities in spermatozoa (20, 21). Based on this information, the male factor should be investigated after spontaneous abortion.

Couples with Environmental Exposure to Endocrine Disruptors

Endocrine-disrupting chemicals (EDCs) impact the endocrine system of animals and humans. There is growing concern about the effects of EDCs in reproductive health and a link between EDC exposure and spontaneous miscarriage is suspected, but studies are limited (22). Environmental exposure to toxic metals has been associated with pregnancy loss and impaired fetal growth, fetal loss, and increased neonatal deaths (23–25).

One of the most common EDCs is bisphenol A (BPA), which is used in manufacturing a wide range of common-use products, including products that are in direct contact with food (plastic bottles, food containers), medical equipment, and paint. Though little is known about the effects of EDC on miscarriage rates, there are several studies suggesting a relationship between exposure and miscarriage or other adverse prenatal outcomes (26–28). Hunt et al. demonstrated that daily oral BPA exposure caused meiotic aneuploidy in female mice (28). In addition, exposure to BPA in the first trimester of pregnancy interfered with cellular growth and affected DNA methylation in humans (29). Therefore, in couples where either partner has known exposure to EDCs, POC chromosomal analysis is indicated.

STATE OF THE ART: WHAT DO WE KNOW TODAY?

Tissue Collection

Once pregnancy loss has occurred, a sample of fetal tissue can be recovered by surgical methods, such as dilation and curettage (D&C) or uterine aspiration. These methods carry risk of maternal cell contamination (MCC) due to maternal and fetal tissue mixing during the procedure. Hysteroembryoscopy (which allows for direct and selective embryo and chorion biopsy) performed before curettage diminishes the likelihood of MCC and avoids the risk of misdiagnosis (30). However, hysteroembryoscopy requires a specialist in the technology and is not common practice in the obstetric field. An alternative to detect the presence of MCC in POC recovered by D&C or aspiration paired with molecular diagnosis of aneuploidies is presented below.

Molecular Approaches for Chromosomal Analysis

Since 2008, emerging technologies have been used as part of a robust and accurate diagnostic approach to assess for aneuploidy of the 22 autosomes and the two sex chromosomes (31). Techniques such as chromosomal comparative genomic hybridization array (aCGH), next-generation sequencing (NGS), bacterial artificial chromosomes (BACs)-on-Beads (BOBs), fluorescence *in situ* hybridization (FISH), multiplex ligation-dependent probe amplification (MLPA), and quantitative fluorescent polymerase chain reaction (QF-PCR) have overcome disadvantages related to conventional cytogenetic techniques in POC after miscarriage, including poor chromosome preparations and failure in culture (32).

Due to recent advances, there has been an increasing trend to perform POC chromosomal assessment using molecular genetics rather than traditional cytogenetic methods, such as Giemsa staining of

TABLE 29.1 Benefits of Molecular Approaches over Conventional Karyotyping for POC Assessment of Chromosomal Abnormalities

CONVENTIONAL KARYOTYPE	*MOLECULAR METHODS*
• Requires *in vitro* cell culture	• Do not require *in vitro* cell culture
• 42% of tests performed are not informative due to tissue degradation	• Results obtained in 99% of tests performed
• 33.3% are false negatives due to MCC	• Rule out false negatives caused by MCC
• Results obtained in 2–4 weeks	• Results obtained in 1 week
• Low-resolution analysis	• Higher resolution than conventional karyotype

Abbreviation: MCC: Maternal cell contamination.
Note: see further refs 18, 19, 22, 49, 50.

metaphase spreads, which relies on cell culture of tissue samples. Molecular methods are more reliable and do not require cell culture, avoiding the possibility of cell culture failure and increasing the proportion of informative results. Cell culture failure is an inherent problem in traditional cytogenetic methods but is not the only problem. Cytogenetic studies cannot distinguish between normal female fetal results and normal female results due to MCC. There is also a high percentage of non-informative results with standard karyotyping due to potential tissue degradation. An additional advantage of molecular techniques is a much lower turn-around time than conventional karyotyping. Finally, resolution is higher with molecular studies. Table 29.1 shows a comparison of conventional karyotype with molecular analysis.

Several groups have evaluated different molecular approaches to POC, including aCGH and, most recently, NGS.

Comparative genomic hybridization (CGH) array

To perform aCGH, DNA extraction must be performed on each type of POC tissue. First, a small portion of tissue is mechanically dilacerated using a scalpel blade. Then, DNA extraction is performed followed by the labeling of POC DNA samples and control DNA with Cy3 and Cy5 fluorophores. Labeled mixes can be combined and hybridized in commercial platforms such as 24sure BAC arrays (Illumina Inc., San Diego, CA, USA) for 6–12 hours. Each probe is specific to a different chromosomal region and occupies a discrete spot on the slide. The different spots are evenly distributed 1 Mb apart. Chromosomal losses or gains are revealed by the color adopted by each spot after hybridization. Fluorescence intensity is detected using a laser scanner and specific software (31, 33).

Single-nucleotide polymorphisms (SNP) array

To perform an SNP array, the tissue sample is rinsed with phosphate-buffered saline (PBS) and separated from maternal decidua and blood using standardized techniques. Next, the tissue is again rinsed in sterile PBS and dissected into small villi/fetal samples (~3 mm^3) and placed in 1.5 mL microfuge tubes. DNA is then extracted from the villi and the maternal blood sample using a QIAamp Circulating Nucleic Acid Kit (Qiagen, Hilden, Germany). Purified DNA is run on an Illumina Cyto 12 SNP microarray according to a standard protocol (34).

Next-generation sequencing (NGS)

NGS is currently the most innovative technology available. Different commercial platforms have a kit developed for aneuploidy, mosaicism, and segmental aneuploidies detection, including:

- *Ion Torrent technology*: Briefly, DNA is amplified, purified, and quantified (a process known as library construction). Then, a clonal amplification (templated preparation) is performed, followed by template enrichment. Finally, the template is loaded onto a sequencing machine. An

Ion Personal Genome Machine (PGM) sequencer sequentially floods a chip with one nucleotide after another. If a nucleotide complements the sequence of the DNA molecule in a particular micro-well, it will be incorporated, and hydrogen ions will be released. The pH of the solution changes in that well and is detected by the ion sensor, essentially going directly from chemical information to digital information (Thermo Fisher Scientific, Waltham, MA, USA). Software aligns the sequence reads to a reference human genome classified in 2 Mb bins and differences in copy numbers for each bin are represented in a plot. With this technology, single and multiple whole-chromosome aneuploidies have been detected in POCs, as well as small deletions/duplications (del/dup) and mosaicism (Ion Reporter 5.4) (35)

- *Illumina technology*: DNA fragmentation is performed using Covaris M220, according to the manufacturer's protocol, and targeted fragment size is set at 180 bp. DNA fragments are used for sequencing library preparation using NEBNext DNA Library Prep Master Mix for Illumina following the NEB Next manual (NEB, E6040S). DNA size in the libraries is checked using an Agilent 2100 Bioanalyzer. A Roche Light Cycler LC480 is employed to evaluate library concentration. Libraries with different indexes are pooled. Sequencing is performed using Illumina PE100 flowcell on the Illumina Hiseq2500 platform. For aneuploid analysis, Illumina flowcells (single end with length 41 bp) are used for sequencing with a low-depth sequencing strategy, about 10 Mb reads per sample (36).

MOLECULAR ANALYSIS OF POC BY NGS

Igenomix Study Design

The following describes an observational study performed at Igenomix that included a total of 2,531 POC samples analyzed using either aCGH (n = 675) or NGS (n = 1,875) (35) (Figure 29.1).

Methodology

The methodology was developed by Igenomix for POC analysis, comparing aCGH and NGS (see Figure 29.2).

Dissection

Multisampling was performed to diminish the chances of MCC. In our laboratories, multisampling comprises three dissections of POC instead of only one. Three dissections make it possible that one or two dissections could belong to the mother or gestational carrier whereas the third belongs to the fetus. This last one would be the one to run on the sequencer for chromosomal analysis and would be able to provide accurate results. In the scenario where only one dissection is performed, if that sample is MCC, there will be a false normal female result if short tandem repeats (STRs; see below) are not performed.

Molecular analysis for aneuploidy testing

The technique to analyze chromosomal aneuploidies was aCGH by Illumina Inc. (San Diego, CA, USA) and NGS by Thermo Fisher Scientific (Waltham, MA, USA) (see previous descriptions of both technologies).

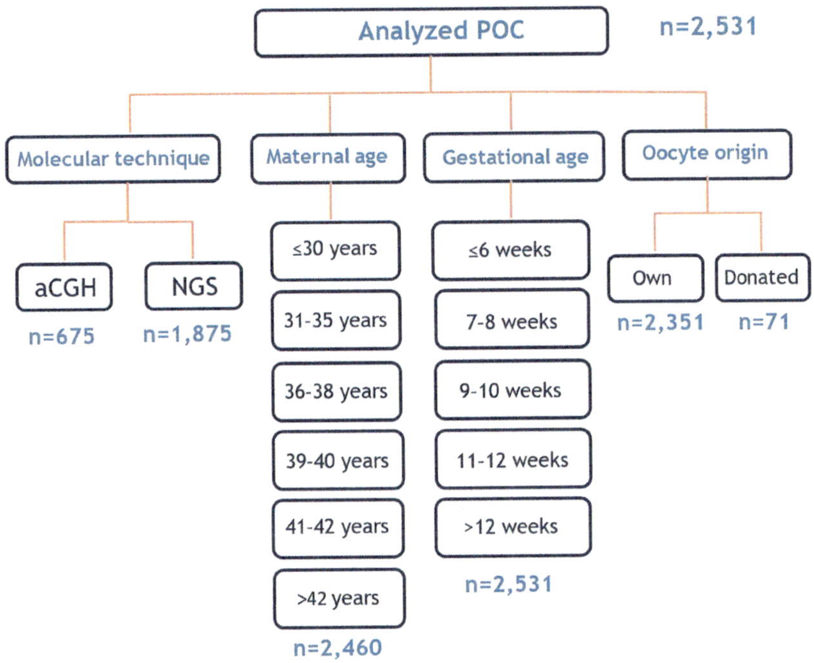

FIGURE 29.1 Study design.

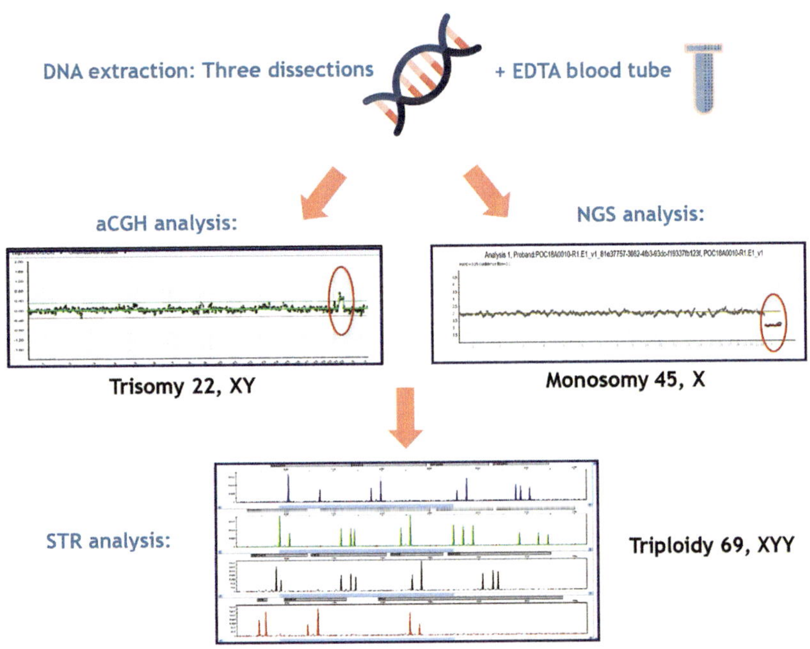

FIGURE 29.2 Methodology: dissection, molecular analysis (aCGH and NGS), and STR analysis.

(a)

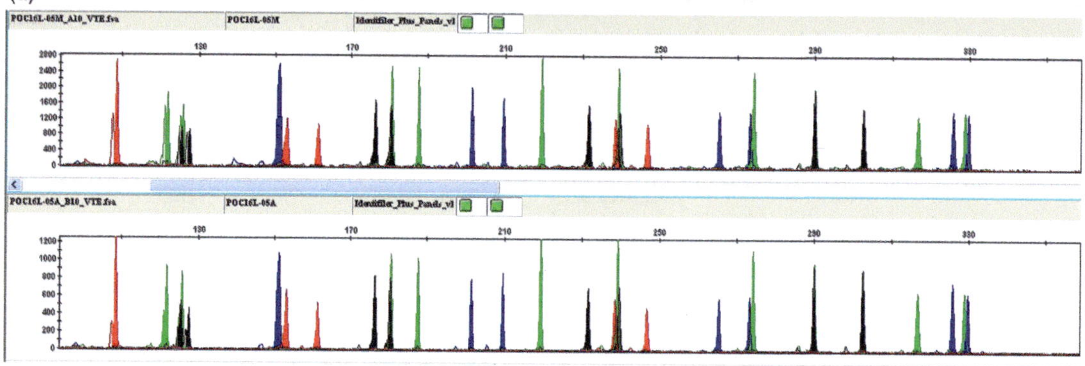

FIGURE 29.3a Maternal cell contamination (MCC). (*Top*) DNA from mother or gestational carrier. (*Bottom*) DNA from fetus. All alleles are coincident between mother and fetus, so that the sample that is supposed to be fetal is not from the fetus but is actually from the mother, resulting in MCC.

(b)

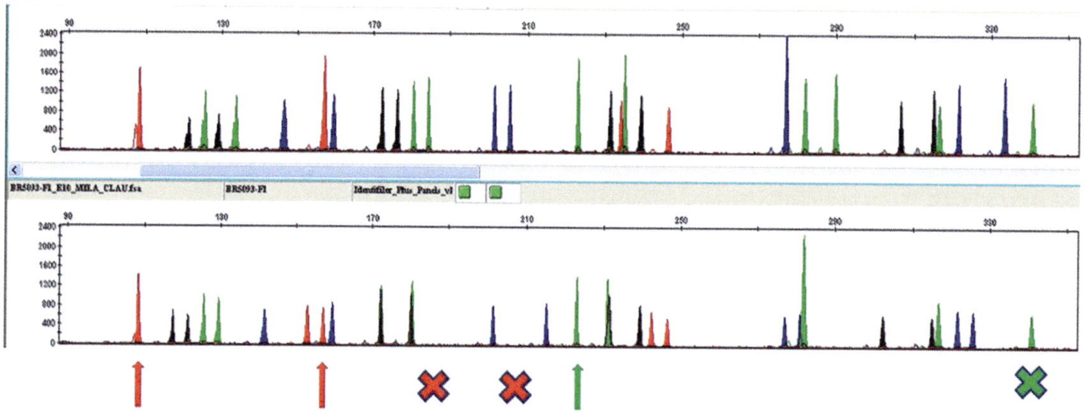

FIGURE 29.3b No MCC (ruled out). (*Top*) DNA from mother or gestational carrier. (*Bottom*) DNA from fetus. Some of the alleles are coincident between mother and fetus (arrows). Some of the alleles are present in the mother but not in fetus (X marks). This means that the sample that is supposed to be fetal is from the fetus and is not contaminated. Sample is determined to be either normal or abnormal after aneuploidy analysis.

Complementary studies to rule out MCC and to identify triploidy

MCC is a major source of misdiagnosis or non-informative results when evaluating POC. A retrospective study showed that over half of the normal 46,XX results in miscarriage specimens were due to MCC (16). Errors like this may lead to misdiagnosis and/or inappropriate counseling recommendations. Accurate results are the only way that a physician and medical team can give appropriate reproductive counseling when a pregnancy loss occurs after natural conception or after ART.

To rule out the risk of misdiagnosis due to MCC, STR can be run in parallel with molecular methods like NGS for aneuploidy detection. DNA from POC samples and gestational carrier DNA are extracted using a QIAamp® DNA Mini and Blood Kit (Qiagen). An AmpFlSTR Identifier Plus Kit PCR Protocol (amp16str) (Thermo Fisher Scientific) is used to run the PCR and subsequent electrophoresis to detect or rule out MCC (Figure 29.3a and 29.3b) and triploidy (Figure 29.4, courtesy of Dr. Ana Cervero).

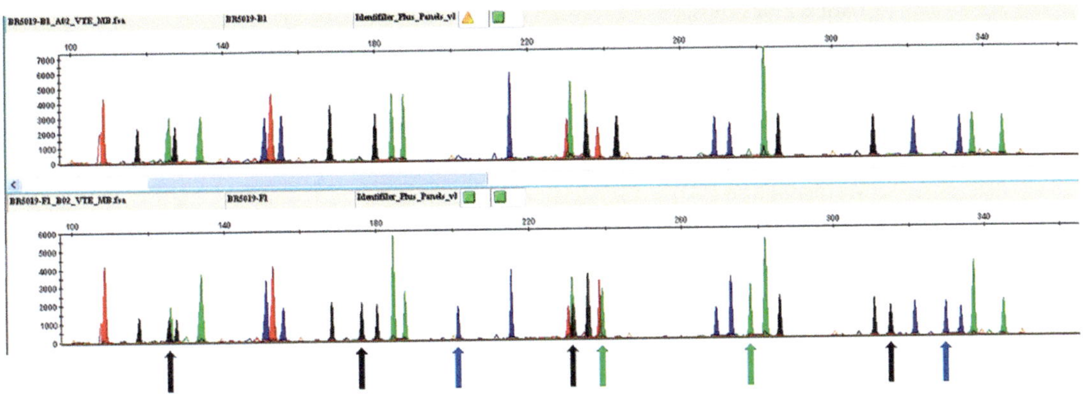

FIGURE 29.4 Polyploidy (triploidy). (*Top*) DNA from mother or gestational carrier. (*Bottom*) DNA from fetus. Extra peaks are observed on the fetus profile (arrows) and not in the maternal profile. This means that the miscarriage was the result of a polyploid (triploid in this case) fetus.

CLINICAL RESULTS

Our group has used NGS/aCGH with STR for aneuploidy detection to analyze more than 2,500 POC cases: 675 by aCGH and 1,875 by NGS. Average maternal age was 35.3 ± 4.9 years and the mean gestational age at which the spontaneous miscarriage occurred was 7.9 ± 2.8 weeks. After three dissections from different locations in the piece of fetal tissue, we reduced the non-informativity rate due to MCC to 12.8%. We were able to provide results (either normal or abnormal) in 86.2% of cases. Notably, three dissections were done per POC sample to establish a lower rate of MCC. We also performed an internal study, taking six or nine samples when the first three were found to be MCC. If the first three dissections proved to be MCC, a larger number of dissections did not change the results. Therefore, when the first three dissections result in MCC, we recommend that no further dissections be performed. We found chromosomal abnormalities in up to 53.5% of cases. As expected, trisomies were the most common abnormality (74.3%) and 7.6% of cases had multiple aneuploidies. Among trisomies, chromosomes 15, 16, and 22 were the most prevalent at 13.3%, 17.6%, and 16.6%, respectively. In addition to trisomies, del/dups were found in 2.3% of cases. Monosomy 45,X was found in more than 9% of cases. Interestingly, we identified monosomy for autosomes in 1.8% of cases. Polyploidy was found in 4.6% of cases. Table 29.2 shows the detailed results.

Results by Maternal Age

We also evaluated the maternal age of women who experienced gestational loss. A total of 2,460 POCs were analyzed, with results stratified by age into five groups: ≤ 30, 31–35, 36–38, 39–40, and ≥ 41 (Figure 29.5).

The percentages of chromosomal abnormalities progressively increased with increasing maternal age. Multiple aneuploidies, in most cases the combination of two chromosomal abnormalities, increased as maternal age increased. The percentage of monosomies for the sex chromosomes appeared to decrease with increasing maternal age. The last type of aneuploidy in which significant differences with respect to maternal age were found after analysis was polyploidies. Polyploidies decreased as the age of the patient increased (Figure 29.6).

TABLE 29.2 Results of STR + 24 Chromosome Analysis (aCGH and NGS). The Percentage of Specific Abnormalities Related to Total Abnormalities Found Are Also Described

	NO. OF CASES	*(%)*
No. of cases processed	2531	---
Maternal age (SD)	35.3 (4.9)	---
Gestational week (SD)	7.9 (2.8)	---
MCC	321	12.8
Informative results	2506	99.0
Normal results	1016	46.5
Abnormal results	1169	53.5
Trisomies	868	74.3
Autosome monosomy	21	1.8
Sex Chr. monosomy	110	9.4
Del/dup	27	2.3
Polyploidy	54	4.6

Abbreviations: SD: Standard deviation; MCC: Maternal cell contamination; Del/dup: Deletion/duplication.

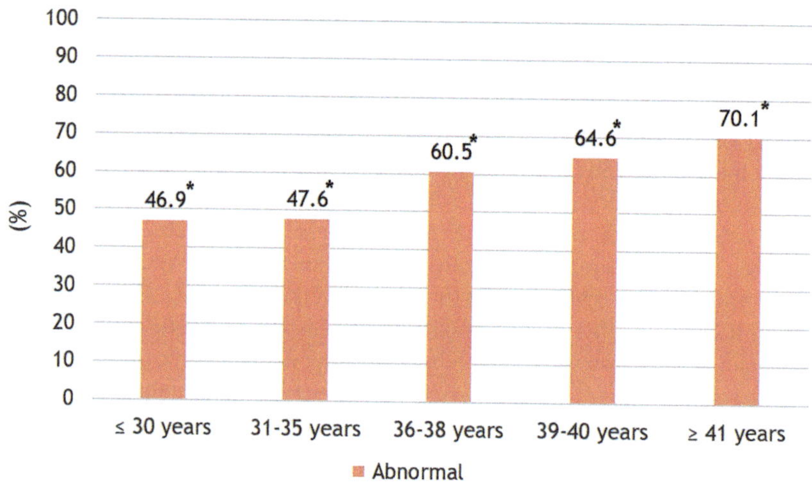

FIGURE 29.5 Percentage of chromosomal abnormalities by maternal age. Data analyzed by Fisher's exact test and Kruskal-Wallis test with Dunn's multiple comparisons test: *, $p < 0.05$.

Results by Gestational Age

Gestational age (in weeks) was evaluated in 2,531 samples divided into groups, according to the gestational week in which the pregnancy loss occurred. The groups were: ≤ 6 weeks, 7–8 weeks, 9–10 weeks, 11–12 weeks, and > 12 weeks.

Decreased percentage of abnormal POCs with increasing gestational age was observed. After the 12th week of gestation, which could be considered as a biological barrier, the percentage of abnormal POCs decreased by half (60.3% vs. 34%). Interestingly, the incidence of monosomies for sex chromosomes increased with gestational age, indicating a lower detrimental effect of sex chromosome aneuploidies in fetal development. Multiple aneuploidies decreased as gestational age increased, and most were seen at gestational age ≤ 6 weeks, due to early fetal arrest.

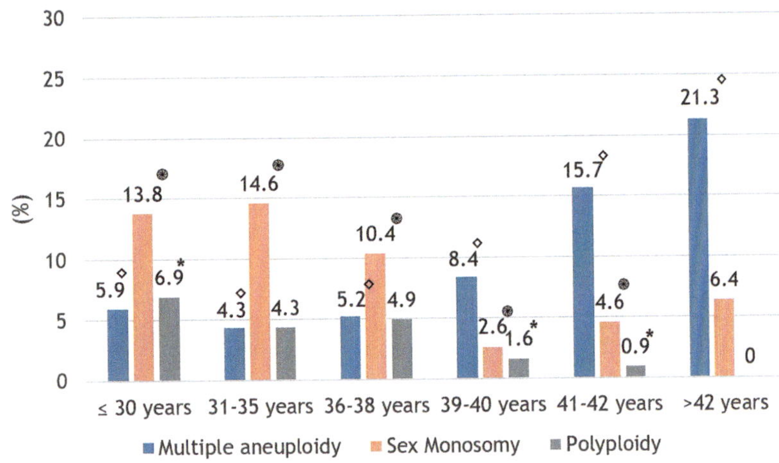

FIGURE 29.6 Percentage of multiple aneuploidies, sex monosomy, and polyploidy by maternal age. Fisher's exact test:*, $p < 0.05$; ●, $p < 0.05$; ◇, $p < 0.05$.

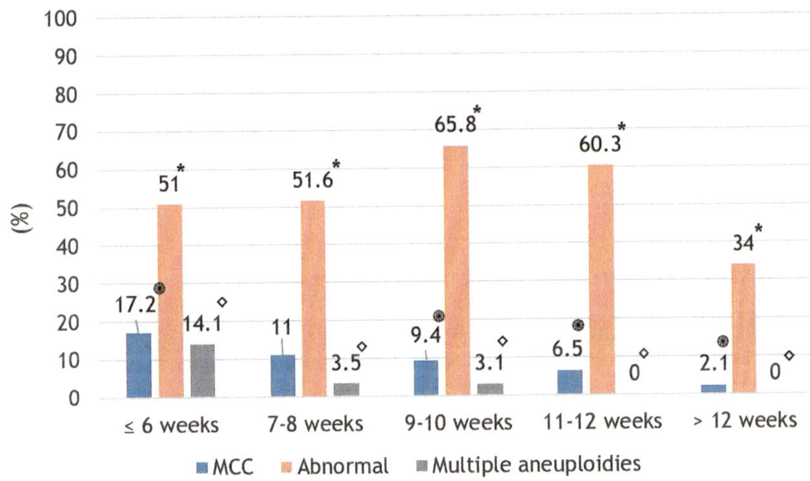

FIGURE 29.7 Percentage of MCC, chromosomal abnormalities, and multiple aneuploidies by gestational age. Fisher's exact test and Kruskal-Wallis test with Dunn's multiple comparisons test: ●, $p < 0.05$; *, $p < 0.05$; ◇, $p < 0.05$.

Gestational age showed an inverse relationship with the percentage of samples with maternal contamination, likely due to the technical difficulty involved in obtaining samples (Figure 29.7).

Results by Oocyte Origin

Results were analyzed in two subgroups: patients who achieved pregnancy with their own oocytes and pregnancies from recipients of donated oocytes. Patients with their own oocytes showed a significant increase in the percentage of abnormal POCs compared with the group of patients with donor oocytes (54.5% vs. 29.7%; $p < 0.001$). When different subtypes of aneuploidies were studied, significant differences were only observed for sex chromosomes (9.1% with own oocytes vs. 36.8% with oocyte donation; $p < 0.001$). In 80% of cases with monosomy for sex chromosomes, the absence of paternal sex chromosomes was thought to be responsible, unlike aneuploidies of autosomes that are mostly of maternal origin (37–40) (Figure 29.8).

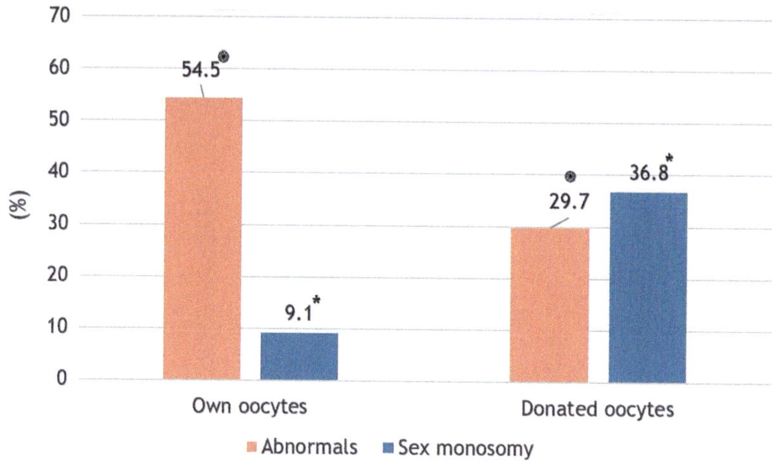

FIGURE 29.8 Percentage of chromosomal abnormalities and sex monosomy aneuploidies by oocyte origin. Patients > 42 years were excluded. Fisher's exact test and Chi-squared test: ⊛, $p < 0.001$; *, $p < 0.001$.

LIMITATIONS AND STRENGTHS OF POC MOLECULAR ANALYSIS

Limitations

Isolation of fetal tissue in the sample collection

One of the important limitations to avoid or rule out MCC and to be able to provide accurate results is tissue collection. General recommendations for proper fetal tissue collection are:

- To take a tissue sample of a minimum size of 3×3 mm (preferably a small but clean sample without blood).
- To remove large blood clots from a tissue sample.
- To use sterile saline solution to wash the samples.
- To place the tissue in a sterile cup with enough saline to cover the sample.
- To perform a blood draw from the mother or gestational carrier before or after D&C for further analysis to rule out MCC or detect polyploidy. Blood should be extracted into an EDTA tube to prevent clotting.

Figure 29.9 shows the appearance of decidual villi in clean tissue (Figure 29.9a) and the appearance of a hematic sample (Figure 29.9b). Ideally, the decidua appears pink, dense, and leaf-like. Once villi are well-rinsed, they tend to float and have a pale appearance and feathers, compared with the decidua.

Mosaicism

Shah et al. published a study comparing three different techniques to analyze POC: traditional cytogenetic analysis, molecular karyotyping using single-nucleotide polymorphism microarrays (SNP), and array comparative genomic hybridization (aCGH).

The authors showed that there was a 33% overall discordant rate between results due to maternal cell contamination, balanced chromosome rearrangements, polyploidy, and placental mosaicism. Regardless

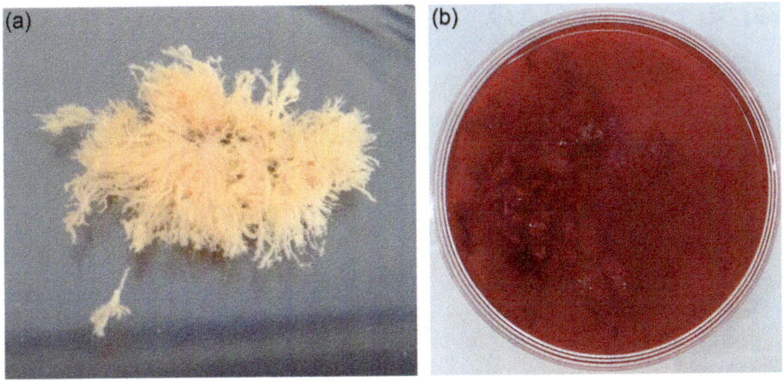

FIGURE 29.9 Tissue sample collection: (a) villi of the decidua in clean tissue; (b) villi of the decidua of a hematic sample.

of the platform utilized, mosaicism was detected in 18% of samples (33). Specifically, confined placental mosaicism (CPM) is characterized by the discrepancy between the chromosomal/genetic makeup of the fetus and the placenta. Approximately 2% of viable pregnancies studied by chorionic villus sampling (CVS) at 9 to 11 weeks of gestation are confined to the placenta (33).

Detection of del/dups below the resolution of the platform

NGS technology improves the detection of del/dups, offering better resolution than conventional karyotyping. Nevertheless, some del/dups below the resolution of the platforms (6 Mb) were not detected and resulted in misdiagnosis (41). Those *de novo* microdeletions represent less than 1% of cases.

Strengths

High rate of informativity

One of the biggest strengths of our study was the high informative rate, with 99% of informative results for all the cases analyzed, at 99.1% for aCGH and 99% for NGS. This proportion of informative results is very important because one of the main disadvantages of conventional karyotyping is that many samples will exhibit growth failure in cell culture (ranging from 10% to 60%) (42–46).

Low rate of MCC

The percentage of samples that presented MCC in our study was 12.8%. Both aCGH and NGS were comparable and did not exhibit significant differences in MCC rate (10.8% with aCGH vs. 13.6% with NGS). Some studies using molecular techniques, such as SNP microarrays, showed that the percentage of MCC exceeded 20% (22%) (34). As mentioned above, our group carried out an internal study (unpublished data) before the clinical program in which DNA extractions were performed in different locations of the received POC samples. In these samples, STR analysis was first performed and compared to the maternal DNA result to assess the minimum number of subsamples necessary to optimize the identification of a subsample without maternal contamination. Such multi-sampling comprises three, six, or up to nine dissections of the POC rather than just one. In this study, we were able to verify that after three subsamples with a MCC profile, it was not necessary to carry out further dissections to see if fetal tissue could be found. Due to multi-sampling, we were able to reduce the percentage of MCC to below 13%.

Shorter turn-around time (TAT)

The turn-around time for diagnosis in the molecular analysis is considerably shorter since chorionic villi can be directly analyzed. This is an advantage over karyotyping, which requires additional steps, such as the preparation of a single-cell suspension, extended cell culture, metaphase conversion, and chromosome staining (36).

Greater cost effectiveness

It is crucial to have a good strategy to achieve not only good results and informativity, but also to reduce the cost to patients. Popescu et al. developed an approach for recurrent pregnancy loss (RPL) evaluation, describing a 24-chromosome microarray of miscarriage tissue (47). In their study, a definite or probable cause of pregnancy loss was identified in the vast majority (95/100; 95%) of RPL patients when a 24-chromosome pair microarray evaluation of POC testing was combined with the standard ASRM RPL workup evaluation at the time of the second or subsequent loss. Only 5/100 (5%) patients, who had a euploid loss and a normal ASRM RPL workup, also had a pregnancy loss without a probable or definitive cause identified. All other losses were explained by an abnormal 24-chromosome microarray analysis of the miscarriage tissue, an abnormal finding on RPL workup, or a combination of both. Interestingly, to save money, the proposed order for testing patient was first the 24-chromosome microarray analysis of the POC, and, if a euploid result was obtained, the ASRM RPL workup was carried out.

CONCLUSIONS IN A NUTSHELL

Molecular techniques, such as NGS, was combined with STR to identify the origin of the analyzed DNA, to facilitate increased efficiency of the analysis of chromosomal abnormalities in POC compared to conventional cytogenetic techniques. These technologies improve the percentage of samples with informative results and decrease the rates of false-negatives and false-positives caused by contamination with DNA of maternal origin.

FUTURE PERSPECTIVES

Molecular POC techniques afford more accurate results and lower contamination rates. Circulating cell-free DNA (cfDNA) may be useful in cases where it is not possible or is very difficult to obtain material from POC, such as in early miscarriages, with degraded material, or pregnancy loss after pharmacological treatment. To this end, our group is developing a non-invasive approach for aneuploidy analysis in clinical miscarriage. A novel approach by Balaguer et al. (48) evaluated maternal cfDNA testing as a feasible alternative to analysis of POC in clinical miscarriage. A total of 12 cases were included in the study and, using molecular POC analysis as standard, the concordance rate between the two studies was 90%. The authors concluded that genome-wide cfDNA testing in the maternal bloodstream is a reliable tool with which to analyze chromosomal aneuploidies in clinical miscarriage. Therefore, cfDNA testing before curettage or pharmacological treatment could be an alternative to POC analysis. Due to small sample size, a larger cohort will be essential to improve cfDNA testing performance.

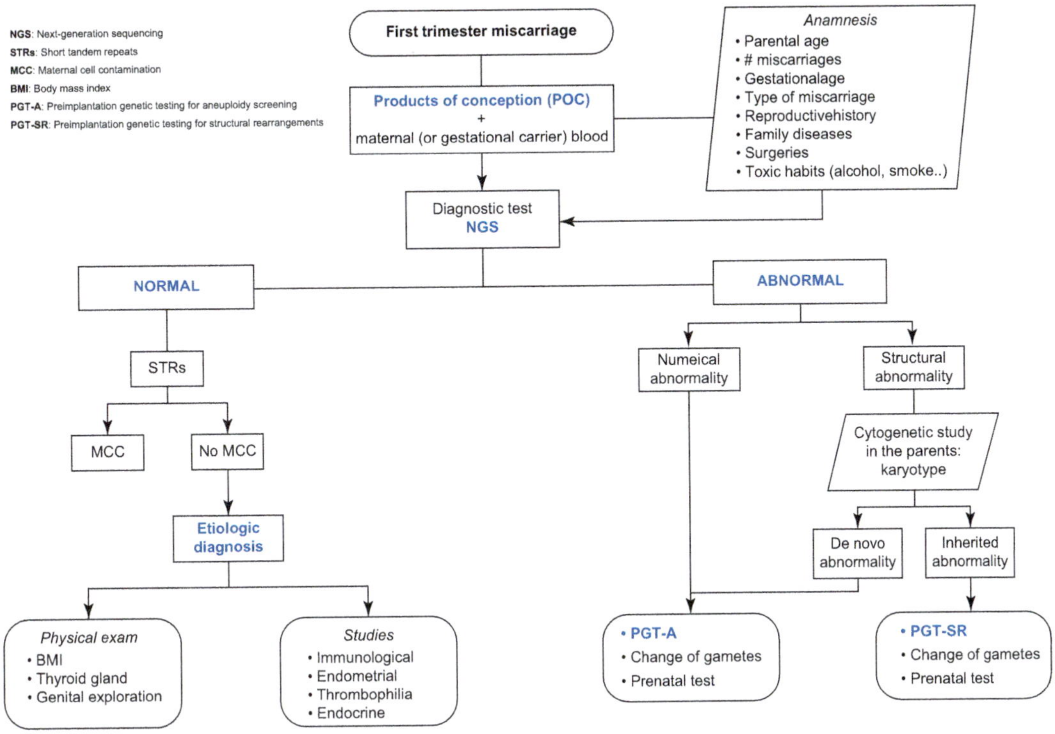

FIGURE 29.10 POC decision-making workflow.

POC DECISION-MAKING WORKFLOW

Following miscarriage, all women (or couples) should have access to support, follow-up, and formal counseling, where necessary. In our lab, a standard operating procedure has been established for decision making in couples who have experienced miscarriage (Figure 29.10). After a first-trimester miscarriage, POC, and maternal or gestational carrier blood are collected. POC is evaluated for 24 chromosome screening by NGS:

- If a normal result is found, we ruled out MCC by performing STR analysis. If the analysis shows no MCC, then we recommend the physician consider other causes of miscarriage.
- If the result is abnormal, there could be a whole-chromosome aneuploidy or a small deletion (loss) or duplication (gain) of a part of the chromosome (del/dup). If an abnormal del/dup is found, we recommend checking if the abnormality is *de novo* or inherited. In some cases, the result could suggest the presence of a balanced chromosome reorganization in one of the partners in the couple. In this case, we recommend performing a karyotype on both partners to establish the risk of miscarriage in further gestations. We also recommend preimplantation genetic testing (PGT) for aneuploidy or for structural rearrangements. Other options could be change of gametes or prenatal test.

REFERENCES

1. Quenby S, et al. *Lancet*. 2021;397(10285):1658–1667.
2. Hassold TJ. *Am J Med Genet*. 1980;32:723–30.

3. Rai R, Regan L. *Lancet.* 2006;368(9535):601–11.
4. Empson M, et al. *Cochrane Database Syst Rev.* 2005;2:CD002859.
5. Kovalevsky G, et al. *Arch Intern Med.* 2004;164(5):558–63.
6. Krabbendam I, et al. *Eur J Obstet Gynecol Reprod Biol.* 2005;118(2):143–53.
7. Devi Wold AS, et al. *Semin Reprod Med.* 2006;24(1):25–32.
8. Arredondo F, et al. *Semin Reprod Med.* 2006;24(1):33–9.
9. Lashen H, et al. *Hum Reprod.* 2004;19(7):1644–6.
10. Pandey MK, et al. *Arch Gynecol Obstet.* 2005;272(2):95–108.
11. Huchon C, et al. *Eur J Obstet Gynecol Reprod Biol.* 2016;201:18–26.
12. Stirrat GM. *Lancet.* 1990;336:673–5.
13. Al-Asmar N, et al. *Fertil Steril.* 2012;98(1):145–50.
14. Warburton D, et al. *Am J Hum Genet.* 2004;75:376–85.
15. Rubio C, et al. *Reprod Biomed Online.* 2009;18(5):687–93.
16. Campana M, et al. *Am J Med Genet.* 1986;24:341–56.
17. Werner M, et al. *J Assist Reprod Genet.* 2012;29(8):817–20.
18. Martínez MC, et al. *Fertil Steril.* 2010;93(1):289–92.
19. Campos-Galindo I, et al. *J Assist Reprod Genet.* 2015;32(5):839–48.
20. Egozcue J, et al. *Placenta.* 2003;24 Suppl B:S62–5.
21. Rubio C, et al. *Hum Reprod.* 2001;16(10):2084–92.
22. Lathi RB, et al. *Fertil Steril.* 2014;102(1):123–8.
23. Rahman A, et al. *Sci Total Environ.* 2016;1:569–570.
24. Buck Louis GM, et al. *Reprod Toxicol.* 2017;2(69):68–74.
25. Ajayi OO, et al. *Afr Health Sci.* 2012;12(2):153–9.
26. Li Q, et al. *Receptors Clin Investig.* 2016;3(3):e1369.
27. Mikołajewska K, et al. *Int J Occup Med Environ Health.* 2015;28(2):209–41.
28. Hunt PA, et al. *Curr Biol.* 2003;13(7):546–53.
29. Basak S, et al. *Reprod Toxicol.* 2018;82:72–79.
30. Ferro J, et al. *Fertil Steril.* 2003;80(5):1260–1264.
31. Rodrigo L, et al. *Biomed Res Int.* 2014;2014:517125.
32. van den Boogaard E, et al. *Hum Reprod.* 2010;25(6):1411–4.
33. Shah M, et al. *Fertil Steril.* 2017;107(4):1028–33.
34. Lathi RB, et al. *Fertil Steril.* 2014;101(1):178–82.
35. Al-Asmar N, et al. Molecular analysis of Products of Conception (POC) provides comprehensive understanding of the chromosomal causes of clinical miscarriage [abstract]. In: *35th Annual Meeting of European Society of Human Reproduction and Embryology*; 2019 June 23–26; Vienna: ESHRE; 2019.
36. Xu J, et al. *Medicine (Baltimore).* 2020;99(5):e18731.
37. Jacobs PA, et al. *Adv Genet.* 1995;33:101–133.
38. Nicolaidis P, et al. *Hum Reprod.* 1998;13:313–319.
39. Hassold T, et al. *Nat Rev Genet.* 2001;2(4):280–91.
40. Hassold T, et al. *Hum Mol Genet.* 2007;16:R203–8.
41. Qi H, et al. *Zhonghua Fu Chan Ke Za Zhi.* 2016;51(2):92–7.
42. Fritz B, et al. *Eur J Hum Genet.* 2001;9:539–47.
43. Lathi RB, et al. *J Assist Reprod Genet.* 2002;19(11):536–8.
44. Stephenson MD, et al. *Hum Reprod.* 2002;17:446–451.
45. Nikitina TV, et al. *Fertil Steril.* 2005;83(4):964–72.
46. Robberecht C, et al. *Prenat Diagn.* 2012;32(10):933–42.
47. Popescu F, et al. *Hum Reprod.* 2018;33(4):579–587.
48. Balaguer Cuenca N, et al. A non-invasive approach for aneuploidy analysis in clinical miscarriages [abstract]. In: *37th Annual Meeting of European Society of Human Reproduction and Embryology*; June 26–July 01; ESHRE; 2021.
49. Zhang T, et al. *BJOG.* 2018; 125(44): 414–420.
50. Al-Asmar Piñar N. 2018. Importancia del estudio cromosómico de los restos abortivos mediante técnicas de secuenciación masiva (NGS) y arrays de CGH para un adecuado consejo reproductivo [Tesis de Doctorado, Universitat de València]. Repositorio Roderic – Universitat de València.

30 Mother and Embryo Cross Communication during Conception

Anna Idelevich, Andrea Peralta, and Felipe Vilella

INTRODUCTION

The prevailing view in the mid-1970s regarded the uterus as a rather passive contributor to periconception events, providing the appropriate temperature, pH, and nutrients—an optimal milieu for embryonic implantation. Over time, this notion changed, and in the 21st century it became apparent that the uterus regulates conception at multiple stages, from sperm viability, where sperm membrane plasma fluidity is maintained, and further on through all stages of pregnancy up to the birth of an offspring. Studies have shown that in cattle and sheep exposure of the embryo to unusual environments prior to blastocyst stage results in the "unusually large offspring" syndrome [1]. These animals had severe organ defects suggesting that the uterine environment provides additional biological support during implantation that is not merely physicochemical.

The explosion of interest in epigenetics and the availability of novel research tools identified important new mechanisms, including microRNA exchange, DNA methylation and others, that play crucial roles in uterus–embryo interactions, imperative to good offspring health [2]. Historically, the first reports indicating cross-communication between gametes, embryos, and the maternal reproductive tract appeared in the scientific literature in the 1990s. The reports demonstrated *de novo* production of proteins in Fallopian tube epithelial cells, when co-cultured with spermatozoa [3, 4]. Although the identity of these proteins was not known at the time, these intriguing reports overruled the dogma that spermatozoa are "inert" cells and showed that the maternal reproductive tract is capable of responding in a molecular fashion to spermatozoan signals. Subsequent studies have shown that the expression of certain genes is upregulated in the fallopian tubes in response to exposure to the spermatozoa, oocytes, and embryo [5–8], using methods such as *in situ* hybridization and microarray. There are several ways by which gametes or embryo can be recognized in the maternal tract and these include pattern recognition receptors and integrins, although our knowledge of these receptors and counterreceptors remains limited. Another means of embryo–mother communication is *via* release of extracellular vesicles carrying information in the form of genetic material, nutrients, and cytokines, which trigger response in the maternal tract wall. The exchange of information occurs at all stages from stem cell differentiation all the way to the implantation, decidualization (differentiation of uterine stromal cells into specialized [decidual] cells), placentation, and

DOI: 10.1201/9781003024941-30

parturition, resulting in birth of the offspring [9–13]. The field of "wireless communication" in pregnancy is rapidly growing and its importance for healthy pregnancy progression can no longer be disputed.

Infertility is common, with ~12% of women in the United States being unable to conceive [14] Only half of women are infertile due to endometriosis and in others the etiology is unknown [15]. Approximately 1.6% of all births in the United States result from *in vitro* fertilization (IVF) [16]. According to the European Society of Human Reproduction and Embryology (ESHRE), more than 8 million babies have been born from IVF, since the world's first in 1978 [17]. A more thorough understanding of such mechanisms is required to enhance pregnancy success rates in fertility clinics, improving assisted reproductive technologies (ARTs), such as IVF. This review attempts to summarize the knowledge of factors implicated in the communication between the embryo and the mother during the conception stage. We start with presenting several important circulating factors, including hormones, cytokines, chemokines, and extracellular vesicles carrying various signaling material, such as microRNA (miRNA). We continue with describing the genetic and epigenetic response to circulating factors in the uterus and placenta. In this chapter, we will cover relevant literature relevant to embryo and mother cross-communication.

EMBRYO IMPLANTATION

Alterations in the periconceptive milieu are necessary to support offspring development and can be viewed as nature's adaptive strategy to changes in the maternal environment. Crosstalk between organs is not unique to reproduction and had been extensively documented in various physiological processes including energy metabolism homeostasis, bone formation, immune response, and others. For example, the immune system helper T cells drive the maturation of immune system B cells to produce antibodies, and this process utilizes neuronal dopamine synthesis machinery and dopamine release into synapses [18]. Neuro-immune crosstalk has been shown to be essential for the maintenance of body homeostasis and involves complex interactions between two or more cell types [19]. Metabolic crosstalk between brain areas, in particular the hypothalamus, and peripheral organs such as fat, bone, pancreas, and muscle has been known for the past two decades and several important neurotransmitters as well as adipokines, such as leptin, were demonstrated to control these processes simultaneously [20–23]. Another example is the vascular endothelium, which regulates not only circulation, platelet function, and pathological conditions such as atherosclerosis, but also whole-body related functions, such as insulin, glucose, and lipid levels and endothelial dysregulation contributes to the development of diabetes mellitus and fatty liver [24, 25].

Pregnancy is a highly complex and tightly regulated process, which relies on multiple communication channels between host—the mother—and a newly developed organism—the baby. Pregnancy begins with fertilization, the union of a haploid (23 chromosomes) sperm with a haploid egg, forming a diploid (46 chromosomes), single-cell zygote, which continues to undergo mitotic divisions as it travels through the fallopian tube toward the uterus, forming a blastocyst. As the zygote travels toward the uterus, it continues to differentiate and forms two layers. The mature blastocyst consists of the outer trophectoderm layer (TE) and the inner cell mass (ICM). At the time of implantation, where the blastocyst adheres to the uterine endometrium, the ICM further differentiates into two distinct cell lineages, the epiblast (EPI) and the primitive endoderm (PE). The EPI gives rise to the fetus, and the PE and the TE give rise to the fetal membrane and the placenta, respectively [9, 10, 26]. Pregnancy relies on blastocyst implantation during a narrow window of uterine receptivity called the window of implantation, with implantation outside this window being associated with spontaneous miscarriages. The uterine sensitivity to implantation-competent blastocysts can be classically divided into three stages: pre-receptive, receptive, and refractory. In humans, the pre-receptive phase occurs after ovulation (7 days; early luteal phase), receptive phase (~7–10 days; mid-luteal phase), followed by the nonreceptive phase (~7–10 to 28–30 days; late-luteal phase), until menstruation ensues [9, 10, 27]. Upon closure of the receptive window, the uterus is a hostile milieu, and the blastocyst cannot implant. The uterus is composed of three major tissue compartments: epithelium, stroma, and myometrium. Since the blastocyst attaches to the luminal epithelium, it is perceived as the crucial site for uterine receptivity, transmitting signals to other compartments [28].

The process of implantation is divided into three stages: apposition, attachment (adhesion), and penetration. Implantation is initiated when the blastocyst meets the uterine wall. The first step is the release from the zone pellucida, a glycoprotein layer surrounding the plasma membrane, a process called "zona hatching." Apposition is the loose connection between the blastocyst and the uterine epithelium. Usually, the apposition occurs in a small crypt in the endometrium, aligning the ICM with the uterine wall. Attachment is the furthest and strongest connection, when trophoblasts adhere and penetrate the endometrium, followed by invasion, embedding the embryo in the endometrium. Eventually, the syncytiotrophoblasts, the protrusions of trophoblast cells, come into contact with maternal blood and form chorionic villi, initiating the formation of the placenta [29].

In 1914, Franklin P. Mall at the Department of Embryology of the Carnegie Institution of Washington and his successor, George L. Streeter, classified human development into 23 stages, or "developmental horizons" [30]. Each stage represents an arbitrary point along the timeline of development, like a "freeze-frame" in a film. In this series of events, stage 1 begins with fertilization, when an oocyte is penetrated by a sperm and ends with intermingling of the paternal and maternal chromosomes of the first mitotic division of the single cell. Stage 2 begins with the division of the zygote into two cells and ends with the appearance of the blastocystic cavity. The embryo is 2–3 days postfertilization and 0.1–0.2 mm in diameter. The DNA increases dramatically with each division, yet the cells become progressively smaller, and the volume of the entire protoplasmic masses decreases. The advanced embryo of 16 blastomeres is called a morula as it resembles a mulberry. During stage 3, the blastocystic cavity first appears in the morula and trophoblast cells develop until the embryo contacts the endometrial lining of the uterus, at embryo postfertilization age of ~4–5 days, measuring 0.1–0.2 mm in diameter. Embryos undergo repeated collapse and expansion before their escape from the zona pellucida and the phenomenon is called blastocyst "breathing." When the blastocyst is ~7–8 days old, hatching occurs, the blastocyst enlarges significantly, the trophoblast develops tight junctions, and small fibrillated trophoblast projections appear at the site where attachment usually occurs. Stage 4 is reserved for blastocyst attachment to the endometrial lining of the uterus. In the areas of contact, the syncytiotrophoblast dissolves the endometrial epithelium. The epithelial cytoplasm disappears before the nuclei, resulting in an accumulation of nuclei in these areas and progressive differentiation of the trophoblast occurs. State 5 is when embryos implant into the uterine mucosa. Two distinct layers are evident in the trophoblast: the syncytiotrophoblast—a thicker outer layer and the cytotrophoblast—a thinner inner layer. Stage 5 embryos are seven to 12 days postfertilization and vary from 0.3 to 1 mm in diameter [30]. These multiple steps involved in embryo implantation are a key feature for successful mammalian pregnancy. In most species, the implantation is superficial, whereas in humans, mice, and only a few other species, the implantation is interstitial and thought to be an evolutionarily inherited feature [31].

To enable embryo implantation, the uterus undergoes major structural changes collectively, known as a process of decidualization. In the field of horticulture, the term "deciduous" means falling off at maturity, referring to trees shedding their leaves in autumn. The endometrium increases in thickness and vascularization and, if no pregnancy occurs, the decidual layer is shed off during menstruation. In the event of pregnancy, the decidual layer expands, becomes filled with glycogen and lipids, and the uterine glands grow in size and number. During decidualization, the uterine stromal cells differentiate and assume characteristic polyhedral (flower-like) shapes. A multitude of small microvillous protrusions, referred to as uterodomes (or pinopodes) develop on the apical surface of the luminal uterine epithelium. These microvilli form a single polyhedral shape, which appears only during the window of implantation, and subsequently inter-digitates with the blastocyst. Uterodomes are characterized by the presence of integrins [32, 33]. The basal and lateral membranes also undergo transformation, specifically at various junctions [28]. Blastocyst attachment to the luminal epithelium overlaps with decidualization of the uterine stroma.

Multiple molecular factors govern the complex dynamic process of implantation, including hormones, cytokines, differentially expressed receptors, and signaling molecules in the uterus, placenta, and embryo, and extracellular vesicles carrying cargo to-and-fro, connecting the developing baby and the mother.

Although significant progress has been made in the field of embryo–mother cross-communication, a lot is still unknown about the specific molecular mechanisms controlling this dialogue. Studies have revealed several histological and biomarker candidates for endometrial receptivity; however, consensus still needs to be established [32, 34]. For example, even the presence of characteristic uterodomes awaits to be proven as a significant biomarker [35, 36]. Another challenge is preoccupation with one potential biomarker, resulting in less effort to explore others [32, 34]. In the following section, we will attempt to cover the cumulative knowledge, focusing on studies in humans and animals.

ENDOCRINE INFLUENCE

The implantation of the embryo in the uterus is among the most finely tuned processes, and two ovarian hormones, estrogen (E2) and progesterone (P4) are placed at the top of the hierarchy of factors regulating the chronological transition between events, supporting continuous interaction between the mother and the developing baby [10, 27]. Ovarian E2 and P4 are crucial for the series of events from uterine receptivity, implantation, decidualization, placentation, and finally birth. Both hormones affect a plethora of growth factors, cytokines, transcription factors, lipid mediators, and cell cycle regulators involved during the course of pregnancy [9, 37]. In mice, ovariectomy on day 4 before preimplantation causes an estrogen surge, and continued P4 treatment places the uterus in a state of quiescence, while maintaining implantation competency which is resumed when estrogen levels peak. This "delayed implantation" model is commonly used in research [38]. Interestingly, although P4 is an absolute requirement for implantation in many species, ovarian estrogen increase is not crucial in subhuman primates [10]. In humans, the 28-to 30-day menstrual cycle begins with menses. The proliferative (follicular) phase is under the influence of rising E2 levels from the growing ovarian follicles, leading to the proliferation of epithelium, stroma, and vascular endothelium to regenerate the endometrium. At midcycle, there is a surge of gonadotropins, namely follicle-stimulating hormone (FSH) and luteinizing hormone (LH), leading to ovulation on day 14. The early secretory (luteal) phase is characterized by thickening of the endometrium and the formation of the corpus luteum from the ruptured follicle and subsequent P4 upsurge in preparation for implantation. Increasing E2 levels, superimposed on rising P4 concentrations, define the window of receptivity. In the absence of a viable embryo, hormone withdrawal and menstruation occur. Conversely, an implanting blastocyst secretes chorionic gonadotropin (hCG) to maintain the corpus luteum, and pregnancy ensues [10].

Robust morphological changes in the endometrium—decidualization—are necessary for blastocyst implantation. Uterine glands invaginate the luminal epithelium and become more irregular with a papillary appearance. At birth, in humans, the glands are sparse and slightly depressed into the stroma. At puberty, they extend toward the myometrium and form a coiled network of tubules [37]. Studies in menstruating animals have demonstrated that P4 treatment during neonatal life impairs gland development, severely affecting fertility [39, 40]. Experimental data also suggest that P4 treatment affects the expression of genes central to the endometrial adenogenesis, including members of the *Wnt* gene family [41–43], the expression and adenogenic role of which have been demonstrated in both glands and stroma [44]. The endometrial glands produce and secrete a cocktail of molecules, including nutrients and growth factors important for embryo survival. For example, uterine glands produce leukemia inhibitory factor (LIF) and vascular endothelial factor (VEF), both under the control of ovarian hormones [45, 46] (see details below). Interestingly, several studies highlighted difference in gland secretion contents between fertile and infertile women, supporting the relevance of these gland secretions to a healthy pregnancy [47, 48]. In addition to changes occurring in the glandular epithelium, major changes occur in the stroma, where cells differentiate and assume polyhedral shape, and this process is also driven by the postovulatory rise in P4 level.

In general, both hormones are necessary for fertility, as mice lacking estrogen receptor a (ERa) are hypoplastic and infertile, as well as mice lacking both progesterone receptor A (PR-A) and progesterone receptor B (PR-B) forms [49, 50]. During pregnancy, the expression pattern of ovarian hormones is dynamic and has compartment-specific functions. PR is transiently expressed in the luminal epithelium just prior to embryo implantation, and at the time of embryo implantation expression is rapidly downregulated in the luminal epithelium, while being upregulated in the stroma, persisting throughout decidualization [51]. Mice with epithelial-specific Esr1–/– show abnormal expression of estrogen-responsive genes and subsequent implantation failure [52]. Mice with epithelial- specific Pgr–/– are also non-responsive to P4 treatment and are infertile due to defects in embryo attachment, stromal cell decidualization, and the inability to cease estrogen-induced epithelial cell proliferation [51, 53]. Mechanistically, the infertility in these mice is attributed to low expression of Indian hedgehog (IHH) and leukemia inhibitory factor (LIF), crucial elements for implantation [51, 54]. IHH is primarily expressed in epithelial cells and interacts with its receptors Patched and Smoothened in the stroma, mediating stromal cell proliferation. LIF is a member of the interleukin-6 family of cytokines shown to be essential for uterine receptivity and implantation, binding to its receptor LIFR and partnering with co-receptor gp130 and signaling through signal transducer and activator of transcription 3 (Stat3), both shown to result in implantation failure upon genetic deletion [55, 56]. Clinically, the role of LIF remains nonconclusive, as LIF administration in a relatively small cohort of hyperstimulated women with multiple etiologies of infertility did not improve pregnancy outcome [57]. Another ovarian hormone-dependent factor thought to be involved in implantation and decidualization is Hand2, and its ablation leads to an infertility phenotype, *via* a mechanism upregulating fibroblast growth factor-extracellular signal regulated kinase (FGF-ERK) signaling [58]. Overall, steroid hormones were shown to affect a plethora of downstream factors crucial for pregnancy progression, acting through bidirectional epithelium–stroma communication channel. Nevertheless, several genes, such as homeobox transcription factor Msx1, were shown to be less dependent on E2 and P4 levels [59]. Genetic studies in mice suggested that Msx1 expression is essential for embryo implantation, and subsequent studies in humans revealed that Msx1 protein levels, normally elevated in the secretory phase endometrium, were significantly reduced in endometrial biopsies obtained from women of infertile couples [60].

Progesterone resistance—a rapidly expanding field of research—has been linked with reduced endometrial receptivity [34, 61]. P4 is anti-inflammatory and induces immune tolerance at implantation. Interference with P4 action by using anti-progestins, such as RU-486, causes pregnancy loss and infertility [62]. Furthermore, early increases in P4 level reduces the success of embryo transfer, even when frozen embryos are normally competent in subsequent transfer. There is a two- to day-day temporal window of P4 exposure, when receptivity is optimal. Overall, data suggest that abnormal P4 exposure or resistance leads to embryo–uterus asynchrony. P4 is also responsible for timely downregulation of ERs, an effect linked to increased expression of integrin aub3, which plays a role in blastocyst attachment to the uterus [34, 63]. Clinically, endometriosis has been associated with progesterone resistance, or non-responsiveness to progesterone signaling, guiding a search for suitable biomarkers that underlie this effect [64].

In addition to E2 and P4, another well-studied hormone affecting pregnancy is human chorionic gonadotropin (hCG). hCG is produced by the trophectoderm of the blastocyst, acting as an antagonist to LH, by maintaining the corpus luteum, and helping ovaries to produce P4 necessary for the establishment and progression of pregnancy. Endometrial epithelial cells respond to hCG by expressing cyclooxygenase-2 (COX2) and prostaglandin E synthase (PGES) through activation of the MAPK signaling and Notch1 pathways, promoting the remodeling of the cytoskeleton and decidualization [65–68]. Subsequently, a decrease in hCG and Notch1 levels are necessary for the completion of decidualization, which is accompanied by an increase in the expression of insulin-like growth factor binding protein-1 (IGFBP1) and prolactin (PRL) [37, 67].

PARACRINE INFLUENCE

The cross-communication between the embryo and the mother during early pregnancy has been studied for several decades, Once thought to be a simple medium for sperm and embryo transport, uterine fluid is now considered an important mode for embryo–mother interactions, allowing delivery and/or exchange of cytokines, extracellular vesicles (EVs), nutrients, and mobile RNAs carrying information "to and fro," coordinating the complex process of pregnancy. In the following sections, we will present a summary of the factors released and acting in the vicinity of the maternal reproductive tract and the developing embryo.

Cytokines

Some of the most important changes in the uterine environment during peri-implantation occur in the immune system. Pregnancy presents an "immunological mystery" because two genetically different individuals coexist. Theoretically, the presentation of paternal histocompatibility antigens by the fetus and placenta should provoke the same kind of tissue rejection response seen following organ transplantation. Nevertheless, the maternal immune system adapts to two counteracting processes: on one hand, it prevents infection and on the other hand, it prevents the immuno-rejection and allows the growth of the semi-allogeneic embryo [69, 70]. Abnormal function of the immune cells in the decidua have resulted in various obstetrical complications, such as recurrent pregnancy loss, preterm delivery, and preeclampsia. Within the maternal innate immune system, decidual natural killer cells (dNK), macrophages, and dendritic cells (DCs) comprise a large proportion of the decidual leukocyte population that play an important role in modulating trophoblast invasion, angiogenesis, and vascular remodeling [70]. Studies have shown that first-trimester human decidual leukocytes are primarily NK cells (~70%) and macrophages (~20%), whereas T cell proportions are more variable (~10–20%), with DCs, B cells, and NKT cells being rare [71]. It is known that the extravillous trophoblast lacks classical major histocompatibility complex (MHC) class I or class II molecules, but still has the capacity to activate maternal immune system and produce anti-paternal HLA antibodies during pregnancy [72]. The dNK cell receptor repertoire is skewed toward HLA recognition in the uterus [73] and has been shown to increase the secretion of inflammatory and pro-angiogenic factors, such as IL-1β, IL-6, IL-8, and TNF-α [74]. A recent study demonstrated the dual role of dNK-mediated HLA cycle in both the mounting cytolytic responses during viral infections and in providing immune tolerance to the fetus and facilitating placental growth [75]. NK cells are also involved in the decidual spiral artery remodeling during pregnancy, into a high-capacitance, low-resistance vessel, which is critical to increase blood flow to the placenta and developing fetus [76].

Another prominent leukocyte type, the macrophages, appear to support a variety of processes throughout the length of the gestation. Macrophages contribute to tissue remodeling, and are strong pathogen sensors, suggesting their central role in the inflammatory response to decidual or placental infection. Generally, macrophages can be divided into two categories: classically activated macrophages (M1) and alternatively activated macrophages (M2). M1 macrophages are involved in inflammatory responses, by secreting IL-12, IL-23, reactive oxygen species, and inducing T helper 1 (Th1)-type responses, whereas M2 macrophages have immunosuppressive qualities, by participating in tissue remodeling and promoting T helper 2 (Th2)-type responses. In essence, M1 is a product of the inducible nitric oxide synthase (iNOS) pathway, whereas M2 is a product of the arginase pathway. These two pathways are antagonistic, and their balance affects the micromilieu of the surrounding tissue. The polarization pattern of decidual macrophages is skewed toward M1 during the peri-implantation period, after which they begin to transition to a mixed M1/M2 profile, when the trophoblasts establish attachment into the endometrial lining and invade the uterine stroma. This mixed polarization pattern remains through the uterine vascular remodeling process to establish an adequate placental-fetal blood supply during the first trimester and the early phase of the second trimester of pregnancy. After the placental development is completed, the decidual macrophages shift toward a predominantly M2 phenotype to prevent rejection of the fetus and allow fetal growth until parturition.

Attempts were made to analyze changes in the levels of signaling molecules in maternal blood throughout the course of pregnancy. Using a liquid chip scanning technology, Zhao et. al. analyzed 30 circulating factors at 14 time points in pregnant rats [12]. The technology is based on flexible Multi-Analyte Profiling (xMAP), integrating colored microspheres, laser technology, fluidics, and the latest mathematical signal processors and computer programming algorithms. The signal molecules showed the greatest changes in the third trimester and moderate changes in the first trimester but were relatively stable during the second trimester. During early pregnancy (days 1–7; equivalent to the first trimester of human pregnancy) the levels of luteinizing hormone (LH) and brain-derived neurotrophic factor (BDNF) increased and decreased, respectively. During this period, sperm-egg binding and fusion formed the fertilized egg, which moved from the fallopian tube to the uterus, sending stimulation signals to the endometrium, preparing it for blastocyst implantation. Compared with pre-pregnancy levels, monocyte chemotactic protein 1 (MCP1), interleukin-10 (IL-10), IL-13, and growth-related oncogene (GRO) were elevated at day 5 (equivalent to the second month of human pregnancy). During that time, there was also maternal T helper 1 (Th1) inhibition and Th2 activation, supporting the involvement of the maternal innate and cellular immune response in fetal development, providing mechanisms inhibiting maternal immune rejection of the fetus. Shifting into mid-pregnancy (days 9–19; equivalent to the second trimester of human pregnancy) results in stabilization of circulating signaling molecules. Growth hormone (GH) and leptin levels increased, promoting muscle growth and fuel anabolism. Th1 and Th2 levels were stable, indicating adjustment and growth of the fetal immune system and reduction in the maternal immune rejection of the fetus, avoiding fetal abortion. During late pregnancy (days 21–23; equivalent to the third trimester of human pregnancy), IL-2, IL-6, IL-12p70m IL-18, Interferon-g (IFN-g), leptin, and GRO levels were significantly increased, whereas levels of adrenocorticotropic hormone (ACTH) and BDNF decreased. At this time, maternal Th1 was rapidly activated, implying immune protection of mother and fetus during childbirth, in preparation for delivery. Previous studies have also shown that IL-2m IL-6, and IL18 are related to uterine expansion. Postpartum was marked by increased vascular endothelial growth factor (VEGF) level, possibly repairing the wounded tissue. Prolactin (PRL) was increased, promoting and maintaining lactation. Over 30 cell types were involved in the intercellular "wireless" communication network. Overall, this study showed common changes in the level of signaling molecules in maternal serum (such as cytokines, chemokines, and hormones), at various time points throughout pregnancy from pre-implantation to post-delivery, in rats. The most pronounced immune responses detected were Th1 transformation to Th2 at day 5, the so-called "Th2 phenomenon," enhancing humoral immunity and inhibiting innate and cellular immunity, to provide fetal protection; and the reverse Th2 transformation to Th1 at day 7, when the fetal heart is fully developed, aiming to activate the innate immunity in the embryo. Several characteristic molecules were detected throughout pregnancy and further investigation is necessary to explore their relevance to human pregnancy.

In humans, several paracrine factors have been shown to be involved in various stages of embryonic implantation and forward to pregnancy progress. Colony-stimulating factor-1 (CSF-1) promotes differentiation of human trophoblast cells into syncytiotrophoblast cells and leads to the production of placental lactogen [77]. Several metalloproteinases have been reported in connection with the invasive ability of the fetal trophoblast, in particular MMP 2 and 9 [78, 79]. Trophoblastic MMPs are regulated in response to IL-1b, tumor necrosis factor alpha (TNFa), IL-1a, macrophage colony-stimulating factor (MCSF), transforming factor b (TGFb), leptin, and endothelial growth factor (EGF), which are secreted from different cells at the fetal–maternal interface. Endometrial extracellular matrix (ECM) remodeling is essential for successful implantation and placentation, and multiple MMPs as well as their substrates are involved in this process. For example, MMP-14 and ADAM10 present in endometrium-derived exosomes act on IL-8, TGFb, CD44, Notch, and its ligand DLL1, promoting their bioactivity [80–83]. A recently published comprehensive review summarizes the roles of MMPs in embryo–mother crosstalk [84].

Extracellular Vesicles

Extracellular vesicles (EVs) are novel players in the field of embryo–mother cross-communication, recently receiving a lot of attention from the scientific community. EVs are enclosed compartments with versatile potential, varying in their biogenesis, size, and cargo [13, 48]. Their small size facilitates trafficking between local sites. EVs activate surface receptors on target cells, merging their membrane contents and releasing cargo. EV cargo—proteins, lipid, genetic material (DNA, RNA, miRNA, and other RNA forms)—reflects the physiological state of the cell or its origin, and this property is exploited in searches for biomarkers for various pathologies, such as cancer [85–87]. Initially, EVs were thought to participate only in the degradation of waste, but recent studies shed light on their diverse properties and roles. Based on their origin and size, EVs are generally subdivided into three classes: apoptotic bodies, microvesicles, and exosomes. The largest are apoptotic bodies (1–5 mm), which form following cytoplasm membrane blebbing in cells undergoing programmed cell death, or apoptosis. Molecular markers include: phosphatidylserine (PS), which serves as an "eat me" signal for phagocytes, although PS has been described in healthy cells as well; thrombospondin; C3b complement protein; VDAC1, a protein forming ionic channels in the mitochondrial membrane; and calreticulin, an endoplasmic reticulum protein [13, 88]. It should be noted that, generally, EV markers are not exclusive, and are often present on more than one EV type. Owing to this overlap, the list of markers indicates proteins enriched in, rather than exclusive to a particular vesicle.

Microvesicles (100–1000 nm) were first described as subcellular material originating from platelets in normal plasma and serum. The molecular markers of microvesicles are ADP-ribosylation factor 6 (ARF6), integrins, selectins, and CD40 ligand. Microvesicles have been studied mainly for their role in blood coagulation and cancer cell-to-cell communication, where they are called oncosomes. Unlike apoptotic bodies and microvesicles, exosomes are small, virus-sized particles (30–150 nm), formed by inward budding of the cytoplasmic membrane. Exosomes are derived from the endolysosomal pathway and represent a more homogeneous population of vesicles than microvesicles. For a long time, they were considered to be nanodust, or dust in electron microscopy. This perception changed dramatically in the past years and their role evolved from debris bins to biologically active particles [89, 90]. The immunomodulatory role of exosomes is the most studied [87, 91], followed by angiogenesis, thrombosis [92], and pathologies, such as cancer [88]. The molecular markers of exosomes include: CD63, CD9, CD81, ALIX, TSG101, flotillin-1, HSC70, and syntenin-1 [13]. Cargo sorting into exosomes involves the endosomal sorting complex required for transport (ESCRT) and other associated proteins.

In general, all EVs were shown to have both biological and pathological roles, serving as messengers communicating information cell-to-cell. EVs participate in the regulation of immune responses, triggering adaptive immune response or suppressing inflammation [93]. EVs confer suppressed immune responses *via* several mechanisms: enhancing the function of T cells, suppressing natural killer (NK) and CD8+ activity, and inhibiting monocyte differentiation into dendritic cells (DCs). The effects on immune activation involve proliferation of hematopoietic stem cells, activation of monocytes, B cells, and NKs. Beyond immunomodulation, EVs contribute to synaptic plasticity, delivering neurotransmitter receptors, playing a role in tissue regeneration following injury, and modulating cell phenotype [85].

In the growing field of embryo–mother cross communication, EVs attracted interest only recently. Data have accumulated showing a key role of EVs at preconception from gamete maturation to implantation and throughout pregnancy [94]. Ng et al. first described the production of EVs by the human endometrial epithelial cell line ECC1 containing a specific subset of micro-RNAs (miRNAs), not detectable in parent cells [95] and later verified to be present in human uterine fluid. Burns et al., demonstrated the presence of EVs in the uterine fluid of pregnant sheep, positive for exosomal markers CD63 and HSP70, containing small RNAs (sRNAs) and miRNAs [96]. Greening et al. demonstrated that the proteome of highly purified exosomes from human endometrial epithelial cells is subject to steroid hormonal regulation—estrogen and progesterone—and varies with the menstrual cycle [97]. Villela et al. performed an elegant study showing internalization of has-miR30d by mouse embryos *via* the trophectoderm, resulting in an indirect overexpression of adhesion-related genes, namely *Itgb3*, *Itga7*, and *Cdh* [98]. In this study, treatment of mouse embryos with miR-30d resulted in increased embryo adhesion [98]. The same group also showed that miR-30d deficiency results in contrasting implantation rate reduction and impaired fetal growth [99], and heterogeneous nuclear ribonucleoprotein C1 (hnRNPC1) has been implicated in the mechanism of cell-to-cell communication [100]. These findings support a model in which maternal endometrial miRNAs act as transcriptomic modifiers of the preimplantation embryo. Analysis of human endometrial liquid biopsy (ELB) material in both natural, and hormonal replacement therapy (HRT) cycles revealed a panel of differentially expressed miRNAs, including members of the miR-30 family [101]. Recently, embryos were shown to "talk back" *via* release of progesterone-induced protein (PIBF) packaged in EVs, modulating the maternal immune response [11, 102]. The presence of EVs in the uterine fluid implies endometrium–embryo crosstalking, although these studies require further, thorough exploration.

EV research is a rapidly growing but still immature field, facing several challenges. Among the notable challenges are a lack of nomenclature for distinct types of EVs based on cellular origin. The terms "exosome" and "microvesicle" have been used interchangeably in many published studies because of as-yet incomplete understanding of EV biogenesis, inconsistencies in purification protocols, and vague characterization [86]. A lot remains unknown regarding the biogenesis, route, and function of EVs in reproductive biology. EVs are considered to be attractive pharmaceutical targets, as well as being capable of direct exploitation as potential therapeutic agents for tissue regeneration and immune response modulation [86].

GENETIC CHANGES—RECEPTORS AND SIGNALING MOLECULES

Uterus

The molecular signature of the uterus at the time of implantation shows elevated expression of a number of factors.

Mucin 1 (MUC 1), a highly glycosylated, polymorphic, mucin-like protein, is secreted by the endometrial luminal epithelium. It is considered to be a "barrier to implantation." In humans, it was shown to be expressed in the luteal phase and pre-implantation phase in a progesterone-dependent manner [103–105]. MUC 1 is more abundant in fertile than in infertile women [106]. In baboons, it was also shown to be progesterone- rather than estrogen-dependent, serving as a marker of the pre-implantation phase [107]. Of interest is a recent study investigating similarity between term pregnancy in marsupials and eutherian mammals, which found that key markers of implantation, including Mucin 1, heparin-binding EGF-like factor (HBEGF), and a range of proinflammatory factors, including IL-6, tumor necrosis factor (TNF), and cyclooxygenase 2 (COX2), are consistent between species and embryo implantation has been suggested. There are transcriptome-wide similarities between the marsupial attachment reaction and implantation in rabbits and humans [108]. Specifically, the marsupial study observed an overlap in the key implantation biomarker osteopontin, which is the only consistent implantation window biomarker identified from five human microarray experiments [109].

Osteopontin (OPN, also known as secreted phosphoprotein 1, SPP1) is a glycosylated phosphoprotein expressed in endometrial epithelium and implicated in attachment and signaling roles at the embryo–epithelium interface [110, 111]. OPN is also a bone-associated protein, produced by several bone cell types, osteoblasts and osteoclasts, and extra-osseous cells, skin, kidney, and lung. Due to differences in post-translational modification, OPNs range in molecular mass from 41 to 75 kDa, OPN was suggested to have cell type-specific structure and function [112, 113]. OPN plays a major role in various physiological processes, including bone remodeling, immune regulation, inflammation, and vascularization, as well as pathologies, such as cancer, with several studies evaluating OPN as a biomarker in tumor progression [112]. Comparative global gene expression studies have illustrated an increase in OPN level following LH surge in the human endometrium [114]; OPN was identified in the vicinity of uterodomes and in decidualizing stroma [115]. Moreover, although OPN-null mice are fertile, they exhibit reduced pregnancy rates

[116]. Mechanistically, OPN interacts with integrins and is classified as a member of the small integrin-binding ligand N linked glycoproteins (SIBLINGs), such as bone sialoprotein (BSP), dentin matrix protein (DMP1), and others. Binding of OPN to integrins activates integrin receptors and cytoskeletal proteins, subsequently promoting focal adhesion in embryo trophectoderm, although the functional significance of this interaction in endometrial receptivity requires further investigation [117].

Integrins, transmembrane glycoproteins with a and b subunits, mediate several processes, including cell-to-cell and cell-to-extracellular matrix (ECM) adhesion. Some integrins are constitutively expressed in the luminal epithelium, whereas others are regulated during the menstrual cycle in a spatial and temporal fashion [28, 118, 119]. Three integrins were discovered that have unique expression patterns which correlated with window of receptivity in women: a1b1, a4b1, and avb3 [32, 33]. Of these, avb3 is the best characterized. The avb3 integrin appears on the apex of luminal and glandular cell surfaces, coinciding with the opening of the window of implantation, and its expression persists into pregnancy with expansion of the decidua [63]. The appearance of avb3 integrin on the apical surface is due to its presence in the subnuclear secretory granules [63]. Expression of the intact heterodimer is regulated by transcription factor HOXA10, the expression of which, together with avb3 integrin, is altered under conditions including adenomyosis, polycystic ovary syndrome, and endometriosis [32]. However, these observations are undermined by several studies showing no reliable link between integrins and fertility [120, 121].

Homeobox protein A10, A11 (HOXA10, HOXA11) play a role in decidualization, and their deletion in mice results in implantation defects [122, 123]. HOXA10-null mice produce normal numbers of embryos that are able to implant in wild-type surrogate mice, whereas wild-type embryos from the surrogate mice cannot implant in the HOXA10-null mice [122, 124]. HOXA11–/– mice show similar phenotype related to implantation [125]. In the human endometrium, HOXA10 is expressed by both epithelial and stromal cells in a menstrual cycle-dependent manner and is regulated by progesterone [126]. Microarray analysis of murine endometrium transfected with HOXA10 cDNA identified 40 downstream target genes including clusterin (Clu), phosphoglycerate 3-dehydrogenase (3-Pgdh), and tumor-associated calcium signal transducer 2 (Tacstd2) [127].

Leptin is an adipokine which regulates energy metabolism. Released from the adipose tissue, leptin travels in circulation until it reaches leptin receptors in the brain, located in hypothalamic nuclei and other sites [20, 128, 129]. Complete absence of leptin (ob/ob mice) is not lethal but results in early-onset obesity, stunted skeletal and brain growth, extreme insulin resistance, hyperphagia, a compromised immune system, and infertility [130]. Interestingly, fertility can be restored in ob/ob mice by exogenous leptin administration, which is characterized by increased LH and FSH levels. However, fertility is not restored by food restriction in these mice, suggesting that leptin affects reproductive system independently of metabolism [131, 132]. Leptin acts on the hypothalamic-pituitary axis to affect steroid hormone release. In pregnant mice and humans, the placenta has also been found to be a major site of leptin expression [130]. Since there is also an increase in maternal leptin levels in adipose tissue, it remains unclear what is the contribution of placenta-derived leptin to embryonic development [133]. In addition to being a site of leptin synthesis, the placenta is a site of abundant leptin receptor expression and it has been proposed to mediate cues regarding energy metabolism status between the mother and the baby [134].

Placenta

The placental interface mediates interaction between the mother and fetus, and efforts were made to analyze changes in placental gene expression [135]. A recent study carried out single-cell transcriptomics on the villous tissue from two human term placentas, obtaining 87 single-cell transcriptomes. These transcriptomes were grouped into three large clusters. The data were further complemented by two transcriptomes of a syncytiotrophoblast, collected from a single placenta by laser microdissection, the transcriptomes of primary undifferentiated endometrial stromal fibroblast (ESF), and the transcriptome of *in vitro* primary differentiated decidual cells. Single-cell data were aligned with tissue-level data to estimate cell origin and the top 25% (2108) most highly expressed genes were found to comprise 80% of the total aligned placental mRNAs. Trophoblast cells at term were concluded to be the most abundant cell type.

Single-cell gene expression profiles were segregated into five clusters, based on combinations of known trophoblast markers (KRT7, KRT8, GCM1, and CYP19A1) and diagnostic genes with > 10-fold higher expression than in uterine and immune cells.

EPIGENETIC CHANGES

Epigenetic regulation is defined as heritable changes in chromosomes without changes in DNA sequence. These changes include DNA methylation, histone modification, and presence of non-coding RNA. In other words, epigenetics covers mechanisms that can perpetuate alternative gene activity states in the context of the same DNA sequence [136]. DNA methylation, the best-characterized epigenetic modification, is achieved by addition of a methyl group ($-CH_3$) to the fifth carbon of the cytosine ring to form 5′-methyl cytosine, during which S-adenosylmethionine acts as the methyl donor. DNA methylation is catalyzed by DNA methyltransferases (DNMTs): DNMT1, DNMT3A, DNMT3B, and DNMT3L. DNMT1 is a methylation-maintaining enzyme responsible for restoring the hemi-methylated sites to full methylation after DNA replication, whereas DNMT3A and DNMT3B are mainly involved in the methylation of new sites, called *de novo* methylation. In humans and other mammals, DNA methylation occurs predominantly on the cytosine base of a cytosine guanine (CG) dinucleotide in the DNA sequence. These CG dinucleotides are often clustered in small stretches of DNA, called CpG islands, which are often associated with promoter regions and at the 5′ end of the gene. In general, CpG islands in gene promoters are "kept free" of methylation, which allows active gene transcription, as their methylation would silence gene expression. However, most CpG sites outside the promoter region (transposons, gene bodies, intergenic regions) are methylated, suggesting a role for methylation in the global maintenance of the genome. The epigenetic modifications can be reversed biochemically, and, many years after the finding of DNMT, the TET family proteins have been shown to be responsible for DNA demethylation [137–140].

Histone modification is another epigenetic mechanism. Histones are basic proteins acting as spools around which DNA winds, packaging it into structural units, called nucleosomes. A histone octamer consisting of two copies of each of the four core histones (H2A, H2B, H3, and H4), around which approximately 146 bp of the DNA winds, comprises a nucleosome. It has been shown that histones are subject to numerous covalent modifications, including methylation, acetylation, phosphorylation, sumoylation, glycosylation, and ubiquitination, at specific tails of selected amino acids. A number of enzymes are involved in this process, including histone methyltransferases (HMTs), acetyltransferases (HATs), kinases, and ubiquitin ligases functioning as writers, as well as erasers, such as histone demethylases, deacetylases (HDACs), and phosphatases, capable of removing modification marks from the histone tails. These modifications impose either transcriptionally repressive or transcriptionally permissive chromatin structures. For instance, histone acetylation usually results in active genes as does the di- or trimethylation of lysine residue 4 in histone H3 (H3K4me2, H3K4me3), whereas H3K9me2/3 and H3K27me3 modifications repress gene expression. In general, unlike DNA methylation, which is believed to confer a more stable and long-term silencing mechanism, various histone modifications seem to exert short-term, flexible regulation important for the plasticity of development [140–144].

DNA sequence variation and epigenetics are inextricably linked. Chromatin states can influence transcription factor binding, and DNA sequence polymorphism influences chromatin states. Furthermore, multiple epigenetic machineries often act together to stabilize heritable states [145, 146]. Recent research revealed multiple examples of environment effect on the epigenetic signature in model organisms and in humans in both physiological and pathological contexts [147–149]. Overall, it becomes apparent that DNA is not the "book of life" and phenotypes depend on specific combinations of genome composition, epigenetic components, and environmental inputs [136].

During early embryo development, the highly methylated paternal genome is hypomethylated by TET enzymes and this reprogramming is meant to erase memory and increase diversity [150]. The cells acquire naïve pluripotency and are primed to exit pluripotency with superior differentiation potential. Tissues and developmental lineages have distinct DNA methylation patterns [151], which arise during early post-implantation and are mediated by DNMT3A and DNMT3B [152], although the precise mechanisms of their generation are unknown. In addition, dynamic changes in histone states during pre-implantation, fertilization, and post-implantation have been documented using advanced technologies, allowing a panoramic view of the landscape of H3K4me3, a histone hallmark for transcription initiation [153, 154]. Now, there is a better understanding of the chromatin architecture, facilitated by a shift from microscopy to a genome-wide chromatin description [155]. Delayed implantation model (ovariectomy before the pre-implantation ovarian E2 surge which is maintained by the injection of P4) studies have shown more heterochromatin in the dormant blastocyst and this dormant state can be mimicked by manipulation of the c-Myc and mTOR pathways [156, 157]. A recent report demonstrated that the expression of inflammatory response genes in the human blastocyst stage is associated with open chromatin accessibility in their regulatory regions [158]. Such studies are of great importance, as zygotic genome activation is different between mice and men, and the dynamics of the chromatin regulatory landscape during human early embryogenesis remains largely unknown.

Epigenetic regulation occurs not only in the embryo, but also in the uterus, increasing uterine receptivity. E2 and P4 are the two main ovarian hormones regulating physiological events during pregnancy. Aberrant DNA methylation of the CpG island in the promoter region of ER or PR has been reported as a marker for endometrial carcinoma, suggesting regulation of the uterus by epigenetic mechanisms [159–161]. Moreover, there is an effect of folate deficiency on the promoter DNA methylation and gene expression of both receptors [162]. Factors critical for uterine receptivity, such as HOXA10 and MUC1, have been shown to be controlled by hormone-dependent DNA methylation, which is associated with infertility in women [140, 163], and DNA methylation changes in the endometrium are correlated with gene expression during the transition from the pre-receptive to the receptive phase in women [164, 165]. Interestingly, comparison of changes in transcriptomes and corresponding DNA methylomes from the same samples revealed an association of DNA methylation and gene expression in a number of loci important in endometrial biology [165], suggesting interplay between hormones and the uterus at the level of the epigenome. Several studies have addressed changes in histone acetylation, methylation, and ubiquitination of the uterus and placenta during various stages of pregnancy [166, 167]. Transcription factors, including Hand2 and Msx2, known to be involved in blastocyst attachment, have been shown to recruit components of epigenetic machinery, such as the polycomb complex [168]. In addition, the small, non-coding RNAs, miRNAs, and circular RNAs (circRNAs), are stably detected in the circulation and EVs. For example, miR-30d is upregulated during the acquisition of receptivity in the endometrium [169], its overexpression inducing changes in the transcriptomic and proteomic of the endometrial epithelium [170], and it is involved in the interaction between the embryo and the uterus, as it is secreted by the uterus and taken up by the embryo [98]. Exosome-associated and free hsa-miR-30d was internalized by embryos *via* the trophectoderm, resulting in an indirect overexpression of genes encoding for certain molecules involved in the murine embryonic adhesion phenomenon—*Itgb3*, *Itga7*, and *Cdh5* [98]. Other miRNA families shown to be important for implantation and potentially mediate embryo–mother dialogue are miR200, Let7, and miR-17-92 clusters [171]. Overall, there is strong evidence that DNA methylation, histone modification, and non-coding RNA components of the epigenetic machinery do not function independently, but rather interact with each other in an interdependent manner.

CONCLUSIONS

For a healthy outcome for the mother and the baby, continuous molecular dialogue is crucial. The language is based on endocrine, paracrine, and autocrine signs. Despite much accumulated knowledge, a lot remains to be elucidated regarding how circulating factors affect genes in the developing fetus and

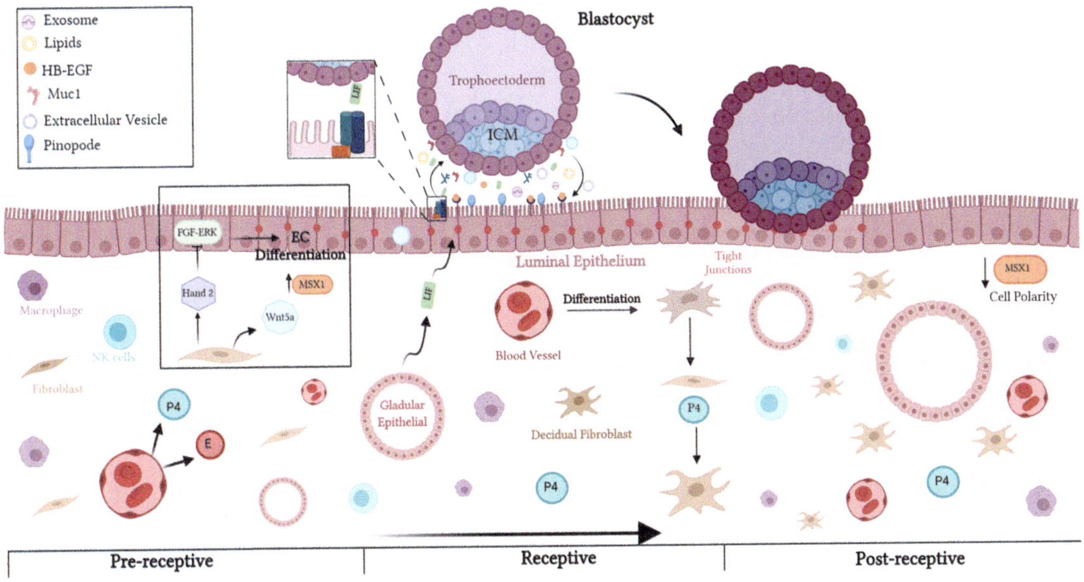

FIGURE 30.1 Schematic representation of the materno–fetal communication during the secretory phase.

how genetic and epigenetic changes are translated into "readable" information. Estrogen, progesterone, and downstream effectors govern differentiation of stroma and remodeling of the endometrium, making it receptive for embryo implantation. The embryo sends a variety of signals to the mother, in the form of EVs carrying miRNA and other material, and the mother responds; see the schematic representation in Figure 30.1 of the crosstalk communication summary during the secretory phase. The exact routes of communication are not well understood and they must be dissected in detail. A better understanding of the physiological mechanisms and the molecules implicated in the mother–baby crosstalk will allow the development of reliable biomarkers for endometrial receptivity, enabling the exploitation of these newly discovered molecules to treat unexplained infertility and to increase healthy birth rate.

REFERENCES

1. Young LE, et al. *Rev Reprod*. 1998;3(3):155–163.
2. Li Y, et al. *Epigenetics*. 2019;14(9):850–876.
3. Ellington JE, et al. *Biol Reprod*. 1993;48(4):851–856.
4. Thomas PG, et al. *Am J Vet Res*. 1995;56(12):1657–62.
5. Lee KF, et al. *Biochem Biophys Res Commun*. 2002;292(2):564–570.
6. Fazeli A, et al. *Biol Reprod*. 2004;71(1):60–65.
7. Long EL, et al. *Biol Reprod*. 2003;69(2):469–474.
8. López-Úbeda R, et al. *PLoS One*. 2015;10(6):e0130128.
9. Dey SK. *J Clin Invest*. 2010;120(4):952–955.
10. Cha J, et al. *Nat Med*. 2012;18(12):1754–1767.
11. Szekeres-Bartho J, et al. *Front Immunol*. 2018;9:2890.
12. Zhao M, et al. *Reprod Biol Endocrinol*. 2019;17(1):40.
13. Simon C, et al. *Endocr Rev*. 2018;39(3):292–332.
14. Jain T, et al. *Fertil Steril*. 2019;111(3):477–488.
15. Miller JE, et al. *Oncotarget*. 2017;8(4):7138–7147.
16. Levine AD, et al. *JAMA*. 2017;317(12):1272–1273.
17. De Geyter C, et al. *Hum Reprod*. 2018;33(9):1586–1601.

18. Qi H. *Nature.* 2017;547(7663):288–290.
19. Jakob MO, et al. *Front Immunol.* 2020;11:308.
20. Karsenty G, et al. *Nature.* 2012;481(7381):314–20.
21. Oury F, et al. *Cell.* 2011;144(5):796–809.
22. Karsenty G, et al. *Cell.* 2016;164(6):1248–56.
23. Idelevich A, et al. *J Clin Invest.* 2018;128(6):2626–2641.
24. Pi X, et al. *Circ Res.* 2018;123(4):477–494.
25. Kim JA, et al. *Circulation.* 2006;113(15):1888–1904.
26. Hirota Y, et al. *Nat Med.* 2010;16(5):529–531.
27. Su RW, et al. *Adv Anat Embryol Cell Biol.* 2015;216:189–213.
28. Murphy CR. *Cell Res.* 2004;14(4):259–267.
29. Díaz-Gimeno P, et al. *Fertil Steril.* 2013;99(2):508–517.
30. Virtual Human Embryo (HEIRLOOM) - The Carnegie Stages of Human Development. [Online]. Available: http://virtualhumanembryo.lsuhsc.edu/HEIRLOOM/Stages/Stages_intro.html. [Accessed: 03-Jun-2020].
31. McGowen MR, et al. *Int J Dev Biol.* 2014;58(2–4):155–161.
32. Lessey BA. *Fertil Steril.* 2011;96(3):522–529.
33. Lessey BA. *J Reprod Immunol.* 2002 55(1–2):101–112.
34. Lessey BA, et al. *Fertil. Steril.* 2017;108(1):19–27.
35. Quinn C, et al. *Fertil. Steril.* 2007;87(5):1015–1021.
36. Usadi RS, et al. *Fertil. Steril.* 2003;79(4):970–974.
37. Massimiani M, et al. *Int J Mol Sci.* 2020;21(1):23.
38. Yuan J, et al. *Nat Commun.* 2018;9(1):603.
39. Gray CA, et al. *Biol Reprod.* 2001;65(5):1311–1323.
40. Cooke PS, et al. *Biol Reprod.* 2012;86(3):63.
41. Hayashi K, et al. *Biol Reprod.* 2011;84(2):308–319.
42. Dunlap KA, et al. *Biol Reprod.* 2011;85(2):386–396.
43. Franco HL, et al. *FASEB J.* 2011;25(4):1176–1187.
44. Mericskay M, et al. *Development.* 2004;131(9):2061–2072.
45. Kojima K, et al. *Hum Reprod.* 1995;10(7):1907–1911.
46. Torry DS, et al. *J Assist Reprod Genet.* 2007;24(7):303–315.
47. Boomsma CM, et al. *Hum Reprod.* 2009;24(6):1427–35.
48. Zhang Y, et al. *Trend Mol Med.* 2017;23(7):604–614.
49. Lydon JP, et al. *Genes Dev.* 1995;9(18):2266–2278.
50. Fernandez-Valdivia R, et al. *Genesis.* 2010;48(2):106–113.
51. Franco HL, et al. *FASEB J.* 2012;26(3):1218–1227.
52. Wipawee W, et al. *Proc Natl Acad Sci U S A.* 2010;107(45):19272–19277.
53. Franco HL, et al. *Semin Cell Dev Biol.* 2008;19(2):178–186.
54. Wang X, et al. *Nat Commun.* 2018;9(1):4421.
55. Song H, et al. *Reproduction.* 2006;131(2):341–349.
56. Sun X, et al. *Mol Endocrinol.* 2013;27(9):1492–1501.
57. Brinsden PR, et al. *Fertil Steril.* 2009;91(4 SUPPL.):1445–1447.
58. Mestre-Citrinovitz AC, et al. *PLoS One.* 2015;10(4):e0124756.
59. Daikoku T, et al. *Dev Cell.* 2011;21(6):1014–1025.
60. Bolnick AD, et al. *Hum Reprod.* 2016;31(9):2042–2050.
61. Vercellini P, et al. *Fertil Steril.* 2016;106(7):1552–1571.e2.
62. Baulieu EE. *Science.* 1989;245(4924):1351–1357,
63. Lessey BA, et al. *J Clin Invest.* 1992;90(1):188–195,
64. Rogers PAW, et al. *Reprod Sci.* 2009;16(4):335–346.
65. Srisuparp S, et al. *Biol Reprod.* 2003;68(2):457–464.
66. Lim H, et al. *Cell.* 1997;91(2):197–208.
67. Afshar Y, et al. *Endocrinology.* 2012;153:2884–96.
68. Banerjee P, et al. *Endocrinology.* 2009;150(9):4326–4337.
69. Velázquez MML, et al. *Anim Reprod Sci.* 2019;206:1–10.
70. Erlebacher A. *Annu Rev Immunol.* 2013;31(1):387–411.
71. Trundley A, et al. *Tissue Antigens.* 2004;63(1):1–12.
72. Lee J, et al. *Am J Reprod Immunol.* 2013;70(2):162–175.
73. Verma S, et al. *Eur J Immunol.* 1997;27(4):979–983.
74. Rajagopalan S, et al. *PLoS Biol.* 2006;4(1):0070–0086.

75. Tilburgs T, et al. *Proc Natl Acad Sci U S A*. 2015;112(43):13312–13317.
76. Moffett A, et al. *J Clin Invest*. 2014;124(5):1872–1879.
77. Strauss, J. F. III, and Lyttle, C. R. *Uterine and Embryonic Factors in Early Pregnancy*. Springer, 2012.
78. Shimonovitz S, et al. *Am J Obstet Gynecol*. 1994;171(3):832–838.
79. Huppertz B, et al. *Cell Tissue Res*. 1997;291(1):133–148.
80. Cuman C, et al. *Reproduction*. 2014;147(3):R75–86.
81. Arici A, et al. *J Clin Endocrinol Metab*. 1998;83(5):1783–1787.
82. Tam EM, et al. *J Biol Chem*. 2004;279(41):43336–43344.
83. Campbell S, et al. *Hum Reprod*. 1995;10(2):425–430.
84. Latifi Z, et al. *J Cell Physiol*. 2018;233(6):4530–4545.
85. Van Niel G, et al. *Nat Rev Mol Cell Biol*. 2018;19(4):213–228.
86. El Andaloussi S, et al. *Nat Rev Drug Discov*. 2013;12(5):347–357.
87. Maia J, et al. *Front Cell Dev Biol*. 2018;6:18.
88. Voloshin T, et al. *Cancer Microenviron*. 2014;7(1–2):11–21.
89. Muller L. *HNO*. 2020;68:56–59.
90. Raposo G, et al. *J Exp Med*. 1996;183(3):1161–1172.
91. Kim HK, et al. *Eur J Cancer*. 2003;39(2):184–191.
92. Janowska-Wieczorek A, et al. *Blood*. 2001;98(10):3143–3149.
93. Chaput N, et al. *Semin Immunopathol*. 2011;33(5):419–440.
94. Machtinger R, et al. *Hum Reprod Update*. 2016;22(2):182–193.
95. Ng YH, et al. *PLoS One*. 2013;8(3):e58502.
96. Burns G, et al. *PLoS One*. 2014;9(3):e90913.
97. Greening DW, et al. *Biol Reprod*. 2016;94(2):38.
98. Vilella F, et al. *Development*. 2015;142(18):3210–3221.
99. Balaguer N, et al. *Am J Obstet Gynecol*. 2019;221(1):46.e1–46.e16.
100. Balaguer N, et al. *Mol Hum Reprod*. 2018;24(8):411–425.
101. Grasso A, et al. *J Clin Endocrinol Metab*. 2020;105(3):dgz146.
102. Pallinger E, et al. *Sci Rep*. 2017;7(2016):1–8.
103. Aplin JD, et al. *Ann NY Acad Sci*. 1994;734(1):103–121.
104. Hey NA, et al. *J Clin Endocrinol Metab*. 1994;78(2):337–342.
105. Wu F, et al. *Reprod Biol Endocrinol*. 2018;16(1):60.
106. Margarit L, et al. *J Clin Endocrinol Metab*. 2010;95(12):5320–5329.
107. Hild-Petito S, et al. *Biol Reprod*. 1996;54(5):939–47.
108. Griffith OW, et al. *Proc Natl Acad Sci U S A*. 2017;114(32):E6566–E6575.
109. Aghajanova L, et al. *Semin Cell Dev Biol*. 2008;19(2):204–211.
110. Berneau SC, et al. *Cells*. 2019;8(5):432.
111. Craciunas L, et al. *Hum Reprod Update*. 2019;25(2):202–223.
112. Hao C, et al. *Int J Mol Med*. 2017;39(6):1327–1337.
113. Christensen B, et al. *J Biol Chem*. 2007;282(27):19463–19472.
114. Kao LC, et al. *Endocrinology*. 2002;143(6):2119–2138.
115. Quenby S, et al. *Reprod Biomed Online*. 2007;14(2):224–234.
116. Weintraub AS, et al. *Pediatr Res*. 2004;55(3):419–424.
117. Davidson LM, et al. *Birth Def Res C*. 2016;108(1):19–32.
118. Singh H, et al. *J Anat*. 2009;215(1):3–13.
119. Bowen JA, et al. *Proc Soc Exp Biol Med*. 2000;223(4):331–343.
120. Creus M, et al. *Hum Reprod*. 1998;13(12):3460–8.
121. Ordi J, et al. *J Clin Endocrinol Metab*. 2003;88(5):2119–2125.
122. Benson GV, et al. *Development*. 1996;122(9):2687–2696.
123. Gendron RL, et al. *Biol Reprod*. 1997;56(5):1097–1105.
124. Satokata I, et al. *Nature*. 1995;374(6521):460–463.
125. Hsieh-Li HM, et al. *Development*. 1995;121(5):1373–1385.
126. Taylor HS, et al. *J Clin Invest*. 1998;101(7):1379–1384.
127. Vitiello D, et al. *Reprod Sci*. 2008;15(5):529–535.
128. Idelevich A, et al. *Bone*. 2018;115:31–42.
129. Nectow AR, et al. *Cell*. 2017;170(3):429–442.e11.
130. Herrid M, et al. *Physiol Res*. 2014;63:543–557.
131. Chehab FF, et al. *Nat Genet*. 1996;12(3):318–320.
132. Mounzih K, et al. *Endocrinology*. 1997;138(3):1190–3.

133. Thomas L, et al. *J Endocrinol.* 2001;169(3):465–476.
134. Hassink SG, et al. *Pediatrics.* 1997;100(1):E1.
135. Pavličev M, et al. *Genome Res.* 2017;27(3):349–361.
136. Cavalli G, et al. *Nature.* 2019;571(7766):489–499.
137. Branco MR, et al. *Nat Rev Genet.* 2012;13(1):7–13.
138. Bird A. *Nature.* 2007;7143:396–398.
139. Weaver ICG, et al. *Nat Neurosci.* 2004;7(8):847–854.
140. Kong S, et al. *Cell Mol Life Sci.* 2019;76(24):4813–4828.
141. Maze I, et al. *Science (80-.).* 2010;327(5962):213–216.
142. Lawrence M, et al. *Trend Genet.* 2016;32(1):42–56.
143. Esteller M. *Nat Rev Genet.* 2007;8(4):286–298.
144. Reik W. *Nature.* 2007;447(7143):425–432.
145. Allshire RC, et al. *Nat Rev Mol Cell Biol.* 2018;19(4):229–244.
146. Von Harsdorf R, et al. *Cell.* 2014;158(1):110–120.
147. De Rubeis S, et al. *Nature.* 2014;515(7526):209–215.
148. Siklenka K, et al. *Science (80-.).* 2006;350(6261):aab.
149. Huypens P, et al. *Nat Genet.* 2016;48(5):497–499.
150. Lee HJ, et al. *Cell Stem Cell.* 2014;14(6):710–719.
151. Ramsköld D, et al. *PLoS Comput Biol.* 2009;5(12):e1000598.
152. Smallwood SA, et al. *Nat Genet.* 2011;43(8):811–814.
153. Zhang B, et al. *Nature.* 2016;537(7621):553–557.
154. Wang C, et al. *Nat Cell Biol.* 2018;20(5):620–631.
155. Ladstätter S, et al. *J Cell Biol.* 2019;218(1):70–82.
156. Bulut-Karslioglu A, et al. *Nature.* 2016;540(7631):119–123.
157. Scognamiglio R, et al. *Cell.* 2016;164(4):668–680.
158. Gao L, et al. *Cell.* 2018;173(1):248–259.e15.
159. Sasaki M, et al. *Cancer Res.* 2001;61(8):3262–6.
160. Sasaki M, et al. *Cancer Res.* 2001;61(1):97–102.
161. Ghabreau L, et al. *Virchows Arch.* 2004;445(2):129–134.
162. Gao R, et al. *Hum Reprod.* 2012;27(9):2756–65.
163. Wu Y, et al. *Am J Obstet Gynecol.* 2005;193(2):371–380.
164. Kukushkina V, et al. *Sci Rep.* 2017;7(1):3916.
165. Houshdaran S, et al. *Mol Endocrinol.* 2014;28(7):1118–1135.
166. Maltepe E, et al. *J Clin Invest.* 2010;120(4):1016–1025.
167. Perrin D, et al. *Oncogene.* 2007;26(17):2518–2524.
168. Wang J, et al. *Dev Cell.* 2011;21(3):575–588.
169. Altmäe S, et al. *Reprod Sci.* 2013;20(3):308–317.
170. Moreno-Moya JM, et al. *Mol Hum Reprod.* 2014;20(6):550–566.
171. Liu W, et al. *Am J Reprod Immunol.* 2016;75(3):263–271.

Perinatal and Pediatric Outcome of Pregnancies Following PGT-M/SR/A

31

Malou Heijligers and Christine de Die-Smulders

INTRODUCTION

Preimplantation genetic testing (PGT; formerly preimplantation genetic diagnosis) was introduced almost three decades ago [1]. Techniques for performing embryo biopsy have changed and developed over time, and currently trophectoderm (TE) biopsy is probably the most commonly used biopsy technique [2, 3]. Applications for PGT are increasing due to the increasing efficiency of the procedure, decreasing costs, and increasing number of genetic diagnoses available due to new techniques like whole-exome sequencing [4].

Studies on the health of offspring following IVF with or without intracytoplasmic sperm injection (ICSI) have been extensively performed, but results are still conflicting and not fully applicable for offspring following PGT [5–7]. The biopsy is an important difference between PGT and regular IVF/ICSI [1, 4]. The safety of the (invasive) biopsy, when considering the effect on embryonic development, but also the reliability of the genetic testing performed and thus the risk of misdiagnosis, should be taken into consideration. In addition, parental factors may be different, especially when comparing couples opting for PGT for monogenic disorders (PGT-M) or PGT for structural rearrangements (PGT-SR) with couples opting for IVF, since in/subfertility is in most cases not an issue for the first two categories. Subfertility itself is considered to be a risk factor for several health problems in the offspring [6, 8]. On the other hand, a proportion of the women opting for PGT-M may suffer from a genetic disorder themselves that could also affect fetal or perinatal outcome. Lastly, growing up in a family with a parent and/or sibling affected with a genetic disorder may also affect the (socio-emotional) development of the offspring [9, 10].

Thus, several factors should be taken into consideration when evaluating the health of offspring following PGT. The risk for adverse obstetric, perinatal, and pediatric outcomes have been studied in short-term follow-up studies and led to recommendations for the follow-up of PGT pregnancies and offspring [5, 9, 11–26]. Long-term follow-up studies, focusing on the health of adolescents and adults born following PGT are still scarce.

DOI: 10.1201/9781003024941-31

DEVELOPMENTAL ORIGINS OF HEALTH AND DISEASE (DOHAD) HYPOTHESIS

The Developmental Origins of Health and Disease hypothesis, formally known as "Fetal Origins of Adult Disease," was proposed in the 1990s [27]. It states that the risk of the development of diseases later in life finds its origin during early fetal development. Exposure to environmental stressors during this critical period may have significant consequences for one's future health, especially for growth and metabolism. Nutritional stress *in utero*, as studied in the Dutch Hunger Winter study, and maternal thalidomide use preventing morning sickness are widely known examples of such stressors [28, 29]. Assisted reproductive technologies (ART), among which PGT is included, are also considered as possible environmental stressors. Natural dynamic conditions present in the genital tract are lost during IVF procedures, whereas cultivation of embryos *in vitro* adds environmental stressors. Sources of stress that may influence preimplantation development are culture media, pH, temperature, oxygen concentration, the use of ICSI, and the use of cryopreservation [30]. The need for a biopsy in PGT in addition to the IVF procedure can prove to be an additional stressor for the development of the embryo. Studies on the influence of the very early embryonic environment also report short-term effects on, for instance, perinatal growth and birthweight [31]. Adverse perinatal growth can subsequently affect long-term outcome.

PREGNANCIES FOLLOWING PGT-M/SR/A

Women who are pregnant following IVF are known to have an increased risk of obstetric complications [6]. Literature on the cause of this increased risk is, however, inconsistent. It is not clear whether this increased risk is a side effect of the IVF procedure, due to parental subfertility, or a combination of both [6, 8]. It is debatable whether these risks are also applicable to pregnancies following PGT. If an increased risk for obstetric complications is secondary to the IVF procedure, this risk will be present in PGT pregnancies as well. On the other hand, the risk will be lower or even absent from pregnancies following PGT when the complications are secondary to parental subfertility. The risk assessment considering the risk due to paternal subfertility is only valid for pregnancies following PGT-M/SR, since PGT for Aneuploidies (PGT-A) is offered as an add-on procedure to regular IVF, and thus to couples with fertility issues.

PGT by itself might increase the risk of obstetric complications due to the need for a cell biopsy. The biopsy, which is necessary to obtain material for genetic analysis, usually takes place at the cleavage stage or the blastocyst stage (TE biopsy) [2, 3, 32, 33]. In the former case, only one or two cells are biopsied from a day 3 embryo, whereas, in the latter case, five to ten cells can be biopsied. Most studies on the safety of PGT include pregnancies following cleavage stage biopsy. These do not report a significantly increased risk of obstetric complications [11–14]. TE biopsy, on the other hand, may increase the risk of obstetric complications, especially placental complications, since cells are removed that are destined to form the placenta. Additionally, TE biopsy can only be performed when followed by cryopreservation of the embryos, which may add an additional risk [15, 25]. Only a few studies have actually reported to date on the risk of obstetric complications following PGT with either cleavage stage or blastocyst stage biopsy, and results are rather inconclusive; the most frequent result is an increased risk of placenta previa and preeclampsia [11–15, 25].

Another known risk of ART is the increased risk of multiple conceptions, which consequently increases the risk of obstetric complications [34]. Over the years, the risk for multiple pregnancies following IVF has fallen due to stringent rules regarding the number of transferred embryos, where single embryo transfer (SET) is highly recommended [35]. Though studies report a higher prevalence of monozygotic twinning after SET in IVF with and without PGT, compared with live births after spontaneous conception [36–39]. The underlying mechanism is not fully understood, but it is hypothesized that the use of zona pellucida manipulating techniques may contribute to this risk [36–38]. Pregnancies following PGT therefore may have an increased risk of monozygotic twinning and accompanying obstetric risks.

PERINATAL OUTCOME

Perinatal outcome of pregnancies following PGT has been the subject of much study [5, 11–13, 16–18]. A major concern regarding PGT was the increased risk of congenital abnormalities due to the embryo biopsy, especially because regular IVF with ICSI is already associated with an increased risk of birth defects [6]. Congenital abnormalities can be detected during pregnancy and after birth. Some of the prenatally detected abnormalities are not compatible with life and lead to intra-uterine fetal death or termination of pregnancy. Reports on congenital abnormalities differentiate between major and minor abnormalities, and between isolated and multiple abnormalities, the latter sometimes being part of a syndrome. Congenital abnormalities are defined as structural, functional, and genetic anomalies. Those that do not have serious medical, functional, or cosmetic consequences are considered minor.

Major congenital abnormality rates in cohorts of children born following PGT differ slightly between studies but are mostly under 4%. Cardiac abnormalities are the most common [5, 11–13, 16–18]. This is in line with the population risk for congenital abnormalities which is about 2–3% [40]. An increased risk of either multiple congenital abnormalities or syndromes has not been reported.

The risk of intra-uterine fetal mortality and/or neonatal mortality was thought to be increased due to PGT. However, in the early years of PGT, the transfer of multiple embryos led to more multiple conceptions. After stratification for these multiples, the risk of perinatal mortality was comparable to the risk in the general population [5, 26]. Recent reports conclude that PGT does not increase the risk of perinatal mortality.

Birth parameters are also an important topic in studies on perinatal outcome. Early studies on birth weight and pregnancy term reported increased risks of both low birth weight and prematurity. A correlation between these outcome measures was seen, with study groups including more premature children also reporting higher incidences of low birthweight [5, 16, 26]. After stratification for multiple pregnancies, the risk was lower and somewhat comparable to the risk following regular IVF. Recent studies on this topic have also concluded that PGT does not increase the risk of either low birth weight or prematurity [5, 12, 13, 41]. These are important results since low birth weight, especially when followed by rapid catch-up growth, is associated with an increased risk of hospital admission after birth and an increased risk for cardiometabolic disorders later in life [42–44]. There is even some evidence that mean birth weight in neonates born following PGT-M/SR is in the higher normal range [5, 16, 17]. Frozen embryo transfer, which is part of the procedure for TE biopsy, is also associated with increased birth weight [25, 45, 46]. The exact underlying mechanism is still not known, but the effect of controlled ovarian stimulation on endometrial receptivity, resulting in abnormal endometrial development in fresh autologous cycles of IVF, might contribute to the better perinatal results in pregnancies following frozen embryo transfer [45, 47].

Overall, data regarding perinatal outcome of pregnancies following PGT are promising and show no increased risk of congenital abnormalities, prematurity, or low birth weight. One should take into account that these conclusions are mostly based on studies regarding PGT with cleavage stage biopsy. The effect of TE biopsy on these outcome measurements seems promising but still needs to be further explored.

MISDIAGNOSIS

Misdiagnosis, defined as either failure of the technical procedure used in PGT or inaccurate or incorrect interpretation of the test results, can occur due to a human error, an intrinsic (embryonic) error, or an extrinsic (technical) error [48]. Misidentification or mislabeling, embryonic mosaicism, parental contamination, or recombination are examples of such errors. Despite the high quality standards in PGT laboratories, misdiagnosis cannot fully be prevented [48, 49]. Misdiagnosis rates for PGT with polymerase chain reaction (PCR) are generally estimated to be about < 1%, whereas misdiagnosis rate for fluorescence *in situ* hybridization (FISH) is slightly higher and generally estimated at < 5% [48]. An evaluation study of PCR-based PGT showed positive effects of performing PGT on two cells instead of one cell, the use of multiplex protocols, and the application of PGT on cells from embryos with good morphology on the validity, robustness, and diagnostic value of PCR-based PGT [49]. The use of TE biopsy makes it possible to assess more embryonic cells and thereby increase diagnostic accuracy [50]. In addition, techniques for genetic analysis will also change in the near future with the implementation of genome-wide haplotyping and karyomapping. These new techniques are probably feasible for all indications, which should improve the laboratory test design process before PGT-M and, as a result of that, the overall reliability of PGT [51–57]. However, evaluation studies on this topic have not been performed extensively and misdiagnosis will remain an inevitable element of PGT [54, 56, 58]. Therefore, invasive prenatal diagnosis (chorionic villus sampling or amniocentesis) is offered in all pregnancies following PGT [59, 60]. Limited literature is available but it is clear that only a minority of couples opt for invasive prenatal testing following PGT [5, 58, 61].

Misdiagnosis can be categorized as either adverse (severe) or benign. Adverse misdiagnoses include embryo transfers leading to a severe adverse event for the patient, for instance, the birth of an affected child or termination of an affected pregnancy. An example of benign misdiagnosis is the transfer of an unaffected carrier embryo of a recessive disorder although it was thought to be free of the mutation and thus not a carrier. The latter has no direct health consequences but is relevant for the child's future reproductive risks. Misdiagnosis can also lead to an early miscarriage or involve embryos which were not transferred and therefore had been missed [48, 49, 58]. Data regarding misdiagnosis have been only rarely reported and may not reflect the true misdiagnosis rate since fetal/pediatric follow-up and genetic analysis of abortion material or untransferred embryos is not routinely performed [4]. PGT-M is increasingly performed for late-onset disorders, like hereditary cancer and Huntington's disease. In these cases, it seems particularly unlikely that couples would opt for invasive prenatal testing [58, 61]. As a result, most PGT offspring have not yet reached adult age, so misdiagnosis could still be undetected [58]. Reliable data regarding misdiagnosis rate following PGT are therefore not yet available.

PEDIATRIC OUTCOME

Pediatric outcome in the literature is differentiated between data regarding physical health, including the growth of the child, and the child's development.

Physical Health

As mentioned earlier, rapid catch-up growth is associated with increased risk of certain health problems, like diabetes mellitus type 2, cardiovascular disease, and obesity [42–44]. The results on the growth of children born following PGT (including body mass index (BMI), height, weight, skin folds, waist and head circumference) are in line with the results regarding birth parameters, showing no actual abnormalities in young children [17, 19–22, 24, 62]. These studies also report no measurable impact from PGT on the children's blood pressure. Comparable data regarding teenagers, adolescents, and adults born after PGT are not available yet, the adult data being particularly important to assess any increased risk of cardiometabolic disease, since these disorders are mostly not present before adulthood.

In order to evaluate physical health, data regarding hospital admissions and medical treatment of children following PGT have been collected by some researchers. These results show slightly increased incidence of medical interventions without a common cause. A proposed explanation is the influence of parental concerns, who may consider their child more precious since it was born after a special treatment, in this case PGT. By itself, PGT does not seem to increase the risk of physical health problems [20–22, 24, 62].

Development

Early childhood development involves motor development and neurological functioning, cognitive development, and socio-emotional or behavioral development.

Motor development encompasses fine and gross motor skills and can be evaluated by checking the age at which motor milestones have been reached. Additional neurological evaluation gives insight into mental status, motor and sensory skills, balance and coordination, and reflexes. Some studies on toddlers born following PGT reported slightly lower scores on certain subscales, i.e., locomotion and fine motor function [17–20], though no effect of PGT was evident in follow-up studies with older children up to the age of nine [17, 21–24]. At the moment, it is not clear whether PGT affects motor skills at a young age or whether there was another reason for this finding in some of the studies. Larger sample sizes are necessary in order to differentiate between a treatment effect and other possible causes.

Cognitive development can be assessed by measuring a subject's intelligence quotient (IQ) by administering, for instance, the Kauffman Assessment Battery for Children or the Wechsler Preschool and Primary Scale of Intelligence [63, 64]. Several study groups measured the IQ of children of different ages [9, 18, 20–23, 65–68]. None observed abnormal mean IQ scores and some even reported mean scores for children born following PGT that were slightly above the population average. This may be due to the higher education level of couples opting for PGT, a finding which is in line with reports on the mean education level of couples opting for ART. Additional data on working memory capacity of children born following PGT showed promising results [9]. This is of importance for future perspectives because working memory is a predictive tool for future cognitive and socio-emotional development and academic success [9]. Overall, PGT does not seem to contribute to adverse cognitive developmental outcomes.

Socio-emotional development can be assessed by use of questionnaire filled in by parents, other caregivers, and/or teachers [9, 18, 20, 21, 66–68]. Most studies evaluating this parameter reported no elevated signs of mental or behavioral problems. One study did report some differences between parental and teacher's scores for externalizing problem behavior, though all scores were within the normal range and thus not indicative of problem behavior [9]. The study group concluded that it was unlikely that these results were due to the embryo biopsy. They could not exclude that family background played a causative role, because the study solely included children born following PGT-M/SR, who were thus born into families with a possible psychological burden due to the familiar genetic disorder.

LIMITATIONS

Most studies on the outcome of pregnancies and offspring born following PGT include data from PGT procedures with cleavage stage biopsy, whereas TE biopsy has currently replaced cleavage stage biopsy in most PGT centers. The biopsy techniques differ when considering cultivation time, number and origin of the biopsied cells, and application of freezing techniques. All of these differences might affect the discussed outcome measures. Knowledge about the safety of currently applied PGT techniques, therefore, is lacking. IVF and PGT procedures keep on developing over time and continuing follow-up programs are needed to evaluate the health of the post-PGT offspring.

Up until now, knowledge about the health of older, post-PGT offspring has also been scarce. Many questions considering adolescents and adults born following PGT may have been raised but are not currently being evaluated. Again, follow-up programs are needed to fill this knowledge gap.

CONCLUSIONS

In general, the evidence currently available suggests that PGT is overall a safe procedure when taking into account outcome measures like obstetric complications, perinatal health, and young children's physical health and development. Prenatal testing, in order to validate the genetic analysis, should be offered in all pregnancies following PGT, although additional obstetric check-ups are not routinely advised. Additional check-ups in offspring born following PGT are not necessary evidence based on current results. Techniques and procedures necessary for PGT are, however, developing and quickly changing over time; as a result, follow-up data are often obsolete because the subsequent assessments by health care providers lag behind technical development. Long-term prognosis for PGT offspring is still not clearly defined.

REFERENCES

1. Handyside AH, et al. *Lancet.* 1989;1(8634):347–349.
2. Scott RT Jr, et al. *Fertil Steril.* 2013;100(3):624–630.
3. Lathi RB, et al. *Reprod Biomed Online.* 2012;25(5):504–507.
4. De Rycke M, et al. *Hum Reprod.* 2017;32(10):1974–1994.
5. Heijligers M, et al. *J Assist Reprod Genet.* 2018;35(11):1995–2002.
6. Pandey S, et al. *Hum Reprod Update.* 2012;18(5):485–503.
7. Davies MJ, et al. *N Engl J Med.* 2012;366(19):1803–1813.
8. Messerlian C, et al. *Hum Reprod.* 2013;28(1):125–137.
9. Heijligers M, et al. *Hum Reprod.* 2018;33(11):2150–2157.
10. Stoeckel M, et al. *Fam Syst Health.* 2015;33(4):356–362.
11. Bay B, et al. *Fertil Steril.* 2016;106(6):1363–1369.e1.
12. Desmyttere S, et al. *Hum Reprod.* 2012;27(1):288–293.
13. Hasson J, et al. *Reprod Biomed Online.* 2017;35(2):208–218.
14. Strom CM, et al. *Am J Obstet Gynecol.* 2000;182(6):1629–1632.
15. Zhang WY, et al. *Fertil Steril.* 2019;112(2):283–290.e2.
16. Eldar-Geva T, et al. *Fertil Steril.* 2014;102(4):1016–1021.
17. Natsuaki MN, et al. *World J Pediatr.* 2018;14(6):555–569.
18. Banerjee I, et al. *Reprod Biomed Online.* 2008;16:376–81.
19. Middelburg KJ, et al. *Pediatr Res.* 2010;67:430–4.
20. Middelburg KJ, et al. *Fertil Steril.* 2011;96:165–9.
21. Schendelaar P, et al. *Hum Reprod.* 2013;28:1508–18.
22. Winter C, et al. *Hum Reprod.* 2014;29:1968–77.
23. Kuiper D, et al. *Hum Reprod.* 2018;33(1):147–155.
24. Heijligers M, et al. *Fertil Steril.* 2019;111(6):1151–1158.
25. Jing S, et al. *Fertil Steril.* 2016;106(1):105–112.e4.
26. Liebaers I, et al. *Hum Reprod.* 2010;25(1):275–82.
27. Barker DJ. *BMJ.* 1990;301(6761):1111.
28. Schulz LC. *Proc Natl Acad Sci U S A.* 2010;107(39):16757–16758.
29. Brent RL, et al. *Teratology.* 1988;38(3):241–251.
30. Feuer S, et al. *Birth Defects Res C Embryo Today.* 2012;96(4):299–314.
31. Kleijkers SH, et al. *Hum Reprod.* 2014;29(4):661–669.
32. Harton GL, et al. *Hum Reprod.* 2011;26(1):25–32.
33. Harton GL, et al. *Hum Reprod.* 2011;26(1):33–40.
34. Bricker L. *Best Pract Res Clin Obstet Gynaecol.* 2014;28(2):305–317.
35. Hazekamp J, et al. *Hum Reprod.* 2000;15(6):1217–1219.
36. Haimov-Kochman R, et al. *Fertil Steril.* 2009;92(6):2037.e11–2037.e2.037E17.
37. Hviid KVR, et al. *Hum Reprod Update.* 2018;24(4):468–483.

38. Verpoest W, et al. *Hum Reprod.* 2009;24(11):2945–2950.
39. Ikemoto Y, et al. *Hum Reprod.* 2018;33(11):1984–1991.
40. Boyle B, et al. *Arch Dis Child Fetal Neonatal Ed.* 2018;103(1):F22–F28.
41. Sunkara SK, et al. *Hum Reprod.* 2017;32(2):432–438.
42. Bay B, et al. *Fertil Steril.* 2014;102:1117–1123.
43. Ceelen M, et al. *Hum Reprod.* 2009;24:2788–2795.
44. Guo XY, et al. *Fertil Steril.* 2017;107:622–31.e5.
45. Sha T, et al. *Fertil Steril.* 2018;109(2):330–342.e9.
46. Luke B, et al. *J Assist Reprod Genet.* 2016;33(1):3–8.
47. Shapiro BS, et al. *Fertil Steril.* 2011;96(2):344–348.
48. Wilton L, et al. *Hum Reprod.* 2009;24(5):1221–1228.
49. Dreesen J, et al. *Eur J Hum Genet.* 2014;22(8):1012–1018.
50. Popovic M, et al. *Hum Reprod.* 2018;33(7):1342–1354.
51. Renwick PJ, et al. *Reprod Biomed Online.* 2006;13(1):110–119.
52. Dimitriadou E, et al. *Hum Reprod.* 2017;32(3):687–697.
53. Zamani Esteki M, et al. *Am J Hum Genet.* 2015;96(6):894–912.
54. Masset H, et al. *Hum Reprod.* 2019;34(8):1608–1619.
55. Handyside AH, et al. *J Med Genet.* 2010;47(10):651–658.
56. Konstantinidis M, et al. *Reprod Biomed Online.* 2015;31(3):394–403.
57. Gould RL, et al. *Expert Rev Mol Diagn.* 2017;17(6):611–621.
58. Hardy T. *Prenat Diagn.* 2020;10.1002/pd.5662.
59. Harton G, et al. [*Hum Reprod.* 2012;27(8):2569]. *Hum Reprod.* 2011;26(1):14–24.
60. Practice Committee of Society for Assisted Reproductive Technology; Practice Committee of American Society for Reproductive Medicine. *Fertil Steril.* 2008;90(5 Suppl):S136–S143.
61. Poulton A, et al. *Prenat Diagn.* 2018;38(12):904–910.
62. Desmyttere S, et al. *Early Hum Dev.* 2009;85(12):755–759.
63. Kaufman AS, et al. *Kaufman Assessment Battery for Children*, 2nd ed. Circle Pines: AGS Publishing; 2004.
64. Wechsler D. *Wechsler Preschool and Primary Scale of Intelligence III.* Amsterdam: NCS Pearson; 2009.
65. Nekkebroeck J, et al. *Hum Reprod.* 2008;23:1560–1566.
66. Nekkebroeck J, et al. *Hum Reprod.* 2008;23:1849–1857.
67. Nekkebroeck J, et al. *Hum Reprod.* 2011;27:299–301.
68. Sacks GC, et al. *Child Neuropsychol.* 2015;22:458–471.

32 Genetic Counseling in Assisted Reproductive Technology

Alyssa Snider and Juliana Cuzzi

WHAT IS GENETIC COUNSELING?

Reproductive genetic counseling is the process of providing information to patients about their genetic health status, and that of future offspring and embryos in a way that empowers them to make decisions about their reproductive care. A central tenet of genetic counseling is non directiveness, specifying that informed patients are capable of their own medical decision making with regard to genetic information, and the provider should avoid directing or compelling a patient (Harper, 2010; Weil, 2000). The genetic counselor's role is to facilitate learning and to support decision making by providing information within a carefully constructed psychosocial framework. Psychosocial genetic counseling includes assessment of the patient's emotional state, health literacy, level of education, and social support network. By attending to psychosocial elements, the genetic counselor can ensure the patient's ability to assimilate, retain, and adapt to new information, thus fostering patient-driven decision making and maximizing positive medical outcomes (Weil, 2000).

Many health care providers, including certified genetic counselors, physicians, and nurse practitioners, provide genetic counseling throughout the course of their practice. Health care professionals providing genetic counseling should have special training in patient communications. It is not sufficient for health care professionals to be knowledgeable only about the test or procedure. Genetic information is deeply personal, and decisions related to one's genetic makeup can have far-reaching ramifications for themselves, their relatives, and their embryos. Certified genetic counselors must also work within their scope of practice with respect to their subspecialty and type of employment. Not all genetic counselors have experience in assisted reproductive technology (ART), and patients will sometimes need counseling in a different field of specialty. Appropriate referral to a genetic counselor with applicable expertise and work setting is imperative.

GENETIC TESTS IN ART

Carrier screening is a genetic test typically performed on both members of a reproductive couple. Carrier screening aims to identify recessive and X-linked variants in a healthy population of patients. If both members

DOI: 10.1201/9781003024941-32

of the reproductive couple are carriers of variants in the same recessive gene, or if a female is a carrier of an X-linked variant, there is a risk of passing on a genetic condition to offspring. Identifying such risk enables a couple to consider various reproductive options, including spontaneous conception followed by prenatal diagnosis or an ART approach utilizing preimplantation genetic testing for monogenic disorders (PGT-M) to select embryos identified to be at reduced risk of developing the condition prior to pregnancy. Some genetic conditions and specific variants are more common in certain ethnic populations, and so ethnicity-based carrier screening panels were historically offered to reproductive couples. Because of advancing technology and the opportunity to perform PGT-M on embryos obtained through *in vitro* fertilization (IVF), many infertility clinics now utilize pan-ethnic expanded carrier screening as part of their standard work-up.

Preimplantation genetic testing for aneuploidy (PGT-A) is an assessment of chromosome copy number in preimplantation embryos. Euploid embryos have two copies of every chromosome, 1 through 22, and either two X chromosomes or one X and one Y chromosome. The genetic instructions for growth and development are encoded in hundreds of genes located on each chromosome, and so having the appropriate number of chromosomes is crucial for embryo viability. Aneuploid embryos have one or more extra or missing chromosomes and are typically non-viable. PGT-A is typically performed on a biopsy of about 3–10 cells from the trophectoderm of a blastocyst stage embryo. Assessing the chromosome copy number of this representative sample provides information about the reproductive potential of each embryo, enabling decisions to be made about which embryos to prioritize. Ideally, a single euploid embryo with the highest reproductive potential can be selected for transfer.

Preimplantation genetic testing for structural rearrangements (PGT-SR) is similar to PGT-A in that it assesses chromosome copy number variation in preimplantation embryos. Unlike PGT-A, PGT-SR is designed to assess, with special attention, segmental and complete imbalances of the chromosomes involved in a known translocation or inversion in embryos conceived from a carrier patient or partner. PGT-SR typically cannot detect balanced translocations or differentiate them from euploid embryos lacking the balanced translocation. Embryos conceived from a healthy carrier parent and carrying the balanced rearrangement would be expected to develop normally.

Preimplantation genetic testing for monogenic disorders (PGT-M) screens preimplantation embryos for single-gene disorders. PGT-M assesses embryos for the inheritance of a disease-associated variant, or of a combination of variants, and facilitates the prioritization of embryos with reduced risk for transfer in an IVF cycle. Knowledge about the genetic status of an embryo can be obtained even before a pregnancy occurs. Decisions about embryo transfer can be made to avoid the birth of a child with a genetic disorder.

Unlike with prenatal diagnosis, with PGT-M, a test must be custom designed for each reproductive couple such that the at-risk and low-risk haplotypes can be defined. Determining the haplotypes involves assessing a set of genetic linkage markers shared among affected and/or unaffected first-degree relatives. The linkage markers (e.g., short tandem repeats [STRs]) flank the disease-causing variant within and around the gene. Linkage markers shared between the patient and an affected/carrier relative define the at-risk haplotype; genetic markers shared between the patient and an unaffected/non-carrier relative define the low-risk haplotype. Where possible, embryo testing involves assessing a combination of linkage markers as well as the disease-associated variant directly. If direct testing of the variant is not possible, PGT-M testing can often be carried out using linkage markers only, provided that the necessary relatives are available to provide samples for linkage analysis. The reason for this elaborate testing methodology is the limited number of cells that can be biopsied from a preimplantation embryo. Such a small sample number precludes many of the robust technologies available to directly detect the disease-causing variant in prenatal or pediatric/adult samples. Utilizing linkage analysis in combination with direct detection of the disease-causing variant, where possible, increases the accuracy of genetic testing applied to the extraordinarily small sample that can be obtained from preimplantation embryos.

Product of conceptions (POC) is a test that analyzes fetal tissue from a miscarriage to evaluate if the pregnancy loss was caused by a chromosomal aneuploidy. Tissue is ideally collected by a clinician but can also be collected at home by the patient. Maternal cell contamination (MCC) is a frequent complication of POC testing, potentially leading to false negative results; however, if the technology utilized includes molecular analysis of polymorphic markers, fetal DNA can be distinguished from maternal DNA, and

MCC can be identified and ruled out. If MCC is confirmed, this would equate to an uninformative result for the fetal testing. Most aneuploidy identified in pregnancy losses is due to sporadic events, but occasionally warrants parental karyotyping to evaluate for the presence of a structural chromosomal rearrangement.

Sperm aneuploidy test (SAT) assesses representative chromosomes for aneuploidy in a sperm sample to better understand the likelihood of abnormal male-factor aneuploidy rates. The assessment characterizes the quality of the sperm but does not facilitate selection of euploid sperm for fertilization. If the SAT test identifies abnormal aneuploidy rates in sperm, the use of a sperm donor can be considered.

Endometrial receptivity analysis (ERA) is a test that assesses the best timing of embryo transfer for successful implantation. Identifying the perfect moment at which the embryo and endometrium are synchronized for implantation can significantly improve a patient's chance of pregnancy. For some patients, transferring embryos earlier or later than typical can enable a pregnancy that would otherwise not be possible. Although this test assesses gene expression levels, it is not considered a genetic test in the classical sense, and so is not discussed in this chapter.

TYPES OF GENETIC COUNSELING

A genetic evaluation and risk assessment aims to determine if there is a genetic cause of infertility and if there are any genetic risk factors that should be addressed at the pre-conception state. A medical, reproductive, and family history can be gathered to assess the partners in a couple for genetic conditions associated with infertility or other concerning genetic risk factors. For example, a personal and/or family history of recurrent miscarriage and/or a child born with a chromosomal imbalance or unspecified genetic disorder might warrant cytogenetic analysis to evaluate the partner concerned for the presence of a balanced translocation. The identification of a balanced translocation can explain the history of infertility and subsequently guide genetic testing for embryos through the IVF process. The family history is assessed for any symptoms which could underlie a heritable single-gene disorder for which PGT-M might be possible. Expanded carrier screening and PGT options can be reviewed. Genetic counseling can include the collection of such reproductive and family history, the interpretation of the family history, and the coordination of any necessary genetic testing.

Pre-test counseling is the process of informing patients about a genetic test prior to having the test performed, to allow informed consent. Genetic testing should never be compulsory, and pre-test counseling therefore includes the opportunity for the patient to elect or decline the test (Uhlman et al., 2009). Pre-test counseling includes a discussion of the benefits of a test as well as the risks and limitations. This discussion ensures the patient is well informed when providing consent. Anticipatory guidance is an essential part of pre-test counseling. A patient should be informed about the potential outcomes of testing as well as how those outcomes could affect their medical care (Uhlman et al., 2009). Anticipatory guidance provides patients with the opportunity to explore how they might react to a given test result or outcome. Such guidance reduces shock, confusion, anxiety, and test regret upon the disclosure of results. Pre-test counseling sets expectations for a patient's medical management after the test has been completed.

Post-test counseling includes a review of a patient's genetic testing results, including a discussion of accuracy, a discussion of reproductive options, and guidance on next steps. Post-test counseling can occur upon disclosure of results, during a subsequent conversation, or even after medical interventions have taken place. Post-test counseling typically provides a greater level of detail, focusing on a patient's specific results. Patients who received effective pre-test counseling generally have a better understanding of their results upon disclosure, and the emotional impact of receiving disappointing results can be mitigated. However, even if pre-test counseling was provided, patients may not have understood or retained crucial information. Most patients are still surprised when they experience negative outcomes that they were counseled were less likely or rare. It may be beneficial to assess a patient's understanding and review pertinent information in the context of their results and experience.

Genetic counselors practice in different work settings (Snider, Isley, et al., 2021). For example, some are employed by fertility centers, and some are employed by genetic testing laboratories. The ART field is somewhat unique in the frequent use of genetic counselors employed by genetic testing laboratories. Although ART genetic counselors employed by an IVF center and those employed by a PGT laboratory share a similar subspecialty, their roles are not entirely the same. Clinic-based genetic counselors are best suited for family and reproductive history gathering and interpretation. They may be best equipped to counsel patients about genetic risks affecting their personal health status. Laboratory-based genetic counselors will have greatest expertise in the genetic testing offered by that laboratory. They will be able to counsel patients on the technology utilized, limitations specific to the technology, and laboratory protocol and reporting policies. Laboratory-based counseling may not be adapted to the policies of the ordering provider's office. IVF clinic-based genetic counselors may be better equipped to synthesize test results between various laboratories. Outside of the ART subspecialty, there are genetic counselors involved in prenatal, pediatric, and cancer care, just to name a few. Appropriate referrals and careful hand-off between providers can greatly benefit a patient's medical management.

GENETIC CAUSES OF INFERTILITY

A young and healthy couple has an approximately 25% chance of becoming pregnant with each cycle (Shah et al., 2003), and infertility is a common health problem, affecting about 15% of couples who wish to start a family (Gardner and Amor, 2018). With the progress of scientific research, an increasing number of genetic causes associated with infertility are being identified. Sporadic aneuploidy in embryos is a widely recognized cause of failed implantation and miscarriage. The frequency of embryonic aneuploidy increases with maternal age, and indeed, advanced maternal age (AMA) is one of the most frequently diagnosed causes of infertility. Aneuploidy of the sex chromosomes is compatible with life and causes syndromes that can affect fertility in adults. Healthy individuals can also carry balanced structural rearrangements in their chromosomes that can disrupt gametogenesis and result in embryos with genetic imbalances. Unfortunately, most of the genetic and molecular factors involved in infertility remain unknown, and about 20% of cases are unexplained (Uehara et al., 2001). In the following sections, we will discuss chromosomal rearrangements that can affect both men and women and the most common genetic causes of both male and female infertility.

Balanced Chromosomal Rearrangements

Although carriers of a balanced chromosomal rearrangement are generally healthy, the rearrangement can cause difficulties conceiving or carrying a child to term. The most common balanced structural rearrangements are translocations, in which segments from two chromosomes dissociate and then rejoin to the reciprocal chromosome, and inversions, where a segment of a single chromosome disassociates but then rejoins in the opposite orientation. Robertsonian translocations are unique types of translocations that involve the acrocentric chromosomes 13, 14, 15, 21, and 22. The chromosomes fuse at the centromere, resulting in a reduction in the total number of chromosomes and the benign loss of the redundant stalks and satellites at the p arm of both chromosomes. A balanced chromosomal rearrangement can lead to difficulty conceiving, spontaneous miscarriage, or the birth of a child with multiple congenital anomalies, depending on the degree of the imbalance. Although balanced in the carrier, structural rearrangements may be unbalanced in the gametes, resulting in developmental abnormalities. In some individuals, the structural rearrangement can impair meiosis and block gametogenesis, especially male spermatogenesis. The disruption of spermatogenesis is variable, even among male relatives carrying the same chromosomal rearrangement. Approximately 1/500 individuals in the general population carry a balanced reciprocal translocation, and approximately 1/1000 individuals carry a balanced Robertsonian translocation. This frequency is likely to be higher in the infertile population.

Cytogenetic analysis is recommended for couples with a history of repeated pregnancy loss, since approximately 50% have balanced reciprocal translocations, 24% have Robertsonian translocations, 12% have sex chromosome mosaicism, and some have other, less frequent, chromosomal changes (Tharapel et al., 1985).

Female Infertility

Advanced maternal age

The high incidence of age-related meiotic errors during oogenesis makes advanced maternal age an important risk factor for early miscarriage and infertility (Grande et al., 2012). Most aneuploidies are incompatible with life. Each chromosome contains many genes essential for normal growth and development, and so a healthy pregnancy cannot occur when an embryo does not have the appropriate chromosome number. The majority (50%–75%) of sporadic losses result from random gains or losses in chromosome number, with a progressively higher incidence as maternal age increases (Sugiura-Ogasawara et al., 2012; Werner et al., 2012). In addition to being correlated with an increased risk of miscarriage, maternal age is also correlated with decreased implantation rates and with the risk of producing offspring affected by chromosomal syndromes, such as Down, Edwards, Patau, Turner, and Klinefelter syndromes (Hunt and Hassold, 2002).

After 35 years of age, women begin to experience a significant decrease in the quality of their oocytes, especially regarding chromosomal aneuploidies. Many scientific theories try to explain the mechanism of ovarian aging and the influence of a woman's age on chromosomal non-disjunction during female gametogenesis, but none have been fully accepted. While embryonic aneuploidies are mainly caused by errors in maternal meiosis, aneuploidy can also be paternally derived, resulting from meiotic errors in spermatogenesis. However, the incidence of aneuploidy does not seem to be correlated with paternal age.

45,X (Turner syndrome)

Monosomy of chromosome X, mosaic monosomy X, and structural variants involving the disruption of Xp underlie Turner syndrome, which is found in approximately 1 in 4000 female births (Gersen and Keagle, 2013). Any woman presenting with primary amenorrhea, especially if she is of short stature, should be evaluated for Turner syndrome. Mosaic female karyotypes (45,X/46,XX) are compatible with ovarian activity and the production of female gametes, although an association with low ovarian reserve and premature ovarian insufficiency has been observed. Additionally, the risk of miscarriage is elevated in women with mosaic karyotypes (Calanchini et al., 2020; Sybert and McCauley, 2004)

47,XXX (Triple X syndrome)

The karyotype 47,XXX is seen in 1 out of every 1000 females born. In 95% of cases, the extra X chromosome originates from the mother and is associated with advanced maternal age (Hassold et al., 1996). Most women with 47,XXX have normal height, weight, and intellectual capability. They have normal prepubertal development and fertility, but develop early menopause, with onset around 30 years of age. Karyotypes with four or more X chromosomes have also been reported. Because the dosage of non-inactivated chromosome X genes contributes to the clinical characteristics of this syndrome, clinical features become more significant with the number of X chromosomes present in the patient's karyotype (Tartaglia et al., 2010).

Primary ovarian insufficiency associated with fragile X syndrome

Expansion of the CGG triplet repeat in the *FMR1* gene is responsible for fragile X syndrome. Women who carry a premutation in this gene do not have fragile X syndrome but can have a unique set of symptoms associated with a premutation, including primary ovarian insufficiency (POI). Women who carry the full mutation in this gene can have a diagnosis of fragile X syndrome, although, because of skewed X-inactivation and the presence of a fully functioning copy of the *FMR1* gene on their other X chromosome, may manifest a milder form of the disease than males who inherit a full mutation. Clinical trials report an incidence of the *FMR1* premutation in as many as 14% to 20% of women with familial POI and 2% to 5% of women with sporadic POI (Torrealday et al., 2017).

Male Factor Infertility

Male infertility can be associated with a wide variety of other factors. In more than half of infertile men, the cause of infertility is unknown (idiopathic). The initial diagnosis of infertility in men is made by analyzing a semen sample. It is estimated that approximately 30% of men seeking assisted reproduction treatment are either oligospermic or azoospermic for no apparent reason. The etiology of low sperm quality is known in less than 50% of cases. Considering that around 15% of men with nonobstructive azoospermia and 4% of men with moderate oligozoospermia are known to be carriers of chromosomal anomalies, karyotyping should be the first genetic test carried out in such patients (Jungwirth et al., 2012).

Approximately 10% of men who do not have sperm in their seminal samples have a Y chromosome deletion. This deleted region corresponds to the chromosomal portion where the azoospermia factor (AZF) locus, the Yq11 region, is located. In this region of the long arm of the Y chromosome, there are four non-overlapping subregions: AZFa, AZFb, AZFc, and AZFd. Each of these areas can be associated with a specific testicular histology, and many genes have been described in these regions. Most commonly, the family of genes deleted in patients with azoospermia (DAZ) is located in AZFc (Krausz et al., 2018).

Karyotype 47,XXY (Klinefelter syndrome)

Klinefelter syndrome affects 1 in every 1000 male children and is found in about 3% of infertile males (Nussbaum et al., 2004). In patients with non-obstructive azoospermia, the frequency is as high as 1 in 7 (Punab et al., 2017). The severity of this syndrome is influenced by the presence or absence of a mosaic 46,XY cell line in individuals' karyotype. Men with Klinefelter syndrome are typically azoospermic; observing spermatozoa in the ejaculate has been described mainly in men with mosaicism. Men with non-mosaic 47,XXY are often sterile whereas men with mosaic karyotypes (47,XXY/46,XY) may have functional germ cells, although with reduced reproductive capacity (Aksglaede and Juul, 2013). Despite reports of increased aneuploidy rates in sperm, an increased risk of having chromosomally abnormal children has not been shown (Gardner and Amor, 2018; Maiburg et al., 2012; Nussbaum et al., 2004).

Karyotype 47,XYY

The 47,XYY syndrome, also known as Jacobs syndrome, is found in 0.1% of the general male population. Most men with 47,XYY are usually fertile (Nussbaum et al., 2004). In the literature, variable meiotic behavior is described among 47,XYY men. Whereas some investigators report no increased incidence of aneuploidy in sperm, others have described a very small but significant increase of diploidy or an increase in both sex chromosome aneuploidy and diploidy (Gardner and Amor, 2018; Rodrigo et al., 2010). Wong et al. (2008) studied germ cells and sperm from a 47,XYY infertile patient, showing the presence of mosaic pachytene cells 46XY/47,XYY leading to an increase in sex and autosome disomy. No discernibly increased risk of having chromosomally abnormal children has been documented (Gardner and Amor, 2018; Nussbaum et al., 2004).

Male infertility associated with the CFTR gene

Variants in the *CFTR* gene are responsible for cystic fibrosis, which is the most common autosomal recessive disease in Caucasians with a carrier frequency of 1 in 25 individuals. At least 95% of men affected with cystic fibrosis are infertile due to congenital bilateral absence of the vas deferens (CBAVD) (Nussbaum et al., 2004). The *CFTR* gene encodes a transmembrane protein that acts in the formation of the ejaculatory duct, seminal vesicle, vas deferens, and the distal two-thirds of the epididymis. Genetic mutation of the *CFTR* gene is one of the primary causes of CBAVD. Genetic studies have shown that 80–97% of patients who present with isolated CBVAVD have a *CFTR* variant, and 63–83% of these carry variants in both alleles (de Souza et al., 2018). Individuals who have CBAVD and one known *CFTR* variant may have a second, unidentified *CFTR* variant, or the condition may not have direct correlation with cystic fibrosis.

Gestational Losses Associated with Gene Mutations

Many diseases with monogenic inheritance are associated with pregnancy loss, among which are hemoglobinopathies and thrombophilias. Unlike most numerical chromosome errors, which typically impact development in the first trimester or even prior to pregnancy, single-gene disorders can cause developmental issues that present in the later stages of pregnancy, causing second- or third-trimester losses or stillbirths.

Alpha-thalassemia major, or Hb Bart disease, is a hemoglobinopathy that can cause fetal loss. Hemoglobin is formed by a tetramer of α-globin and β-globin molecules. Changes in the stoichiometric ratios of α-globin and β-globin molecules impact the formation of hemoglobin and the delivery of oxygen to the fetus during pregnancy. The α-globin gene region contains two tandem α-globin genes on each allele, for a total of four copies in a typical individual. Hb Bart disease, caused by deletions in all four α-globin copies (–/–), results in severe intrauterine hypoxia and hydrops fetalis. The α-thalassemia trait genotype –/αα is common among Southeast Asians and is associated with a 25% risk of passing on Hb Bart disease if both members of the reproductive couple are carriers. Reproductive genetic counseling should be recommended for couples with a family history of thalassemia.

Gene mutations that cause thrombophilia, including factor V Leiden, prothrombin G2021A (factor II), and antithrombin, protein C, and protein S deficiencies, are associated with sporadic and recurrent miscarriage (Venkatesh et al., 2014). Thrombophilias are a heterogeneous group of disorders that predispose the individual to thromboembolism, gestational complications, and fetal loss. Although the mechanism involved in this disease is poorly understood and may involve oligogenic interactions, mutations in genes essential for blood clotting can result in damage to the chorionic vascularization and utero-placental circulation, leading to thrombosis and, consequently, spontaneous miscarriage.

Multiple single genes have been implicated in euploid pregnancy loss, including *RYR1*, *CHRNA1*, and *DYNC2H1*, among others (Colley et al., 2019). Others, including *NLRP7* and *KHDC3L*, have been associated with recurrent diploid molar pregnancies (Carey et al., 2015). Loss of heterozygosity, including that caused by consanguineous unions, can increase the risk of recessive monogenic disease, potentially having an impact on fetal development. Expanded carrier screening should be performed to assess the risk of single gene disorders. The inclusion of any genes of interest should be investigated, and other targeted or recurrent pregnancy loss panel can be explored as appropriate.

MODES OF INHERITANCE

Sporadic aneuploidy is a random occurrence and refers to aneuploidy that is not inherited. Errors during maternal meiosis are the most frequent cause of sporadic aneuploidy in embryos. Chromosomal rearrangements, such as translocations and inversions, can be passed on from generation to generation in the balanced form and can also result in imbalances, causing developmental abnormalities in offspring. Recurrent aneuploidy of the acrocentric chromosomes can be caused by a Robertsonian translocation carried by one member of the reproductive couple. Recurrent aneuploidy, especially recurrent segmental aneuploidy, can be caused by a balanced reciprocal translocation or inversion in a carrier parent.

Mendelian inheritance refers to single-gene disorders, where mutation(s) in a single nuclear gene causes a genetic disorder (Nussbaum et al., 2004). Other modes of inheritance include mitochondrial, where mutation occurs on the mitochondrial genome, and multifactorial, where multiple genetic, environmental, and random factors combine to cause a disease. Autosomal dominant, autosomal recessive, and X-linked inheritance patterns will be described here because these are the class of diseases for which PGT-M can be offered.

Autosomal dominant inheritance is when mutation in one allele of an autosomal gene is sufficient to cause disease. An affected individual typically has one non-functional allele and one functional allele. Either allele can be passed on to subsequent generations, resulting in a 50% recurrence risk for each offspring. Dominant conditions can sometimes be seen in a family, passed down from generation to generation and shared among siblings. Sometimes, dominant conditions are caused by a de novo gene change, when an individual can be the first in the family to be affected.

Autosomal recessive inheritance is when mutation in both alleles of an autosomal disease is necessary to cause disease. An affected individual has two non-functional alleles, one inherited from each carrier parent. An individual carrying a single variant in a recessive gene is unaffected with the condition, but when two individuals are carriers, there is a 25% chance that both variants will be passed on, resulting in an affected child. There may be no family history of a recessive condition when a child is found to be affected. The variant can be passed down generation to generation without ever causing disease until, by random chance, both members of a reproductive couple are carriers, and future offspring are at risk of being affected. Carrier screening can therefore be an effective tool in detecting such carrier couples.

X-linked inheritance is when a disease-associated mutation occurs on the X chromosome. Males are generally more severely affected than females. Some X-linked conditions are so severe in males that they are lethal. Some X-linked conditions are considered dominant, and some are considered recessive. Regardless of dominant or recessive classification, the severity of symptoms in females varies widely, depending on which of the two X chromosomes become inactivated. When a female is a carrier of an X-linked condition, there is a 50% chance of passing on the variant. Daughters who inherit the variant would also be carriers, and sons who inherit the variant would be affected. Affected males would pass on the variant along with the X chromosome to all daughters, who would therefore be carriers, and would pass on the Y chromosome to all sons, who would therefore be unaffected.

GENETIC COUNSELING FOR PGT-A

Indication for PGT-A

PGT-A is available to any patient who has embryos obtained from IVF to biopsy. This is because any embryo, regardless of maternal age or other factors, has a chance of being aneuploid. Because the risk of embryonic aneuploidy is positively associated with advancing maternal age, PGT-A may have the greatest clinical benefit for women who are in older reproductive age categories. Indeed, some studies have shown no significant increases in the ongoing pregnancy rate for women under 35 years old who pursued PGT-A compared with control patients who transferred embryos based on morphology alone (Munné et al., 2019). Many IVF centers are offering PGT-A to all patients undergoing IVF treatment. Patients may elect to pursue testing that helps prioritize embryos for transfer, reduces the risk of miscarriage due to aneuploidy, reduces the risk of a child being born with an aneuploid syndrome, and generally gives a sense of greater control over a process that is otherwise so subject to random chance.

Indication for PGT-SR

In contrast to PGT-A, PGT-SR is available only to patients who carry a balanced rearrangement such as a Robertsonian translocation, reciprocal translocation, or inversion. PGT-SR is not available or indicated for couples with normal karyotypes or couples who have not had cytogenetic analysis. PGT-SR can be offered for translocation carriers regardless of age due to the increased incidence of imbalances due to the translocation. Larger translocations are more likely to result in implantation failure or early miscarriage, whereas smaller translocations have greater potential to sustain development to term and result in a child born with significant medical complications. PGT-SR is appropriate under either circumstance. Laboratories require the patient's karyotype for review and must accept each case for testing. The laboratory will assess the potential imbalances that could result in gametes from balanced rearrangement carriers. Depending on the sizes of the segments in potential imbalances, some cases are rejected due to technical limitations. The unbalanced translocation results from various segregation patterns during meiosis. In contrast, the unbalanced inversion results from recombination events during meiosis, the frequency of which depends on the size of the inverted segments. Inversions can also have increased incidence of unbalanced gametes, but, because imbalances result from recombination events rather than from chromosome segregation alone, the increased rate of unbalances is more variable compared with translocation cases. For example, the inverted segment of the common chromosome 9 inversion, inv(9)(p11q13), is small enough that recombination is rare, an increased incidence of unbalanced gametes has not been observed, and the inversion is considered benign (Anton et al., 2002). PGT-SR is not indicated for this inversion. Recombinant chromosomes for the inv(3)(p25q21), on the other hand, have been observed in 31% of gametes, and PGT-SR is clearly of clinical utility (Gardner and Amor, 2018).

Pre-Test Counseling for PGT-A

Pre-test counseling for PGT-A need not require an extensive reproductive history review since the test can be offered to all IVF patients. However, a patient's family, medical, and reproductive history should be reviewed to evaluate any potential genetic causes of infertility, as described above. A history for which PGT-SR or PGT-M would be indicated should be ruled out. A patient with a history of recurrent miscarriage or with a child born with a chromosomal imbalance should have karyotyping to rule out an underlying chromosomal rearrangement. NGS-based technologies can be performed on embryos regardless of the karyotypic status of the patient or partner; however, SNP-based technologies may be incompatible with testing embryos conceived from individuals who are themselves aneuploid (e.g., from a male with Klinefelter syndrome). The family history should be reviewed to screen for any genetic condition potentially associated with a single-gene disorder. The results of any carrier screening performed should be available prior to IVF treatment in case PGT-M may be utilized.

The age of the egg source should be used in determining the risk of embryonic aneuploidy. In most cases, this will be the female patient pursuing IVF treatment. If donor eggs are being used, the age of the egg donor will determine the chance of embryonic aneuploidy. Aneuploidy rates can be found in some published literature, can be requested from the PGT laboratory, or collected from an individual fertility center. In Figure 32.1, we provide age-related aneuploid rates from approximately 200,000 blastocysts tested from Igenomix USA.

Deviations from statistical expectations are possible and even likely with the statistically small numbers of embryos typically tested following an individual IVF cycle. The possibility of having all embryos reported as aneuploid should be conveyed to the patients prior to testing. Patients should be aware of the number of embryos expected for each IVF cycle, which will affect the likelihood that at least one euploid embryo will be obtained and embryo transfer can take place. In Table 32.1 are data from approximately 40,000 cycles, including about 200,000 blastocysts tested from Igenomix USA, showing the chance of having at least one euploid embryo per cycle.

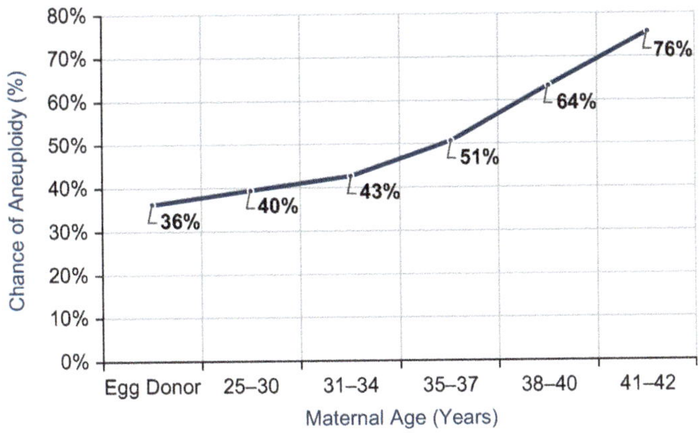

FIGURE 32.1 Age-related aneuploid rates from approximately 200,000 blastocysts tested from Igenomix USA.

TABLE 32.1 Chance of Having at Least One Euploid Embryo Per Cycle

MATERNAL AGE (YEARS)	*PROBABILITY OF 1 + NORMAL EMBRYO PER IVF CYCLE*
25–30	91%
31–34	89%
35–37	83%
38–30	68%
41–42	51%

The risks of PGT-A largely center on the biopsy procedure. Studies show that blastocyst biopsy has minimal impact on embryonic viability (Cimadomo et al., 2016; Treff et al., 2013); however, clinic-specific survival rates should be provided where possible.

The limitations of PGT-A should also be reviewed. Importantly, PGT-A does not test for "everything," and a healthy baby cannot be guaranteed. PGT-A is, in general, a chromosome copy number screen. Genetic conditions not caused by changes in chromosome copy number cannot be assessed with PGT-A. This includes, but is not limited to, single-gene disorders and multifactorial conditions. Birth defects, which are most often not genetic, cannot be tested with PGT-A.

One of the most relevant limitations of PGT-A is the accuracy of the test, typically quoted by laboratories at 98%. Results depend on the ploidy of the cells that were biopsied and which serve as a representative sample of the embryo. The cells of the inner cell mass, which will develop to become the fetus, cannot be tested without damage to the embryo. Mosaic embryos are those that have two or more cell lines with different chromosomal counts. Mosaicism in an embryo cannot be excluded, even when the PGT-A technology allows for the detection of mosaicism in the biopsy. Therefore, false-positives and false-negatives are possible.

Pre-Test Counseling for PGT-SR

Imbalances due to the chromosomal rearrangement may be isolated or may be in combination with age-related spontaneous aneuploidies. Therefore, when quoting aneuploidy rates for PGT-SR patients, one must consider both the age-related risk as well as the risk of unbalanced chromosomal rearrangements.

Potential PGT-A Results

Euploid embryos

Embryos with no chromosomal imbalances detected are reported to be euploid. Monosomies, trisomies, and segmental aneuploidies above the detection threshold have been ruled out, within the limitations of the test. These embryos have the highest chance of implanting, developing to term, and resulting in the birth of a healthy baby. Ongoing pregnancy and live birth rates for euploid embryos vary but can be quoted generally between 50–70% and 45–65%, respectively. Success rates should be quoted by the fertility center where possible. Over half of all first-trimester miscarriages are caused by aneuploidy; thus, the risk of miscarriage is also reduced when transferring euploid embryos. The residual risk of miscarriage following the transfer of a euploid embryo is between 7 and 15%.

Indeterminate

Embryos with indeterminate results did not receive a reliable result and can be considered for transfer in a manner similar to untested embryos. Depending on embryo quality, rebiopsy may be considered.

Non-viable whole-chromosome aneuploidy

Embryos with non-mosaic, whole-chromosome aneuploidy are typically not viable. Transfer of such embryos would most likely fail to implant or would miscarry in the first trimester (Scott et al., 2012; Tiegs et al., 2021). As such, most aneuploidies are not associated with a recognized syndrome. This point can be helpful to explain to patients who might otherwise turn to the internet in a fruitless effort to research syndromes that don't exist.

Viable autosomal trisomies

There are three viable autosomal trisomies associated with recognized syndromes: trisomy 21 (Down syndrome), trisomy 18 (Edwards syndrome), and trisomy 13 (Patau syndrome). Embryos reported to have trisomy 21 have an increased risk of miscarriage. When detected prenatally, pregnancies with trisomy 21 have only about a 24% chance of developing to term. Individuals with Down syndrome have varying degrees of cognitive impairment and birth defects, including characteristically dysmorphic facial features and heart defects. Down syndrome affects approximately 1 in 800 individuals in the general population.

Trisomy 18, also known as Edwards syndrome, is a significant chromosome condition occurring in approximately 1/7,500 births. Approximately 95% of pregnancies identified to have trisomy 18 miscarry. Babies born with trisomy 18 typically have intrauterine growth restriction and an increased risk of multiple congenital anomalies, including heart defects. Affected babies usually cannot live for more than a few months.

Trisomy 13, also known as Patau syndrome, is associated with several life-threatening medical problems. More than 95% of pregnancies identified to have trisomy 13 miscarry. Trisomy 13 is observed in only about 1/20,000 births and affected babies usually cannot live for more than a few weeks or months. Babies with trisomy 13 typically have multiple congenital anomalies, including heart defects and brain or spinal cord abnormalities.

Viable sex chromosome aneuploidies

Sex chromosome aneuploidies include Turner (XO) syndrome, triple X (XXX) syndrome, Klinefelter (XXY) syndrome, and Jacobs (XYY) syndrome.

Despite Turner syndrome being a recognized syndrome affecting approximately 1/4,000 female births, less than 1% of embryos with XO are capable of developing to term. Those that do develop to term are likely mosaic (Gersen and Keagle, 2013). Individuals with Turner syndrome have a range of symptoms, including a risk of cardiac and renal defects and infertility. Intelligence is typically normal; however, about 10% require special education (Nussbaum et al., 2004). Women with Turner syndrome frequently are of short stature, have a webbed neck, and premature ovarian failure.

Aneuploidy of 47,XXY is compatible with life and causes a genetic syndrome called Klinefelter syndrome, found in the human population at an incidence of 1/1,000 male births. Despite being compatible with life, approximately 50% of pregnancies identified to have 47,XXY miscarry. Klinefelter syndrome is not phenotypically recognizable at birth. Individuals with Klinefelter syndrome may appear taller than average and produce less testosterone, which can affect development during puberty and fertility. Children with Klinefelter syndrome may have learning disabilities and delayed speech and language development.

Trisomy X is compatible with life and causes a genetic syndrome called Triple X syndrome. Triple X syndrome is common in the human population with an incidence of 1/1000 female births. Most pregnancies identified to have Triple X syndrome are able to develop to term. Triple X syndrome is not phenotypically recognizable at birth. Affected individuals may appear taller than average and have learning disabilities. Fertility is usually normal.

XYY syndrome is not phenotypically recognizable at birth. Affected individuals have normal intelligence but may have an increased risk of learning disabilities and behavioral problems. Adults may appear taller than average, and fertility is expected to be normal. Early reports that the condition is associated with aggression or psychopathological behavior have been discounted as being due to ascertainment bias.

Segmental aneuploidy

Embryos with a segmental aneuploidy have gained or lost part of a chromosome. The resolution of PGT-A is such that the potential gains and losses are expected to be large enough to affect global development, rather than microdeletions or microduplications associated with specific viable syndromes. Typically, multiple genes are included in PGT-A-detected segmental aneuploidies. Therefore, segmental aneuploidies detected on PGT-A could compromise implantation, result in developmental abnormalities and miscarriage, or cause the birth of a child with multiple congenital anomalies.

Rarely are segmental imbalances, which are large enough to have been picked up by PGT-A, inherited from a healthy adult. Deletions and duplications can be mitotic in origin, and therefore, mosaic in an embryo, regardless of whether biopsied cells were mosaic for that imbalance (Girardi et al., 2020).

Mosaic aneuploid results

Mosaicism is the finding of two cell lines, with differing chromosomal counts, within an embryo. PGT-A assesses for chromosome copy number change in the biopsied cells as a group rather than in each individual cell. Embryo biopsies that are found to have an intermediate copy number change are reported as mosaic. The level of mosaicism is based on bioinformatic calling rather than an exact count of the number of aneuploid or euploid cells. The level of mosaicism in a biopsy may not represent the level of mosaicism in the embryo. The embryo may contain a similar proportion of mosaic cells, a pure population of euploid cells, a pure population of aneuploid cells, or any proportion of euploid and aneuploid cells in between. Given these limitations, when an embryo is reported as mosaic, embryos should be considered to be putatively mosaic.

Emerging data show that mosaic embryos have reduced potential to implant, develop to term, or result in a healthy liveborn baby compared with euploid embryos; however, mosaic embryos do have an increased potential to result in a seemingly healthy pregnancy compared to embryos reported as non-mosaic aneuploid (Fragouli et al., 2017; Greco et al., 2015; Munné and Wells, 2017; Victor et al., 2019; M. Viotti et al., 2018; Manuel Viotti et al., 2021). Taking into consideration several different studies, mosaic embryos have between a 15–45% chance of resulting in an ongoing pregnancy or live birth. Embryos with lower-level mosaicism are likely to have higher reproductive potential than embryos with higher level mosaicism, and some studies have suggested that the reproductive potential of low level mosaicism may be nearly as high as for embryos reported as euploid (Capalbo et al., 2021; Rubino et al., 2018). Others have documented no differences between the reproductive potential of low and high mosaic categories (Munné & Wells, 2017; Victor et al., 2019). One must take into consideration the mosaic calling thresholds, which vary considerably among studies. The type of mosaic aneuploidy also appears to be a factor in the reproductive potential of putatively mosaic embryos. Several studies have shown that embryos reported as mosaic for segmental aneuploidies have the highest reproductive potential (Fragouli et al., 2017; Viotti et al., 2018; Viotti et al., 2021). Others have shown evidence that embryos reported as having mosaicism for several chromosomes have the lowest reproductive potential (Munné et al., 2017; Viotti et al., 2021). Researchers have submitted several opinions on prioritization of mosaic embryos, none of which are universally agreed upon (*CoGEN Statement - IVF-Worldwide*, n.d.; Cram et al., 2019; Grati et al., 2018; Practice Committee and Genetic Counseling Professional Group [GCPG], 2020; Viotti et al., 2021).

If an embryo is mosaic, it is possible that euploid cells within the embryo will progress and non-viable, aneuploid cells will not, resulting in a healthy child. Progression of aneuploid cells in a pregnancy or in isolated tissue or organs in a liveborn child is possible, but rare (Greco et al., 2015; Kahraman et al., 2020; Viotti et al., 2021). There are many case reports of pregnancies and liveborn babies affected with different mosaic aneuploidies and shown to have multiple congenital anomalies and/or neurological differences. Although the studies provide valuable information about clinical features that may be associated with a specific mosaic aneuploidy, they document naturally conceived pregnancies and babies, rather than pregnancies resulting from a mosaic embryo transfer. This distinction is important when counseling using an evidence-based approach (Besser et al., 2021).

Predictive value of aneuploid results

PGT-A is considered a screening test, rather than a diagnostic testing. Screening test performance measures, such as positive and negative predictive values, are unavailable for PGT-A due to the lack of a diagnostic testing that can be performed on embryos as a comparison. The positive predictive values of the different PGT-A results are likely to vary but cannot be accurately measured.

Using outcome data to calculate predictive values of PGT-A is one worthy approach, however imperfect. A few non-selection studies have been performed to measure the predictive value of euploid and aneuploid PGT-A results based on ongoing pregnancy/delivery rates as the reference point (Tiegs et al., 2021; Werner et al., 2015). These studies have shown that the predictive value of a euploid result was approximately 65%, correlating positively with the probability of an ongoing pregnancy/delivery; the predictive value of an aneuploid result was nearly 100%, owing to their finding that no aneuploid transfer resulted in an ongoing pregnancy/delivery. These studies were not sufficiently powered to assess predictive values of each chromosome aneuploidy, segmental aneuploidy, or varying numbers of chromosomes in a complex or chaotic result.

High aneuploidy rates

Patients often wonder if any conclusions can be drawn about their health status or fertility based on the PGT-A results. Typically, the answer is no. Even when there is a higher rate of aneuploidy than would be expected based on the age-related risks of aneuploidy, this is most likely due to random chance, rather than to anything unusual about the couple. With the statistically small number of embryos that can typically be obtained from a single or even multiple IVF cycles, it is generally not surprising to see results with all aneuploid embryos. The chance of aneuploidy for each embryo in subsequent cycles would be independent of the results of previous cycles.

Patterns of aneuploidy in PGT-A results

Occasionally, striking patterns of aneuploidy recurring on particular chromosomes can be seen. In these cases, the patient or partner could be a carrier of a balanced rearrangement (Snider, Darvin, et al., 2021).

This is particularly true if the recurring aneuploidies are segmental deletions or duplications with similar breakpoints seen in more than one embryo. The hallmark of unbalanced gametes from a translocation carrier is a pair of segmental imbalances corresponding to each chromosome of the reciprocal translocation. Robertsonian translocations, especially the der(13;14) translocation, can result in gametes with complete aneuploidies of acrocentric chromosomes. These patterns can be more difficult to assess because whole-chromosome imbalances are indistinguishable from age-related, sporadic aneuploidy when assessed with PGT-A. In all cases, cytogenetic analysis through karyotyping is necessary to confirm the balanced rearrangement in the egg and sperm source.

Prenatal Testing for Pregnancies Conceived Following PGT-A

Prenatal testing should be offered to patients who have had PGT-A and transfer of a euploid embryo. PGT-A should not be considered a replacement for prenatal care and testing. PGT-A is a screening tool and can give false-negative results due to the limitation that testing is performed on a representative trophectoderm sample. The inner cell mass is not directly assessed, and embryonic mosaicism cannot be ruled out.

Prenatal testing can be grouped into two broad categories of diagnostic testing and non-invasive screening. Diagnostic testing includes chorionic villus sampling (CVS) and amniocentesis, and accuracy exceeds 99%. The sample type is amenable to various technologies and specialized testing, including chromosomal microarray (CMA), cytogenetics, FISH analysis, uniparental disomy (UPD) analysis, methylation studies, microdeletion/duplication assessment, and targeted testing for single-gene disorders. CVS can be performed around 11–13 weeks of pregnancy, and amniocentesis can be performed beginning at around 15 weeks of pregnancy. CVS and amniocentesis require sampling of the chorionic villi or amniotic fluid, respectively, and are procedures that involve slightly increased risks of miscarriage. Non-invasive prenatal testing (NIPT) requires a maternal blood draw and is thus not associated with increased risk of miscarriage. NIPT is a screening test with a chance of false-positives and false-negatives, and typically only screens chromosomes associated with common, viable aneuploidies. Non-invasive testing is very popular among those who have undergone extensive fertility treatment to achieve a pregnancy. For some patients, however, the increased accuracy and comprehensiveness of the diagnostic tests are paramount.

There are no formal recommendations for the type of prenatal testing that should take place following transfer of an embryo reported as mosaic; however, a committee from the American Society for Reproductive Medicine's (ASRM's) Genetic Counseling Professional Group issued an option statement outlining points to consider when offering prenatal testing to patients who are pregnant following transfer of an embryo reported as mosaic (Practice Committee and Genetic Counseling Professional Group [GCPG], 2020). Persisting mosaicism following transfer of a mosaic embryo has been documented in a single case report (Kahraman et al., 2020), but this phenomenon seems to be rare, leading some experts to argue that the need for diagnostic testing does not outweigh the potential risks. Cell-free fetal DNA circulating in maternal serum originates from placental cells, which are derived from the embryonic trophectoderm. Similarly, CVS collects cells from placental villi, derived from embryonic trophectoderm. Thus, amniocentesis is the only test that assesses fetal cells directly. NIPT, therefore, offers a non-invasive approach but is also a limited screening test that assesses placental tissues. CVS is a comprehensive diagnostic test, and normal/euploid results can be very reassuring, but it is also limited in that it cannot rule out confined placental mosaicism (CPM) and carries slight procedure-associated risks. Amniocentesis has the advantage of being a comprehensive and diagnostic test that assesses fetal cells directly, but also carries slight procedure-associated risks and is only available in the second trimester. Patients should be counseled about the pros and cons of these prenatal tests and empowered to make the most appropriate decision for their situation.

GENETIC COUNSELING FOR PGT-M

Indication for PGT-M

PGT-M can be a powerful tool in reducing the risk of genetic disease in offspring prior to pregnancy. PGT-M was historically developed to screen for severe genetic conditions that could affect the life of an individual. With the evolution of the genetic testing landscape, including the emergence of expanded carrier screening, PGT-M is now performed for a variety of genetic disorders with a wide spectrum of severity. In the infertile population, for example, some IVF patients are having embryos biopsied for PGT-A, and PGT-M is performed as a secondary or add-on test for very mild conditions. Some conditions for which PGT-M is performed have reduced penetrance, adult onset, and/or variable expressivity. In the United States, the use of PGT-M for such conditions, although requiring additional consideration and counseling, is considered ethically permissible by the American Society of Reproductive Medicine and is generally supported by genetic counselors (Daar et al., 2018; Porto et al., 2021). The use of PGT-M for these conditions may be federally regulated in European and other countries.

When determining whether PGT-M testing is indicated, one must recognize that not all genetic variants cause severe disease, and some genetic variants are benign. An understanding of the potential severity and variability of the genetic condition is crucial when deciding upon PGT-M. For some individuals, having a child with a relatively mild genetic disorder is deemed intolerable, while others are more accepting of genetic variability and disease. Infertile patients may have difficulty creating a sufficient number of euploid embryos, and, if faced with the alternative of having no biological child, might find having a biological child with a mild genetic disorder preferable. PGT-M might have limited clinical utility for patients in such situations.

The specific variant(s) must also be considered. A variant of unknown significance (VUS) in a gene associated with disease might not be disease causing. Patients carrying a VUS might elect for PGT-M to reduce risk but might also ethically decline PGT-M given the uncertain pathogenicity of the VUS. Variant classification may change over time as relevant cases are added to genetic databases. Current variant classification should be sought before determining whether PGT-M should be offered. Performing PGT-M for benign or likely benign variants would be limited to no clinical utility and could even be a detriment. Furthermore, some specific combinations of variants are benign, even when the same variants are pathogenic when in combination with other variants. Therefore, the provider must review the interpretation of each variant and, for recessive conditions, each combination of variants, before determining whether PGT-M is appropriate.

On the other hand, providers should freely and equitably mention the possibility of PGT-M to patients with a family history of a genetic disease or to reproductive couples with positive carrier screening results. What may be considered mild to one individual or family might have a significant impact on another, and providers should be cautious about making subjective determinations on a patient's behalf.

Considerations When Offering PGT-M

There are several important points to convey to patients when raising the possibility of PGT-M. The first is that PGT-M is not always possible for every genetic condition or every patient. The genetic variant(s) underlying the condition must be known for testing to be accurately performed on embryos. If a patient is affected with a genetic condition, but no variant has been identified, PGT-M is usually not possible. If genetic testing has not previously identified the underlying genetic change associated with the condition, affected patients may need to undergo genetic testing before the possibility of PGT-M can even be evaluated. Genetic testing is a process that can take time, and, in some cases, a causative variant may never be confirmed. Even after the genetic variant has been identified, it is crucial to set expectations for test development, or "probe development." The additional set-up time required can be unexpected and discouraging, especially for patients battling infertility. The need for DNA samples from first-degree relatives should be raised in early discussions. There may be precious opportunities to obtain and bank DNA samples from prenatal testing or from relatives with terminal illness. Additionally, it should be understood that PGT-M testing is never a guaranteed option until the development of the probe has been completed. PGT-M test development should take place prior to a patient's IVF treatment, unless extraordinary circumstances dictate otherwise.

Prior to performing PGT-M, both the physician and the patient should have an understanding of how the test results will be utilized in embryo transfer decisions. There is often an expectation that embryos with reduced risk will be produced, in which case, the decision to transfer these embryos is clear. In the unfortunate outcome that all embryos are found to be at risk of developing the genetic condition, it is important that the reproductive options available to a patient are clear. Such options might include pursuing additional IVF cycles with PGT-M testing, pursuing additional IVF cycles without PGT-M testing, or transferring embryos with known genetic risk. For some conditions, transferring embryos with known genetic risk has serious ethical considerations. If the clinic's policy precludes transferring at-risk embryos, this should be clearly communicated to the patient prior to testing.

Lastly, patients should clearly understand that PGT-M is a selection tool, not a gene editing tool. With today's technology, embryos cannot be "cured" of a genetic disease, they can only be selected against. PGT-M should not be considered a replacement for prenatal diagnosis. Conceiving spontaneously and pursuing prenatal diagnosis is an alternative.

Pre-Test Counseling for PGT-M

Patients should be counseled about the recurrence risks of the condition for which they are pursuing testing. This corresponds to the inheritance pattern of the condition: 50% for autosomal dominant conditions and 25% for autosomal recessive conditions. For X-linked conditions, the appropriate recurrence risks to quote will depend on whether the female carriers are generally expected to be affected: 50% for X-linked conditions classified as dominant and 25% for X-linked conditions classified as recessive. Even when the condition is considered to be classically recessive, patients should be cautioned about the chance of female carriers manifesting the condition due to skewed X-inactivation.

Many patients whose embryos are biopsied for PGT-M take advantage of the availability of PGT-A to further select embryos with the highest reproductive potential. Some patients pursuing PGT-M will have infertility as the primary indication to pursue IVF, whereas others pursue IVF with the primary purpose of testing embryos for the single-gene disorder. Both groups of patients, but especially those in the latter group, may be surprised by the high incidence of aneuploidy and the potential complexity of PGT-A results. It is therefore essential not to gloss over the age-related risks of aneuploidy for PGT-M patients who are also pursuing PGT-A. The chance of having embryos that are both euploid and at reduced risk for the genetic condition can be calculated by multiplying the age-related aneuploidy rate with the recurrence risk of the single-gene disorder and subtracting this number from 1.

Statistical expectations may not be realized in PGT-M results. The fractions of at-risk and low-risk euploid and aneuploid embryos can be greatly skewed by random chance. The possibility of having no embryos available for transfer must be communicated in advance.

If IVF is pursued for the primary purpose of facilitating PGT-M testing, the potential risks of IVF treatment must be factored into a patient's decision to proceed with PGT-M. Similar to PGT-A, the risks of embryo biopsy are thought to be low; however, the IVF center performing the biopsy should be consulted for any clinic-specific risk estimates. The risks and limitations of PGT-A should be well explained to patients, even if pursuing PGT-M is their primary focus.

The limitations of PGT-M should be reviewed during pre-test counseling. PGT-M cannot detect single-gene disorders other than the condition for which PGT-M was requested and for which a probe was developed. PGT-M assesses embryos for the inheritance of only the variant(s) reported for the patient and/

or partner and cannot screen embryos for unidentified variants that may be associated with the condition. Misdiagnosis could result if a separate, unidentified, or unreported variant is responsible for the condition. PGT-M is not a diagnostic tool, and its accuracy is standardly quoted at 98%, allowing for a chance of false-positive or false-negative results. PGT-M does not screen for birth defects.

PGT-M Test Results

Test results are typically grouped into black and white categories of *low-risk* (negative for the variant or combination of variants screened) and *high-risk* (positive for the variant/variants screened). Low-risk embryos testing negative for the indicated condition will be identified and selected for transfer. At-risk embryos testing positive for the condition will be selected against or deprioritized. For dominant conditions, the results are either positive or negative for the variant. For recessive conditions, results are non-carriers, carriers, or affected, where carriers and non-carriers are expected to be unaffected with the genetic condition. For X-linked conditions where the egg source is a carrier, results are either positive or negative for males, or carrier or non-carrier for females. Grey results are possible, such as carrier females of many X-linked conditions and embryos positive for a VUS. Ideally, the clinical management of such embryos will have been discussed in advance.

Ideally, test results will be informative. Test results are sometimes complicated by allelic dropout, where the DNA at the locus of interest fails to amplify and cannot be interpreted. If allelic dropout affects all the linkage markers and the variant itself, the embryo will be reported as uninformative. Fortunately, allelic dropout usually affects only one or two linkage markers, and the information obtained from the remaining linkage markers is sufficient to provide a reliable diagnosis. Indeed, increasing the chance of an informative result is the purpose of using multiple linkage markers in PGT-M analysis. Rebiopsy and re-testing can be considered for embryos with uninformative results due to allelic dropout.

Prenatal Diagnosis for Pregnancies Conceived Following PGT-M

Prenatal diagnosis should be offered to patients who are pregnant following transfer of a PGT-M-tested embryo. The accuracy of prenatal diagnosis is superior to that of PGT-M. Whereas PGT-M is standardly quoted at 98% accuracy, the accuracy of prenatal diagnosis following amniocentesis or CVS exceeds 99%. Clinics and governmental bodies may regulate the availability of prenatal diagnosis for some genetic conditions. Non-invasive prenatal testing, or NIPT, is a powerful and non-invasive tool to screen pregnancies for Down syndrome and other chromosome copy number disorders; however, NIPT is, to date, unable to reliably diagnose single-gene disorders.

Transfer of Embryos with Positive PGT-M Results

Most patients set out to pursue PGT-M with the intention of transferring low-risk embryos that test negative for the condition. Unfortunately, the outcome where no embryo is reported as low risk for the condition (and euploid if PGT-A was performed) is not uncommon. Depending on the condition, individuals may consider transfer of embryos with the at-risk allele(s). For some patients, an embryo testing positive for an adult-onset or mild genetic disorder may be their only chance of having a biological child. The position statement from ASRM suggests that "valid and reasoned arguments exist to support provider decisions to either assist in transferring genetically anomalous embryos or to decline to assist in such transfers" (Daar et al., 2017). In making this difficult decision, patients should be counseled on the variability of the condition, especially if they themselves are affected, potentially causing a personal bias on the likely severity of the condition. The patients should also consider the emotional and ethical burden of knowingly bringing a child into existence with a known risk of a genetic condition.

REFERENCES

Aksglaede, L., et al. (2013). *European Journal of Endocrinology, 168*(4), 67–76.

Anton, E., et al. (2002). *Cytogenetic and Genome Research, 97*(3–4), 149–154.

Besser, A. G., et al (2021). *Fertility and Sterility*, 1–5.

Calanchini, M., et al. (2020). *Fertility and Sterility, 114*(1), 144–154.

Capalbo, A., et al. (2021). *MedRxiv, 2021.*

Carey, L., et al. (2015). *Translational Pediatrics, 4*(2), 181–188.

Cimadomo, D., et al. (2016). *BioMed Research International, 2016.*

CoGEN Statement - IVF-Worldwide. (n.d.). Retrieved August 8, 2021, from https://ivf-worldwide.com/cogen/general/cogen-statement.html

Colley, E., et al. (2019). *Human Reproduction Update, 25*(4), 452–472.

Cram, D. S., et al. (2019). *Reproductive BioMedicine Online, 39*, e1–e4.

Daar, J., et al. (2017). *Fertility and Sterility, 107*(5), 1130–1135.

Daar, J., et al. (2018). *Fertility and Sterility, 109*(6), 989–992.

de Souza, D. A. S., et al. (2018). *Andrology, 6*(1), 127–135.

Fragouli, E., et al. (2017). *Human Genetics, 136*(7), 805–819.

Gardner, R. J. M., et al. (2018). *Gardner and Sutherland's Chromosome Abnormalities and Genetic Counseling* (Vol. *1*). Oxford University Press.

Gersen, S. L., & Keagle, M. B. (2013). *The Principles of Clinical Cytogenetics.* Springer.

Girardi, L., et al. (2020). *American Journal of Human Genetics, 106*(4), 525–534.

Grande, M., et al. (2012). *Human Reproduction, 27*(10), 3109–3117.

Grati, F. R., et al. (2018). *Reproductive BioMedicine Online, 36*(4), 442–449.

Greco, E., et al. (2015). *The New England Journal of Medicine, 373*(21), 2089–2090.

Harper, P. (2010). *Practical Genetic Counseling* (7th ed.). Edward Arnold Ltd.

Hassold, T., et al. (1996). *Environmental and Molecular Mutagenesis, 28*(3), 167–175.

Hunt, P. A., et al. (2002). *Science, 296*(5576), 2181–2183.

Jungwirth, A., et al. (2012). *European Urology, 62*(2), 324–332.

Kahraman, S., et al. (2020). *Human Reproduction, 35*(3), 727–733.

Krausz, C., et al. (2018). *Expert Review of Molecular Diagnostics, 18*(4), 331–346.

Maiburg, M., et al. (2012). *Fertility and Sterility, 98*(2), 253–260.

Munné, S., et al. (2017a). *Fertility and Sterility, 108*(1), 62–71.e8.

Munné, S., et al. (2017b). *Fertility and Sterility, 107*(5), 1085–1091.

Munné, S., et al. (2019). *Fertility and Sterility, 112*(6), 1071–1079.e7.

Nussbaum, R. L., et al. (2004). *Thompson & Thompson Genetics in Medicine* (7th ed.). Saunders Elsevier.

Porto, A., et al. (2021). *Journal of Genetic Counseling*, June, 1–8.

Practice Committee and Genetic Counseling Professional Group (GCPG). (2020). *Fertility and Sterility, 114*(2), 246–254.

Punab, M., et al. (2017). *Human Reproduction, 32*(1), 18–31.

Rodrigo, L., et al.(2010). *Fertility and Sterility, 94*(4), 1380–1386.

Rubino, P., et al. (2018). *Fertility and Sterility, 110*(4), e406–e407.

Scott, R. T., et al. (2012). *Fertility and Sterility, 97*(4), 870–875.

Shah, K., et al. (2003). *Reproduction, 126*:13–25.

Snider, A. C., et al. (2021a). *F&S Reports, 2*(1), 72–79.

Snider, A. C., et al. (2021b). *F&S Reports, 2*(1), 80–87.

Sugiura-Ogasawara, M., et al. (2012). *Human Reproduction, 27*(8), 2297–2303.

Sybert, V., et al. (2004). *The New England Journal of Medicine, 351*(12), 1227–1238.

Tartaglia, N. R., et al. (2010). *Orphanet Journal of Rare Diseases, 5*(8).

Tharapel, A. T., et al. (1985). *BJOG: An International Journal of Obstetrics & Gynaecology, 92*(9), 899–914.

Tiegs, A. W., et al. (2021). *Fertility and Sterility, 115*(3), 627–637.

Torrealday, S., et al. (2017). *Open Peer Review Premature Ovarian Insufficiency* [version 1; referees: 3 approved].

Treff, N. R., et al. (2013). *Fertility and Sterility, 100*(3), 624–630.

Uehara, S., et al. (2001). *Fertility and Sterility, 76*, 908–914.

Uhlman, W. R., et al. (2009). *A Guide to Genetic Counseling* (2nd ed.). Wiley-Blackwell.

Venkatesh, T., et al. (2014). *Application of Clinical Genetics, 7*, 235–243.

Victor, A. R., et al. (2019). *Fertility and Sterility, 111*(2), 280–293.
Viotti, M., et al. (2018). *Fertility and Sterility, 110*(4), e420.
Viotti, Manuel, et al. (2021). *Fertility and Sterility, 115*(5), 1212–1224.
Weil, J. (2000). *Psychosocial Genetic Counseling*. Oxford University Press.
Werner, M. D., et al. (2015). *Fertility and Sterility, 104*(3), e12–e13.
Werner, M., et al. (2012). *Journal of Assisted Reproduction and Genetics, 29*(8), 817–820.
Wong, E. C., et al. (2008). *Human Reproduction, 23*(2), 374–378.

Index

A

ABC, *see* ATP-binding cassette
Abnormal meiosis, 46
Abnormal reproductive tract microbiome, 240–244
ACMG, *see* American College of Medical Genetics and Genomics
ACOG, *see* American College of Obstetricians and Gynecologists
ACTH, *see* Adrenocorticotropic hormone
Addition algorithm, 110
Adrenocorticotropic hormone (ACTH), 351
Adult-onset disorders, 33–34, 311
Advanced maternal age (AMA), 25, 40, 76, 106, 125–126, 371, 372
Advanced Therapy Medicinal Product (ATMP), 273
AF, *see* Amniotic fluid
Allele dropout (ADO), 124
Alpha-thalassemia, 374
Altered meiosis, 45, 48
AMA, *see* Advanced maternal age
American College of Medical Genetics and Genomics (ACMG), 31, 306
American College of Obstetricians and Gynecologists (ACOG), 31, 311
Amniotic fluid (AF), 25, 158, 289, 301, 302, 310, 381
Aneuploidy, 106
- morphokinetics
  - barriers to access, 112–114
  - early stages of embryonic development, 107–108
  - later stages of embryonic development, 108–109
  - model risk and algorithms, 110–112
  - pronuclei, 107
- testing, 160

Array comparative genomic hybridization (aCGH), 117, 118, 154, 161
Array hybridization, 38
Array single nucleotide polymorphism (aSNP), 154
Asherman's syndrome (AS), 267
- CD133+ BMDSCS, 269–273
- regenerative medicine, 273
- stem cells, 268–269

Assisted hatching (AH), 131, 161
Assisted reproduction techniques (ART), 64, 100, 103, 244, 346, 362, 368
Ataxia telangiectasia (ATM), 33
ATP-binding cassette (ABC), 268
Autosomal dominant inheritance, 375
Autosomal recessive inheritance, 375
Azoospermia, 36, 41, 51, 142, 373

B

Bacterial artificial chromosome (BAC), 123, 332
Bacterial vaginosis (BV), 226, 238, 248
Balanced chromosomal rearrangements, 54, 371–372
Balanced structural chromosome rearrangements, 138
- inversions, 141
- translocations, 138–140

Base editors, 210
BDNF, *see* Brain-derived neurotrophic factor
Binary Alignment Map (BAM), 5
Bioinformatic tools, 5, 40
Blastocyst, 87, 103–104, 109, 143, 347
Blastocyst-stage biopsy, 120–121
Blastomere, 120, 180, 206
Body mass index (BMI), 284, 296, 364
Bone marrow-derived stem cells (BMDSCs), 268
Bone sialoprotein (BSP), 354
Brain-derived neurotrophic factor (BDNF), 351
Bridge amplification, 278
BSP, *see* Bone sialoprotein
Bulk vs. single cells, 12
BV, *see* Bacterial vaginosis

C

CAKUT, *see* Congenital anomalies of the kidneys and urinary tract
Cancer-associated fibroblasts (CAFs), 21
Carrier screening (CS), 37, 40, 368
CBAVD, *see* Congenital bilateral absence of the vas deferens
CCR, *see* Complex chromosomal rearrangements
CCT, *see* Comprehensive chromosome testing
CE, *see* Chronic endometritis
Cell free DNA (cfDNA), 124, 133, 161
- clinical results, 170
- culture conditions, IVF laboratory, 165–166
- embryonic culture medium, 161–165
- NGS, 168–169
- ploidy analysis, 163–164
- SBM vs. WB, 163
- strengths and limitations, 170–171
- WGA, 168–169

Cell-free fetal DNA (cffDNA), 284, 304, 381
Centromere-specific multicolor fluorescence *in situ* hybridization (cenM-FISH), 43
Centromeric (CEN), 144
Centromeric probes (CEP), 53
CFTR, *see* Cystic fibrosis transmembrane conductance regulator
CG, *see* Cytosine guanine
CGH, *see* Comparative genomic hybridization
Chaotic mosaic, 152
Chorionic villus sampling (CVS), 25, 151, 158, 284, 302, 341, 364, 381
Chromatin immunoprecipitation experiments (ChIP-seq), 99
Chromatin remodeling, 98
Chromosomal abnormalities, 24, 54
- chorionic villus structure, 327

clinical outcome, 55–56
embryonic outcomes, 56
genetic counseling, 56–57
limitations, 57
offspring, 56
origin, 322–324
sex chromosome monosomy, 324–325
sperm
clinical outcome, 55
embryonic outcomes, 56
offspring, 56
strengths, 57
tetraploidy, 326–327
triploidy, 325–326
trisomy, 325
Chromosomal inversions, 29
Chromosomal mosaicism, 130
Chromosomal rearrangements, 136
balanced and unbalanced, 137
Chromosomal translocation
abnormal liveborns, 28
cumulative likelihood of livebirths, 29
frequency and management of carriers, 27–28
theoretical vs. empirical risks, 28
Chromosome fingerprinting, 80–81
Chronic endometritis (CE), 240, 247
diagnosis
endometrial culture, 249
histology and immunohistochemistry, 249–250
hysteroscopy, 250
molecular microbiome, 250
etiopathogenesis, 247–249
intrauterine infusion of antibiotics, 252–253
pathologies, 253
transplantation of vaginal bacteria, 252
treatment, 251–252
Circulating tumor cells (CTC), 281
Cleavage-stage biopsy, 117, 120, 144
Cleavage-stage human embryos, 154
Clinical genetics, 7–10
CNV, *see* Copy number variants
Colonization resistance, 240
Colony-stimulating factor-1 (CSF-1), 351
Community state types (CST), 227, 238
Comparative genomic hybridization (CGH), 29, 73, 122, 196, 316, 333
Competitive exclusion, 240
Complex chromosomal rearrangements (CCR), 140, 208
Comprehensive chromosomal screening (CCS), 118, 122
Comprehensive chromosome testing (CCT), 154
Congenital abnormalities, 149, 288, 363
Congenital anomalies of the kidneys and urinary tract (CAKUT), 308
Congenital anomaly, 24
Congenital bilateral absence of the vas deferens (CBAVD), 26, 374
Copy number variants (CNV), 6, 33, 276, 285
Cord blood, 289
Cryopreservation, 129
CS, *see* Carrier screening
CST, *see* Community state types
CTC, *see* Circulating tumor cells
CTS, *see* Circulating tumor cells
CVS, *see* Chorionic villus sampling
Cyclic adenosine monophosphate (cAMP), 261
Cyclooxygenase-2 (COX2), 349
Cystic fibrosis transmembrane conductance regulator (CFTR), 13
Cytogenetics, 136
Cytokines, 350–351
Cytosine guanine (CG), 355

D

Data Coordination Platform (DCP), 14
DC, *see* Dendritic cells
Decidualization
barriers to access, 264
blastocyst invasion, 259
challenge, 264–265
infertility and pregnancy complications, 257–260
late secretory menstrual phase, 256–257
methodology, 261
PE, 260–261
research results, 261–264
strengths and limitations, 264
Decidual natural killer cells (dNK), 13, 350
Delayed implantation, 348
Dendritic cells (DCs), 350, 352
De novo methylation, 355
Dentin matrix protein (DMP), 354
DET, *see* Double embryo transfer
Developmental horizons, 347
Developmental origins of health and disease (DOHAD) hypothesis, 362
Digenic effect, 38
Digital PCR (dPCR), 293
D-loop, 176
DMP, *see* Dentin matrix protein
DNA methylation, 92–93
development in a mouse, 94
embryonic reprogramming, 95
reprogramming, 93–95
DNA methyltransferase (DNMT), 62, 93
DNA sequencing technologies, 278
dNK, *see* Decidual natural killer cells
DNMT, *see* DNA methyltransferase
DOHAD, *see* Developmental origins of health and disease hypothesis
Double embryo transfer (DET), 127
Double-stranded DNA breaks (DSBs), 205
Down syndrome, 56, 72
Droplet single-cell assay for transposase-accessible chromatin using sequencing (dscATAC-seq), 17–18
DSB, *see* Double-stranded DNA breaks
Dysbiosis, 240

E

Early embryo
Ca^{2+} signaling and fertilization, 181
metabolism, 179–180
oxidative stress, 180
ECM, *see* Endometrial extracellular matrix
ECS, *see* Expanded carrier screening

EDC, *see* Endocrine-disrupting chemicals
Edwards syndrome, 378
EGA, *see* Embryonic genome activation
EGF, *see* Endothelial growth factor
ELB, *see* Endometrial liquid biopsy
Elective SET (eSET), 129
Elective single embryo transfers (eSET), 121
EMA, *see* European Medicines Agency
Embryo biopsy strategies
  blastomere, 120
  first and second polar body biopsy, 119–120
  trophectoderm, 120–121
Embryo implantation, 346–348
Embryonic genome activation (EGA), 92
Embryo transfer (ET), 252
Endocrine-disrupting chemicals (EDC), 332
Endogenous retroviral elements (ERV), 99
Endometrial extracellular matrix (ECM), 351
Endometrial liquid biopsy (ELB), 353
Endometrial receptivity analysis (ERA), 213–214, 370
  accuracy, 214–215
  challenging, 215–216
  clinical protocols, 216–217
  decision tree, 223
  endometriosis, 221
  first IVF cycle, 221–222
  interpreting results, 217
  limitations, 222
  pathological conditions, 220–221
  reproducibility, 214–215
Endometrial stromal fibroblast (ESF), 354
Endometriosis, 221
Endosomal sorting complex required for transport (ESCRT), 352
Endosymbiotic theory, 173–174
Endothelial growth factor (EGF), 351
Endothelial progenitor cells (EPC), 267, 269
Epigenetic regulation, 356
Epigenetics, 62
  assistive reproductive technologies, 68–69
  chromatin structure, 63
  CpG islands, 68
  DNA methylation, 62
  dysfunction, 69
  histone modification and retention, 62–63
  male infertility
    chromatin modification, 64–65
    histone modification and retention, 64
    molecular diagnostics, 67–68
    non-coding RNA, 65
    RNA methylation, 65
    transgenerational disease, 68
  mechanisms and marks, 62
  methylation of RNA, 63
  non-coding RNA, 63
  transgenerational inheritance, 65–67
Epimutation, 65
eQTL, *see* Expression quantitative trait loci
ERA, *see* Endometrial receptivity analysis
ERV, *see* Endogenous retroviral elements
Erythroblastosis fetalis, 188
ESCRT, *see* Endosomal sorting complex required for transport
eSET, *see* Elective SET
ESF, *see* Endometrial stromal fibroblast
ESHRE, *see* European Society of Human Reproduction and Embryology
ET, *see* Embryo transfer
Ethnicity-based carrier screening, 30–31
Etiopathogenesis, 247–249
European Medicines Agency (EMA), 271
European Society of Human Reproduction and Embryology (ESHRE), 346
Exome sequencing (ES), 306
  analysis and interpretation of, 306–307
  clinical use of, 308
  medical management, 308–310
  proband vs. trio sequencing, 308
  technical limitations of, 307–308
Exons, 29
Expanded carrier screening (ECS), 32–33, 37
Expression quantitative trait loci (eQTL), 279
Extracellular vesicles (EVs), 351–353
Extravillous trophoblasts (EVTs), 259

F

FACS, *see* Fluorescent activated cell sorting
False-positive rate (FPR), 284
Fanconi anemia (FAN), 33
FCT, *see* First-trimester screening
Female infertility
  advanced maternal age, 372
  45, X (Turner syndrome), 372
  47, XXX (Triple X syndrome), 372
  ovarian insufficiency, 372–373
Female translocation, 146
Fertilization, 92
FET, *see* Fresh embryo transfer
Fetal exome sequencing
  current clinical experience, 308
  ES
    analysis and interpretation of, 306–307
    clinical use of, 308
    medical management, 308–310
    proband vs. trio sequencing, 308
    technical limitations of, 307–308
  incidental and secondary findings, 310
  NGS, 306
  prenatal results interpretation, 310
  technical challenges, 310
Fetal fraction (FF), 284
FFPE, *see* Formalinfixed paraffin-embedded
Fibroblast growth factor-extracellular signal regulated kinase (FGF-ERK), 349
First-trimester screening (FCT), 287
Flexible Multi-Analyte Profiling (xMAP), 350–351
Fluorescence in situ hybridization (FISH) technique, 7, 52, 121–122, 196
Fluorescent activated cell sorting (FACS), 15
Follicle-stimulating hormone (FSH), 348
Formalinfixed paraffin-embedded (FFPE), 280
47,XYY syndromes, 53
FPR, *see* False-positive rate
Fresh embryo transfer (FET), 120, 363
FSH, *see* Follicle-stimulating hormone

G

G-banding karyotype, 7
G-CSF, *see* Granulocyte-CSF
Gene mutations, 374
Genetic risk
birth defects, 26
chromosomal inversions, 29
chromosomal translocation, 27
family history, 25–26
newborn screening, 30
primary and secondary prevention, 26
rationale for determining, 24–25
recurrent aneuploidy, 26–27
single-gene (monogenic) disorders, 29–30
Genome activation, 98–100
Genome-wide association study (GWAS), 87
Genomic data, 6–7
Genomic diagnostic process, 3
Genomic index, 198
Genomic profiling technologies, 11
Germline genome editing (GE), 204
Gestational losses, 374
Granulocyte-CSF (G-CSF), 270
Green fluorescent protein (GFP), 269
Growth-related oncogene (GRO), 351
GWAS, *see* Genome-wide association study

H

Hamster oocytes, 51
HAT, *see* Histone acetyltransferases
HBEGF, *see* Heparin-binding EGF-like factor
HCA, *see* Human Cell Atlas
HD, *see* Huntington's disease
HDR, *see* Homology directed repair
Hematopoietic stem cells (HSC), 269
Hemoglobinopathies, 24
Hemolytic disease, 188
Heparin-binding EGF-like factor (HBEGF), 353
Hereditary leiomyomatosis, 279
Heterogeneous nuclear ribonucleoprotein C1 (hnRNPC1), 352
High-grade serous ovarian cancers (HGSOC), 20
Histology (HIS), 248
Histone acetyltransferases (HATs), 355
Histone methyltransferases (HMTs), 355
Histone modification asymmetry, 96–97
Homeobox protein, 354
Homologous recombination (HR), 205
Homology directed repair (HDR), 205
HSC, *see* Hematopoietic stem cells
Human aneuploidy
aging features, 85
embryo development, 77
embryonic arrest, 87–88
error type in meiosis I, 79
female meiosis, 85
GWAS, 87
historical context, 72–73
maternal chromosome errors, 77–79
mitotic errors, 85–87
oocytes and preimplantation embryos, 74–77
recombination patterns, 83
reverse segregation, 80–81
SAC, 85
time-lapse imaging, 81–83
U-shaped curve, 73–74
Human Cell Atlas (HCA), 11
bulk vs. single cells, 12
cell types, 13
comprehensive understanding, 14
Human chorionic gonadotropin (hCG), 349
Human endometrial stromal cells (hESCs), 256
Human ovary, 14
Human pachytene spermatocyte, 44
Huntington's disease (HD), 187
Hysteroscopy (HYS), 248, 250

I

Illumina technology, 334
Immunohistochemistry (IHC), 248, 249
Immunological mystery, 350
Imprinting control regions (ICRs), 96
Inducible nitric oxide synthase (iNOS) pathway, 350
Inerolysin, 229
Infertility, intracytoplasmic sperm injection (ICSI), 51
Inflammatory state of the endometrium (IISE), 248
Inner cell mass (ICM), 92, 121, 153, 163, 346
Insulin growth factor binding protein 1 (IGFBP-1), 256
Insulin-like growth factor binding protein-1 (IGFBP1), 349
Interchromosomal effect (ICE), 142
Intracytoplasmic sperm injection (ICSI), 165, 180, 205
Intrauterine adhesions (IUAs), 267
Intrauterine fetal death (IUFD), 289
Intrauterine fetal growth restriction (IUFGR), 288
Intrauterine growth restriction (IUGR), 258
In vitro fertilization (IVF), 69, 117, 138
Ion Personal Genome Machine (PGM), 334
Ion Torrent technology, 333
Isoimmunization, 188

J

Jacobs syndrome, 373

K

Karyomapping, 189–190
Karyotyping, 136
Klinefelter syndrome, 26, 55, 72, 373, 379
Knowledge arena, 6–7

L

Laboratory-based counseling, 371
Leiomyomas (LM)/leiomyosarcomas (LMS)
barriers to access, 281
clinical diagnostic tools, 279–280
differential diagnosis, 277–278
liquid biopsy, 282
NGS, 278–279
strengths and limitations, 280
Leishman stain
chromosome identification, 46

meiotic figures, 45
spermatogenic cells, 44
Leptin, 354
Leukemia inhibitory factor (LIF), 348, 349
Live-birth rates (LBR), 117
Locus-specific (LSI), 144
Long non-coding RNAs (lncRNAs), 63
Long Q-T syndrome (LGT), 34
Long terminal repeat (LTR), 97
Luteinizing hormone (LH), 348, 351

M

Macrophage colony-stimulating factor (MCSF), 351
Major histocompatibility complex (MHC), 350
MALBAC, *see* Multiple annealing and looping-based amplification cycles
Male factor infertility, 127
Male infertility, 47
chemotherapy and radiotherapy treatments, 55
chromosomal positioning, 48–49
chromosomopathy, 55
epigenetics
chromatin modification, 64–65
histone modification and retention, 64
molecular diagnostics, 67–68
non-coding RNA, 65
RNA methylation, 65
transgenerational disease, 68
genetic counseling
CFTR gene, 373–374
Karyotype 47, XXY (Klinefelter syndrome), 373
impaired meiosis in testicular analysis, 54
impaired sperm parameters, 54
interpretation of results, 48
limitations of technique, 48
repetitive implantation failure, 55
RPL, 55
Maternal cell contamination (MCC), 332, 336, 369
Maternal-to-zygotic transition (MZT), 92
Materno–fetal communication, 357
MCC, *see* Maternal cell contamination
MCSF, *see* Macrophage colony-stimulating factor
MDA, *see* Multiple displacement amplification
Medium-chain acyl-CoA dehydrogenase (MCAD), 30
MeioMapping, 84
Meiosis, 42, 43
abnormal meiosis, 46
altered meiosis, 45
anomalies, 45–47
low chiasmata count, 46–47
meiotic arrest, 47
methods, 43–45
normal meiosis, 45
presence of univalents, 47
vs. infertile males, 47
Meiotic recombination, 42
Mendelian disorders, 1–2
Mendelian inheritance, 375
Metaphase CGH (mCGH), 122
M-FISH, *see* Multiplex fluorescence *in situ* hybridization protocols
MHC, *see* Major histocompatibility complex
Microbiome, 225
Microbiota, 225
Microtubule organizing centers (MTOC), 85
Miscarriage, 330
ART treatments, 331
CGH, 333
chromosomal analysis, 332–333
EDS, 332
gestational age, 338–339
greater cost effectiveness, 342
high rate of informativity, 341
low rate of MCC, 341
male factor, 331–332
maternal age, 337–338
NGS, 333–334
oocyte origin, 339–340
POC molecular analysis
detection of del/dups, 341
isolation of fetal tissue, 340
mosaicism, 340–341
recurrent, 331
shorter TAT, 342
SNP, 333
tissue collection, 332
Misdiagnosis, 363–364
Mitochondria
DNA characteristics, 175–176
early embryo
Ca^{2+} signaling and fertilization, 181
metabolism, 179–180
oxidative stress, 180
embryonic implantation, 182–183
functions, 175
inheritance and bottleneck selection, 176
morphological changes, 178–179
nucleoid proteins, 178
origin, 173–174
replication/mtDNA replication, 176–178
stress sensor, 181–182
Mitochondrial DNA (mtDNA), 174
Monocyte chemotactic protein 1 (MCP1), 351
Monogenic disease, 202–203
clinical research, 207–208
CRISPR-Cas9, 204–207
embryo selection, 203–204
ethics, 210
GE, 204
genome editing, 208
germline genome editing, 210
off-target consequences, 209–210
preimplantation embryos, 208–209
Morphokinetics
barriers to access, 112–114
early stages of embryonic development, 107–108
later stages of embryonic development, 108–109
model risk and algorithms, 110–112
pronuclei, 107
Morphology vs. morphokinetics, 103–105
Morula, 347
Mosaicism, 151
classifications, 152
clinical outcomes, 157
detection tools, 153–154

diagnosis, 154–155
embryo, 152–153
mitotic errors, 152
non-selection trial, 156–157
origin, 151–152
prenatal diagnosis, 157–158
treatment, 155–156
MTOC, *see* Microtubule organizing centers
Mucin 1 (MUC 1), 353
Multiple annealing and looping-based amplification cycles (MALBAC), 122
Multiple displacement amplification (MDA), 122
Multiplex fluorescence *in situ* hybridization protocols (M-FISH), 44
MZT, *see* Maternal-to-zygotic transition

N

National Society of Genetic Counselors (NSGC), 31
Negative predictive value (NPV), 291
Next-generation sequencing (NGS), 1, 2, 7, 117, 118, 333–334
NIPD, *see* Non-invasive prenatal diagnosis
NLD, *see* Non-Lactobacillus dominated
NOD-like receptors (NLRs), 229
Non-homologous end joining (NHEJ), 205
Non-invasive PGT-A (niPGT-A), 121, 161
Non-invasive prenatal diagnosis (NIPD), 285
applications, 305
barriers to access, 311
clinical results, 305
de novo variants, 303
genetic counseling, 312
methodology, 303
strengths and limitations, 306
testing, 302
whole-genome cfDNA analysis, 304–305
Non-invasive prenatal testing (NIPT), 25, 158
analysis of common aneuploidies, 287
barriers to access, 296
clinical trials, 285–286
confined placental mosaicism, 294
decision tree, 298
false-negative results, 296
maternal factors, 295–296
monogenic disorders, 292–293
RAT, 288–289
SA, 289–292
SCAs, 287–288
SNP-based, 287
strengths and limitations, 293–294
targeted genome sequencing, 286
VT, 294–295
WGS-based NIPT, 286
Non-Lactobacillus dominated (NLD), 235
Non-receptive endometrium, 217
Normal meiosis, 45
Normozoospermic men, 53
NPV, *see* Negative predictive value
NSGC, *see* National Society of Genetic Counselors
Nucleosome remodeling histone deacetylation complex (NuRD), 96
Nucleosomes, 355

O

Oligoasthenozoospermic males, 54
Oligozoospermia, 51
Oncosomes, 352
Oocyte-derive transcripts, 100
Orphan Drug Designed (ODD) therapy, 273
Osteopontin (OPN), 353
Oxidative stress, 175

P

Pan-ethnic screening, 31–32
DNA-based expanded, 31–32
genetic counseling, 32–33
Paracentric inversions, 141
Patau syndrome, 378
PCR, *see* Polymerase chain reaction
Pericentric inversions, 141
Personal genomics, 10
PGES, *see* Prostaglandin E synthase
PGT, *see* Preimplantation genetic testing
PGT for structural rearrangements (PGT-SR), 361
cognitive development, 365
early childhood development, 365
limitations, 365
misdiagnosis, 363–364
perinatal outcome, 363
physical health, 364–365
pregnant following IVF, 362–363
socio-emotional development, 365
Phosphatidylserine (PS), 352
Placenta, 354–355
POC, *see* Products of conception
Polar body (PB), 119
Polygenic disease
comparison of disease risk, 200
comparison of PGT platforms, 197
DNA sequence-based predictor, 195
future perspectives, 199
limitations and barriers to access, 198–199
machine learning algorithms, 196
method and implementation, 197–198
Polymerase chain reaction (PCR), 196, 303, 364
Post-fertilization, 97–98
Post-test counseling, 370
PPROM, *see* Preterm premature rupture of membranes
Precocious separation of sister chromatids (PSSC), 79
Pre-conception testing, 26
Predivision, 79
Preeclampsia (PE), 258, 260–261
Pregnancy loss, 315
Preimplantational genetic testing (PGT-A), 106–107
Preimplantation genetic testing (PGT), 25, 154, 196
Preimplantation genetic testing for aneuploidy (PGT-A), 28, 117, 369
clinical outcomes, 128
cost effectiveness, 130
cryopreservation, 129
history, 118
invasive biopsy, 131
miscarriage and pregnancy loss, 132
moral status of the embryo, 131

mosaicism, 130–131
non-invasive, 133
pregnancy, 128–129
reproductive autonomy, 131–132
SET, 129
testing methodologies
array CGH, 123
CGH, 122–123
FISH, 121–122
implantation and pregnancy rates, 127
NGS, 124
SNP, 123–124
Preimplantation genetic testing for monogenic (PGT-M), 185, 369
cancer predisposition, 187
clinical results, 190–191
diagnostic methods
fragment analysis, 189
karyomapping, 189–190
PCR, 189
WGA amplification, 189
HLA matching, 187–188
isoimmunization, 188
late-onset diseases, 187
limitations and barriers to access, 191–192
NGS, 190
single gene disorders, 186
Preimplantation genetic testing for structural rearrangements (PGT-SR), 136, 142–143, 369
chromosomal abnormalities, 145–146
clinical outcome, 146–147
genetic diagnostic technologies, 146
methodology
array/NGS-based, 144–145
FISH-based, 144, 145
reproductive medicine, 146
strengths and limitations, 147–149
technical approaches, 143
Preterm birth (PTB), 258
Preterm premature rupture of membranes (PPROM), 241
Pre-test counseling, 370
Primitive endoderm (PE), 346
Products of conception (POC), 330, 369
aneuploidy testing, 334–336
decision-making workflow, 343
dissection, 334
Igenomix study design, 334
MCC, 336–337
miscarriage
detection of del/dups, 341
isolation of fetal tissue, 340
mosaicism, 340–341
Progesterone receptor (PR), 215–216
Prognosis patients, 127
Prolactin (PRL), 256, 351
Prophase, 42
Prostaglandin E synthase (PGES), 349
Protamine, 97–98
PSSC, *see* Precocious separation of sister chromatids

Q

Quantitative polymerase chain reaction (qPCR), 154, 196, 254

R

Randomized clinical trials (RCTs), 122, 158
Rare autosomal trisomies (RATs), 285
Reactive oxygen species (ROS), 175
Real-time quantitative PCR (RT-qPCR), 118
Receptive endometrium, 217
Reciprocal translocations, 27, 138, 146
Recurrent implantation failure (RIF), 126–127, 138, 217–220, 252
Recurrent pregnancy loss (RPL), 55, 126, 138
Re-endothelialization, 269
Relative haplotype dosage (RHDO) method, 292
Relative mutation dosage (RMD) method, 292, 303
Reproductive genetic counseling, 368
balanced chromosomal rearrangements, 371–372
cytogenetic analysis, 372
female infertility
advanced maternal age, 372
45, X (Turner syndrome), 372
47, XXX (Triple X syndrome), 372
ovarian insufficiency, 372–373
infertility, 371
male factor infertility
CFTR gene, 373–374
Karyotype 47, XXY (Klinefelter syndrome), 373
PGT-A, 375
aneuploidy, 380–381
euploid embryos, 378
high aneuploidy rates, 380
indeterminate, 378
mosaic aneuploid results, 379–380
non-viable whole-chromosome aneuploidy, 378
predictive value of aneuploid results, 380
prenatal testing, 381
segmental aneuploidy, 379
viable autosomal trisomies, 378
viable sex chromosome aneuploidies, 378–379
PGT-M
considerations, 382–383
indication, 382
prenatal diagnosis, 384
pre-test counseling, 383–384
test results, 384
transfer of embryos, 384
PGT-SR, 376
pre-test counseling for PGT-A, 376–377
pre-test counseling for PGT-SR, 377
types, 370–371
Reproductive medicine
in vitro models, 21
single-cell atlases
disease, 20–21
healthy tissue, 19–20
Ribonucleoprotein (RNP), 205
Ribosomal RNA (rRNA), 235
RNA-guided programmable endonuclease (RGEN), 205
Robertsonian translocations, 138
Roche sequencers, 4

S

Screening tests, 8
Second-generation sequencing methods, 278

Secreted phosphoprotein 1 (SPP1), 353
Secretory leukocyte protease inhibitor (SPLI), 229
Severe male factor (SMF), 127
Sex chromosome aneuploidies (SCAs), 285
Sex chromosome monosomy, 324–325
Short tandem repeats (STRs), 188
Sickle-cell trait, 31
Side Population (SP), 268
Single-cell atlases
  disease, 20–21
  healthy tissue, 19–20
Single-cell RNA sequencing (scRNA-seq), 214
Single-cell sequencing
  droplet-based combinatorial indexing, 17–18
  droplet-based protocols, 15–17
  plate-based combinatorial indexing, 17
  plate-based protocols, 15, 16
  RNA-seq, 15
  spatial location, 18–19
Single-cell transcriptomics, 19
Single embryo transfer (SET), 103, 129, 362
Single-gene disorder screening, 36
  clinical validation reports, 39
  limitations, 39–40
Single nucleotide polymorphisms (SNP), 123–124, 188, 196
  array, 333
Single-nucleotide variants (SNVs), 5, 276
Single stranded guide RNA (sgRNA), 205
Small integrin-binding ligand N-linked glycoproteins (SIBLINGs), 354
Small supernumerary marker chromosomes (sSMCs), 142
Smooth muscle tumors of uncertain malignant potential (STUMP), 277
Society for Assisted Reproductive Technology (SART), 125
Spent blastocyst media (SBM), 133
Spent blastocyst medium (SBM) vs. trophectoderm (TE) vs. polar body (PB) biopsy, 162
Sperm
  aneuploid, 52–55
  chromosome abnormalities, 53
    clinical outcome, 55
    embryonic outcomes, 56
    offspring, 56
  genetic counseling, 56–57
  limitations, 57
  strengths, 57
Sperm aneuploidy test (SAT), 370
Spermatagonial stem cells (SSC), 65
Spermatogenesis, 51
Sperm RNA elements (SRE), 65
Sperm washing, 207
Spindle assembly checkpoint (SAC), 85
Splitpool cycle, 16
Split-pool ligation-based transcriptome sequencing (SPLiT-seq), 17
Spontaneous abortions (SABs), 315
  chromosome abnormalities, 317, 320–322, 327–328
  maternal age, 322
  maternal contamination, 317–320
  methodological approaches, 316–317
  tissue sample, 315–316
Sporadic aneuploidy, 374–375
Structural variations (SVs), 6
Subtelomere labeling, 43, 144
Syngamy, 92

T

Tay-Sachs disease, 24, 31
Ten-eleven translocation (TET), 62, 94
Tetrahydrofolate (THF), 175
Tetraploidy, 326–327
The Cancer Genome Atlas (TCGA), 21, 278
Thermo Fisher sequencers, 4
Third-generation sequencing technologies, 3
Three-dimensional fluorescent *in situ* hybridization (3D-FISH), 48
Th2 phenomenon, 351
Thymine DNA glycosylases (TDG), 95
Time-lapse systems, 105
Time of pronuclei fading (tPNf), 107
Toll-like receptors (TLRs), 229
Topologically associated domains (TADs), 98, 99
Transcription-activator-like effector nucleases (TALENs), 205
Transcription factor (TF), 96
Transforming factor b (TGFb), 351
Triple X syndrome, 372, 379
Triploidy, 325–326
Trisomy, 325
Trophectoderm (TE) biopsy, 92, 104, 119–121, 151, 346
Tumor necrosis factor (TNF), 353
Turn-around time (TAT), 306, 310
Turner syndrome, 72, 372

U

Unbalanced structural chromosome rearrangements, 142
Uniform aneuploidies, 151
Uniparental disomy (UPD), 124, 288
Unique molecular identifier (UMI), 17
Uterine-derived cells (UDCs), 269
Uterine microbiome, 235
  embryo signaling and implantation, 239–240
  endometrial dysbiosis, 243–244
  endometrial microbiome, 236–238
  implications for the uterus, 240
  peri-conceptional period, 239
  phylogeny vs. taxonomy, 236
  potential therapies, 243–244
  uterine pathophysiology, 241–243
  vaginal microbiome, 238–239
Uterus, 353–354

V

Vaginal microbiota
  characterization, 225–226
  composition, 227–228
  host response, 228–229
  molecular-based profiling, 226–227
  pathogen colonization, 228–229
  reproductive health outcomes, 229–232
Vagitypes, 227
Vanishing twin (VT), 284, 294–295
Variant call format (VCF), 5
Variant interpretation, 38

Variant of uncertain significance (VUS), 6, 33, 39, 307
Vascular endothelial factor (VEF), 348
Vascular endothelial growth factor (VEGF), 351
Vasculogenesis, 269
Visual microscope-based techniques, 7

W

Whole-exome sequencing (WES), 2
Whole-genome amplification (WGA), 122, 144, 196
Whole-genome sequencing (WGS), 2
Wild-type (WT) alleles, 292
Window of implantation (WOI), 213
Wireless communication, 346

X

X-linked inheritance, 375

Z

Zinc finger nucleases (ZFNs), 205
Zona hatching, 347
Zygotic genome activation (ZGA), 92